Fundamental Theories of Physics

Volume 217

Series Editors

Henk van Beijeren, Utrecht, The Netherlands

Philippe Blanchard, Bielefeld, Germany

Bob Coecke, Oxford, UK

Dennis Dieks, Utrecht, The Netherlands

Bianca Dittrich, Waterloo, ON, Canada

Ruth Durrer, Geneva, Switzerland

Roman Frigg, London, UK

Christopher Fuchs, Boston, MA, USA

Domenico J. W. Giulini, Hanover, Germany

Gregg Jaeger, Boston, MA, USA

Claus Kiefer, Cologne, Germany

Nicolaas P. Landsman, Nijmegen, The Netherlands

Christian Maes, Leuven, Belgium

Mio Murao, Tokyo, Japan

Hermann Nicolai, Potsdam, Germany

Vesselin Petkov, Montreal, QC, Canada

Laura Ruetsche, Ann Arbor, MI, USA

Mairi Sakellariadou, London, UK

Alwyn van der Merwe, Greenwood Village, CO, USA

Rainer Verch, Leipzig, Germany

Reinhard F. Werner, Hanover, Germany

Christian Wüthrich, Geneva, Switzerland

Lai-Sang Young, New York City, NY, USA

The international monograph series "Fundamental Theories of Physics" aims to stretch the boundaries of mainstream physics by clarifying and developing the theoretical and conceptual framework of physics and by applying it to a wide range of interdisciplinary scientific fields. Original contributions in well-established fields such as Quantum Physics, Relativity Theory, Cosmology, Quantum Field Theory, Statistical Mechanics and Nonlinear Dynamics are welcome. The series also provides a forum for non-conventional approaches to these fields. Publications should present new and promising ideas, with prospects for their further development, and carefully show how they connect to conventional views of the topic. Although the aim of this series is to go beyond established mainstream physics, a high profile and open-minded Editorial Board will evaluate all contributions carefully to ensure a high scientific standard.

Sibel Başkal · Young Suh Kim · Marilyn E. Noz

Theory and Applications of the Poincaré Group

Second Edition

 Springer

Sibel Başkal
Department of Physics
Middle East Technical University
Ankara, Türkiye

Young Suh Kim
Department of Physics
University of Maryland
College Park, USA

Marilyn E. Noz
Department of Radiology
New York University
New York, NY, USA

ISSN 0168-1222   ISSN 2365-6425 (electronic)
Fundamental Theories of Physics
ISBN 978-3-031-64378-1   ISBN 978-3-031-64376-7 (eBook)
https://doi.org/10.1007/978-3-031-64376-7

This Springer imprint is published by the registered company Springer Nature Switzerland AG
The registered company address is: Gewerbestrasse 11, 6330 Cham, Switzerland

If disposing of this product, please recycle the paper.

Preface to the Second Edition

The First Edition of this book was published in 1986, and that book consisted of interpretations of Eugene Wigner's paper entitled *On Unitary Representations of the Inhomogeneous Lorentz Group* published in the Annals of Mathematics in 1939 (Vol. 40, pages 149-204). This paper of Wigner's is applicable to internal space-time symmetries of elementary particles in Einstein's Lorentz-covariant world.

In addition to its energy and momentum, the particle has internal variables. Of special interest, is the spin variable corresponding to the internal angular momentum. Wigner's 1939 paper clearly tells us how this spin variable appears as the necessary internal dynamical variable in Einstein's Lorentz-covariant world.

When Wigner wrote his paper in 1939, all particles were point particles. He did not consider the non-zero size of particles; as, protons and neutrons were point particles when Wigner wrote this paper.

However, high energy physics experiments first conducted in the 1950's, concluded that the proton was not a point particle, but had an extended distribution. Gell-Mann's paper published in 1964 in Physics Letters entitled *A schematic model of baryons and mesons* (Vol. 8, pages 214-215) told us that the proton and the neutron and indeed all hadrons in the rest state are quantum bound states of more fundamental particles called *quarks*. This bound state, together with the minimal time-energy uncertainty relation, has a space-time extension in the Lorentz-covariant world. This Lorentz-covariant picture allows us to look at fast-moving protons, i.e., with a speed close to that of light. This leads to all the peculiarities observed in the parton picture, detailed by Feynman in a paper entitled *Very High-Energy Collisions of Hadrons* published in Physics Letters in 1969 (Vol. 23, pages 1415-1417).

While the First Edition was based mostly on the papers published by Young S. Kim and and Marilyn E. Noz on applications of Wigner's 1939 papers on particle physics, three major improvements have been made in this Second Edition:

1. While Kim and Noz continued their efforts on applications of Wigner's 1939 paper, a younger author named Sibel Başkal joined the research team after Kim and Noz published their First Edition in 1986. This new edition contains the research results published by Başkal, Kim, and Noz.

2. If not Planck's hypothesis of light being emitted in discrete units of energy, it is the invention and operating principles of lasers that is considered to be the birth of quantum optics. It provides a substantiation for the fundamental characteristics of quantum physics as coherence and entanglement, that has either directly or incidentally contributed to the emergence and development of quantum information and computing, a new field anticipated to revolutionize how we approach science. This new edition shows us that Wigner's 1939 paper provides a natural language for this flourishing subject.

3. In 1963, Paul A. M. Dirac published a paper entitled *A Remarkable Representation of the 3 + 2 de Sitter Group* in the Journal of Mathematical Physics (Vol. 20, pages 901-909) in which he showed that the quantum mechanics of two coupled oscillators leads to the symmetries of the Lorentz group with three space dimensions and two time variables. In this new edition, it is shown that one of those time variables can be converted into translations in the Lorentzian space of three space dimensions with one time coordinate. Thus, it is possible to derive Einstein's $E = m c^2$ from quantum mechanics. This is consistent with Dirac's life-long ambition to combine quantum mechanics with special relativity.

This new edition contains four new chapters, two of which are consistent with Dirac's aim to combine the important developments in physics in the twentieth century, namely quantum mechanics and special relativity. Moreover, these new chapters also discuss various aspects of classical and quantum optics, that are now understood to be interrelated. Most of the original chapters have been updated, either with new material added or in some instances reinterpretation of the original. The order of the chapters has been rearranged to create a more cohesive presentation. The original purpose of the First Edition, namely to present examples to which physics students and researchers can relate, has not been altered.

This book still remains as a teaching tool which is directed toward those who desire a deeper understanding of group theory in terms of examples applicable to the physical world and/or of the physical world in terms of the symmetry properties which can best be formulated in terms of group theory. Those who are interested in the relationship between group theory and physics will find it instructive. In particular, those engaged in high-energy physics and foundations of quantum mechanics will find this book rich in illustrative examples of relativistic quantum mechanics.

Ankara, Türkiye
Collge Park, MD, USA
New York, NY USA

Sibel Başkal
Young Suh Kim
Marilyn E. Noz

Acknowlegement: We would like to thank Professor Emeritus Gerald Q. Maguire Jr. of KTH Royal Institute of Technology for redrawing several figures for this book.

Preface to the First Edition

Special relativity and quantum mechanics, formulated early in the twentieth century, are the two most important scientific languages and are likely to remain so for many years to come. In the 1920's, when quantum mechanics was developed, the most pressing theoretical problem was how to make it consistent with special relativity. In the 1980's, this is still the most pressing problem. The only difference is that the situation is more urgent now than before, because of the significant quantity of experimental data which need to be explained in terms of both quantum mechanics and special relativity.

In unifying the concepts and algorithms of quantum mechanics and special relativity, it is important to realize that the underlying scientific language for both disciplines is that of group theory. The role of group theory in quantum mechanics is well known. The same is true for special relativity. Therefore, the most effective approach to the problem of unifying these two important theories is to develop a group theory which can accommodate both special relativity and quantum mechanics.

As is well known, Eugene P. Wigner is one of the pioneers in developing group theoretical approaches to relativistic quantum mechanics. His 1939 paper on the inhomogeneous Lorentz group laid the foundation for this important research line. It is generally agreed that this paper was somewhat ahead of its time in 1939, and that contemporary physicists must continue to make real efforts to appreciate fully the content of this classic work.

Wigner's 1939 paper is also a fundamental contribution in mathematics. Since 1939, in order to achieve a better understanding of Wigner's work, mathematicians have developed many concepts and tools, including little groups, orbits, groups containing Abelian invariant subgroups, induced representations, group extensions, group contractions and expansions. These concepts are widely discussed in many of the monographs and textbooks in mathematics [Segal (1963), Gel'fand et al. (1966), Hermann (1966), Gilmore (1974), Mackey (1978), and many others].

Indeed, the mathematical research along this line has been extensive. It is therefore fair to say that there is at present an imbalance between mathematics and physics in the sense that there are not enough physical examples to enrich the theorems mathematicians have developed. The main purpose of this book is to reduce the gap between mathematical theorems and physical examples.

This book combines in a systematic manner numerous articles published by the authors primarily in the American Journal of Physics and lecture notes prepared by the authors over the past several years. It is intended mainly as a teaching tool directed toward those who desire a deeper understanding of group theory in terms of examples applicable to the physical world and/or of the physical world in terms of the symmetry properties which can best be formulated in terms of group theory. Both graduate students and others interested in the relationship between group theory and physics will find it instructive. In particular, those engaged in high-energy physics and foundations of quantum mechanics will find this book rich in illustrative examples of relativistic quantum mechanics.

For numerous discussions, comments, and criticisms while the manuscript was being prepared, the authors would like to thank S. T. Ali, L. C. Biedenharn, J. A. Brooke, W. E. Caswell, J. F. Carinena, A. Das, D. Dimitroyannis, P. A. M. Dirac, G. N. Fleming, H. P. W. Gottlieb, O. W. Greenberg, M. Haberman, M. Hamermesh, D. Han, W. J. Holman, T. Hubsch, P. E. Hussar, S. Ishida, P. B. James, T. J. Karr, S. K. Kim, W. Klink, R. Lipsman, G. Q. Maguire, V. I. Man'ko, M. Markov, S. H. Oh, S. Oneda, E. F. Redish, M. J. Ruiz, G. A. Snow, D. Son, L. J. Swank, K. C. Tripathy, A. van der Merwe, D. Wasson, A. S. Wightman, E. P. Wigner, and W. W. Zachary. The chapters of this book on massless particles are largely based on the series of papers written by one of the authors (YSK) in collaboration with D. Han and D. Son.

Contents

List of Figures

List of Tables

Chapter 1
Overview

Abstract One of the most fruitful and still promising approaches to unifying quantum mechanics and special relativity has been the covariant formulation of quantum field theory. An important starting point is Wigner's 1939 paper and the physical ideas presented in his 1931 book. This early work was overshadowed by the success of quantum field theory until the 1950's. However, with the discovery of the extended charge distribution of protons leading to the quark theory for all hadrons, the difficulties with using quantum field theory for quarks confined in bound states became apparent along with the difficulties of reconciling quantum mechanics with special relativity. Paul A. M. Dirac devoted much of his professional life to this important task. In his attempt to construct a *relativistic dynamics of atom* contained in his 1949 article, Dirac emphasizes that the task of constructing a relativistic dynamics is equivalent to constructing a representation of the Poincaré group, also known as the inhomogeneous Lorentz group. In the 1950's, the limitations of the present form of quantum field theory became apparent. Currently, there are two different opinions on the difficulty of using field theory in dealing with bound-state problems or systems of confined quarks.

One of the most fruitful and still promising approaches to unifying quantum mechanics and special relativity has been and still is the covariant formulation of quantum field theory. The role of Wigner's work on the Poincaré group (also known as the inhomogeneous Lorentz group) and the physical ideas presented in his 1931 book on atomic spectra [1, 2] in quantum field theory is nicely summarized in the fourth paragraph of the introductory article by V. Bargmann et al. in the commemorative issue of Reviews of Modern Physics in honor of Wigner's 60th birthday [3] which concludes with the sentences:

> Those who had carefully read the preface of Wigner's great 1939 paper on relativistic invariance and had understood the physical ideas in his 1931 book on group theory and

atomic spectra were not surprised by the turn of events in quantum field theory in the 1950's. A fair part of what happened was merely a matter of whipping quantum field theory into line with the insights achieved by Wigner in 1939.

It is important to realize that quantum field theory has not been and is not at present the only theoretical machine with which physicists attempt to unify quantum mechanics and special relativity. Paul A. M. Dirac devoted much of his professional life to this important task. In his attempt to construct a *relativistic dynamics of atom* [4] using *Poisson brackets* Dirac emphasizes that the task of constructing a relativistic dynamics is equivalent to constructing a representation of the Poincaré group.

Dirac's form of relativistic quantum mechanics had been overshadowed by the success of quantum field theory throughout the 1950's. However, with Hofstader's [5] discovery of the extended charge distribution of protons leading to the quark theory in the 1960's [6, 7, 8] for all hadrons, the difficulties with using quantum field theory when it was necessary to deal with quarks confined permanently inside hadrons, the limitations of the present form of quantum field theory became apparent. Currently, there are two different opinions on the difficulty of using field theory in dealing with bound-state problems or systems of confined quarks. One of these regards the present difficulty merely as a complication in calculation. According to this view, we should continue developing mathematical techniques which will someday enable us to formulate a bound-state problem with satisfactory solutions within the framework of the existing form of quantum field theory. The opposing opinion is that quantum field theory is a model that can handle only scattering problems in which all particles can be brought to free-particle asymptotic states. According to this view, we have to make a fresh start for relativistic bound-state problems, possibly starting from Dirac's 1949 paper.

We contend that these two opposing views are not mutually exclusive. Bound-state models developed in these two different approaches should have the same space-time symmetry. It is quite possible that independent bound-state models, if successful in explaining what we see in the real world, will eventually complement field theory. One of the purposes of this book is to discuss a relativistic bound-state model built in accordance with the principles laid out by Wigner in 1939 [1] and Dirac in 1949 [4], which can explain basic hadron features observed in high-energy laboratories.

Another important development in modern physics is the extensive use of gauge transformations in connection with massless particles and their interactions. Wigner's 1939 paper has the original discussion of space-time symmetries of massless particles. However, it was only recently recognized that gauge-dependent electromagnetic four-potentials form the basis for a finite-dimensional non-unitary representation of one little group of the Poincaré group. This enables us to associate gauge degrees of freedom with the degrees of freedom left unexplained in Wigner's work. Hence it is possible to impose a gauge condition on the electromagnetic four-potential to construct a unitary representation of the photon polarization vectors.

The organization of this book is identical to that of Wigner's original paper, but the emphasis will be different. In discussing representations of the Poincaré group for free particles, we use the method of little groups, as is summarized in Table 1.1.

Table 1.1: Wigner's little groups discussed in this book.

P: Four-momentum	Subgroup of $O(3, 1)$	Subgroup of $SL(2, c)$
Timelike~Massive: $M^2 > 0$	$O(3)$-like subgroup of $O(3, 1)$: hadrons	$SU(2)$-like subgroup of $SL(2, c)$: electrons
Null~Massless: $M^2 = 0$	$E(2)$-like subgroup of $O(3, 1)$: photons	$E(2)$-like subgroup of $SL(2, c)$: spin-1/2
Spacelike~Imaginary: $M^2 < 0$	$O(2, 1)$-like subgroup of $O(3, 1)$	$Sp(2)$-like subgroup of $SL(2, c)$
$P = 0$	$O(3, 1)$	$SL(2, c)$

Wigner observed in 1939 that Dirac's electron has an $SU(2)$-like internal space-time symmetry. However, quarks and hadrons were unknown at that time. By discussing Dirac's form of relativistic bound-state quantum mechanics, which starts from the representations of the Poincaré group, it is possible to study the $O(3)$-like little group for massive particles. Since Dirac's form leads to hadron wave functions which can describe fairly accurately the distribution of quarks inside hadrons, a substantial portion of hadron physics can be incorporated into the $O(3)$-like little group for massive particles.

As for massless particles, Wigner showed that their internal space-time symmetry is locally isomorphic to the Euclidian group or $E(2)$ in two-dimensional space. However, Wigner did not explore the content of this isomorphism, because the physics of the translation-like transformations of this little group was unknown in 1939. Neutrinos were known only as *Dirac electrons without mass*, although photons were known to have spins either parallel or anti-parallel to their respective momenta. We now know the physics of the degrees of freedom left unexplained in Wigner's paper. Much more is also known about neutrinos today than in 1939. It is also possible to discuss internal space-time symmetries of massless particles starting from Wigner's $E(2)$-like little group.

The $O(2, 1)$-like little group could explain internal space-time symmetries of particles which move faster than light. Since these particles are not observable, this little group is not of immediate physical interest. However, the mathematics of this group has been and is still being discussed extensively in the literature. We shall discuss the mathematical aspect of this and other little groups in this book. It is of interest to note in particular that $O(2, 1)$ is isomorphic to the two-dimensional symplectic group or $Sp(2)$, which is playing an increasingly important role in all branches of physics.

Also included in this book are discussions of hadron phenomenology. The above-mentioned $O(3)$-like little group for hadrons and the bound-state model based on this concept will be meaningful only if they can describe the real world. We shall discuss hard experimental data and curves describing mass spectra, form factors, the parton model, and the jet phenomenon. We shall then show that a simple harmonic oscillator [9, 4] can produce results which can be compared with the relevant experimental data.

Although group theory has been well-known to mathematicians for almost two centuries, we owe Wigner for the propagation of group theory to quantum mechanics. In his book *Inward Bound: Of Matter and Forces in the Physical World*, Abraham Pais writes [10]:

> So, Wigner told me, he went to consult his friend the mathematician Johnny Von Neumann. Johnny thought a few moments and told him that he should read certain papers by Frobenius and by Schur which he promised to bring the next day. As a result Wigner's paper on the case of general n (no spin), was ready soon [11] and was submitted in November 1926. It contains an acknowledgment to Von Neumann, and also the following phrase: *There exists a well-developed mathematical theory which one can use here: the theory of transformation groups which are isomorphic with the symmetric group (the group of permutations).*
>
> Thus did group theory enter quantum mechanics.

Here, we give very basic definitions and concepts about the old venerable group theory and Lie algebras and give a more complete discussion of these important topics in Appendices A and B respectively. It is recommended that those readers who are less familiar with these topics, read the Appendices before starting the text.

1.1 Basic Concepts in Group Theory

It often happens that the three-dimensional rotation group is the physicists' first exposure to group theory when studying quantum mechanics. In connection with the quark model, group theory has, since the 1960's, become an indispensable tool in theoretical physics. Because it describes the fundamental space-time symmetries in the four-dimensional Minkowski space, the Poincaré group occupies an important place in the recent trend to study space-time coordinate transformations and in constructing explicit representations of non-compact groups. In particular, those which are neither simple nor semisimple.

A set of elements g can form a group G if there is a group operation (sometimes referred to as *multiplication*) which is defined for any two elements a and b of G such that:

1. Closure: for any a and b in G, $a \cdot b$ is in G, where $\cdot$ is the group operation generally referred to as group multiplication.
2. Associative law: $(a \cdot b) \cdot c = a \cdot (b \cdot c)$, where c is also in G.
3. Identity: there exists a unique identity element e such that $e \cdot a = a \cdot e$ for all a in G.
4. Inverse: for every a in G, there exists an inverse element, denoted a^{-1}, such that $a^{-1} \cdot a = a \cdot a^{-1} = e$.

The order of G is determined by the number of elements in the group, which can be either finite or infinite. Some examples of groups with which the reader might be familiar are:

1. The group S_3 which consists of permutations of three objects.
2. The group $O(3)$, namely the three-dimensional rotation group without space inversions, sometimes referred to as the *proper rotation group*.

The order of S_3 is six, while the order of $O(3)$ is infinite.

If the group multiplication is commutative the group is called Abelian. The group consisting of rotations around the origin on the xy-plane or $O(2)$ is Abelian. The group of translations in three-dimensional space is also Abelian. S_3 and $O(3)$ are non-Abelian groups.

Two groups are isomorphic if there is a one-to-one mapping F from one group (G) to another group (H) that preserves group multiplication. The groups are *homomorphic* if the mapping is onto but not one-to-one. The three-dimensional rotation group is homomorphic to the group $SU(2)$ which consists of two-by-two unitary matrices which are unimodular (the determinant is equal to one).

Another example which will play an important role throughout this book is the two-dimensional Euclidean group, $E(2)$. In this book we'll use the coordinate system (t, z, x, y). This is used to represent the Lorentz group $O(3, 1)$ in Minkowski space. In matrix notation, this transformation takes the form:

$$\begin{pmatrix} 1 \\ 1 \\ x \\ y \end{pmatrix} = \begin{pmatrix} 1 & 0 & 0 & 0 \\ 0 & 1 & 0 & 0 \\ 0 & u & \cos\theta & -\sin\theta \\ 0 & v & \sin\theta & \cos\theta \end{pmatrix} \begin{pmatrix} 1 \\ 1 \\ x_0 \\ y_0 \end{pmatrix}. \tag{1.1}$$

The determinant of the above transformation matrix is 1. Like matrices representing the three-dimensional rotation group, the matrix in Eq. (1.1) also contains three parameters. While being similar to the rotation group in some aspects, $E(2)$ also shares many characteristics with the Poincaré group. Indeed, this group will play a very important role both in illustrating mathematical theorems and in constructing representations of the Poincaré group.

A set of elements H, contained in a group G, forms a subgroup of G, if all elements in H satisfy group operations. Even permutations in S_3 form a subgroup. Rotations around the z-axis form a subgroup of the rotation group.

If H is a subgroup of G, the set gH with g in G is called the left coset of H, while Hg is called the right coset of H. A right coset of H in general is not identical to the left coset. There are certain subgroups H in which every right coset is a left coset. Subgroups H having this property are called invariant subgroups. The set of even permutations in S_3 is an invariant subgroup. The group of rotations around the z-axis is not an invariant subgroup of the three-dimensional rotation group.

If a group does not contain invariant subgroups, it is called a simple group. $O(3)$ is a simple group. If a group contains invariant subgroups which are not Abelian, it is called a semisimple group. S_3 is a semisimple group. $E(2)$ contains an Abelian invariant subgroup, and is therefore neither simple nor semisimple.

If p is a point in the vector space X, the maximal subgroup $G(p)$ of G which leaves p invariant, i.e., $G^{(p)}p = p$, is called the *little group* of G at p. In the four-dimensional (t, z, x, y) space, rotations around the z-axis form the $O(2)$-like little group of $O(3)$ at $(0, 1, 0, 0)$. This little group is conjugate to the little group at $(0, 0, 1, 0)$ which consists of rotations around the x-axis.

The set of points V that can be reached through the application of G on a single point p in X is called the *orbit* of G at p. Two orbits are either identical or disjoint. The orbit of $O(3)$ at $(0, R, 0, 0)$ is the surface of the sphere with radius R centered around the origin. The orbit of $O(2)$ at the point $(x = a, y = 0)$ on the xy-plane is the circumference of a circle with radius a.

A set of linear transformation matrices homomorphic to the group multiplication of G is called a *representation* of the group. The word *homomorphic* is appropriate here because the matrices need not have a one-to-one correspondence with the group elements. In addition, there can be more than one set of matrices forming a representation of the group. For instance, the three-dimensional rotation group can be represented by two-by-two, three-by-three, or n-by-n matrices, where n is an arbitrary integer. If a representation does not contain subrepresentations, it is said to be *irreducible*. Schur's Lemma, derived in Appendix A, gives a practical method for determining if a given representation is irreducible.

1.2 Basic Concepts of Lie Groups and Lie Algebras

Lie groups are groups that are also smooth manifolds. Additionally, Lie groups can be defined as the connected component of a continuous group which can continuously be connected to the identity element. Lie groups have continuous parameters, and give rise to Lie algebras. In general, the number of parameters of a group is not necessarily equal to the dimension of the space where the coordinates are defined. For instance, there are semidirect product of groups as in $E(2)$, whose number of parameters is larger than the number of coordinates.

Conversely, to any finite-dimensional Lie algebra over real or complex numbers, there is a corresponding connected Lie group unique up to finite coverings. Due to the correspondence between Lie groups and Lie algebras, it is possible to study the structure and classification of Lie groups in terms of their Lie algebras.

Lie algebra is a vector space together with an operation called the Lie bracket

$$\left[X_\sigma, X_\rho\right] = X_\sigma X_\rho - X_\rho X_\sigma \tag{1.2}$$

which is an alternating bilinear map, such that the Jacobi identity

$$[X_\rho, [X_\sigma, X_\tau]] + [X_\sigma, [X_\tau, X_\rho]] + [X_\tau, [X_\rho, X_\sigma]] = 0 \tag{1.3}$$

holds.

If X_{σ_i}, with $\{\sigma_i\} = \{\sigma, \rho, ...\}$, are the generators of the Lie algebra then

$$[X_\sigma, X_\rho] = iC^\tau_{\sigma\rho}X_\tau, \tag{1.4}$$

where $C^\tau_{\sigma\rho}$ are called the structure constants.

From the Lie algebra the Lie group can be obtained by the exponential map. If X is any complex valued square matrix then the exponential of X is

$$\exp X \equiv e^X = \sum_{n=0}^{\infty} \frac{1}{n!} X^n. \tag{1.5}$$

With this definition of the exponential map of matrices, the group element corresponding to any parameter σ is obtained by

$$G(\sigma) = e^{-i\sigma X}. \tag{1.6}$$

From the Lie group one can go back to its algebra by differentiation

$$i\frac{d}{d\sigma}G(\sigma)\bigg|_{\sigma=0} = X_\sigma. \tag{1.7}$$

An infinitesimal differential generator for the Lie algebra takes the form

$$X_\sigma = -i \sum_{k=1}^{n} a^k_\sigma(x^j) \frac{\partial}{\partial x^k} \tag{1.8}$$

where $a^k_\sigma(x^j)$ are matrix elements to be determined. The corresponding matrix is not necessarily invertible.

As an example let us again consider the group $E(2)$ whose transformation matrix takes the form as in (1.1). This group is the semidirect product of the rotation group and the Euclidean vector space. There are two coordinate variables $\{x^j\} = \{x, y\}$ and three group parameters $\{\sigma\} = \{\theta, u, v\}$. The generators take the form

$$X_\theta(x^j) = -i\left(x\frac{\partial}{\partial y} - y\frac{\partial}{\partial x}\right),$$

$$X_u(x^j) = -i\frac{\partial}{\partial x}, \qquad X_v(x^j) = -i\frac{\partial}{\partial y} \tag{1.9}$$

and they satisfy the commutation relations

$$[X_\theta(x^j), X_u(x^j)] = iX_v(x^j), \qquad [X_\theta(x^j), X_v(x^j)] = -iX_u(x^j), \tag{1.10}$$
$$[X_u(x^j), X_v(x^j)] = 0.$$

The structure constants can be read off as

$$C^v_{\theta u} = 1, \qquad C^u_{\theta v} = -1, \qquad \text{and} \qquad C^\theta_{uv} = 0. \tag{1.11}$$

Conversely, the above structure constants completely determine the Lie brackets of all the elements of the Lie algebra of the $E(2)$ group.

1.3 Plan of the Book

Chapter 2 contains a pedagogical elaboration of Wigner's original work on the Poincaré group which occupies an important place in physics. Therefore we discuss the representations of the Poincaré group and note that the Poincaré group is non-compact and is neither simple nor semisimple. We study first the four-by-four Lorentz transformation matrices and the resulting Lie algebra. We then study the orbits and little groups of the Lorentz group, in preparation for constructing representations of the Poincaré group, the little group decomposition of the Poincaré group, and the Casimir operators for each little group.

Chapter 3 contains a discussion of the groups $SU(2)$ and $SL(2, c)$ and their correspondence with the rotation group $O(3)$ and the Lorentz group $O(3, 1)$. The group $SL(2, c)$, the covering group of the Lorentz group, is more elaborate from a mathematical point of view. On the other hand, since $SL(2, c)$ can be represented by two by two matrices with complex parameters, calculations are relatively easier compared to the four-by-four matrices of the Lorentz group. The transformation properties of spinors with regard to the $SL(2, c)$ group is given and their correspondence between the components of a four-vector is made evident. It is also shown that $E(2)$-like groups, which leave the four-momentum of relativistic massless particles invariant, form a subgroup of $SL(2, c)$. The spinor space of the Dirac equation is presented, and the symmetries of the Dirac equation both for massive and massless particles are discussed. It is shown that, thanks to those symmetry properties, a plane wave solution to the Dirac equation can be found easily.

Chapter 4 discusses the principles of group contraction. We explain first how the squeeze transformation can be used and that it is possible a straight line can be produced from a function on the xy-plane. We give a summary of Wigner's $E(2)$ little group and the cylindrical group, and contract the $O(3)$ group into both the $E(2)$ and cylindrical groups. The Lorentz and the Galilei transformations are discussed as well as the $O(2, 1)$ Lorentz group. Additionally, we show how we can use the squeeze transformation to contract $O(2, 1)$ into the $E(2)$ and cylindrical groups and the Lorentz group $O(3, 1)$ into the Galilean group.

In Chap. 5, the covariant harmonic oscillator formalism is discussed as a mathematical device useful in constructing space-time solutions of the commutator equations for Dirac's relativistic bound-state quantum mechanics. It is pointed out that the harmonic oscillator formalism is useful also in explaining basic hadron features we observe in the real world. The relation between the $O(4)$, the moving $O(4)$ and the harmonic oscillator coordinate systems is also included.

Chapter 6 shows that the harmonic oscillator formalism indeed satisfies all the requirements for Dirac's form of relativistic quantum mechanics, and therefore that the formalism is consistent with the established rules of quantum mechanics and special relativity. Included is a discussion of the c-number uncertainly principle and of Dirac's light-cone coordinate system. The material here is based mainly on Dirac's papers from 1927, 1945, and 1949 [12, 13, 9, 4].

Chapter 7 first presents two coupled harmonic oscillators, the Lorentz covariance associated with them and then Dirac's coupled harmonic oscillators. In addition, the origins of the $O(3,2)$ and $O(3,3)$ Lorentz groups are elucidated together with a discussion of the squeezed states of light, the symmetries of both Dirac's two oscillator system and of Dirac's γ matrices. Finally, we contract $O(3,2)$ into the Poincaré group. The material presented here is based mainly on Dirac's 1963 paper [14].

In Chap. 8, the groups $O(2,1)$, $SU(1,1)$, $SL(2,r)$, and $Sp(2)$ are revisited. In particular, $O(2,1)$ is locally isomorphic to $Sp(2)$, $SU(1,1)$, $SL(2,r)$, and has a rich mathematical content. We discuss these groups in some detail in this chapter. The importance of the group $Sp(2)$ is well-known in the Hamiltonian formulations of both classical and quantum mechanics. It has also found a myriad of applications in relatively new research areas such as quantum information and computing. The correspondence of $SU(1,1)$ with $SL(2,r)$ or $Sp(2)$, consisting of real two-by-two matrices, is given. In this setting, the geometry of $SL(2,r)$ or $Sp(2)$ is made clear. Wigner, Bargmann, and Iwasawa decompositions are discussed within the context of little groups. It is shown that the study of $O(2,1)$ is that of the Legendre functions. Complex angular momentum is introduced and physical applications are demonstrated. We also outline the steps in examining the unitary irreducible representations of $SU(1,1)$.

Next, Chap. 9, presents representations of the $O(3,1)$ Lorentz group which has become particularly important since the appearance of the papers of Bargmann (1947) and of Harish-Chandra (1947) [15, 16]. The homogeneous Lorentz group plays the central role in studying Lorentz transformation properties of quantum mechanical state vectors and operators. We shall study $O(3,1)$ not as a little group but as the symmetry group for the process of orbit completion.

Chapter 10 deals with the internal space-time symmetries of massless particles. Among massless particles photons are particularly significant due to their involvement in the development of quantum mechanics, as well as in quantum electrodynamics. One of the difficulties associated with studying photons has been that the four-vector form for the photon wave function is not unique, depending on gauge degrees of freedom which are not measurable. The little group for massless particles is $E(2)$-like. We first examine the $E(2)$-like little group as an infinite-momentum/zero-mass limit of the $O(3)$-like little group for massive particles and the $E(2)$-like little group for photons. The four-momentum of relativistic particles is altered by Lorentz transformations, whose rotations are represented by unitary matrices while boosts, by non-unitary matrices. Therefore, we consider the possibility of constructing a unitary representation by combining the boost and gauge transformations. Both finite and infinite-dimensional representations of the $E(2)$ group are considered. We study also the transformation property of Maxwell's equations as a further illustrative example of spinor representations. Additionally, we study the little group for massless spin-1/2 particles.

Chapter 11 discusses the one and two dimensional representations of Heisenberg's uncertainty relations. We then show that the one- and two-dimension step-up and step-down operators can be combined into Hermitian matrices, the later one of which, when off-diagonal elements are added, can be linearly combined with the ten generators for Dirac's two oscillator system leading to the $(3 + 2)$ de Sitter or $O(3,2)$ Lorentz group, which is isomorphic to $Sp(4)$. The $O(3,2)$ group can then be contracted to the Poincaré group with four translation generators corresponding to the four-momentum in the Lorentz-covariant world. This Lorentz-covariant four-momentum is known as Einstein's $E = mc^2$.

Chapters 12 and 13 deal with various applications to hadron phenomenology of the harmonic oscillator formalism developed in Chap. 5 and Chap. 6. Since Hofstadter's discovery in 1955 [5], it has been known that the proton or hadron is not a point particle, but has a space-time extension. This idea is compatible with the basic concept of the quark model in which hadrons are quantum bound states of quarks having space-time extensions. At present, this concept is totally consistent with all qualitative features of hadron phenomenology. It is widely believed that quark motions inside the hadron generate Rydberg-like mass spectra. It is also believed that fast-moving extended hadrons are Lorentz-deformed, and that this deformation is responsible for peculiarities observed in high-energy experiments such as the parton and jet phenomena. The question is how to describe all these covariantly.

In Chap. 12, we show first that the $O(3)$-like little group is the correct language for describing covariantly the observed mass spectra and the space-time symmetry of confined quarks, and then show that the harmonic oscillator model describes the mass spectra observed in the real world. This is illustrated with the comparison of the results of a harmonic oscillator based mass formula for baryons with data from high energy experiments. We then discuss mesons and give a summary of known mesons derived from high energy experiments. In conclusion we give a summary of particles, such as bosons, which are not hadrons and present the current state of experimental knowledge about them.

In Chap. 13, we point out first that the space-time extension of the hadron and the Lorentz-Dirac deformation thereof are responsible for behavior of hadron form factors. We present a comprehensive review of theoretical models constructed along this line, and compare the calculated form factors with experimental data. Second, we discuss in detail the peculiarities in Feynman's parton picture which are universally observed in high-energy hadron experiments. According to the parton model, the hadron consisting, at rest, of a finite number of quarks appears as a collection of an infinite number of partons when the hadron moves very rapidly. Since partons appear to have properties which are different from those of quarks, the question arises whether quarks are partons. While this question cannot be answered within the framework of the present form of quantum field theory, the harmonic oscillator formalism provides a satisfactory resolution of this paradoxical problem. Third, the proton structure function is calculated from the Lorentz-boosted harmonic oscillator wave function, and is compared with experimental data collected from electron and neutrino scattering experiments. Detailed numerical analyses are presented. Fourth, it is shown that Lorentz-Dirac deformation is responsible also for formation of hadron jets in high-energy experiments in which many hadrons are produced in the final state. Both qualitative and quantitative discussions are presented.

We present in Chap. 14 the Poincaré sphere and study as well the Stokes parameters and the decoherence of light. It is noted that the $O(3, 2)$ group contains two Lorentz subgroups. The change in the determinant in one Lorentz subgroup can be compensated by the other. It is thus possible to describe the decoherence process as well as the radius variation of the Poincaré sphere, as a symmetry transformation in the $O(3, 2)$ space. It is shown also that these two coupled Lorentz groups provide a concrete example of Feynman's rest of the universe. The two-by-two matrix representations of the energy-momentum four-vectors in the two Lorentz subgroups of the $O(3, 2)$ are also related through the sum of their determinants.

In Chap. 15 we discuss the role of the Lorentz group in classical optics. First we study the Jones vector formalism for studying the polarization properties of the two component electric field vector. A general form for the attenuator is introduced, and thereof attenuations, phase-shifts, and the rotation of the polarization axes are reformulated. Their combined effects are investigated and it is shown that they amount to a two-by-two representation of the six-parameter $SL(2, c)$ group. In connection with these combined effects, the Wigner rotation is also reviewed. Ray transfer $(ABCD)$ matrices, are introduced and their equi-diagonalization process is explained. It is shown how the results of this process facilitate computations when a large number of cycles is required to account for the behaviour of a beam subjected to the effects of a series of optical elements. It is also noted that the Wigner and Bargmann decompositions can be explained in terms of the decomposition properties of the $ABCD$ matrix. This matrix as a general element of the $Sp(2)$ group can be composed by assembling some specific combinations of lens and translation matrices. Concrete physical examples of $ABCD$ matrices governing the propagation of rays in a camera, in a laser cavity, in multi-layers, and in an optically active medium are presented. It is observed that they serve as analog computers for the essential features of Wigner's little groups dictating the internal space–time symmetries of relativistic particles.

In Appendix A we give a basic discussion of group theory. The treatment given here is not meant to be complete. However, we organize the material in such a way that the reader can acquire a basic introduction to and understanding of group theory. We start from the basic concept of groups, and give a discussion of subgroups, cosets, and invariant subgroups as well as of equivalence classes, orbits, and little groups. This is followed by a presentation of group representations and representation spaces. Included as well are the properties of matrices and Schur's lemma.

In Appendix B we concentrate our discussion on continuous groups. We start from the basic concepts of Lie groups and then give the basic theorems associated with them. Next we discuss the properties of Lie algebras and Lie groups. We present without proof some further theorems concerning Lie groups.

The scope of the Poincaré group is much wider than this book can formally cover. Hence, many illustrative exercises and problems are included in most Chapters.

References

1. E.P. Wigner, On Unitary Representations of the Inhomogeneous Lorentz Group, The Annals of Mathematics **40**(1), 149–204 (1939). DOI 10.2307/1968551. URL http://www.jstor.org/stable/1968551?origin=crossref

2. E.P. Wigner, *Group Theory: And its Application to the Quantum Mechanics of Atomic Spectra* (Academic Press, New York, NY, USA, 1959). ISBN 978-0127505503. (Originally published as: Gruppentheorie und ihre Anwendung auf die Quantenmechanik der Atomspektren, Springer Verlag, Braunscheig, Germany 1931.)

3. V. Bargmann, M.L. Goldberger, S.B. Treiman, J.A. Wheeler, A. Wightman, To Eugene Paul Wigner on his Sixtieth Birthday, Reviews of Modern Physics **34**(4), 587–591 (1962). DOI 10.1103/RevModPhys.34.587. URL https://link.aps.org/doi/10.1103/RevModPhys.34.587

4. P.A.M. Dirac, Forms of Relativistic Dynamics, Reviews of Modern Physics **21**(3), 392–399 (1949). DOI 10.1103/RevModPhys.21.392. URL https://link.aps.org/doi/10.1103/RevModPhys.21.392

5. R. Hofstadter, R.W. McAllister, Electron Scattering from the Proton, Physical Review **98**(1), 217–218 (1955). DOI 10.1103/PhysRev.98.217. URL https://link.aps.org/doi/10.1103/PhysRev.98.217

6. M. Gell-Mann, *The Eightfold Way: A theory of strong interaction symmetry*, vol. TID-12608;CTSL-20 (California Institute of Technology, Synchrotron, Laboratory, Pasadena, CA, 1961). DOI 10.2172/4008239

7. M. Gell-Mann, Symmetries of Baryons and Mesons, Physical Review **125**(3), 1067–1084 (1962). DOI 10.1103/PhysRev.125.1067. URL https://link.aps.org/doi/10.1103/PhysRev.125.1067

8. M. Gell-Mann, A Schematic Model of Baryons and Mesons, Physics Letters **8**(3), 214–215 (1964). DOI 10.1016/S0031-9163(64)92001-3. URL http://linkinghub.elsevier.com/retrieve/pii/S0031916364920013

9. P.A.M. Dirac, Unitary Representations of the Lorentz Group, Proceedings of the Royal Society A: Mathematical, Physical, and Engineering Sciences **183**(994), 284–295 (1945). DOI 10.1098/rspa.1945.0003. URL http://rspa.royalsocietypublishing.org/cgi/doi/10.1098/rspa.1945.0003

10. A. Pais, *Inward bound: of matter and forces in the physical world*, reprint edn. (Clarendon Press, Oxford, Enland, 2002). ISBN 9780198519973. (Originally published 1986; reprinted with corrections 1994.)

11. E. Wigner, Über nicht kombinierende Terme in der neueren Quantentheorie. II. Teil, Zeitschrift für Physik **40**(11-12), 883–892 (1927). DOI 10.1007/BF01390906. URL http://link.springer.com/10.1007/BF01390906

12. P.A.M. Dirac, The Quantum Theory of the Emission and Absorption of Radiation, Proceedings of the Royal Society A: Mathematical, Physical, and Engineering Sciences **114**(767), 243–265 (1927). DOI 10.1098/rspa.1927.0039. URL http://rspa.royalsocietypublishing.org/cgi/doi/10.1098/rspa.1927.0039

13. P.A.M. Dirac, The Quantum Theory of Dispersion, Proceedings of the Royal Society A: Mathematical, Physical, and Engineering Sciences **114**(769), 710–728 (1927). DOI 10.1098/rspa.1927.0071. URL http://rspa.royalsocietypublishing.org/cgi/doi/10.1098/rspa.1927.0071

14. P.A.M. Dirac, A Remarkable Representation of the 3 + 2 de Sitter Group, Journal of Mathematical Physics **4**(7), 901–909 (1963). DOI 10.1063/1.1704016. URL http://aip.scitation.org/doi/10.1063/1.1704016

15. V. Bargmann, Irreducible Unitary Representations of the Lorentz Group, The Annals of Mathematics **48**(3), 568–640 (1947). DOI 10.2307/1969129. URL http://www.jstor.org/stable/1969129?origin=crossref

16. Harish-Chandra, Infinite irreducible representations of the Lorentz group, Proceedings of the Royal Society of London. Series A. Mathematical and Physical Sciences **189**(1018), 372–401 (1947). DOI 10.1098/rspa.1947.0047. URL https://royalsocietypublishing.org/doi/10.1098/rspa.1947.0047

Chapter 2
Theory of the Poincaré Group

Abstract Wigner introduced group theory to physics in 1927 and since then it has been an essential tool in almost all branches of theoretical physics. The study of space-time coordinate transformations and the construction of explicit representations of non-compact groups, particularly fundamental space-time symmetries in the four-dimensional Minkowski space is now of great interest. The Poincaré group occupies an important place here. The purpose of this book is to discuss the representations of the Poincaré group which is non-compact and which is neither simple nor semisimple. We study first the four-by-four Lorentz transformation matrices and the resulting Lie algebra. We then study the orbits and little groups of the Lorentz group, in preparation for constructing representations of the Poincaré group, the little group decomposition of the Poincaré group, and the Casimir operators for each little group. The problem of constructing unitary representations of Lorentz transformations is discussed as well as the finite-dimensional representations, which are non-unitary and commonly used in quantum field theory. We also discuss space and time reflections applicable to the representations frequently used in physics, with a particular emphasis on little groups.

Since 1927, the time when Wigner introduced group theory in the realm of quantum mechanics, it has become an essential tool in almost all branches of theoretical physics [1]. Just to name a few we may observe the use of group theory in the covariant formulation of quantum electrodynamics (QED) [2], the quark model of quantum chromodynamics (QCD) [3], and also in supersymmetry and string theory [4]. As is manifested in the study of gauge theories of gravitation [5, 6], unified theories of gravity such as the Kaluza-Klein theory [7], classical and quantum optics [8], we are becoming more interested in studying space-time coordinate transformations and in constructing explicit representations of non-compact groups, particularly fundamental space-time symmetries in the four-dimensional Minkowski space. The Poincaré

S. Başkal et al., *Theory and Applications of the Poincaré Group*, Fundamental Theories of Physics 217, https://doi.org/10.1007/978-3-031-64376-7_2

group occupies an important place in this trend. The purpose of this book is to discuss the representations of the Poincaré group which is non-compact and which is neither simple nor semisimple.

The two-dimensional Euclidean group $E(2)$, apart from being interesting in its own right, plays a very important role both in illustrating mathematical theorems and in constructing representations of the Poincaré group as $E(2)$ and the Poincaré group have many characteristics in common.

The $E(2)$ group consists of rotations and translations in the xy-plane, resulting in the linear transformation:

$$\begin{aligned} x &= x_0 \cos\theta - y_0 \sin\theta + u \,, \\ y &= x_0 \sin\theta + y_0 \cos\theta + v \,. \end{aligned} \tag{2.1}$$

This is the rotation of the coordinate system by angle θ followed by the translation of the origin to (u, v). In matrix notation,

$$\begin{pmatrix} 1 \\ 1 \\ x \\ y \end{pmatrix} = \begin{pmatrix} 1 & 0 & 0 & 0 \\ 0 & 1 & 0 & 0 \\ 0 & u & \cos\theta & -\sin\theta \\ 0 & v & \sin\theta & \cos\theta \end{pmatrix} \begin{pmatrix} 1 \\ 1 \\ x_0 \\ y_0 \end{pmatrix}, \tag{2.2}$$

where $x^\mu = (t, z, x, y)$ is used as the four-dimensional Minkowskian coordinate system.

The determinant of the above transformation matrix is 1, and the inverse transformation takes the form

$$\begin{pmatrix} 1 \\ 1 \\ x_0 \\ y_0 \end{pmatrix} = \begin{pmatrix} 1 & 0 & 0 & 0 \\ 0 & 1 & 0 & 0 \\ 0 & u' & \cos\theta & \sin\theta \\ 0 & v' & -\sin\theta & \cos\theta \end{pmatrix} \begin{pmatrix} 1 \\ 1 \\ x \\ y \end{pmatrix}, \tag{2.3}$$

where

$$u' = -(u \cos\theta + v \sin\theta) \,, \qquad v' = u \sin\theta - v \cos\theta \,. \tag{2.4}$$

Considering Eq. (1.7), one can obtain the regular representation of $E(2)$

$$X_\theta = \begin{pmatrix} 0 & 0 & 0 & 0 \\ 0 & 0 & 0 & 0 \\ 0 & 0 & 0 & -i \\ 0 & 0 & i & 0 \end{pmatrix}, \quad X_u = \begin{pmatrix} 0 & 0 & 0 & 0 \\ 0 & 0 & 0 & 0 \\ 0 & i & 0 & 0 \\ 0 & 0 & 0 & 0 \end{pmatrix}, \quad X_v = \begin{pmatrix} 0 & 0 & 0 & 0 \\ 0 & 0 & 0 & 0 \\ 0 & 0 & 0 & 0 \\ 0 & i & 0 & 0 \end{pmatrix} \tag{2.5}$$

in matrix form.

Furthermore, the matrix $[a^k_\sigma(x^j)]$ of Eq.(1.8) can be found as:

$$[a^k_\sigma(x^j)] = \begin{pmatrix} -y & 1 & 0 \\ x & 0 & 1 \end{pmatrix} \tag{2.6}$$

then the generators take the form

$$X_\theta(x^j) = -i\left(x\frac{\partial}{\partial y} - y\frac{\partial}{\partial x}\right),$$

$$X_u(x^j) = -i\frac{\partial}{\partial x}, \quad X_v(x^j) = -i\frac{\partial}{\partial y}. \tag{2.7}$$

Both Eq. (2.5) and Eq. (2.7) satisfy the commutation relations:

$$[X_\theta(x^j), X_u(x^j)] = iX_v(x^j), \quad [X_\theta(x^j), X_v(x^j)] = -iX_u(x), \tag{2.8}$$
$$[X_u(x^j), X_v(x^j)] = 0.$$

The details of these calculations are in Appendix B.

We use the two-dimensional Euclidean group extensively, for the following reasons:

(**a**) Transformations of the $E(2)$ group are mathematically simple. They can be explained in terms of the two-dimensional geometry which can be sketched on a piece of paper.
(**b**) Like the Poincaré group, $E(2)$ is non-compact and contains an Abelian invariant subgroup.
(**c**) The $E(2)$ group is isomorphic to the little group of the Poincaré group for relativistic massless particles, i.e. it is neither simple nor semisimple.
(**d**) Like the three-dimensional rotation group, $E(2)$ is a three-parameter Lie group. It can be obtained by contracting $O(3)$.

The Poincaré group is the group of inhomogeneous Lorentz transformations, namely Lorentz transformations followed by space-time translations. In order to study this group, we have to understand first the group of Lorentz transformations [9, 10], the group of translations, *and* how these two groups are combined to form the Poincaré group.

As far as the coordinate transformations are concerned, the study of representations is a matter of constructing four-by-four and five-by-five matrices for the Lorentz and Poincaré groups respectively. However, we are interested in the symmetry problems associated with quantum states and operators. We are particularly interested in Lorentz-transformation properties of Hilbert spaces and in operators acting on a Hilbert space.

To attack these problems, we use the original method of Wigner based on the little groups which leave the four-momentum of a given free particle invariant [11]. We should realize however that the mathematical techniques available to today's physicists are much richer and far more diverse than those of 1939. We should note in particular the following points:

(**a**) While the Poincaré group is a continuous group, Wigner did not employ the mathematical techniques available for Lie groups in his 1939 paper. The fact that the generators can serve useful purposes was not recognized until the appearance of the paper of Bargmann and Wigner [12].

(b) The concept of covering group is widely known by physicists these days. The covering group for the Lorentz group is $SL(2, c)$ and is double-valued. For this reason, we do not have to repeat the lengthy argument Wigner gave on the double-valuedness of the representation.

(c) In his 1939 paper, Wigner was mostly interested in constructing representations of the little groups of the Poincaré group, although the concept of orbit completion was also clearly spelled out. These days, the study of relativistic particles includes calculations of orbit completion. The word *orbit completion*, in practical terms, means the calculation of Lorentz boosts.

(d) In 1939, electrons and photons were known to exist. Wigner's 1939 paper makes contact with the physical world through the representation for electrons. We now have a better understanding of photons and neutrinos. In addition, we now have quarks, hadrons, and vector mesons. It is therefore necessary to extend Wigner's original form of the representation theory to include explanations for these new particles.

(e) In 1939, there did not exist a covariant formulation of quantum field theory. While Wigner was primarily interested in unitary representations in his original 1939 paper, finite-dimensional non-unitary representations play important roles in quantum field theory. It is therefore of interest to add discussions of non-unitary representations when we study representations of the Poincaré group.

In spite of the development of quantum field theory in which finite-dimensional non-unitary representations are commonly used, Wigner's original proposal to construct unitary representations is still the most fundamental issue. We have not forgotten this original proposal, and shall discuss in detail the method of constructing concrete models in this Chapter and continue the discussion is Chaps. 5 and 6. Chap. 10 includes a discussion of the unitary representation of the photon polarization vectors.

In Sect. 2.1, we study the four-by-four Lorentz transformation matrices and the resulting Lie algebra. In Sect. 2.2, we study the orbits and little groups of the Lorentz group, in preparation for constructing representations of the Poincaré group. In Sect. 2.3, the little group decomposition of the Poincaré group is carried out, and the Casimir operator is constructed for each little group.

Section 2.4 deals with the problem of constructing unitary representations of Lorentz transformations applicable to wave functions. It is shown that the problem reduces to that of finding a Hilbert space consisting of functions square integrable over the spatial and time-like coordinates.

We discuss in Sect. 2.5 finite-dimensional representations commonly used in quantum field theory. The little group and the corresponding orbit completion are discussed for each type of field. In Sect. 2.6, we discuss space and time reflections applicable to the representations commonly used in physics, with a particular emphasis on the little groups. Section 2.7 consists of exercises and problems.

2.1 Group of Lorentz Transformations

We shall use throughout this book the four-vector notation:

$$x_\mu = (t, -z, -x, -y) \quad \text{and} \quad x^\mu = (t, z, x, y) \,. \tag{2.9}$$

The reordering of the coordinates as above proves to be convenient, as things go in many cases we can ignore the x, y coordinates and work only with the t, z coordinates, without loss of account. The coordinate z is distinctive since in special relativity it is common practice to boost along that direction. Also in optics, z is chosen to be the optical axis.

The conversion from x^μ to x_μ is achieved through the relation:

$$x_\mu = \eta_{\mu\nu} x^\nu \tag{2.10}$$

where $\eta_{00} = -\eta_{11} = -\eta_{22} = -\eta_{33} = 1$ while all other $\eta_{\mu\nu}$ are zero. It is often convenient to use the matrix:

$$(\eta_{\mu\nu}) = \begin{pmatrix} 1 & 0 & 0 & 0 \\ 0 & -1 & 0 & 0 \\ 0 & 0 & -1 & 0 \\ 0 & 0 & 0 & -1 \end{pmatrix} \,. \tag{2.11}$$

We shall use the notation x_i to cover the last three space-like components of the four-vector and use x_0 for the first time-like component:

$$x^0 = x_0 = t, \quad x_i = -x^i = (-z, -x, -y) \,. \tag{2.12}$$

The lower case Greek and Roman letters will be used for the four-vector and three-component spatial vectors, respectively.

The Lorentz group consists of all real four-by-four matrices A satisfying the condition:

$$A\eta A^T = \eta, \tag{2.13}$$

where A^T is the transpose of A. The A matrix leaves the quantity $(t^2 - z^2 - x^2 - y^2)$ invariant. This group is therefore often called $O(3, 1)$. If a^μ_ν are the elements of the matrix A, where μ and ν are the row and column indices respectively, the above equation can be written as

$$\eta_{\mu\nu} a^\mu_\sigma a^\nu_\lambda = \eta_{\sigma\lambda} \,. \tag{2.14}$$

This expression contains ten conditions on the sixteen elements of the four-by-four matrix A. Therefore, the Lorentz transformation matrix contains six independent parameters.

Since the determinant of A^T is the same as that of A, the condition of Eq. (2.13) leads to

$$\det(A) = \pm 1. \tag{2.15}$$

The Lorentz group with $\det(A) = \pm 1$ is called $O(3, 1)$. Furthermore, if we take the $\sigma = \lambda = 0$ component of Eq. (2.14),

$$|a_0^0|^2 = 1 + \sum_{i=1}^{3} |a_0^i|^2 . \tag{2.16}$$

Thus

$$|a_0^0| \geq 1. \tag{2.17}$$

Transformations with positive a_0^0 are called orthochronous Lorentz transformations. The group of orthochronous Lorentz transformations with $\det(A) = 1$ is called the *proper Lorentz group*.

The proper Lorentz group is an invariant subgroup of the general Lorentz group. $O(3, 1)$ can therefore be expanded in terms of the cosets of the proper Lorentz group. The cosets in this case form a quotient group isomorphic to the four-element group consisting of the following matrices:

$$I(++) = \begin{pmatrix} 1 & 0 \\ 0 & 1 \end{pmatrix}, \qquad I(+-) = \begin{pmatrix} 1 & 0 \\ 0 & -1 \end{pmatrix},$$

$$I(-+) = \begin{pmatrix} -1 & 0 \\ 0 & 1 \end{pmatrix}, \qquad I(--) = \begin{pmatrix} -1 & 0 \\ 0 & -1 \end{pmatrix}. \tag{2.18}$$

These matrices represent the four group of type B discussed in Appendix A. The $I(++)$ component represents the proper Lorentz group, and is the identity. The upper-left element of the above matrix specifies the space-inversion property. If the element is (-1), then the spatial coordinates are to be inverted. The lower-right element is for the time reflection property. If (-1), the time variable is to change its sign. The sign of the determinant of the above matrix is the same as that of the four-by-four Lorentz transformation matrix.

The matrices which perform rotations on the three spatial coordinates form the $O(3)$-like rotation subgroup of the proper Lorentz group. The coordinate transformation matrix takes the reduced form

$$A_r(\theta, \phi, \psi) = \begin{pmatrix} 1 & 0 & 0 & 0 \\ 0 & & 3-by-3 & \\ 0 & & Rotation & \\ 0 & & Matrix & \end{pmatrix}, \tag{2.19}$$

where ψ is the amount of rotation, and the angle variables θ and ϕ specify the direction of the rotation axis. As is well known, the rotation group has three independent parameters.

The boost matrix along the z-direction takes the form

$$A_b(z, \eta) = \begin{pmatrix} \cosh\eta & \sinh\eta & 0 & 0 \\ \sinh\eta & \cosh\eta & 0 & 0 \\ 0 & 0 & 1 & 0 \\ 0 & 0 & 0 & 1 \end{pmatrix}. \tag{2.20}$$

A pure boost along the (θ, ϕ)-direction is achieved by

$$A_b(\theta, \phi, \eta) = A_r(\theta, \phi, 0) A_b(z, \eta) [A_r(\theta, \phi, 0)]^{-1}. \tag{2.21}$$

A pure boost also has three independent parameters.

While rotation matrices are orthogonal:

$$(A_r)^{-1} = (A_r)^T, \tag{2.22}$$

it is easy to prove using Eq. (2.21) that pure boost matrices are symmetric:

$$A_b = (A_b)^T. \tag{2.23}$$

A multiplication of two A_r matrices results in another A_r. However, it is easy to check that a multiplication of two different A_b matrices does not give another A_b matrix. Instead,

$$A_b A_{b'} = A_r A_{b''}. \tag{2.24}$$

The most general transformation matrix for the proper Lorentz group can be decomposed into the form:

$$A = A_r A_b, \tag{2.25}$$

where both A_r and A_b have their respective three independent parameters. The proper Lorentz group is therefore a six-parameter group.

The transformation operators for the proper Lorentz group can be written in exponential form as

$$A = \exp\left[-i \sum_{i=1}^{3} (\theta_i J_i + \eta_i K_i)\right], \tag{2.26}$$

where J_i and K_i are the generators of rotations and proper Lorentz boosts respectively. They form six independent generators of the Lorentz group. When applied to the four-vector x^μ, namely the column vector $(t, z, x, y)^T$, the generators take the matrix form:

$$J_1 = \begin{pmatrix} 0 & 0 & 0 & 0 \\ 0 & 0 & 0 & i \\ 0 & 0 & 0 & 0 \\ 0 & -i & 0 & 0 \end{pmatrix}, \qquad K_1 = \begin{pmatrix} 0 & 0 & i & 0 \\ 0 & 0 & 0 & 0 \\ i & 0 & 0 & 0 \\ 0 & 0 & 0 & 0 \end{pmatrix},$$

$$J_2 = \begin{pmatrix} 0 & 0 & 0 & 0 \\ 0 & 0 & -i & 0 \\ 0 & i & 0 & 0 \\ 0 & 0 & 0 & 0 \end{pmatrix}, \qquad K_2 = \begin{pmatrix} 0 & 0 & 0 & i \\ 0 & 0 & 0 & 0 \\ 0 & 0 & 0 & 0 \\ i & 0 & 0 & 0 \end{pmatrix},$$

$$J_3 = \begin{pmatrix} 0 & 0 & 0 & 0 \\ 0 & 0 & 0 & 0 \\ 0 & 0 & 0 & -i \\ 0 & 0 & i & 0 \end{pmatrix}, \qquad K_3 = \begin{pmatrix} 0 & i & 0 & 0 \\ i & 0 & 0 & 0 \\ 0 & 0 & 0 & 0 \\ 0 & 0 & 0 & 0 \end{pmatrix}. \qquad (2.27)$$

In general, the generators of the proper Lorentz groups applicable to an arbitrary function of the coordinate variables are

$$J_i = -i\varepsilon_{ijk}x^j \left(\partial/\partial x^k \right),$$
$$K_i = i \left[t \left(\partial/\partial x^i \right) - x_i \left(\partial/\partial t \right) \right], \qquad (2.28)$$

with $x^i = (z, x, y)$ and $i = 3, 1, 2$. These generators satisfy the following commutation relations:

$$\left[J_i, J_j \right] = i\varepsilon_{ijk}J_k, \qquad \left[J_i, K_j \right] = i\varepsilon_{ijk}K_k, \qquad \left[K_i, K_j \right] = -i\varepsilon_{ijk}J_k. \qquad (2.29)$$

The first commutator indicates that transformations generated by J_i form a rotation subgroup of the proper Lorentz group. The second and third commutators imply that boosts alone do not form a group, and that a multiplication of two pure boosts will result in a multiplication of a boost and a rotation. The rotation subgroup is not an invariant subgroup of the Lorentz group.

The six generators given in Eq. (2.28) can be combined into the covariant form

$$L_{\mu\nu} = i \left\{ x_\mu \left(\partial/\partial x^\nu \right) - x_\nu \left(\partial/\partial x^\mu \right) \right\}, \qquad (2.30)$$

where

$$J_i = \frac{1}{2}\varepsilon_{ijk}L_{jk} \qquad \text{and} \qquad K_i = L_{oi} . \qquad (2.31)$$

From the transformation properties of x^μ and x_μ and those of their derivatives, it is clear that $L_{\mu\nu}$ transforms like a second-rank tensor:

$$L'_{\mu\nu} = a_\mu^\rho a_\nu^\sigma L_{\rho\sigma} \qquad (2.32)$$

under the Lorentz transformation. The commutation relations for $L_{\mu\nu}$ are [12]:

$$[L_{\mu\nu}, L_{\rho\sigma}] = i(\eta_{\mu\rho}L_{\nu\sigma} + \eta_{\nu\sigma}L_{\mu\rho} - \eta_{\mu\sigma}L_{\nu\rho} - \eta_{\nu\rho}L_{\mu\sigma}) . \qquad (2.33)$$

2.2 Orbits and Little Groups of the Proper Lorentz Group

According to the definition given in Sect. A.3 of Appendix A, the little group of the proper Lorentz group consists of the subset $A^{(p)}$ of the proper transformation matrices satisfying the condition

$$A^{(p)}p = p, \tag{2.34}$$

where p is a column four-vector. Since the above relation is a linear form, the little group for p is the same as that for αp, where α is an arbitrary real number.

Again, according to the definition given in Sect. A.4 of Apendix A, the orbit of p is the *surface* in the four-dimensional p space on which the quantity:

$$p^2 = p_\mu p^\mu = \pm m^2 \tag{2.35}$$

is constant. There is a one-to-one correspondence between this surface and the coset space $A/A^{(p)}$.

For the proper Lorentz group, there are six such orbits in the p space. The first two orbits consist of two time-like surfaces with positive m^2, and with positive and negative values of p_0 respectively. The third and fourth orbits are the forward and backward light cones respectively. The fifth orbit is a space-like surface with negative values of m^2. The sixth orbit is the origin in the p space where $p_0 = p_1 = p_2 = p_3 = 0$. These orbits are illustrated in Figure 2.1.

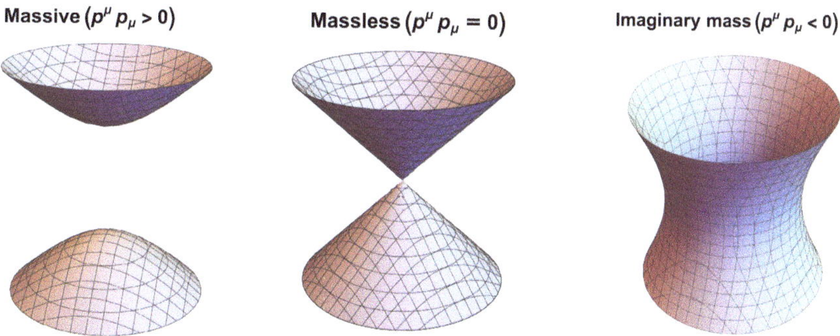

Fig. 2.1: Orbits of the Lorentz group for all possible four-momenta. There are six orbits. The four-momentum can be time-like ($p^\mu p_\mu > 0$), light-like ($p^\mu p_\mu = 0$), space-like ($p^\mu p_\mu < 0$), respectively, or zero which is represented here by the vertex of the double-cone.

These orbits enable us to study all possible little groups by studying the little group for one representative four-vector on each orbit. For the first and second orbits, we can consider the little group as the group of transformations which leave the four-vector $(\pm p_0, 0, 0, 0)$ invariant. The little groups for the third and fourth orbits can be represented by the null vector, which leave $(\pm\omega, \omega, 0, 0)$ invariant. The little group representing the the fifth orbit leaves invariant the four-vector $(0, q, 0, 0)$. The little group for the sixth orbit leaves only the zero-vector invariant.

The task of finding representations of the little groups is now reduced to solving the linear equations:

$$
\begin{pmatrix} a_0^0 & a_1^0 & a_2^0 & a_3^0 \\ a_0^3 & a_1^3 & a_2^3 & a_3^3 \\ a_0^1 & a_1^1 & a_2^1 & a_3^1 \\ a_0^2 & a_1^2 & a_2^2 & a_3^2 \end{pmatrix} \begin{pmatrix} p^0 \\ p^3 \\ p^1 \\ p^2 \end{pmatrix} = \begin{pmatrix} p^0 \\ p^3 \\ p^1 \\ p^2 \end{pmatrix}.
$$
(2.36)

for each representative p subject to the condition $\eta_{\sigma\rho} a_\mu^\sigma a_\nu^\rho = \eta_{\mu\nu}$. Because this matrix equation imposes three additional conditions if the column vector is non-zero, the little groups for the first five orbits have three independent parameters. There are no additional conditions for the sixth orbit. Let us study each little group carefully.

(a) For the first and second orbits, the column vector $(\pm m, 0, 0, 0)$ should remain invariant. From the above matrix equation and from the relation given in Eq. (2.16), it is clear that $a_i^0 = a_0^i = 0$. This condition can be written explicitly as

$$
\begin{pmatrix} 1 & 0 & 0 & 0 \\ 0 & a_3^1 & a_2^3 & a_3^3 \\ 0 & a_1^1 & a_2^1 & a_3^1 \\ 0 & a_1^2 & a_2^2 & a_3^2 \end{pmatrix} \begin{pmatrix} \pm m \\ 0 \\ 0 \\ 0 \end{pmatrix} = \begin{pmatrix} \pm m \\ 0 \\ 0 \\ 0 \end{pmatrix}.
$$
(2.37)

The little group in this case is therefore the rotation subgroup of the proper Lorentz group.

(b) For the third and fourth orbits, we have to consider Lorentz transformations which leave the four-vectors $(\pm\omega, \omega, 0, 0)$ invariant. Let us study first the third orbit represented by $(\omega, \omega, 0, 0)$. It is clear that rotations around the z-axis will leave this four-vector invariant. The rotation matrix in this case takes the form

$$
R(\phi) = \begin{pmatrix} 1 & 0 & 0 & 0 \\ 0 & 1 & 0 & 0 \\ 0 & 0 & \cos\phi & -\sin\phi \\ 0 & 0 & \sin\phi & \cos\phi \end{pmatrix}.
$$
(2.38)

In addition, as is shown in Exercise 1 in Sect. 2.7, the representative four-vector remains invariant under the transformation [11, 13]:

$$T(u, v) = \begin{pmatrix} 1 + (u^2 + v^2)/2 & -(u^2 + v^2)/2 & u & v \\ (u^2 + v^2)/2 & 1 - (u^2 + v^2)/2 & u & v \\ u & -u & 1 & 0 \\ v & -v & 0 & 1 \end{pmatrix}. \tag{2.39}$$

This matrix has the algebraic property:

$$T(u, v) = T(u, 0)T(0, v) = T(0, v)T(u, 0),$$
$$T(u_2, v_2)T(u_1, v_1) = T(u_2 + u_1, v_2 + v_1), \tag{2.40}$$

which is identical to that of the translation subgroup of $E(2)$.

The most general form of the three-parameter four-by-four matrix representing this little group can be written as

$$D(u, v, \phi) = T(u, v)R(\phi). \tag{2.41}$$

The D matrix has the algebraic property:

$$D(u_2, v_2, \phi_2)D(u_1, v_1, \phi_1) = T(u', v')R(\phi_2 + \phi_1), \tag{2.42}$$

where

$$u' = u_2 + u_1 \cos \phi_2 - v_1 \sin \phi_2,$$
$$v' = v_2 + u_1 \sin \phi_2 + v_1 \cos \phi_2. \tag{2.43}$$

This algebraic property is identical to that for the F matrices representing the $E(2)$ group which were discussed in Appendix A.

The calculations are essentially the same for the fourth orbit represented by $(-\omega, \omega, 0, 0)$. The R matrix remains the same. The T matrix in this case is the transpose or the Hermitian conjugate of the T matrix given in Eq. (2.39) which represents an Abelian group. Therefore, the algebraic property of the new D matrix is the same as that of the original D matrix for the third orbit given in Eq. (2.41). Indeed, the little groups for the third and fourth orbits are isomorphic to $E(2)$.

(c) For the fifth orbit, the little group is represented by the three-parameter matrices which leave $(0, q, 0, 0)$ invariant. Here again the rotation around the z-axis will satisfy the requirement. Boosts along the x- and y-axes will also leave the representative four-vector invariant. These boosts together with the rotation around the z-axis form a three-dimensional Lorentz group operating on two spatial dimensions and one time-like dimension. This group is often called $O(2, 1)$.

(d) The sixth orbit consists of only the origin. The six-parameter matrices of the proper Lorentz group will leave this point invariant.

Let us summarize this section by writing down the generators of the little groups and the commutation relations. For the first and second orbits consisting of time-like surfaces in the four-vector spaces, the little group is generated by J_1, J_2 and J_3, and their commutation relations are given in Eq. (2.29).

For the third orbit consisting of the forward light cone, the little group is generated by J_3 given in Eq. (2.27) and by N_1 and N_2 where

$$N_1 = \begin{pmatrix} 0 & 0 & i & 0 \\ 0 & 0 & i & 0 \\ i & -i & 0 & 0 \\ 0 & 0 & 0 & 0 \end{pmatrix}, \qquad N_2 = \begin{pmatrix} 0 & 0 & 0 & i \\ 0 & 0 & 0 & i \\ 0 & 0 & 0 & 0 \\ i & -i & 0 & 0 \end{pmatrix}. \tag{2.44}$$

These matrices are the infinitesimal limits of the $T(u, 0)$ and $T(0, v)$ matrices outlined in Eq. (2.39) and Eq. (2.40). From the matrices given in Eq. (2.27), it is clear that

$$N_1 = K_1 - J_2 \qquad \text{and} \qquad N_2 = K_2 + J_1 . \tag{2.45}$$

These operators have the commutation relations:

$$[N_1, N_2] = 0, \qquad [J_3, N_1] = iN_2 , \qquad [J_3, N_2] = -iN_1 . \tag{2.46}$$

These commutation relations are identical to those for the $E(2)$ group. If the N operators in the above equations are replaced by their respective Hermitian conjugates, the commutation relations remain invariant. The little group for the fourth orbit is generated by J_3 and the Hermitian conjugates of N_1 and N_2.

The little group for the fifth orbit consisting of space-like surfaces is generated by J_3, K_1, and K_2.

$$[K_1, K_2] = -iJ_3 , \qquad [K_1, J_3] = -iK_2 , \qquad [K_2, J_3] = iK_1 . \tag{2.47}$$

As was pointed out before, the little group for the sixth orbit consisting only of the origin is the proper Lorentz group generated by the matrices given in Eq. (2.27).

These little groups will be discussed in more detail in later Chapters in connection with representations of the Poincaré group. In the meantime, we should keep in mind the fact that the local isomorphism does not necessarily mean that groups having the same Lie algebra are identical. Clearly $SU(2)$ is not identical to $O(3)$. The little group for massless particles has the same Lie algebra as that of $E(2)$, but they are quite different groups. We shall use the word *like* to specify the local isomorphism. For instance, $SU(2)$ is an $O(3)$-*like* subgroup of $SL(2, c)$. The little group for massless particles is an $E(2)$-*like* subgroup of the Lorentz group.

2.3 Representations of the Poincaré Group

The study of the Poincaré group starts with the group of inhomogeneous Lorentz transformations on the four-dimensional Minkowski space in the form

$$x'^\mu = a^\mu_\nu x^\nu + b^\mu . \tag{2.48}$$

The matrix performing the above linear transformation is not difficult to construct. Let us consider the five-by-five matrices of the form

$$
A = \begin{pmatrix} & 4-by-4 & & 0 \\ & Lorentz & & 0 \\ & transformation & & 0 \\ & matrix & & 0 \\ 0 & 0 & 0 & 1 \end{pmatrix}, \qquad
B = \begin{pmatrix} 1 & 0 & 0 & 0 & b^0 \\ 0 & 1 & 0 & 0 & b^3 \\ 0 & 0 & 1 & 0 & b^1 \\ 0 & 0 & 0 & 1 & b^2 \\ 0 & 0 & 0 & 0 & 1 \end{pmatrix}. \tag{2.49}
$$

applicable to the five-element column vector $(t, z, x, y, 1)^T$. Then the the inhomogeneous Lorentz transformation of Eq. (2.48) can be written as

$$
x' = B(b)A(a)x, \tag{2.50}
$$

where x, x', a, and b are abbreviations of the their respective quantities with indices. This matrix is indeed the representation of the inhomogeneous Lorentz transformation.

The A matrix has the same properties as the four-by-four Lorentz transformation matrices which we discussed in Sects. 1 and 2. The B matrix representing translations in the four-dimensional Minkowski space is Abelian and is an invariant subgroup of the Poincaré group:

$$
B(b')B(b) = B(b' + b),
$$
$$
A(a)B(b) = B(aba^{-1})A(a). \tag{2.51}
$$

The group BA therefore has the algebraic property:

$$
[B(b')A(a')][B(b)A(a)] = B(b' + a'ba'^{-1})A(a'a). \tag{2.52}
$$

Indeed, the Poincaré group is a semidirect product of the Lorentz group $A(a)$ and the translation group $B(b)$.

Because B is an invariant subgroup, matrices of the form AB, instead of BA, are equally adequate for representing the group. This form of representation is convenient when we study space-time symmetries of a relativistic particle with a given four-momentum p, whose state vector has the momentum dependence in the form

$$
\exp(\pm i p_\mu x^\mu). \tag{2.53}
$$

From this expression and also from the five-by-five matrix form for B, it is clear that the translation is generated by

$$
P_\mu = i\left(\frac{\partial}{\partial x^\mu}\right) = i\left(\frac{\partial}{\partial x^0}, -\frac{\partial}{\partial x^i}\right). \tag{2.54}
$$

From the discussions given in Sects. 2.1 and 2.2, we can write down the most general form for the generators of the Lorentz transformation:

$$M_{\mu\nu} = L^*_{\mu\nu} + L_{\mu\nu},\tag{2.55}$$

where the $L^*_{\mu\nu}$ part is applicable to coordinate variables, while $L_{\mu\nu}$ is applicable to all other variables which are not affected by $L^*_{\mu\nu}$. In the case of massive Dirac particles, the $L_{\mu\nu}$ operators are to generate Lorentz transformations on the Dirac spinor. In the case of photons, we have to construct $L_{\mu\nu}$ applicable to the polarization vector. In the case of particles with a finite size and with internal motion of constituent particles, the $L_{\mu\nu}$ operator has to generate Lorentz transformations on the internal coordinate variables. In this case, the internal coordinate variables are independent of the particle coordinates.

With this in mind, we can write the commutation relations among the generators of the Poincaré group as

$$\begin{aligned}
[P_\mu, P_\nu] &= 0\\
[M_{\mu\nu}, P_\rho] &= i(\eta_{\mu\rho}P_\nu - \eta_{\nu\rho}P_\mu)\\
[M_{\mu\nu}, M_{\rho\sigma}] &= i(\eta_{\mu\rho}M_{\nu\sigma} - \eta_{\nu\rho}M_{\mu\sigma} + \eta_{\mu\sigma}M_{\rho\nu} - \eta_{\nu\sigma}M_{\rho\mu}).
\end{aligned}\tag{2.56}$$

The fact that P_μ does not commute with $M_{\mu\nu}$ is a clear indication that the Poincaré group is not a direct product of the Lorentz and translation groups.

In constructing irreducible representations of the group, it is convenient to construct, from the generators of the group, the maximal set of operators which commute with all the generators. These are the Casimir operators discussed in Sect. B.3 of Appendix B. In the case of the Poincaré group, we can show after a lengthy but straightforward calculation [Problem 5 in Section 7] that the Casimir operators are

$$P^2 = P_\mu P^\mu \qquad\text{and}\qquad W^2 = W_\mu W^\mu,\tag{2.57}$$

where

$$W_\mu = \frac{1}{2}\varepsilon_{\mu\nu\alpha\beta}P^\nu M^{\alpha\beta}.\tag{2.58}$$

W_μ transforms like a four-vector but is constrained to satisfy

$$P_\mu W^\mu = 0.\tag{2.59}$$

The four-vector W therefore has only three independent components. Furthermore, it commutes with P_μ:

$$[P_\mu, W_\nu] = 0.\tag{2.60}$$

When the system is a free-particle state, the operator P^2 determines the (mass)^2 of the particle. Because of Eq. (2.60), P_μ in Eqs. (2.57) and (2.58) can be replaced by its eigenvalue p_μ. We are then led to suspect that the three independent components of W_μ are the generators of the little group at p^μ. Let us examine this point for each orbit.

(i) For the first and second orbits where $p^\mu = (\pm m, 0, 0, 0)$, $W_0 = 0$, and

$$W_i = \pm(m/2)\varepsilon_{ijk}M_{jk},\tag{2.61}$$

where m is the particle mass. Clearly three W_i's generate an $O(3)$-like little group with its spin oriented in an arbitrary direction. The $L_{ij}{}^*$ part of M_{ij} has no effect on the system, and therefore can be dropped. The L_{ij} part represents the internal angular momentum.

We are already quite familiar with spin-1/2 electrons where the internal angular momentum comes from the spinor representation of the rotation group. For the case of massive vector mesons, represented by a four-vector A^μ with the constraint $p_\mu A^\mu = 0$, the L_{ij} matrices should be those of the $O(3)$-like subgroup of the proper Lorentz group. They are four-by-four matrices. If the particle is a relativistic extended hadron with internal quark coordinate y_i, then the L_{ij} operators take the form

$$L_{ij} = i[y_i(\partial/\partial y_j) - y_j(\partial/\partial y_i)] . \tag{2.62}$$

The internal space-time symmetry of the extended hadron will be studied in detail in Chaps. 5, 6, 12, and 13.

(ii) The third and fourth orbits are specified by the four-vectors $p^\mu = (\pm\omega, \omega, 0, 0)$. Here again the L_{ij}^* part drops out, and

$$\begin{aligned}
W_1 &= \omega(L_{02} - L_{23}) , \\
W_2 &= \omega(L_{01} - L_{31}) , \\
W_3 &= W_0 = \omega L_{12} .
\end{aligned} \tag{2.63}$$

W_0 in this case is redundant. It is clear that the first three W operators are proportional to the generators of the $E(2)$-like little groups discussed in Sect. 2.2. The massless particle most familiar to us is of course the photon, which is commonly represented by the four-vector potential A^μ. The photon is known to have spin 1 parallel or antiparallel to its momentum. A^μ depends on gauge parameters which do not produce any measurable effect. Can the symmetry of the little group explain all these? The matrices of the little group generated by W_i operators applicable to A^μ are the four-by-four matrices $D(u, v, \phi)$ given in Eq. (2.41).

There are also interesting stories about massless particles which have spin-1/2. The basic difference between photons and massless spin-1/2 particles is that the latter are polarized whereas the photon spin can be both parallel and antiparallel. Does the symmetry of the little group play any role in the polarization of massless spin-1/2 particles? This problem, together with other massless particles including photons, will be discussed in Chap. 10.

(iii) For the fifth orbit specified by the four-vector $p^\mu = (0, q, 0, 0)$, the W operators are

$$W_1 = qL_{02}, \quad W_2 = qL_{10}, \quad W_3 = 0, \quad W_0 = qL_{12}. \tag{2.64}$$

The three non-vanishing generators are proportional to $J_3 = L_{12}$, $K_1 = L_{02}$, and $K_2 = L_{10}$. These generators satisfy the commutation relations for the $O(2, 1)$ group given in Eq. (2.47).

(iv) The sixth orbit is specified by a four-momentum whose components vanish. The W operators do not exist in this case. Since this null four-vector is invariant under Lorentz transformations, there is no need for orbit completion.

The little groups are summarized in the Table 1.1 given in the Chap. 1. They are subgroups of $O(3, 1)$ and $SL(2, c)$. For the case of time-like and light-like p^μ, their representations describe internal space-time symmetries of observed free particles. Although we do not see immediate physical applications of the little groups for the space-like and point-like orbits, they are of mathematical interest and should be included in the discussion of the Poincaré group.

The $O(3)$-like little group for massive particles is compact and can be studied with the techniques developed for the three-dimensional rotation group. The $O(2, 1)$-like little group is non-compact and its representations are either unitary and infinite-dimensional or non-unitary and finite-dimensional. The same is true for representations of $O(3, 1)$. The $E(2)$-like little group for massless particles is expected to have the same algebraic property as $E(2)$. As was discussed in Appendix B, the $E(2)$ group is non-compact, but it has an Abelian invariant subgroup. We shall see how this affects its representations in Chap. 10.

Let us summarize the method of constructing representations of the Poincaré group. This group is a ten-parameter Lie group and is a semi-direct product of the four-parameter space-time translation group and the six-parameter group of Lorentz transformations. We are interested here only in those representations which are relevant to physics. The fundamental contribution Wigner made in 1939 [11] was to observe that the little group which leaves invariant the four-momentum of a given relativistic particle describes the internal space-time symmetry. In the case of electrons, for instance, the $O(3)$-like little group describes the spin angular momenta of electrons.

Indeed, Wigner's little group serves as the starting point for a more ambitious program for studying representations of the Lorentz group. After constructing representations of the little group for the four-momentum of a given particle, we have to study their Lorentz transformation properties as the four-momentum traces its orbit. This process of orbit completion is necessarily the study of a non-compact group, whose finite-dimensional representations are not unitary, and whose unitary representations are infinite-dimensional.

The little group is definitely a subgroup of the Lorentz group. Indeed, the above-mentioned orbit completion leads to the idea of constructing representations of a given group starting from those of its subgroups. The representation constructed in this manner is called the induced representation.

When we construct representations of the Poincaré group by taking a semi-direct product of the space-time translation group and the Lorentz group, it is an induced representation. In Appendix A, it was noted that the representations of $E(2)$ can be constructed from those of $O(2)$ and the two-dimensional translation group. This construction is also an induced representation. From the mathematical point of view, the study of the Poincaré group consists of constructing induced representations. The mathematics of induced representation is a very rich subject [14, 15]. However, in this book, we shall restrict ourselves to the examples which explain the physical world.

2.4 Lorentz Transformations of Wave Functions

In Sects. 2.2 and 2.3, we have been preoccupied with the little groups of the Poincaré group whose transformations leave the four-momentum of the free particle invariant. As we noted in Sect. 2.2 we still have to *complete the orbit* of each little group. In order to discuss fully the relativistic space-time symmetry, we have to study the effect of Lorentz transformations which change the four-momentum.

We already know that the Lorentz scalars such as P^2 and W^2 should remain invariant under Lorentz transformations. Since, however, we are dealing with quantum mechanics, we have to study the transformation properties of the Hilbert space and of the operators applicable to the Hilbert space. Indeed, this was the ultimate purpose of Wigner's original paper [11]. The 1948 paper of Bargmann and Wigner contains the following paragraphs [12]:

The wave functions, ψ, describing the possible states of a quantum mechanical system form a linear vector space V which, in general, is infinite dimensional and on which a positive definite inner product (ϕ, ψ) is defined for any two wave functions ϕ and ψ (i.e., they form a Hilbert space). The inner product usually involves an integration over the whole configuration or momentum space and, for particles of higher spin, a summation over the spin indices.

If the wave functions in question refer to a free particle and satisfy relativistic wave equations, there exists a correspondence between the wave functions describing the same state in different Lorentz frames. The transformations considered here form the group of all *inhomogeneous* Lorentz transformations (including translations of the origin in space and time). Let $\psi_{\ell'}$ and ψ_ℓ be the wave functions of the same state in two different Lorentz frames ℓ' and ℓ, respectively. Then $\psi_{\ell'} = U(L)\psi_\ell$, where $U(L)$ is a linear unitary operator which depends on the Lorentz transformation L leading from ℓ to ℓ'.

Bargmann and Wigner continue:

By proper normalization, U is determined by L up to a factor ± 1. ... Moreover, the operator U form a single- or double-valued representation of the inhomogeneous group, i.e., for a succession of two Lorentz transformations L_1 and L_2, we have

$$U(L_1 L_2) = \pm U(L_1)U(L_2) . \tag{1}$$

Since all Lorentz frames are equivalent for the description of our system, it follows that, together with ψ, $U(L)\psi$ is also a possible state viewed from the original Lorentz frame ℓ. Thus, the vector space V contains, with every ψ, all transforms $U(L)\psi$, where L is any Lorentz transformation.

A modern interpretation of the double-valuedness of Bargmann and Wigner's Eq. (1) is that $SL(2, c)$ is the universal covering group for the group of Lorentz transformations. In his original paper [11], Wigner was primarily interested in unitary representations of $U(\Lambda)$ with positive definite metric:

$$\psi^n_{\alpha'} = \sum_k u^n_k(\alpha', \alpha)\psi^k_\alpha, \tag{2.65}$$

with

$$\sum_k |u^n_k(\alpha', \alpha)| = 1, \tag{2.66}$$

where the indices n and k specify quantum states. The summation is expected to extend from zero to infinity for unitary representations. Indeed, the construction of representations satisfying the above requirement is one of the primary purposes of this book.

In order that the representation be unitary, the generators of Lorentz transformations have to be Hermitian. We are familiar with the case in which the rotation operators are Hermitian. However, attempts to construct finite-dimensional matrices lead to non-unitary representations. The last hope therefore is in the *differential forms of the generators* given in Eq. (2.28). Here again there is no problem with the generators of rotations. In order that the boost operators be Hermitian, the operators K_i should satisfy the condition

$$(\phi, K_i\psi) = (K_i\phi, \psi). \tag{2.67}$$

K_i contains a derivative with respect to time. For this reason, the wave functions to which K_i is applicable have to be square integrable in the time variable as well as the spatial variables.

It is not common in quantum mechanics to construct functions square-integrable in the time variable, because there is no Hilbert space associated with this variable. For this reason, we are tempted to consider functions which do not depend on time. In this case, the time derivative in K_i disappears, and K_i is still Hermitian. The resulting unitary representations have been studied in detail by Naimark [16]. However, functions which do not depend on time in one Lorentz frame are not independent of time in other frames. It is therefore essential that we understand the physics of this time dependence before making any attempt to construct representations.

Another complication in constructing the above-mentioned unitary representations is that we are discussing here a free particle with a definite four-momentum. If this is the case, is there any non-trivial space-time distribution? Are there space-time variables which are other than those conjugate to the components of the four-momentum?

By definition, the problem of finding unitary representations of the Lorentz transformation is that of constructing a square-integrable Hilbert space in both space and time variables. We shall discuss in Chap. 5 a concrete model which satisfies this criterion. This problem is much deeper than a simple mathematical construction. The issue is how to construct a localized probability distribution which can be Lorentz-transformed. We shall discuss in Chap. 6 the physical interpretation of the mathematical construction given in Chap. 5. The phenomenological aspects of this formalism will be discussed in Chaps. 12 and 13.

Let us return to the question of the dimensionality of representations. As was noted before, the summation given in Eq. (2.65) is an infinite sum indicating that unitary representations of non-compact groups are infinite-dimensional. If the dimension is to be finite, as in the case of most of the particles in quantum field theory, the representation is not unitary. The question then is whether this lack of unitarity is the final word for the representations we use in field theory.

As we discussed in Sect. A.4 of Appendix A, the word unitarity is associated with the positive definite mctric. As was also pointed out there, the concept of pseudo-unitarity or pseudo-orthogonality plays an important role in both mathematics and physics. In the pseudo-unitary representation, the metric is not positive definite, and some components of the vector space have negative metric.

The four-by-four Lorentz transformation matrix applicable to four-vectors constitute a pseudo-orthogonal representation. The four-by-four transformation matrix applicable to Dirac spinors belong to a pseudo-unitary representation. The Lorentz transformation matrix applicable to the photon four-potential is of course a pseudo-orthogonal matrix. However, in the case of photon four-potentials, it is possible to eliminate the time-like component imposing a gauge condition in a Lorentz-invariant manner. In this case, the transformation matrix becomes unitary. This photon problem will be discussed in Chap. 10. We shall study in Sect. 2.6 of the present Chapter some of the non-unitary representations which commonly appear in the existing literature.

In his original paper [11], Wigner was interested in constructing unitary representations. Theories based on finite-dimensional non-unitary representations are also consistent with Wigner's original spirit, if physical interpretations can be given to the pseudo-unitary representation in question.

2.5 Lorentz Transformations of Free Fields

In quantum field theory, most of the particles and fields have finite components and are Lorentz-transformed by finite-dimensional non-unitary matrices, with the metric which is not positive definite. We explained in Sect. 2.4 why the discussions of these representations does not destroy Wigner's original purpose of constructing unitary representations.

The particles most familiar to us through our experience with quantum field theory are scalar or pseudoscalar mesons, photons, vector mesons, electrons, and neutrinos. The Lorentz transformation properties of these fields based on finite-dimensional matrices have been discussed extensively in the literature. We are interested here in the little groups for these fields.

Let us discuss first the free scalar field which satisfies the homogeneous Klein-Gordon equation. The solution of the Klein-Gordon equation takes the form

$$\phi_p = e^{\pm i p_\mu x^\mu} . \tag{2.68}$$

This solution is in an eigenstate of the four-momentum. If we boost this scalar particle in such a way that it will have the four-momentum p', then p' is obtained through an active Lorentz transformation:

$$p'^{\mu} = a^{\mu}_{\nu} p^{\nu}. \tag{2.69}$$

The transformed solution takes the form

$$\phi_{p'} = e^{\pm i p'_{\mu} x^{\mu}}. \tag{2.70}$$

Because the exponent is a Lorentz scalar, we can make arbitrary Lorentz transformations on p and x simultaneously. If we make a transformation which is inverse to that of Eq. (2.69), then the above expression can be written as

$$\phi_{p'} = e^{\pm i p_{\mu} x'^{\mu}} \tag{2.71}$$

where
$$x^{\mu} = a^{\mu}_{\nu} x'^{\nu}. \tag{2.72}$$

The transformation of x to x' is the inverse of that of the transformation for p in Eq. (2.69). This is a passive transformation. Indeed, we can achieve the same purpose by making an active transformation on p or by making a passive transformation on the space-time coordinate. The scalar particles have no internal space-time structure subject to transformations of the little group.

Let us next consider electromagnetic fields. In spite of the well-known inconvenience associated with gauge degrees of freedom, the four-potential form of electromagnetic fields is still the most useful representation in all branches of physics, especially in quantum field theory. For free photons, the solution of the homogeneous wave equation becomes

$$A^{\mu}(x) = A^{\mu} e^{\pm i p_{\mu} x^{\mu}}. \tag{2.73}$$

The coordinate dependence is only through the exponential form. The Lorentz-transformation property of this exponential form is the same as that for the scalar field. As for the four-vector A^{μ}, we have to apply an active Lorentz transformation on it.

The complication in using the four-vector form is that not all the components are independent. There are in fact only two independent components. For instance, if the momentum of the photon is in the z-direction, we can let the first and second components of A^{μ} vanish :

$$A^{\mu} = (0, 0, A_x, A_y) \tag{2.74}$$

or equivalently impose the conditions:

$$p_{\mu} A^{\mu} = 0 \quad \text{and} \quad \mathbf{p} \cdot \mathbf{A} = 0. \tag{2.75}$$

The first of the above conditions is Lorentz-invariant, but the second is not. For this reason, the Lorentz transformation of the photon four-vector is more complicated than that for the momentum four-vector.

However, we should note that the four-by-four matrix of the little group applicable to the photon case is given in Eq. (2.39). Although this matrix does not change the four-momentum, it changes A^μ. Thus, there is a reason to expect that the little group matrix will play a role in making the conditions of Eq. (2.75) covariant. This problem will be discussed in Chap. 10.

The four-vector representation of the free massive vector particle can also be written as Eq. (2.73), with the subsidiary condition:

$$p_\mu A^\mu = 0 \, . \tag{2.76}$$

Unlike the massless case, there are three independent components. If the particle is at rest, then the first component of A^μ vanishes. The little group in this case is generated by J_1, J_2 and J_3 given in Eq. (2.27). If the particle gains a velocity along the z-direction, we have to perform an active Lorentz boost on the A^μ four-vector using the boost matrix

$$B(\eta) = e^{-i\eta K_3} = \begin{pmatrix} \cosh\eta & \sinh\eta & 0 & 0 \\ \sinh\eta & \cosh\eta & 0 & 0 \\ 0 & 0 & 1 & 0 \\ 0 & 0 & 0 & 1 \end{pmatrix} \, . \tag{2.77}$$

The little group for this moving vector meson is generated by J_1', J_2', and J_3', where

$$J_i' = B(\eta)J_i[B(\eta)]^{-1} \, . \tag{2.78}$$

These generators also satisfy the commutation relations for the three-dimensional rotation group:

$$[J_i', J_j'] = i\varepsilon_{ijk}J_k' \, . \tag{2.79}$$

Free spin-1/2 Dirac particle wave functions satisfy the Dirac equation:

$$\left(i\gamma^\mu \frac{\partial}{\partial x^\mu} + m \right) \psi(x) = 0 \, . \tag{2.80}$$

This equation has four linearly independent solutions. The Dirac matrices satisfy the algebraic relations:

$$\gamma_\mu\gamma_\nu + \gamma_\nu\gamma_\mu = 2\eta_{\mu\nu} \, . \tag{2.81}$$

The smallest matrices satisfying the above commutation relations are four-by-four. We can write them as

$$\gamma_i = \begin{pmatrix} 0 & \sigma_i \\ -\sigma_i & 0 \end{pmatrix}, \tag{2.82}$$

$$\gamma_0 = \begin{pmatrix} 0 & I \\ I & 0 \end{pmatrix}, \qquad \gamma_5 = \begin{pmatrix} I & 0 \\ 0 & -I \end{pmatrix}. \tag{2.83}$$

This form of the Dirac matrices is called the Weyl representation.

In terms of these γ matrices, we can construct the generators of Lorentz transformations:

$$L_{\mu\nu} = (i/4)[\gamma_\mu, \gamma_\nu]. \tag{2.84}$$

In the Weyl representation, this quantity can be written as

$$J_i = \frac{1}{2}\varepsilon_{ijk}L_{jk} = \frac{1}{2}\begin{pmatrix} \sigma_i & 0 \\ 0 & \sigma_i \end{pmatrix}, \tag{2.85}$$

$$K_i = L_{0i} = \frac{1}{2}\begin{pmatrix} i\sigma_i & 0 \\ 0 & -i\sigma_i \end{pmatrix}. \tag{2.86}$$

The form of J_i in the above expression is in block diagonal form. The boost generators K_i are also in block diagonal form. However, the first and second blocks of the boost generators have the opposite sign to each other. In order to see whether this is acceptable, let us go back to $SL(2, c)$ discussed in Sect. B.6 of Appendix B (Problem 7). The generators of $SL(2, c)$ satisfy the same Lie algebra as that for $O(3, 1)$ given in Eq. (2.29). These commutation relations of Eq. (2.29) remain unchanged when the boost operators change their sign. For this reason, the above rotation and boost operators satisfy the same commutation relations as those for $O(3, 1)$ or $SL(2, c)$. Furthermore, they can be combined into the covariant form

$$[L_{\mu\nu}, L_{\rho\sigma}] = i(\eta_{\mu\rho}L_{\nu\sigma} - \eta_{\nu\rho}L_{\mu\sigma} - \eta_{\mu\sigma}L_{\rho\nu} + \eta_{\nu\sigma}L_{\mu\rho}). \tag{2.87}$$

The little group for the Dirac particle is generated by three J_i operators of Eq. (2.85) when the particle is at rest. If the particle gains a velocity along the z-direction, the generators of the little group become

$$J'_i = B(\eta)J_i\left[B(\eta)^{-1}\right], \tag{2.88}$$

where $B(\eta)$ is the boost matrix and takes the form

$$B(\eta) = \begin{pmatrix} e^{\eta/2} & 0 & 0 & 0 \\ 0 & e^{-\eta/2} & 0 & 0 \\ 0 & 0 & e^{-\eta/2} & 0 \\ 0 & 0 & 0 & e^{\eta/2} \end{pmatrix}. \tag{2.89}$$

If the mass of the Dirac particle vanishes, the little group is generated by J_3 in Eq. (2.85), and N_1 and N_2:

$$N_1 = K_1 - J_2, \qquad N_2 = K_2 + J_1, \tag{2.90}$$

where $K_{1,2}$ and $J_{1,2}$ are four-by-four matrices given in Eqs. (2.85) and (2.86). However, in this case, the Dirac equation becomes separated, and the upper two components of the Dirac spinor becomes completely decoupled from the lower components. Therefore, we can choose either the upper component or lower component, by choosing the eigenvalue of γ_5 to be $+1$ or -1 respectively. Accordingly, the matrices representing the little group become two-by-two.

We shall continue the discussion of the symmetry of the Dirac equation in Chap. 3.

2.6 Discrete Symmetry Operations

We have so far studied representations of the Poincaré group and its little groups without paying much attention to the fact that there are also discrete symmetries associated with this group. As was noted in Sect. 2.1, if the space and time reflections are taken into account, the Lorentz group has four branches. In addition, we have to note the fact that wave functions and operators in quantum mechanics are complex, and the results of calculations to be compared with experiment remain unchanged if all mathematical symbols in the formalism are replaced by their complex conjugates. Thus the study of discrete symmetry should include space reflection, time reflection, and complex conjugation.

The primary purpose of studying these discrete symmetry operations is to understand properties of elementary particles under space inversion (P, parity), charge conjugation (C), and time reversal (T). We expect that the momentum will reverse its direction under P and T, and that the angular momentum will change its sign under C and T.

We shall examine these operations using the generators of Lorentz transformations. The above-mentioned discrete symmetry operations will in general change the signs of the generators. The rotation operators correspond to physically measurable dynamical quantities. Thus if they change their sign, the direction of the angular momentum is reversed. However, there is no measurable quantity associated with the boost operators. Furthermore, the change in sign of the boost generators does not necessarily mean that the direction of motion is changed.

In order to examine this point more effectively, let us rewrite the commutation relations for the J_i and K_i operators:

$$[J_i, J_j] = i\varepsilon_{ijk}J_k, \qquad [J_i, K_j] = i\varepsilon_{ijk}K_k, \qquad [K_i, K_j] = -i\varepsilon_{ijk}J_k. \qquad (2.91)$$

It is necessary to change J_i to $-J_i$ when J_i changes direction. However, the commutation relations remain invariant even when the boost generators change sign. Thus the change in sign of K_i alone does not lead to any definite physical conclusion.

In order to study the effect of the discrete symmetry operation on the boost generator, we have to see the final form of the transformation matrix. If a massive particle at rest is boosted along one direction, it will move along that direction. If it is boosted along the opposite direction, it will move in the opposite direction.

Let us examine first the effect of the discrete symmetry operations on the representations based on the differential forms of the generators given in Eq. (2.28). It is quite clear that all the generators will change sign under complex conjugation. However, complex conjugation is not the same as the Hermitian conjugation. The angular momentum operators remain invariant under space and time reflections, while the boost operators change their sign under space reflection or time reflection. These are summarized in Table 2.1.

Table 2.1: Properties of the differential forms of J_i and K_i under discrete symmetry operations.

	Space reflection	Time reflection	Complex conjugation	Hermitian conjugation
For J_i:	J_i	J_i	$-J_i$	J_i
For K_i:	$-K_i$	$-K_i$	$-K_i$	K_i (unitary) $-K_i$ (non-unitary)

Because the generators are pure imaginary, the transformation matrices are real and remain invariant under complex conjugation. Thus the parity operation is achieved by space-reflection. Charge conjugation is simply a complex conjugation. Time reversal is time reflection followed by complex conjugation.

The above-mentioned symmetry operations are applicable to bosons with integer spin in which the differential forms of the generators are used, including photons, vector mesons, and gravitons. Pseudoscalar mesons have an intrinsic parity.

In addition, we would expect that these symmetry operations will be applicable to the unitary representation which we will discuss in Chap. 5. As is seen in Table 2.1, the Hermitian conjugation in the unitary representation is different from that for the non-unitary representation .

Let us next discuss massive Dirac particles. If we use for the γ matrices the expressions given in Eqs. (2.82) and (2.83) and for the generators the expressions in terms of the γ matrices given in Eqs. (2.85) and (2.86), space inversion or P is achieved through

$$\begin{aligned}
(\gamma_0)J_i(\gamma_0)^{-1} &= \quad J_i\,, \\
(\gamma_0)K_i(\gamma_0)^{-1} &= -K_i
\end{aligned} \tag{2.92}$$

followed by space reflection. The parity operation will leave the direction of the spin invariant, but will reverse the direction of the momentum.

Charge conjugation or C requires sign changes in both the rotation and boost generators. This is done through

$$\begin{aligned}
(\gamma_2)J_i^*(\gamma_2)^{-1} &= -J_i\,, \\
(\gamma_2)K_i^*(\gamma_2)^{-1} &= -K_i\,.
\end{aligned} \tag{2.93}$$

This will lead to the spin flip and change in the energy state.

The time reversal or T operation requires a change in sign of J_i:

$$(\gamma_1, \gamma_3)J_i^*(\gamma_1, \gamma_3)^{-1} = -J_i,$$
$$(\gamma_1, \gamma_3)K_i^*(\gamma_1, \gamma_3)^{-1} = K_i. \tag{2.94}$$

This will lead to the reversal in the direction of the angular momentum as well as in the direction of the momentum.

It is now evident that, under the successive operations of PCT, each generator is brought back to itself, or PCT is the identity operation. We often call this effect the PCT theorem applicable to massive fermions. These effects are summarized in Table 2.2. In Table 2.2, we used the following forms for the rotation and boost matrices respectively:

$$R(\mathbf{n}, \alpha) = \exp(-i\alpha\,\mathbf{n} \cdot \mathbf{J}) \tag{2.95}$$

and

$$B(\mathbf{n}, \eta) = \exp(-i\eta\,\mathbf{n} \cdot \mathbf{K}) \tag{2.96}$$

and took into account the fact that complex conjugation is required for C and T.

Table 2.2: Effects of P, C, and T on the generators of Lorentz transformation.

	P	C	T	PCT
J_i	J_i	$-J_i$	$-J_i$	J_i
$R(\mathbf{n}, \alpha)$	R	R	R	R
K_i	$-K_i$	$-K_i$	K_i	K_i
$B(\mathbf{n}, \eta)$	B^{-1}	B	B^{-1}	B

It is easy to study the effect of P, C, and T on little groups for massive particles. For massive Dirac particles, the three-parameter group is generated by J_i which corresponds to the spin of the particle. The direction of the spin remains unchanged under parity operation. However, it changes sign under time reversal and charge conjugation.

For massless Dirac particles, there is no Lorentz frame in which the particle is at rest, and the generators of the little group are linear combinations of J_i and K_i. For this reason, the generators of the little group undergo more changes than a simple change in sign. This will lead to a non-trivial result, which we shall discuss in Chap. 10.

Likewise, the generators of the little group for photons will undergo non-trivial transformations under discrete symmetry operations. However, while neutrinos share some properties with electrons, photons share properties with massive vector mesons.

Unlike photons, there is a Lorentz frame in which the vector meson is at rest. We can study P, C, and T operations in this Lorentz frame. It is important to realize that these operations are not Lorentz-invariant. For example, $I(+, +)$ and $I(-, -)$

matrices in Eq. (2.18) are Lorentz invariant, but $I(-,+)$ and $I(+,-)$ are not. If we boost the $I(+,-)$ matrix along the z-direction,

$$B_z(\eta)I(+,-)B_z(-\eta) = \begin{pmatrix} \cosh\eta & \sinh\eta \\ \sinh\eta & \cosh\eta \end{pmatrix} \begin{pmatrix} 1 & 0 \\ 0 & -1 \end{pmatrix} \begin{pmatrix} \cosh\eta & -\sinh\eta \\ -\sinh\eta & \cosh\eta \end{pmatrix}$$

$$= \begin{pmatrix} \cosh(2\eta) & -\sinh(2\eta) \\ \sinh(2\eta) & -\cosh(2\eta) \end{pmatrix}. \tag{2.97}$$

The above form is quite different from the rest-frame expression for $I(+,-)$.

Finally, for unitary representations in which both the rotation and boost generators are Hermitian, we will have to deal with the differential forms for these operators. It is clear from Tables 2.1 and 2.2 that P is simply a space reflection in this case. C is complex conjugation, and T is complex conjugation followed by time reflection. Since all the generators are imaginary, the transformation matrices are real and are unaffected by complex conjugation.

2.7 Exercises and Problems

Exercise 1. The translation-like matrix of the E(2)-like little group for massless particles is given in Eq. (2.39). Show that this matrix is a product of one rotation matrix and two boost matrices.

In order to solve this problem, we have to construct non-trivial four-by-four transformation matrices which leave the four-momentum invariant. Consider a massless particle moving along the z-direction, whose four-momentum without loss of generality, can be expressed as $(1, 1, 0, 0)$. A Lorentz boost along the x-axis, followed by a rotation around the y-axis can be implemented, which will bring it back to the z-axis. A boost along the z-axis will bring the momentum to its original condition. In matrix notation, these operations can be written as a product of three matrices [17]:

$$T_x(u) = \begin{pmatrix} \frac{1}{2}\omega(2-u^2) & -\frac{1}{2}\omega u^2 & 0 & 0 \\ -\frac{1}{2}\omega u^2 & \frac{1}{2}\omega(2-u^2) & 0 & 0 \\ 0 & 0 & 1 & 0 \\ 0 & 0 & 0 & 1 \end{pmatrix} \begin{pmatrix} 1 & 0 & 0 & 0 \\ 0 & 1/\omega & u & 0 \\ 0 & -u & 1/\omega & 0 \\ 0 & 0 & 0 & 1 \end{pmatrix} \begin{pmatrix} \omega & 0 & \omega u & 0 \\ 0 & 1 & 0 & 0 \\ \omega u & 0 & \omega & 0 \\ 0 & 0 & 0 & 1 \end{pmatrix}. \tag{2.98}$$

where

$$\omega = (1 - u^2)^{-1/2}. \tag{2.99}$$

In the above expression, the first matrix is a boost along the z-direction, the second a rotation around the y-axis, and the third a boost along the x-direction. The multiplication of the above three matrices is straightforward, and

$$T_x(u) = \begin{pmatrix} 1 + \frac{1}{2}u^2 & -\frac{1}{2}u^2 & u & 0 \\ \frac{1}{2}u^2 & 1 - \frac{1}{2}u^2 & u & 0 \\ u & -u & 1 & 0 \\ 0 & 0 & 0 & 1 \end{pmatrix}. \tag{2.100}$$

The above procedure is physically meaningful only when u^2 is smaller than one. If $|u|$ is greater than one, w in Eq. (2.99) becomes pure imaginary. However, the above $T_x(u)$ has the property:

$$T_x(u_1 + u_2) = T_x(u_1)T_x(u_2). \tag{2.101}$$

We can thus divide an arbitrarily large u into u/n, where n is an integer large enough to make $|u/n|$ smaller than one, and repeat the same process n times.

By carrying out a similar procedure on the yz-plane, we can get another four-by-four T matrix:

$$T_y(v) = \begin{pmatrix} 1 + \frac{1}{2}v^2 & -\frac{1}{2}v^2 & 0 & v \\ \frac{1}{2}v^2 & 1 - \frac{1}{2}v^2 & 0 & v \\ 0 & 0 & 1 & 0 \\ v & -v & 0 & 1 \end{pmatrix}. \tag{2.102}$$

Then it is straightforward to show that T_x and T_y matrices commute. Thus the matrices:

$$T(u, v) = T_y(v)T_x(u) \tag{2.103}$$

form a two-parameter Abelian subgroup of the little group.

Exercise 2. Consider a Lorentz boost along the x-direction followed by another along the y-direction. Calculate the resulting Lorentz transformation matrix. Does the resulting matrix represent a pure boost or a boost times a rotation?

Since the z-direction is not affected by these transformations, we can use the vector (t, x, y), and three-by-three matrices to perform boosts. Let us start with the boost matrix along the x-direction:

$$B_x(\xi) = \begin{pmatrix} \cosh \xi & \sinh \xi & 0 \\ \sinh \xi & \cosh \xi & 0 \\ 0 & 0 & 1 \end{pmatrix}. \tag{2.104}$$

We can obtain a boost matrix along the ϕ-direction by rotating the above matrix:

$$B_\phi(\lambda) = R(\phi)B_x(\xi)[R(\phi)]^{-1}, \tag{2.105}$$

where

$$R(\phi) = \begin{pmatrix} 1 & 0 & 0 \\ 0 & \cos \phi & -\sin \phi \\ 0 & \sin \phi & \cos \phi \end{pmatrix}. \tag{2.106}$$

Since the inverse of the rotation matrix is its transpose, boost matrices are symmetric.

Let us next consider the boost matrix along the y-direction:

$$B_y(\eta) = \begin{pmatrix} \cosh\eta & 0 & \sinh\eta \\ 0 & 1 & 0 \\ \sinh\eta & 0 & \cosh\eta \end{pmatrix}. \tag{2.107}$$

Then the boost along the x-direction followed by another boost along the y-direction can be represented by the three-by-three matrix $B_y(\eta)B_x(\xi)$. However, this matrix is not symmetric. For this reason, the resulting matrix cannot be a pure boost matrix. On the other hand, we can still consider the form [18]:

$$B_y(\eta)B_x(\xi) = B_\phi(\lambda)R(\alpha). \tag{2.108}$$

Here, $R(\alpha)$ is appropriately called the Wigner rotation matrix and α is the Wigner rotation angle [19]. The computation of the above matrix equation is straightforward. There are three parameters to determine. There are however nine matrix elements to compare. The result is

$$\cosh\lambda = (\cosh\xi)\cosh\eta,$$
$$\tan\phi = (\coth\xi)\sinh\eta,$$
$$\tan\frac{\alpha}{2} = \tanh\frac{\xi}{2}\tanh\frac{\eta}{2}. \tag{2.109}$$

When the amount of boost is the same for both x and y-directions: $\xi = \eta$,

$$\cosh\lambda = (\cosh\xi)^2, \qquad \tan\phi = \cosh\xi, \qquad \tan\frac{\alpha}{2} = \left(\tanh\frac{\xi}{2}\right)^2. \tag{2.110}$$

This means that, when ξ is very small,

$$\lambda = \sqrt{2}\xi, \quad \phi = 45°, \quad \text{and} \quad \alpha = \xi^2/2. \tag{2.111}$$

This result is consistent with what we expect from non-relativistic kinematics. In fact the nonvanishing ξ is the Thomas effect which manifests itself in atomic spectra.

On the other hand, when ξ is very large,

$$\lambda = 2\xi, \quad \phi = 90°, \quad \text{and} \quad \alpha = 90°. \tag{2.112}$$

This result is somewhat unexpected. However, we should be able to see this effect in connection with massless particles.

A more interesting problem would be two successive Lorentz boosts along arbitrary directions. We can handle this problem using the mathematical technique given here. However, it is not an efficient method to work with nine different equations to determine three parameters. For this reason, we shall discuss a more efficient method in Chap. 8.

Problem 1. Consider two four-vectors V and W and their "inner" product

$$(V, W) = (V_\mu)^* W^\mu. \tag{2.113}$$

Show that, if they are orthogonal to each other: $(V, W) = 0$, and if one of them is time-like, then the other is space-like. See Section 4B of Wigner's paper [11].

Problem 2. The four-by-four Lorentz transformation matrix has four eigenvalues and four eigenvectors. Show that, if λ is an eigenvalue, $1/\lambda$, λ^* and $1/\lambda^*$ are also eigenvalues. Show that the eigenvectors V_1 and V_2 corresponding respectively to λ_1 and λ_2 are orthogonal if $\lambda_1^* \lambda_2 \neq 1$ See Section 4B of Wigner's paper [11].

Problem 3. Show that the complex eigenvalue of the Lorentz transformation matrix has unit modulus, and that the real eigenvalue is positive. See Section 4B of Wigner's paper [11].

Problem 4. Wigner's paper [11] contains a proof that the group of Lorentz transformations is simple. Use Cartan's criterion (see: Appendix B, Sect. B.5) to show that this group is a simple group.

Problem 5. Show that Casimir operators for the Lorentz group are

$$C_1 = \frac{1}{2} M^{\mu\nu} M_{\mu\nu} \qquad \text{and} \qquad C_2 = \frac{1}{4} \varepsilon_{\mu\nu\alpha\beta} M^{\mu\nu} M_{\alpha\beta} \tag{2.114}$$

and show that the operators given in Eq. (2.57) are the Casimir operators for the Poincaré group. Show also that C_1 and C_2 for the Lorentz group cannot serve as the Casimir operators for the Poincaré group.

Problem 6. Solve the matrix equation in Eq. (2.36) starting from the most general form of the real four-by-four matrix with sixteen elements. Use the condition that it be a Lorentz transformation matrix to reduce the number of independent parameters from sixteen to six. For the little groups for non-vanishing four-momenta, solve the matrix equation to reduce further the number of independent parameters from six to three [Halpern and Branscomb [20], Halpern [21]].

Problem 7. Write down the Dirac equation for an electron in an external electromagnetic field. Using this equation, show that the operations on the Dirac matrices in Eqs. (2.92), (2.93), and (2.94) are parity, charge conjugation, and time reversal respectively.

References

1. E. Wigner, Über nicht kombinierende Terme in der neueren Quantentheorie. II. Teil, Zeitschrift für Physik **40**(11-12), 883–892 (1927). DOI 10.1007/BF01390906. URL http://link.springer.com/10.1007/BF01390906
2. R.P. Feynman, *Quantum electrodynamics*. Advanced book classics (CRC Press, Boca Raton London New York, 2018). ISBN 9780201360752
3. H. Fritzsch, *Murray Gell-Mann: Selected Papers, World Scientific Series in 20th Century Physics*, vol. 40 (World Scientific, 2010). DOI 10.1142/7101. ISBN 97898128368479789812836854. URL https://www.worldscientific.com/worldscibooks/10.1142/7101

4. E. Witten, Anti-de Sitter space and holography, Advances in Theoretical and Mathematical Physics **2**(2), 253–291 (1998). DOI 10.4310/ATMP.1998.v2.n2.a2. URL http://www.intlpress.com/site/pub/pages/journals/items/atmp/content/vols/0002/0002/a002/

5. L. O'Raifeartaigh, *Group Structure of Gauge Theories*, 1st edn. (Cambridge University Press, 1986). DOI 10.1017/CBO9780511564031. ISBN 9780521347853,9780521252935, 9780511564031. URL https://www.cambridge.org/core/product/identifier/9780511564031/type/book. (Published on line 2011.)

6. M. Blagojevic, F.W. Hehl (eds.), *Gauge theories of gravitation: a reader with commentaries* (Imperial College Press, London, UK, 2013). ISBN 9781848167261. (Distributed by World Scientific, Singapore; Hackensack, NJ USA; OCLC: 679937453.)

7. R. Kerner, Generalization of the Kaluza-Klein theory for an arbitrary non-abelian gauge group, Annales de l'institut Henri Poincaré. Section A, Physique Théorique **9**(2), 143–152 (1968). URL http://www.numdam.org/item?id=AIHPA_1968__9_2_143_0

8. S. Başkal, Y. Kim, M. Noz, *Mathematical Devices for Optical Sciences.* (IOP Publishing, Bristol, UK, 2019). ISBN 978-0-7503-1612-5. URL https://dx.doi.org/10.1088/2053-2563/aafe78. (OCLC: 1034620988.)

9. S. Başkal, Y.S. Kim, M.E. Noz, *Physics of the Lorentz Group (Second Edition): Beyond high-energy physics and optics* (IOP Publishing, Bristol, UK, 2021). DOI 10.1088/978-0-7503-3607-9. ISBN 978-0-7503-3607-9. URL https://iopscience.iop.org/book/978-0-7503-3607-9

10. M. Carmeli, S. Malin, *Representations of the rotation and Lorentz groups: an introduction.* Lecture notes in pure and applied mathematics; No. 16 (M. Dekker, New York, NY USA, 1976). ISBN 978-0-8247-6449-4

11. E.P. Wigner, On Unitary Representations of the Inhomogeneous Lorentz Group, The Annals of Mathematics **40**(1), 149–204 (1939). DOI 10.2307/1968551. URL http://www.jstor.org/stable/1968551?origin=crossref

12. V. Bargmann, E.P. Wigner, Group Theoretical Discussion of Relatistic Wave Equations, Proc. Nat. Acad. Sci. (USA) **34**(5), 211–223 (1948). DOI 10.1073/pnas.34.5.211. URL https://doi.org/10.1073/pnas.34.5.211

13. E.P. Wigner, Relativistische Wellengleichungen, Zeitschrift für Physik **124**(7-12), 165–184 (1948). DOI 10.1007/BF01668901. URL http://link.springer.com/10.1007/BF01668901

14. G.W. Mackey, *Unitary group representations in physics, probability, and number theory.* The advanced book program (Addison-Wesley, Redwood City, CA, USA, 1989). ISBN 978-0-201-51009-6

15. G. Mackey, *Induced Representations of Groups and Quantum Mechanics* (W. A. Benjamin, Inc./Editore Boringhieri, Boston, MA, USA, 1968). (ASIN B001DOM62C.)

16. M. Naimark, Linear Representation of the Lorentz Group, Usp.Mat. Nauk **9**, 19–93 (1954). ISBN 9781483169170. (Naimark M A 1957 Linear Representation of the Lorentz Group Am. Math. Soc. Transl. Ser. 2 6 379–458 Engl. transl.; Naimark M A Linear Representations of the Lorentz Group International Series of Monographs in Pure and Appliead Mathematics vol 63, First Edition 1964, reprinted 2014, series editor: Farahat, H. K., Oxford UK: Pergamon, Engl. transl.)

17. D. Han, Y.S. Kim, Little group for photons and gauge transformations, American Journal of Physics **49**(4), 348–351 (1981). DOI 10.1119/1.12509. URL http://aapt.scitation.org/doi/10.1119/1.12509

18. D. Han, Y.S. Kim, D. Son, Decomposition of Lorentz transformations, Journal of Mathematical Physics **28**(10), 2373–2378 (1987). DOI 10.1063/1.527773. URL http://aip.scitation.org/doi/10.1063/1.527773

19. D. Han, Y.S. Kim, D. Son, Eulerian parametrization of Wigner's little groups and gauge transformations in terms of rotations in two–component spinors, Journal of Mathematical Physics **27**(9), 2228–2235 (1986). DOI 10.1063/1.526994. URL http://aip.scitation.org/doi/10.1063/1.526994

20. F.R. Halpern, E. Branscomb, *Wigner's Analysis of the Unitary Representations of the Poincaré Group.* (U.S. Atomic Energy Commission, Lawrence Radiation Laboratory, Berkeley, CA,

USA, 1965). URL https://babel.hathitrust.org/cgi/pt?id=mdp.39015077327065&view=1up&seq=6
21. F.R. Halpern, *Special Relativity and Quantum Mechanics* (Prentice-Hall, Englewood Cliffs, NJ USA, 1968). ISBN 9780138271135

Chapter 3
Theory of Spinors

Abstract The groups $SU(2)$ and $SL(2, c)$ are revisited and their correspondence with the rotation group $O(3)$ and the Lorentz group $O(3, 1)$ is given. The group $SL(2, c)$ is the covering group of the Lorentz group, therefore is more elaborate from a mathematical point of view. On the other hand, since they can be represented by two by two matrices with complex parameters, calculations are relatively easier compared to the four-by-four matrices of the Lorentz group. The transformation properties of spinors with regard to the $SL(2, c)$ group is given and their correspondence between the components of a four-vector is made evident. Finding all possible irreducible representations of $SU(2)$ is sketched out. It is also shown that $E(2)$-like groups, which leave the four-momentum of relativistic massless particles invariant, form a subgroup of $SL(2, c)$. The spinor space of the Dirac equation is presented, and the symmetries of the Dirac equation both for the massive and the massless cases are discussed. It is shown that, thanks to those symmetry properties, a plane wave solution to the Dirac equation can be found easily.

One easy way for physicists to understand group theory is in terms of coordinate transformations. Indeed, as we did in Chap. 2, the study of the Lorentz group naturally starts with the group of coordinate transformation matrices operating on four-component Minkowski vectors. The question then is whether those four-by-four matrices are the smallest matrices having the algebraic properties of the proper Lorentz group [1]. The answer to this question is *No*.

The three-dimensional rotation group is locally isomorphic to $SU(2)$ generated by three Hermitian traceless two-by-two matrices. Thus it is quite natural for us to look for six two-by-two matrices which satisfy the same set of commutation relations as that for the Lorentz group. Indeed, there are six traceless complex two-by-two matrices satisfying this requirement. They are called the generators of $SL(2, c)$. We have already begun discussion of this group in connection with the Dirac equation

© The Author(s), under exclusive license to Springer Nature Switzerland AG 2024
S. Başkal et al., *Theory and Applications of the Poincaré Group*, Fundamental Theories of Physics 217, https://doi.org/10.1007/978-3-031-64376-7_3

in Chap. 2. The purpose of this Chapter is to study the general properties of the $SL(2, c)$ group.

There are several advantages in studying this subject. First, there are particles in nature whose space-time symmetry is that of $SL(2, c)$. Second, $SL(2, c)$ is a covering group for the Lorentz group, and therefore more fundamental from a mathematical point of view. Third, since every value of spin can be regarded as a multiple of 1/2, the theory of spinors together with spin addition mechanisms will constitute a theory of angular momenta. Fourth, because we are dealing with two-by-two matrices, calculation of Lorentz transformations is much easier than when dealing with four-by four matrices. Finally, in $SL(2, c)$, the boost generators are simply i times the rotation generators. For this reason, we can discuss this group with three generators, instead of six, by complexifying the group parameters.

In Sect. 3.1, we discuss $SL(2, c)$ as the covering group of the group of Lorentz transformations. Section 3.2 gives a summary of Wigner's energy momentum four-vector matrices. We study in Sect. 3.3 subgroups of $SL(2, c)$, particularly those which correspond to the little groups discussed in Chap. 2. $SU(2)$ is the $O(3)$-like subgroup of $SL(2, c)$. Since this group is discussed extensively in the literature, we give in Sect. 3.4 a summary of what is already known to most readers.

The generators of $SL(2, c)$ satisfy the same set of commutation relations as those for $O(3, 1)$. We are of course interested in knowing more than this local property. Section 3.5 deals with the problem of constructing explicitly the four-by-four Lorentz transformation matrix from the two-by-two matrices of $SL(2, c)$, and with the problem of constructing four-vectors from $SL(2, c)$ spinors. In physics, the representation of $SL(2, c)$ appears most often in the form of the Dirac equation. For this reason, in Sect. 3.6, we discuss in detail the symmetry of the Dirac equation. In Sect. 3.7, we present some of the items not discussed in Sects. 3.1 to 3.6 in the form of exercises and problems.

3.1 SL(2, c) as the Covering Group of the Lorentz Group

Let us consider the group generated by the following six two-by-two matrices:

$$J_i = (1/2)\sigma_i \quad \text{and} \quad K_i = (i/2)\sigma_i, \tag{3.1}$$

where σ_i are the Pauli spin matrices defined as

$$\sigma_1 = \begin{pmatrix} 0 & 1 \\ 1 & 0 \end{pmatrix}, \quad \sigma_2 = \begin{pmatrix} 0 & -i \\ i & 0 \end{pmatrix}, \quad \text{and} \quad \sigma_3 = \begin{pmatrix} 1 & 0 \\ 0 & -1 \end{pmatrix}. \tag{3.2}$$

The above operators take the following explicit forms:

$$J_1 = \begin{pmatrix} 0 & 1/2 \\ 1/2 & 0 \end{pmatrix}, \qquad K_1 = \begin{pmatrix} 0 & i/2 \\ i/2 & 0 \end{pmatrix},$$

$$J_2 = \begin{pmatrix} 0 & -i/2 \\ i/2 & 0 \end{pmatrix}, \qquad K_2 = \begin{pmatrix} 0 & 1/2 \\ -1/2 & 0 \end{pmatrix},$$

$$J_3 = \begin{pmatrix} 1/2 & 0 \\ 0 & -1/2 \end{pmatrix}, \qquad K_3 = \begin{pmatrix} i/2 & 0 \\ 0 & -i/2 \end{pmatrix}. \tag{3.3}$$

These matrices satisfy the commutation relations:

$$[J_i, J_j] = i\varepsilon_{ijk}J_k, \qquad [J_i, K_j] = i\varepsilon_{ijk}K_k, \qquad [K_i, K_j] = -i\varepsilon_{ijk}J_k. \tag{3.4}$$

These relations are identical to those for the proper Lorentz group. We tabulate the the two-by-two $SL(2, c)$ matrices and their corresponding four-by-four matrices of the Lorentz group in Table 3.1.

There is therefore every reason to expect that the group of two-by-two matrices of the form

$$W = \exp\left[-i \sum_{i=1}^{3} (\theta_i J_i + \eta_i K_i)\right] \tag{3.5}$$

will have the same algebraic properties as the proper Lorentz group. This group is often called $SL(2, c)$. Because the generators are traceless, the determinant of W is one. $SL(2, c)$ is therefore an unimodular group. Since the K_i matrices are not Hermitian, $SL(2, c)$ is not a unitary group.

We note in Eq. (3.4) that the Lie algebra for the generators is invariant under the sign change of the boost operators K_i. In the case of $O(3, 1)$ which was discussed in Chap. 2 the sign of the boost generators K_i is unambiguously defined in terms of the space and time variables. However, in the present case of $SL(2, c)$, we have to consider both signs. Since the sign change can be performed easily, we shall choose the positive sign given in Eq. (3.3) unless otherwise required.

Let us look at the expression of Eq. (3.5). Since K_i is just i times J_i, W of Eq. (3.5) can be written as

$$W = \exp\left[-\frac{i}{2}(\zeta_1 \sigma_1 + \zeta_2 \sigma_2 + \zeta_3 \sigma_3)\right] \tag{3.6}$$

with

$$\zeta_i = \theta_i + i\eta_i. \tag{3.7}$$

This means that, when needed, we can study $SL(2, c)$ with three generators and three complex parameters.

In order to see the correspondence with the matrices of Lorentz transformations, let us note first that the three-parameter subgroup generated by J_i is already familiar to us from the rotation group. This is the $SU(2)$ group homomorphic to $O(3)$. We know how the matrices of the $SU(2)$ group change the direction of the spin operators.

With this point in mind, let us consider the two-by-two coordinate matrix

$$X = Ix_0 + \sigma_i x_i = \begin{pmatrix} t + z & x - iy \\ x + iy & t - z \end{pmatrix}. \tag{3.8}$$

where I is the two-by-two unit matrix. Then

$$\det(X) = t^2 - z^2 - x^2 - y^2. \tag{3.9}$$

Therefore unimodular transformations on the above matrix are Lorentz transformations.

Let us consider the transformation

$$X' = WXW^\dagger. \tag{3.10}$$

Since W and $W^\dagger$ are unimodular, $\det(X)$ is preserved. If W is restricted to the $SU(2)$ subgroup W_r generated by the three J_i matrices, then W_r is unitary and does not affect the t component in Eq. (3.8). The effect of a W_r transformation on the $\sigma_i x_i$ is a rotation in the three-dimensional (z, x, y) space. The rotation by ψ around the (θ, ϕ)-direction is represented by

$$W_r(\theta, \phi, \psi) = \exp\left[-i\frac{\psi}{2}\mathbf{n}\cdot\boldsymbol{\sigma}\right] = I\cos\frac{\psi}{2} - i\mathbf{n}\cdot\boldsymbol{\sigma}\sin\frac{\psi}{2}, \tag{3.11}$$

where $\mathbf{n}$ is the unit vector along the (θ, ϕ)-direction.

Let us next consider the effect of $W_b(z, \eta) = \exp[-i\eta K_3]$. The explicit form of this matrix is

$$W_b(z, \eta) = \begin{pmatrix} e^{\eta/2} & 0 \\ 0 & e^{-\eta/2} \end{pmatrix}. \tag{3.12}$$

The effect of this matrix on X is clearly a boost along the z-axis:

$$z' = z\cosh\eta + t\sinh\eta, \qquad t' = z\sinh\eta + t\cosh\eta. \tag{3.13}$$

The boost along the (θ, ϕ)-direction is achieved through

$$W_b(\theta, \phi) = W_r(\theta, \phi, 0)W_b(z, \eta)[W_r(\theta, \phi, 0)]^\dagger. \tag{3.14}$$

As in the case of Eq. (2.25) in Chap. 2 for the proper Lorentz group, the most general form for the W matrix is

$$W = W_r(\theta, \phi, \psi)W_b(\theta, \phi). \tag{3.15}$$

Once we write down the above expression for two-by-two matrices containing six-parameters, we are led to the question of whether there is a one-to-one correspondence between a given matrix of $SL(2, c)$ and a four-by-four matrix in the proper Lorentz group. The answer to this question is *No*. If we require that the W matrix be the two-by-two unit matrix when all the parameters are zero, the rotation parameter ψ is allowed to vary from 0 to 4π for the W matrices while the corresponding parameter for the four-by-four A matrices for the proper Lorentz group takes values between 0 and 2π. For this reason, for each A, there are two W matrices, namely W and $-W$ which produce the same effect on the transformation defined in Eq. (3.10). $SL(2,c)$ is simply connected and is therefore the universal covering group for the proper Lorentz group.

Table 3.1: Generators and transformation matrices of $SL(2,c)$. The last column gives the corresponding four-by-four transformation matrices of the Lorentz group $O(3,1)$, that are applicable to the Minkowski space of (t,z,x,y).

Generators	Two-by-two	Four-by-four
$J_1 = \frac{1}{2}\begin{pmatrix} 0 & 1 \\ 1 & 0 \end{pmatrix}$	$\begin{pmatrix} \cos(\theta/2) & i\sin(\theta/2) \\ i\sin(\theta/2) & \cos(\theta/2) \end{pmatrix}$	$\begin{pmatrix} 1 & 0 & 0 & 0 \\ 0 & \cos\theta & 0 & \sin\theta \\ 0 & 0 & 1 & 0 \\ 0 & -\sin\theta & 0 & \cos\theta \end{pmatrix}$
$K_1 = \frac{1}{2}\begin{pmatrix} 0 & i \\ i & 0 \end{pmatrix}$	$\begin{pmatrix} \cosh(\lambda/2) & \sinh(\lambda/2) \\ \sinh(\lambda/2) & \cosh(\lambda/2) \end{pmatrix}$	$\begin{pmatrix} \cosh\lambda & 0 & \sinh\lambda & 0 \\ 0 & 1 & 0 & 0 \\ \sinh\lambda & 0 & \cosh\lambda & 0 \\ 0 & 0 & 0 & 1 \end{pmatrix}$
$J_2 = \frac{1}{2}\begin{pmatrix} 0 & -i \\ i & 0 \end{pmatrix}$	$\begin{pmatrix} \cos(\theta/2) & -\sin(\theta/2) \\ \sin(\theta/2) & \cos(\theta/2) \end{pmatrix}$	$\begin{pmatrix} 1 & 0 & 0 & 0 \\ 0 & \cos\theta & -\sin\theta & 0 \\ 0 & \sin\theta & \cos\theta & 0 \\ 0 & 0 & 0 & 1 \end{pmatrix}$
$K_2 = \frac{1}{2}\begin{pmatrix} 0 & 1 \\ -1 & 0 \end{pmatrix}$	$\begin{pmatrix} \cosh(\lambda/2) & -i\sinh(\lambda/2) \\ i\sinh(\lambda/2) & \cosh(\lambda/2) \end{pmatrix}$	$\begin{pmatrix} \cosh\lambda & 0 & 0 & \sinh\lambda \\ 0 & 1 & 0 & 0 \\ 0 & 0 & 1 & 0 \\ \sinh\lambda & 0 & 0 & \cosh\lambda \end{pmatrix}$
$J_3 = \frac{1}{2}\begin{pmatrix} 1 & 0 \\ 0 & -1 \end{pmatrix}$	$\begin{pmatrix} e^{i\phi/2} & 0 \\ 0 & e^{-i\phi/2} \end{pmatrix}$	$\begin{pmatrix} 1 & 0 & 0 & 0 \\ 0 & 1 & 0 & 0 \\ 0 & 0 & \cos\phi & -\sin\phi \\ 0 & 0 & \sin\phi & \cos\phi \end{pmatrix}$
$K_3 = \frac{1}{2}\begin{pmatrix} i & 0 \\ 0 & -i \end{pmatrix}$	$\begin{pmatrix} e^{\eta/2} & 0 \\ 0 & e^{-\eta/2} \end{pmatrix}$	$\begin{pmatrix} \cosh\eta & \sinh\eta & 0 & 0 \\ \sinh\eta & \cosh\eta & 0 & 0 \\ 0 & 0 & 1 & 0 \\ 0 & 0 & 0 & 1 \end{pmatrix}$

3.2 Wigner Energy-Momentum Four-Vector Matrices

A Minkowski four-vector can be expressed by a two-by-two matrix as in Eq. (3.8), and a $SL(2, c)$ matrix of the form of Eq. (3.15) performs Lorentz transformations on this vector. Similarly, the energy-momentum four-vector can be expressed by a two-by-two matrix as

$$P = \begin{pmatrix} E + p_z & p_x - i p_y \\ p_x + i p_y & E - p_z \end{pmatrix},$$ (3.16)

whose determinant is

$$E^2 - p_z^2 - p_x^2 - p_y^2$$ (3.17)

and is equal to m^2, 0, or $-m^2$, depending on the nature of the mass of the particle. If a massive particle moves only along the z-direction, this matrix becomes

$$\begin{pmatrix} E + p & 0 \\ 0 & E - p \end{pmatrix}.$$ (3.18)

If this particle is at rest then it takes the form

$$P_m = m \begin{pmatrix} 1 & 0 \\ 0 & 1 \end{pmatrix}.$$ (3.19)

On the other hand, $E = p$ for massless particles, while for tachyon particles, the mass is imaginary. Thus, for all these cases there are three distinct two-by-two matrix representations proportional to

$$P_m = \begin{pmatrix} 1 & 0 \\ 0 & 1 \end{pmatrix}, \qquad P_0 = \begin{pmatrix} 1 & 0 \\ 0 & 0 \end{pmatrix}, \qquad P_{im} = \begin{pmatrix} 1 & 0 \\ 0 & -1 \end{pmatrix}$$ (3.20)

where m is conveniently factored out. These matrices are collectively called Wigner four-vector-matrices and denoted by $P_w = \{P_m, P_0, P_{im}\}$. For all P_w matrices

$$e^{-\frac{i}{2}\sigma_3} = Z(\phi) = \begin{pmatrix} e^{i\phi/2} & 0 \\ 0 & e^{-i\phi/2} \end{pmatrix}$$ (3.21)

satisfies

$$P_w = Z(\phi) P_w Z(\phi)^\dagger.$$ (3.22)

More generally, consider the matrices that leave the energy-momentum four-vector-matrices invariant

$$P = W P W^\dagger.$$ (3.23)

These W matrices dictate the internal space-time symmetries of the particle, and they constitute Wigner's little group. In particular, for P_w of Eq.(3.20) we have

$$P_w = W_i P_w W_i^\dagger.$$ (3.24)

Corresponding Wigner matrices W_i are tabulated in Table 3.2. We note that, these are unimodular matrices with real elements, thus they belong to the group $Sp(2)$. This group will be discussed more thoroughly in Sect. 8.1 of Chap. 8.

Table 3.2: The two-by-two matrix representation of the Wigner energy-momentum four-vectors. Their determinants are positive, zero, and negative for massive, massless, and imaginary-mass particles, respectively. Their corresponding Wigner transformation matrices leave four-vector matrices invariant, thus belong to Wigner's little group [2].

Particle mass	Wigner four-vector P_w	Wigner transformation matrix W_i
Massive	$\begin{pmatrix} 1 & 0 \\ 0 & 1 \end{pmatrix}$	$\begin{pmatrix} \cos(\theta/2) & -\sin(\theta/2) \\ \sin(\theta/2) & \cos(\theta/2) \end{pmatrix}$
Massless	$\begin{pmatrix} 1 & 0 \\ 0 & 0 \end{pmatrix}$	$\begin{pmatrix} 1 & -\gamma \\ 0 & 1 \end{pmatrix}$
Imaginary-mass	$\begin{pmatrix} 1 & 0 \\ 0 & -1 \end{pmatrix}$	$\begin{pmatrix} \cosh(\lambda/2) & \sinh(\lambda/2) \\ \sinh(\lambda/2) & \cosh(\lambda/2) \end{pmatrix}$

3.3 Subgroups of SL(2, c)

As we noted in Appendix A, $SL(2, c)$ consists of non-singular two-by-two matrices of the form:

$$W = \begin{pmatrix} a & b \\ c & d \end{pmatrix}, \qquad \text{with} \qquad (ad - bc) = 1, \tag{3.25}$$

where $a, b, c,$ and d are complex numbers. This matrix is applicable to spinors of the form

$$x = \begin{pmatrix} u \\ v \end{pmatrix}. \tag{3.26}$$

Like the proper Lorentz group, $SL(2, c)$ is a simple group, i.e., it has no invariant subgroups. We are interested in those subgroups which are locally isomorphic to the little groups of the Lorentz group discussed in Chap. 2, and we call for simplicity those subgroups the little groups of $SL(2, c)$.

Among the subgroups of $SL(2, c)$, we would expect that the $O(3)$-like little group corresponding to the time-like orbit is generated by J_1, J_2, and J_3. The $E(2)$-like little groups for the forward and backward light-like orbits are generated by J_3, N_1 and N_2, where

$$N_1 = K_1 - J_2, \qquad N_2 = K_2 + J_1, \qquad J_3 \qquad (3.27)$$

and their Hermitian conjugates respectively. The $O(2, 1)$-like little group for space-like orbit is generated by J_3, K_1 and K_2. The $O(3, 1)$ group corresponding to the point-like orbit confined to the origin is $SL(2, c)$ itself.

In obtaining these little groups, the easiest and quickest method would be to use the above-mentioned generators. However, in order to study the properties of $SL(2, c)$ more effectively, we use a somewhat inefficient way of obtaining subgroups by imposing additional conditions on the parameters $a, b, c,$ and d.

First, we can put the restriction that the above matrix be unitary. Then this subgroup is $SU(2)$ generated by three J_i matrices satisfying the commutation relations for the three-dimensional rotation group. In this case, the elements of the matrix of Eq. (3.25) are

$$|a|^2 + |b|^2 = 1, \qquad |c|^2 + |d|^2 = 1, \qquad a^*b + c^*d = 0, \qquad (3.28)$$

and consequently the two-by-two matrix of Eq. (3.25) becomes

$$\begin{pmatrix} a & b \\ -b^* & a^* \end{pmatrix}, \qquad \text{with} \qquad aa^* + bb^* = 1. \qquad (3.29)$$

This matrix has three independent real parameters. When applied to the spinor of Eq. (3.26), this group preserves the norm

$$|u|^2 + |v|^2. \qquad (3.30)$$

$SU(2)$ is indeed a unitary group. $SU(2)$ is a simple group and has no invariant subgroups. It has a one-parameter unitary subgroup generated by J_3 of Eq. (3.3). Since this group plays the pivotal role in the rotation and Lorentz groups, we shall discuss the representations for it in detail in Sect. 3.4.

Second, the matrices with real elements form a subgroup. This subgroup is called $SL(2, r)$. This group is equivalent to the group of matrices of the form:

$$M = \begin{pmatrix} a & b \\ b^* & a^* \end{pmatrix}, \qquad \text{with} \qquad (aa^* - bb^*) = 1. \qquad (3.31)$$

This group is generated by J_3, K_1, and K_2, and is locally isomorphic to $O(2, 1)$. The above matrix satisfies the condition:

$$M^\dagger J M = J, \qquad (3.32)$$

where

$$J = \begin{pmatrix} 1 & 0 \\ 0 & -1 \end{pmatrix} \qquad (3.33)$$

and therefore the transformation of the spinor of Eq. (3.26) preserves the quantity:

$$|u|^2 - |v|^2 \,. \tag{3.34}$$

For this reason, this subgroup is sometimes called $SU(1, 1)$. The group of two-by-two real unimodular matrices satisfying the condition of Eq. (3.32) is sometimes known as the real symplectic group of dimension 2 or $Sp(2, r)$ [3]. This group which is called $SL(2, r)$, $SU(1, 1)$, or $Sp(2, r)$ is also simple. It has two one-parameter subgroups. One is the unitary group generated by J_3, and the other is a non-unitary group generated by K_1.

Third, let us see whether $SL(2, c)$ has a subgroup consisting of triangular matrices of the form

$$F = \begin{pmatrix} a & b \\ 0 & 1/a \end{pmatrix} . \tag{3.35}$$

The determinant of this matrix is 1. If a and b are allowed to be complex, there are four parameters. In fact this subgroup is generated by J_3, K_3, N_1 and N_2, where

$$N_1 = \begin{pmatrix} 0 & i \\ 0 & 0 \end{pmatrix}, \qquad N_2 = \begin{pmatrix} 0 & 1 \\ 0 & 0 \end{pmatrix} \tag{3.36}$$

and the expressions for J_3 and K_3 are given in Eq. (3.3). These four operators satisfy the commutation relations [Problem 15 in Sect. 3.7]:

$$[J_3, N_1] = iN_2 , \quad [J_3, N_2] = -iN_1 , \quad [N_1, N_2] = 0 , \tag{3.37}$$

$$[K_3, N_1] = -N_2 , \quad [K_3, N_2] = N_1 , \quad [K_3, J_3] = 0 . \tag{3.38}$$

Among the above six commutation relations, the first three form a closed algebraic relation. This means that matrices of the form

$$D(b_1, b_2, \phi) = \begin{pmatrix} e^{-\phi/2} & b_1 - ib_2 \\ 0 & e^{\phi/2} \end{pmatrix} \tag{3.39}$$

represent a three-parameter subgroup generated by J_3, N_1 and N_2. As we did in Appendix B, we call this group $W(2)$.

The fundamental algebraic relations for the generators of $SL(2, c)$ given in Eq. (3.5) remain invariant under the sign change in K_i. Under this sign change, the N operators become their Hermitian conjugates. Thus the resulting matrices are of the lower triangular form. The algebraic properties of the lower triangular matrices are the same as those for the upper triangular matrices.

Let us go back to the upper triangular matrix of Eq. (3.39). We can decompose this matrix as

$$D(b_1, b_2, \phi) = T(b_1, b_2)R(\phi) . \tag{3.40}$$

where

$$T(b_1, b_2) = \begin{pmatrix} 1 & b_1 - ib_2 \\ 0 & 1 \end{pmatrix},$$

$$R(\phi) = \begin{pmatrix} e^{-i\phi/2} & 0 \\ 0 & e^{i\phi/2} \end{pmatrix}, \tag{3.41}$$

where b_1 and b_2 are real parameters. $T(b_1, b_2)$ can further be decomposed as

$$T(b_1, b_2) = T(b_1, 0)T(0, b_2) = T(0, b_2)T(b_1, 0). \tag{3.42}$$

Furthermore, it can be shown that $D(b_1, b_2, \phi)$ is a semi-direct product of $R(\phi)$ and $T(b_1, b_2)$ as in the case of the $E(2)$ group discussed in Appendix A and Chap. 2.

Another interesting property of the N matrices is that they cannot be brought to a diagonal form by any similarity transformation. They however have the property that

$$(N_1)^2 = (N_2)^2 = 0. \tag{3.43}$$

Thus the exponentiation takes the following simple form:

$$T(b_1, b_2) = \exp(-i[b_1 N_1 + b_2 N_2]) = \begin{pmatrix} 1 & b_1 - ib_2 \\ 0 & 1 \end{pmatrix}. \tag{3.44}$$

As in the instance of the $E(2)$ group, $T(b_1, b_2)$ is an invariant subgroup of $W(2)$. Since $W(2)$ a subgroup of $SL(2, c)$, $T(b_1, b_2)$ is also a subgroup of $SL(2, c)$, but is not an invariant subgroup of $SL(2, c)$. $SL(2, c)$ does not have any invariant subgroup.

3.4 SU(2)

The $SU(2)$ group has been studied exhaustively in the literature in connection with the rotation group, and it is unnecessary to give another full-fledged treatment of this group. We mention in this section only those features which will be useful in our later discussions of $SL(2, c)$, starting from what we already know.

Let us discuss first its connection with $O(3)$. It is possible to construct three-by-three rotation matrices from the elements of the two- by-two matrix given in Eq. (3.29) [Problem 5 in Sect. 3.7]. The rotation matrix applicable to the three dimensional space of (z, x, y) takes the form

$$\begin{pmatrix} aa^* - bb^* & 2Re(aa^*) & -2Im(aa^*) \\ -2Re(ab) & Re(a^2 - b^2) & -Im(a^2 + b^2) \\ -2Im(ab) & Im(a^2 - b^2) & Re(a^2 + b^2) \end{pmatrix}. \tag{3.45}$$

The elements of the above three-by-three matrix are quadratic and homogeneous in the elements of the two-by-two matrix of Eq. (3.25). Therefore, to a given three-by-three matrix of $O(3)$, there correspond two two-by-two matrices. One is the expression given in Eq. (3.45), and the other is the negative of Eq. (3.29). This is of course the manifestation of the fact that the correspondence between $SU(2)$ and $O(3)$ is two-to-one.

The two-by-two matrix of Eq. (3.25) can be constructed from the generators of $SU(2)$, namely J_1, J_2, and J_3. Then it is convenient to use the spinor notation α, β where

$$\alpha = \begin{pmatrix} 1 \\ 0 \end{pmatrix}, \qquad \beta = \begin{pmatrix} 0 \\ 1 \end{pmatrix}, \tag{3.46}$$

which are the eigenstates of J_3 with the eigenvalues $\pm 1/2$ respectively. The total angular momentum for these spinors is 1/2. The application of the two-by-two matrix of Eq. (3.25) results in a rotation of the spinor. For instance, the matrix

$$\exp(-i\theta J_2) = \begin{pmatrix} \cos\frac{\theta}{2} & -\sin\frac{\theta}{2} \\ \sin\frac{\theta}{2} & \cos\frac{\theta}{2} \end{pmatrix} \tag{3.47}$$

rotates the spin in the z-direction to the direction which makes the angle θ in the zx-plane [4, 5]. The corresponding matrix in the $O(3)$ group applicable to the Cartesian vector (z, x, y) is

$$\begin{pmatrix} \cos\theta & \sin\theta & 0 \\ -\sin\theta & \cos\theta & 0 \\ 0 & 0 & 1 \end{pmatrix}, \tag{3.48}$$

where we have used the sign convention of the passive transformation, since we will be dealing with functions of the coordinates rather than coordinates themselves.

It is often more convenient to use the spherical vector:

$$\begin{pmatrix} z \\ -\frac{1}{\sqrt{2}}(x + iy) \\ \frac{1}{\sqrt{2}}(x - iy) \end{pmatrix} = \begin{pmatrix} 1 & 0 & 0 \\ 0 & -\frac{1}{\sqrt{2}} & -\frac{i}{\sqrt{2}} \\ 0 & \frac{1}{\sqrt{2}} & -\frac{i}{\sqrt{2}} \end{pmatrix} \begin{pmatrix} z \\ x \\ y \end{pmatrix} \tag{3.49}$$

which is obtained from the Cartesian form by a unitary transformation. The rotation matrix applicable to this spherical vector is

$$\begin{pmatrix} \cos\theta & -\frac{1}{\sqrt{2}}\sin\theta & \frac{1}{\sqrt{2}}\sin\theta \\ \frac{1}{\sqrt{2}}\sin\theta & \frac{1}{2}(\cos\theta - 1) & -\frac{1}{2}(1 + \cos\theta) \\ -\frac{1}{\sqrt{2}}\sin\theta & -\frac{1}{2}(\cos\theta + 1) & \frac{1}{2}(\cos\theta - 1) \end{pmatrix} \tag{3.50}$$

which is obtained from Eq. (3.48) through the conjugate transformation with the unitary matrix in Eq. (3.49):

$$\begin{pmatrix} 1 & 0 & 0 \\ 0 & -\frac{1}{\sqrt{2}} & -\frac{i}{\sqrt{2}} \\ 0 & \frac{1}{\sqrt{2}} & -\frac{i}{\sqrt{2}} \end{pmatrix} \begin{pmatrix} \cos\theta & \sin\theta & 0 \\ -\sin\theta & \cos\theta & 0 \\ 0 & 0 & 1 \end{pmatrix} \begin{pmatrix} 1 & 0 & 0 \\ 0 & -\frac{1}{\sqrt{2}} & \frac{1}{\sqrt{2}} \\ 0 & -\frac{i}{\sqrt{2}} & -\frac{i}{\sqrt{2}} \end{pmatrix}. \tag{3.51}$$

It is possible to add the angular momenta of two spin-1/2 particles by taking the direct product of their spinors. For particles 1 and 2, the sums of the spin operators are

$$J_i = J_i^{(1)} + J_i^{(2)} \, . \tag{3.52}$$

If we add spins of two spin-1/2 particles, the total spin is either zero or 1. If the total angular momentum is zero, the spin wave function is

$$|j = 0, m = 0\rangle = \frac{1}{\sqrt{2}}(\alpha_1\beta_2 - \beta_1\alpha_2) \, , \tag{3.53}$$

where the subscripts 1 and 2 are for particles 1 and 2 respectively. The rotation matrix applicable to this one-dimensional representation is 1. On the other hand, if the total angular momentum is 1, then the spin wave functions take the form

$$|1, \quad 1\rangle = \alpha_1\alpha_2 \, ,$$

$$|1, \quad 0\rangle = \frac{1}{\sqrt{2}}(\alpha_1\beta_2 + \beta_1\alpha_2) \, ,$$

$$|1, -1\rangle = \beta_1\beta_2 \, . \tag{3.54}$$

The rotation matrix applicable to this set of wave functions is identical to that for the spherical vector of Eq. (3.49) [Problem 7 in Sect. 3.7].

We started above with two spin-1/2 particles. If we take a direct product of two wave functions, it can be decomposed into one symmetric and one antisymmetric combination. The symmetric wave functions form a representation space for $j = 1$, while j for the antisymmetric combination is 0. Because this symmetry property is invariant under rotations, this method is extensively used in constructing representations of all finite groups.

Let us next see how this symmetry method works for the three-particle case. J_i is now a sum of all three spin operators.

$$J_i = J_i^{(1)} + J_i^{(2)} + J_i^{(3)} \, . \tag{3.55}$$

The resulting angular momentum is either 3/2 or 1/2. Because each particle has two possible states, there are eight wave functions for this three particle system [Problem 7 in Sect. 3.7]. They are, for $j = 3/2$,

$$\left|\frac{3}{2}, \frac{3}{2}\right\rangle = \alpha_1\alpha_2\alpha_3 \, ,$$

$$\left|\frac{3}{2}, \frac{1}{2}\right\rangle = (1/\sqrt{3})(\alpha_1\alpha_2\beta_3 + \alpha_1\beta_2\beta_3 + \beta_1\alpha_2\alpha_3) \, ,$$

$$\left|\frac{3}{2}, -\frac{1}{2}\right\rangle = (1/\sqrt{3})(\alpha_1\beta_2\beta_3 + \beta_1\alpha_2\beta_3 + \beta_1\beta_2\alpha_3) \, ,$$

$$\left|\frac{3}{2}, -\frac{3}{2}\right\rangle = \beta_1\beta_2\beta_3 \, . \tag{3.56}$$

These wave functions are totally symmetric. There cannot be a totally antisymmetric
wave function because each spinor has only two possible spin states. We can next
construct two wave functions which are symmetric in the first two indices:

$$\left|\frac{1}{2}, \frac{1}{2}\right\rangle = (1/\sqrt{6})(\alpha_1\alpha_2\beta_3 + \alpha_1\beta_2\alpha_3 - 2\beta_1\alpha_2\alpha_3)\,,$$

$$\left|\frac{1}{2}, -\frac{1}{2}\right\rangle = (1/\sqrt{6})(\beta_1\beta_2\alpha_3 + \beta_1\alpha_2\beta_3 - 2\alpha_1\beta_2\beta_3)\,. \tag{3.57}$$

There are two additional wave functions antisymmetric in the first two indices.

$$\left|\frac{1}{2}, \frac{1}{2}\right\rangle = (1/\sqrt{2})(\alpha_1\alpha_2\beta_3 - \alpha_1\beta_2\alpha_3)\,,$$

$$\left|\frac{1}{2}, -\frac{1}{2}\right\rangle = (1/\sqrt{2})(\beta_1\beta_2\alpha_3 - \beta_1\alpha_2\beta_3)\,. \tag{3.58}$$

All four of these wave functions should be orthogonal to the totally symmetric
wave functions. We shall discuss the symmetry properties of the three-particle wave
functions in detail in Chap. 12. What is important is that, if we rotate each of the
spinors in the wave functions, the resulting transformation is exactly like that of the
one spinor with total spin-1/2 [Problem 8 in Sect. 3.7].

The rotation matrix applicable to the above spin-1/2 wave functions is the same
as the one for α and β. The rotation matrix applicable to the $j = 3/2$ wave functions
are four-by-four. How can we construct this rotation matrix?

For a given value of j which is either an integer or half integer, we start with the
diagonal $(2j + 1)$-by-$(2j + 1)$ matrix J_z:

$$J_z = \begin{pmatrix} j & 0 & 0 & 0 & 0 \\ 0 & j-1 & 0 & 0 & 0 \\ 0 & 0 & * & 0 & 0 \\ 0 & 0 & 0 & -j+1 & 0 \\ 0 & 0 & 0 & 0 & -j \end{pmatrix}. \tag{3.59}$$

Then J_x and J_y can be constructed from the commutation relations:

$$[J_i, J_j] = i\varepsilon_{ijk}J_k\,. \tag{3.60}$$

In practice, we take the combinations:

$$J_\pm = J_x \pm iJ_y \tag{3.61}$$

and write commutation relations as

$$[J_\pm, J_z] = \pm J_z\,, \qquad [J_+, J_-] = iJ_z\,, \tag{3.62}$$

We then construct the matrices using the relations

$$J_z \,|j,m\rangle = |m|j,m\rangle \, ,$$

$$J_\pm \,|j,m\rangle = [(j \mp m)(j \pm m + 1)]^{1/2}\,|j,m \pm 1\rangle \, . \tag{3.63}$$

We often couple a spin-1/2 wave function with a wave function of an arbitrary integer angular momentum ℓ, which is represented by the spherical harmonics $Y_\ell^m(\theta, \phi)$. The resulting total angular momentum is either $j = \ell + 1/2$ or $j = \ell - 1/2$. The wave functions in this case are

$$\Phi_{j=\ell+1/2}^m = \begin{pmatrix} \sqrt{(\ell + m + 1/2)/(2\ell + 1)}\,Y_\ell^{m-1/2}(\theta, \phi) \\ \sqrt{(\ell - m + 1/2)/(2\ell + 1)}\,Y_\ell^{m+1/2}(\theta, \phi) \end{pmatrix} \, ,$$

$$\Phi_{j=\ell-1/2}^m = \begin{pmatrix} -\sqrt{(\ell - m + 1/2)/(2\ell + 1)}\,Y_\ell^{m-1/2}(\theta, \phi) \\ \sqrt{(\ell - m + 1/2)/(2\ell + 1)}\,Y_\ell^{m+1/2}(\theta, \phi) \end{pmatrix} \, . \tag{3.64}$$

The point of the above discussion is that there are many different ways to add angular momenta. However, they all result in irreducible representations or invariant vector space s with total angular momentum j. The rotation matrix applicable to a given j is $(2j + 1)$-by-$(2j + 1)$. For instance, the rotation matrix applicable to one spin-1/2 particle is the same as the one for the spin-1/2 state constructed from three spin-1/2 particles. For this reason, it is sufficient to study the rotation matrix which is applicable to the maximum value of j obtainable from a given number of spin-1/2 particles. If, for instance, we are interested in constructing representations for $j = 5/2$, it is sufficient to calculate the rotation matrix for five spin-1/2 particles whose spins are added to give the maximum value of the total angular momentum.

In order that the spinors give the maximum value of the total angular momentum, they should form a totally symmetric combination. We are thus interested in finding all irreducible representations of $SU(2)$ by constructing the representations of totally symmetric products. Starting from u and v and transforming according to Eq. (3.47), we consider the set of products [6, 7]:

$$f_m = \frac{u^{j+m}v^{j-m}}{\sqrt{(j + m)!(j - m)!}} \tag{3.65}$$

with $m = -j, -j + 1, \ldots, j - 1, j$, where j is integral or half-integral. This form satisfies the relation

$$\sum_m |f_m|^2 = \sum_m \frac{|u^{j+m}v^{j-m}|^2}{(j + m)!(j - m)!} = (|u|^2 + |v|^2)^j/(2j)! \, . \tag{3.66}$$

This sum is independent of m, and will be invariant under transformations which preserve $[|u|^2 + |v|^2]$. We are therefore led to consider whether the above form will serve as the basis vector for representations of $SU(2)$.

With this point in mind, let us consider the operators:

$$J_3 = \frac{1}{2}\left(u\frac{\partial}{\partial u} - v\frac{\partial}{\partial v}\right),$$

$$J_+ = u\frac{\partial}{\partial v}, \qquad J_- = v\frac{\partial}{\partial u}. \tag{3.67}$$

These operators satisfy the commutation relations given in Eq. (3.62). Consequently, these operators satisfy the relations of Eq. (3.63):

$$J_3 f_j^m = m f_j^m,$$

$$J_\pm f_j^m = \sqrt{(j \mp m)(j \pm m + 1)} f_j^{m\pm1}. \tag{3.68}$$

The form of Eq. (3.65) forms the basis for a $(2j + 1)$-dimensional representation of $SU(2)$.

Next, let us consider the transformation of the spinor of Eq. (3.26) by the $SU(2)$ matrix given in Eq. (3.29). For a fixed j, the homogeneous polynomials f_m are transformed among themselves by the linear transformation:

$$u' = au + bv, \qquad v' = -b^*u + a^*v. \tag{3.69}$$

Thus

$$R(a, b) f_m = \sum_m D_{mm'}^j(a, b) f_{m'} \tag{3.70}$$

where

$$D_{mm'}^j(a, b) = \sum_\mu \frac{\sqrt{(j + m)!(j - m)!(j + m')!(j - m')!}}{k!(j + m - k)!(j - m' - k)!(m' - m + k)}$$

$$\times \quad a^{j+m-k}(a^*)^{j-m'-k} b^k(-b^*)^{m'-m+k}. \tag{3.71}$$

The above formula gives a complete calculation of all irreducible representations of the rotation group. For example, for $j = 1/2$, we have $f_{1/2} = u$ and $f_{-1/2} = v$. We can calculate each element in the two-by-two matrix which rotates the spinor around the y-axis by θ using the above formula. The result of this calculation will confirm the matrix given in Eq. (3.47). For $j = 1$, we have $f_1 = u^2/\sqrt{2}$, $f_0 = uv$, $f_{-1} = v^2/\sqrt{2}$ for the three-dimensional representation. Here there are nine matrix elements for the three-by-three matrix. We can check this calculation for the rotation around the y-axis and compare with the expression given in Eq. (3.50).

3.5 SL(2, c) Spinors and Four-Vectors

Let us go back to the commutation relations of Eq. (3.4) for the generators of $SL(2, c)$. While it is not possible to change the sign of J_i without changing the commutation relations, the boost operators K_i can take two different signs. For $J_i = \frac{1}{2}\sigma_i$, the boost generators K_i can be either iJ_i or $-iJ_i$. We shall continue to use α and β for the spinor space in which $K_i = iJ_i$. However, we shall use $\dot\alpha$ and $\dot\beta$ for the spinor space in which the boost generators are $\dot K_i = -iJ_i$. Under rotations generated by J_i, the spinors in the undotted and dotted spaces behave in the same manner as is described in Sect. 3.4. The boost operators applicable to the dotted space are the inverse of those applicable to the undotted space.

We often wonder why the Dirac matrices are four-by-four instead of two-by-two. The reason is very simple, as was discussed in Chap. 2 and will be studied further in Sect. 3.6 of the present Chapter, the Dirac wave function is a direct sum of both undotted and dotted spinors.

The next question is whether two spin-1/2 states can be combined to give a spin-1 state. This possibility within the framework of $SU(2)$ and $O(3)$ is widely discussed in quantum mechanics textbooks. This case is also mentioned in Sect. 3.4. In the $SU(2)$ case, the symmetric product gives a vector and the antisymmetric combination becomes a scalar. The question then is whether we can extend this procedure to the $SL(2, c)$ regime to include Lorentz boosts [8].

Let us start with the spinor combinations for massive particles at rest, and consider two spin-1/2 particles. Let us assume that particle 1 and particle 2 are boosted by $K_i = \pm i\sigma_i$ respectively. Then this spinor combination $\alpha_1\dot\alpha_2$ is invariant under boosts along the z-direction, and is like $-(x + iy)$ under rotations. Thus we are led to write:

$$\alpha_1\dot\alpha_2 = (0, 0, -1, -i) \,. \tag{3.72}$$

Likewise, for other combinations, we can consider

$$\beta_1\dot\beta_2 = (0, 0, 1, -i) \,, \qquad \alpha_1\dot\beta_2 = (1, 1, 0, 0) \,, \qquad \beta_1\dot\alpha_2 = (-1, 1, 0, 0) \,. \tag{3.73}$$

Indeed, $\beta_1\dot\beta_2$ is like $(x - iy)$, and $\frac{1}{2}(\alpha_1\dot\beta_2 + \beta_1\dot\alpha_2)$ and $\frac{1}{2}(\alpha_1\dot\beta_2 - \beta_1\dot\alpha_2)$ are like t and z under rotations in which undotted and dotted spinors are transformed in the same manner.

Under the boost along the z-direction, $\alpha_1\dot\alpha_2$ and $\beta_1\dot\beta_2$ remain invariant, $\frac{1}{2}(\alpha_1\dot\beta_2 \pm \beta_1\dot\alpha_2)$ behave like t and z respectively. As for the x-direction, the boost operator is

$$B_x(\eta) = B_{1x}^{(+)}(\eta)B_{2x}^{(-)}(\eta) \,, \tag{3.74}$$

where $B_{1x}^{(+)}(\eta)$ and $B_{2x}^{(-)}(\eta)$ are the boost operators applicable to particle 1 and 2 respectively. The boost operators applicable to the first and second states are

$$B_x^{(\pm)}(\eta) = \begin{pmatrix} \cosh(\eta/2) & \pm\sinh(\eta/2) \\ \pm\sinh(\eta/2) & \cosh(\eta/2) \end{pmatrix} \,, \tag{3.75}$$

respectively. The effect of this boost on $\alpha_1\dot\alpha_2$ is

$$B_x(\eta)(-\alpha_1\dot\alpha_2) = -\alpha_1\dot\alpha_2 \cosh^2(\eta/2) + \beta_1\dot\beta_2 \sinh^2(\eta/2)$$
$$+\frac{1}{2}(\alpha_1\dot\beta_2 - \beta_1\dot\alpha_2)\sinh(\eta)$$
$$= (\sinh\eta, 0, \cosh\eta, i). \tag{3.76}$$

This result is identical to that of boosting $-(x + iy)$. We can carry out similar calculations for $B_x(\eta)\beta_1\dot\beta_2$, $B_x(\eta)\alpha_1\dot\beta_2$, and $B_x(\eta)\beta_1\dot\alpha_2$ [Problem 14 in Sect. 3.7]. As for the boost along the y-direction, we know how to rotate this system of spinors around the z-axis by $90°$.

The above procedure of constructing four-vectors from $SL(2, c)$ spinors will be useful when we study gauge transformations for massless particles in Chap. 10. We are then led to the question of what happens to the combination of two undotted spinors or two dotted spinors. This question will be discussed in Sect. 10.6 of Chap. 10.

In the meantime, our immediate interest is in constructing the real four-by-four Lorentz transformation matrix from the complex elements of a, b, c, d of the two-by-two matrix given in Eq. (3.25) representing $SL(2, c)$. The problem of converting the two-by-two $SU(2)$ matrix of Eq. (3.29) to the three-by-three rotation matrix of Eq. (3.45) is widely discussed in textbooks. The question here is whether it is possible to use the same procedure in the $SL(2, c)$ case.

If we go back to Eq. (3.8), the calculation is straight-forward. It is a matter of solving the linear equations resulting from Eq. (3.8) using the explicit form of Eq. (3.25):

$$\begin{pmatrix} t' + z' & x' - iy' \\ x' + iy' & t' - z' \end{pmatrix} = \begin{pmatrix} a & b \\ c & d \end{pmatrix} \begin{pmatrix} t + z & x - iy \\ x + iy & t - z \end{pmatrix} \begin{pmatrix} a^* & c^* \\ b^* & d^* \end{pmatrix}. \tag{3.77}$$

The result of the above calculation is the following four-by-four Lorentz transformation matrix [Problem 9 in Sect. 3.7]:

$$\begin{pmatrix} \frac{1}{2}(aa^* + bb^* + cc^* + dd^*) & \frac{1}{2}(aa^* - bb^* + cc^* + dd^*) & Re(ab^* + cd^*) & -Im(ab^* + cd^*) \\ \frac{1}{2}(aa^* + bb^* - cc^* - dd^*) & \frac{1}{2}(aa^* - bb^* - cc^* + dd^*) & Re(ab^* - cd^*) & -Im(ab^* - cd*) \\ Re(ac^* + bd^*) & Re(ac^* - bd^*) & Re(ad^* + bc^*) & -Im(ad^* - bd^*) \\ Im(ac^* + bd^*) & Im(ac^* - bd^*) & Im(ad^* + bc^*) & Re(ad^* - bd^*) \end{pmatrix} \tag{3.78}$$

acting on (t, z, x, y).

This procedure is the same as taking the direct product of the spinors α and β to which the transformation matrix W of Eq. (3.5) is applicable and those for the complex conjugate of W. If we use the notation α^* and β^* as the spinors to which W^* is applicable, the elements of the matrix of Eq. (3.77) can be identified as [9]

$$\alpha_1\alpha_2^* = (t + z), \qquad \alpha_1\beta_2^* = (x - iy),$$
$$\beta_1\alpha_2^* = (x + iy), \qquad \beta_1\beta_2^* = (t - z). \tag{3.79}$$

The expressions are quite different from combinations given in Eqs. (3.72) and (3.73) for undotted and dotted representations. The completion of $SL(2, c)$ through complex conjugation is mathematically convenient [10, 11], especially when we use the conformal form

$$z' = \frac{az + b}{cz + d} \tag{3.80}$$

and the complex conjugate of this expression. However, the dotted representation is more convenient when we deal with the Dirac spinors.

Let us see how these two representations are related to each other. If we write W of Eq. (3.5) as

$$W = \exp\left[-\frac{i}{2}\sum_i(\theta_i\sigma_i + i\eta_i\sigma_i)\right], \tag{3.81}$$

then

$$\dot{W} = \exp\left[-\frac{i}{2}\sum_i(\theta_i\sigma_i - i\eta_i\sigma_i)\right], \tag{3.82}$$

and

$$W^* = \exp\left[-\frac{i}{2}\sum_i(-\theta_i\sigma_i^* + i\eta_i\sigma_i^*)\right]. \tag{3.83}$$

On the other hand, in the standard notation for the Pauli spin matrices

$$\sigma_2\sigma_i^*\sigma_2 = -\sigma_i. \tag{3.84}$$

Therefore,

$$\dot{W} = \sigma_2 W^* \sigma_2 \tag{3.85}$$

and

$$\dot{\alpha} = \sigma_2\alpha^* \qquad \text{and} \qquad \dot{\beta} = \sigma_2\beta^* \tag{3.86}$$

up to a constant factor of unit modulus. We choose the convention:

$$\alpha^* = \dot{\beta} \quad \text{and} \quad \beta^* = -\dot{\alpha}. \tag{3.87}$$

In order to make a connection with Eq. (3.73), we have to make a further adjustment to accommodate the usual passive transformation applicable to the spherical harmonics by reversing the sign of y.

3.6 Symmetries of the Dirac Equation

We discussed the Dirac equation briefly in Chap. 2 as an illustrative example of the representation of the Poincaré group for free particles with internal angular momentum. We noted there that if we use the Weyl representation of the Dirac matrices, the generators of the rotations and boosts take the form [Problem 1 in Sect. 3.7]

$$J_i = \begin{pmatrix} (1/2)\sigma_i & 0 \\ 0 & (1/2)\sigma_i \end{pmatrix} \tag{3.88}$$

$$K_i = \begin{pmatrix} (i/2)\sigma_i & 0 \\ 0 & (-i/2)\sigma_i \end{pmatrix} . \tag{3.89}$$

It was noted in Sect. 3.1 that the boost operators can take two different signs in the $SL(2, c)$ regime. Indeed, the significance of the Dirac equation is that it represents a direct sum of the representations with the two different signs of the boost operators.

It is also very important to note that $\gamma_5 = i\gamma_1\gamma_2\gamma_3\gamma_0$ takes the form

$$\gamma_5 = \begin{pmatrix} I & 0 \\ 0 & -I \end{pmatrix} . \tag{3.90}$$

This means that γ_5 commutes with both the rotation and boost generators, and remains invariant under Lorentz transformations. Furthermore, the eigenvalues of this matrix determine the sign of the boost generators.

With this preparation, let us consider the Dirac equation for a free particle with four-momentum p_μ. For a given sign of p_0, there are two different solutions corresponding to two different signs of the energy:

$$\psi^{(\pm)} = U(\pm p)e^{\pm i\, p_\mu x^\mu} . \tag{3.91}$$

In terms of $U(p)$, the Dirac equation becomes

$$(\gamma_\mu p^\mu - M)U(p) = 0 \tag{3.92}$$

where the explicit forms for the γ matrices are given in Sect. 2.5 of Chap. 2. M is the mass of the particle. We can consider also the equation in which the four-vector p_μ is replaced by $-p_\mu$. This is one way to deal with both signs of the energy, and this sign convention is used frequently in quantum field theory.

However, when we discuss symmetry problems, it is more convenient to use a different convention. This different but equivalent way is to rotate around the y-axis the above-mentioned negative-energy solution by $180°$. Then the sign of the momentum is the same as that of the positive energy solution, while the sign of energy is opposite to that of the positive-energy solution. The positive-energy solution takes the form

$$\psi^{(+)} = U(\mathbf{p})e^{i(\mathbf{p}\cdot\mathbf{x}-Et)} \tag{3.93}$$

and the negative energy solution is

$$\psi^{(-)} = V(\mathbf{p})e^{i(\mathbf{p}\cdot\mathbf{x}+Et)} , \tag{3.94}$$

where

$$E = +\left[p^2 + M^2\right]^{1/2} . \tag{3.95}$$

Therefore, we can obtain the negative-energy solution for the positive- energy solution by simply reversing the sign of E. If we write the Dirac spinor $U(\mathbf{p})$ as

$$U(\mathbf{p}) = \begin{bmatrix} x_+ \\ x_- \end{bmatrix} \tag{3.96}$$

then the above spinor satisfies the matrix equation

$$\begin{bmatrix} -M & E + \boldsymbol{\sigma} \cdot \mathbf{p} \\ E - \boldsymbol{\sigma} \cdot \mathbf{p} & -M \end{bmatrix} \begin{bmatrix} x_+ \\ x_- \end{bmatrix} = 0 . \tag{3.97}$$

The solution of this equation is

$$x_- = \frac{E - \boldsymbol{\sigma} \cdot \mathbf{p}}{M} x_+ \qquad \text{or} \qquad x_+ = \frac{E + \boldsymbol{\sigma} \cdot \mathbf{p}}{M} x_- . \tag{3.98}$$

Let us assume without loss of generality that the momentum of the particle is along the z-direction. The spin along the direction of the momentum is called the helicity [12, 13]. If the spin is parallel (anti-parallel) to the momentum, the helicity is said to be positive (negative). It can be shown that the energy operator commutes with the helicity operator J_3 [Problem 3 in Sect. 3.7]. Since $\boldsymbol{\sigma} \cdot \mathbf{p}$ acting on the helicity state does not change the helicity, we can let $x_\pm$ be proportional to either α (positive helicity) or β (negative helicity). Since there are two possible helicity states for a given sign of E, there are four linearly independent solutions.

For positive p_0 or $p_0 = E$, the positive helicity spinor takes the form

$$U_+(\mathbf{p}) = \begin{bmatrix} \left(\dfrac{E + P}{E - P}\right)^{1/4} \alpha \\ \left(\dfrac{E - P}{E + P}\right)^{1/4} \dot{\alpha} \end{bmatrix} , \tag{3.99}$$

where $P = p_z$, and we used the identity:

$$E - P = \frac{M^2}{E + P}, \tag{3.100}$$

in deriving the above expression from Eq. (3.97). The negative helicity state is

$$U_-(\mathbf{p}) = \begin{bmatrix} \left(\dfrac{E - P}{E + P}\right)^{1/4} \beta \\ \left(\dfrac{E + P}{E - P}\right)^{1/4} \dot{\beta} \end{bmatrix} . \tag{3.101}$$

For negative p_0 or $p_0 = -E$,

$$V_+(\mathbf{p}) = \begin{bmatrix} -\left(\dfrac{E - P}{E + P}\right)^{1/4} \alpha \\ \left(\dfrac{E + P}{E - P}\right)^{1/4} \dot{\alpha} \end{bmatrix} \tag{3.102}$$

for positive helicity. The negative helicity spinor is

$$V_-(\mathbf{p}) = \begin{bmatrix} \left(\frac{E+P}{E-P}\right)^{1/4} \beta \\ -\left(\frac{E+P}{E-P}\right)^{1/4} \dot{\beta} \end{bmatrix}. \tag{3.103}$$

In order to study the symmetry of the little groups, let us start with the spinors for the particle at rest:

$$U_+(0) = \begin{bmatrix} \alpha \\ \dot{\alpha} \end{bmatrix}, \qquad U_-(0) = \begin{bmatrix} \beta \\ \dot{\beta} \end{bmatrix}, \tag{3.104}$$

$$V_+(0) = \begin{bmatrix} -\alpha \\ \dot{\alpha} \end{bmatrix}, \qquad V_-(0) = \begin{bmatrix} \beta \\ -\dot{\beta} \end{bmatrix}. \tag{3.105}$$

For a massive particle, the little group is $SU(2)$ applicable to both upper and lower components of the above spinors. Indeed, the above spinors form the representation space for rotations generated by the spin operators given in Eq. (3.1).

Let us boost along the z-axis the above spinors by applying the boost operators generated by K_3 of Eq. (3.3):

$$B_z(\eta) = \begin{pmatrix} B^{(+)}(\eta) & 0 \\ 0 & B^{(-)}(\eta) \end{pmatrix}, \tag{3.106}$$

where

$$B_z^{(\pm)}(\xi) = \exp\left(\pm\frac{\xi}{2}\sigma_3\right) = \begin{pmatrix} e^{\pm\xi/2} & 0 \\ 0 & e^{\mp\xi/2} \end{pmatrix}, \tag{3.107}$$

with

$$\xi = \sinh^{-1}\left(\frac{P}{E}\right) \quad \text{or} \quad \exp\left(\frac{\xi}{2}\right) = \left(\frac{E+P}{E-P}\right)^{1/4}. \tag{3.108}$$

The result of this calculation is

$$B_z(\eta)U_\pm(0) = U_\pm(P), \qquad B_z(\eta)V_\pm(0) = V_\pm(P). \tag{3.109}$$

We have thus obtained the solutions of the Dirac equation purely from the symmetry considerations. We constructed first the representation space of the $O(3)$-like little group, and then completed the orbit by boosting the system. Thus the solutions of the free-particle Dirac equation, which does not contain any dynamical effects, constitute a manifestation of the space-time symmetry properties.

For the moving Dirac particle, the J_i matrices given in Eq. (3.88) are no longer the generators of the little group. The generators are

$$J_i' = B_z(\xi)J_i B_z^{-1}(\xi), \tag{3.110}$$

where $B_z(\xi)$ is given in Eq. (3.107). These new generators are not Hermitian but have the same algebraic property as that of J_i.

From the spinor expression given in Eqs. (3.99), (3.101), (3.102), and (3.103), it is likely that we can obtain the spinors for massless particles by taking the large-momentum zero-mass limit of those spinors. However, the limiting process is far more complicated than this, because the little group for massless particles is not like $O(3)$ but is like $E(2)$. This problem is discussed in detail in Chap. 10. In the meantime, we shall study the Dirac equation for massless particles as an independent entity.

The Dirac equation for a massless particle can be written as:

$$\begin{bmatrix} \sigma \cdot \mathbf{p} & 0 \\ 0 & -\sigma \cdot \mathbf{p} \end{bmatrix} \begin{bmatrix} x_1 \\ x_2 \end{bmatrix} = p_0 \begin{bmatrix} x_1 \\ x_2 \end{bmatrix}. \tag{3.111}$$

There are two uncoupled equations in the above expression. The basic difference between the above equation and the equation for a massive particle in Eq. (3.92) is that the above form is invariant under the simultaneous sign changes in p_0 and $\mathbf{p}$. We thus expect an additional symmetry.

This additional symmetry comes from the fact that γ_5 commutes with the Hamiltonian operator, and can therefore be simultaneously diagonalized with the energy. The eigenvalue of this matrix is $+1$ or -1, as is specified in Eq. (3.90). This symmetry appears as the polarization of massless spin-1/2 particles. What implication does this $\gamma_5 = \pm 1$ symmetry have in terms of the little groups?

Let us assume again that the massless particle moves along the z-direction, and the particle has a definite helicity. If the energy has the same sign as the momentum, the solution of the above matrix equation is

$$U_+(\mathbf{p}) = \begin{bmatrix} \alpha \\ 0 \end{bmatrix} \quad \text{or} \quad U_-(\mathbf{p}) = \begin{bmatrix} 0 \\ \dot{\beta} \end{bmatrix}. \tag{3.112}$$

If the sign of the energy is opposite to that of the momentum,

$$V_+(\mathbf{p}) = \begin{bmatrix} 0 \\ \dot{\alpha} \end{bmatrix} \quad \text{or} \quad V_-(\mathbf{p}) = \begin{bmatrix} \beta \\ 0 \end{bmatrix}. \tag{3.113}$$

If we rotate the above wave functions by $180°$ around the y-axis, they become identical to those of Eq. (3.111). Thus, unlike the case of massive particles, there are only two independent solutions, and we can restrict ourselves to the solutions of Eq. (3.111).

Within the system of spinors given in Eq. (3.111), we have two ways to construct the massless spin-1/2 particle wave function. The first way is to choose

$$\psi_+(x) = U_+(\mathbf{p})e^{i\omega(z-t)} \tag{3.114}$$

and the second choice is

$$\psi_-(x) = U_-(\mathbf{p})e^{i\omega(z-t)}. \tag{3.115}$$

The eigenvalues of γ_5 for the first and second choices are $+1$ and -1 respectively. The helicity for the first and the second wave functions are positive and negative respectively.

We can obtain the anti-particle wave functions by performing charge conjugation on each of the above wave functions. The charge conjugation of Eq. (3.114) is

$$(\psi_+(x))^C = U_-(\mathbf{p})e^{-i\omega(z-t)}. \tag{3.116}$$

The charge conjugation for the second choice is

$$(\psi_-(x))^C = U_+(\mathbf{p})e^{-i\omega(z-t)}. \tag{3.117}$$

The charge conjugation changes the helicity in both cases, while γ_5 remains invariant.

It is clear from the above analysis that if $U_+(\mathbf{p})$ is the spinor for the particle (anti-particle), then $U_-(\mathbf{p})$ is the spinor for the antiparticle (particle). It is thus sufficient to study only the spinors in Eqs. (3.114) and (3.115). If $\gamma_5 = +1$, then the particle has positive helicity, while the helicity of the anti-particle is negative. If $\gamma_5 = -1$, then it is in the other way around. In the world of spin-1/2 paricles, their polarization characteristic shows consistently $\gamma_5 = -1$ [14]. In order to study the problem of what fundamental space-time symmetry causes the separation of the $\gamma_5 = 1$ state from that of $\gamma_5 = -1$, let us go back to the generators of Lorentz transformations applicable to the four-component spinors given in Eq. (3.111). The boost operator applicable to the lower component is the inverse of that for the upper component.

The little group in this case is generated by J_3, N_1 and N_2, which can be written explicitly as

$$J_3 = \frac{1}{2}\begin{pmatrix} \sigma_3 & 0 \\ 0 & \sigma_3 \end{pmatrix},$$

$$N_1 = K_1 + J_2 = \begin{pmatrix} i\sigma_- & 0 \\ 0 & i\sigma_+ \end{pmatrix}, \qquad N_2 = K_2 - J_1 = \begin{pmatrix} \sigma_- & 0 \\ 0 & \sigma_+ \end{pmatrix}, \tag{3.118}$$

where

$$\sigma_+ = \begin{pmatrix} 0 & 1 \\ 0 & 0 \end{pmatrix} \qquad \text{and} \qquad \sigma_- = \begin{pmatrix} 0 & 0 \\ 1 & 0 \end{pmatrix}. \tag{3.119}$$

These matrices satisfy the commutation relations for the generators of the $E(2)$ group.

Since we demand that the wave function be diagonal in J_3, there is no need to construct the rotation matrix it generates. The four-by-four transformation matrices generated by N_1 and N_2 take the form

$$\begin{aligned} D(u,v) &= \exp\left[-i(uN_1 + vN_2)\right] \\ &= I - iuN_1 - ivN_2 \\ &= \begin{pmatrix} I + (u+iv)\sigma_- & 0 \\ 0 & I + (u+iv)\sigma_+ \end{pmatrix}. \end{aligned} \tag{3.120}$$

The spinor solutions U_+ and U_- given in Eq. (3.112) are invariant under the operation of the above matrix. Therefore, the $\gamma_5 = \pm 1$ symmetry can be translated into the invariance under the transformation of the $E(2)$-like little group.

We noted before in the case of massive particles that the Dirac spinors can be constructed from the symmetry considerations alone starting from the representation space of the little group. The question then is whether this can be done in the massless case. More specifically, let us see whether we can get the result equivalent to $\gamma_5 = \pm 1$ by symmetry property of the little group. We can start with the wave functions diagonal in the J_3 operator:

$$\psi_\pm(x) = W_\pm \exp\left[\pm i\omega(z - t)\right] \tag{3.121}$$

where

$$W_\pm = \begin{pmatrix} A\ x_\pm \\ B\ x_\pm \end{pmatrix}. \tag{3.122}$$

Because all six generators of Lorentz transformations are in block diagonal form, each of the upper and lower components of the above four-spinor forms a representation basis for the $SL(2, c)$ group. The above wave function satisfies the second-order D'Alembertian equation, but is not a solution of the Dirac equation. If we demand that the wave function satisfy the Dirac equation for massless particles, the result is the separation of $\gamma_5 = \pm 1$. The question is whether we can get the same result from the symmetry of the little group without using the Dirac equation.

The four-by-four D matrix of Eq. (3.120) consists of the two-by-two $\sigma_\pm$ matrices. When these step-up and step-down operators are applied to $x_\pm$ contained in the wave function of Eq. (3.121), they produce:

$$\begin{aligned}
\sigma_+ x_+ &= 0, & \sigma_+ x_- &= x_+, \\
\sigma_- x_+ &= x_-, & \sigma_- x_- &= 0.
\end{aligned} \tag{3.123}$$

Thus if we insist on the invariance under $D(u, v)$, W has to be

$$W_-^{(+)} = \begin{pmatrix} \alpha \\ 0 \end{pmatrix} \qquad \text{or} \qquad W_+^{(+)} = \begin{pmatrix} 0 \\ \dot{\beta} \end{pmatrix}. \tag{3.124}$$

Indeed, we can therefore obtain $\gamma_5 = \pm 1$ from the condition that the world be invariant under the $D(u, v)$ transformation.

It can be said further that the polarization of massless spin-1/2 particles is a consequence of the invariance under this D transformation [15]. We know that this transformation can be interpreted in terms of the basic space-time transformations. What then its physical significance? What interpretation does this kind of transformation have in the case of photons? We shall discuss these questions in Chap. 10.

3.7 Exercises and Problems

Exercise 1. Let us go back to the commutation relations of Eq. (3.4). They are invariant under Hermitian conjugation, but not under complex conjugation. Show how the complex conjugated generators of $SL(2, c)$ satisfy the original commutation relations.

From the form of the Pauli spin matrices,

$$(J_1)^* = J_1 , \qquad (J_2)^* = -J_2 , \qquad (J_3)^* = J_3 . \tag{3.125}$$

Thus the commutation relations

$$[(J_i)^* , (J_j)^*] = i\varepsilon_{ijk}(J_k)^* \tag{3.126}$$

are satisfied. As for the boost operators,

$$(K_1)^* = -K_1 , \qquad (K_2)^* = K_2 , \qquad (K_3)^* = -K_3 . \tag{3.127}$$

From Eqs. (3.125) and (3.127), we can construct the commutation relations:

$$[(J_i)^*, (K_j)^*] = i\varepsilon_{ijk}(K_k)^* ,$$
$$[(K_i)^*, (K_j)^*] = -i\varepsilon_{ijk}(J_k)^* . \tag{3.128}$$

This property of $SL(2, c)$ is not shared by $O(3, 1)$.

Let us check some of the transformation properties of the spinor combinations for this complex conjugate representation [11, 9]. Rotations around x- and z-axes are in the same direction as in the original representation, but the rotation around the y-axis is in the opposite direction. Boosts along the x- and z-directions are in the opposite directions, but the boost along the y-axis is in the same direction.

With this point in mind, let us check some of the transformation properties of the spinor combinations given in Eq. (3.79). $\alpha_1\alpha_2^*$ and $\beta_1\beta_2^*$ should be invariant under rotations around the z-axis. If we rotate around the x- or y-axis, the combination $\frac{1}{2}(\alpha_1\alpha_2^* + \beta_1\beta_2^*)$ remains invariant. Therefore this combination can be identified as t, and $\frac{1}{2}(\alpha_1\alpha_2^* - \beta_1\beta_2^*)$ as z. Under rotations around the z-axis, $\alpha_1\beta_2^*$ and $\beta_1\alpha_2^*$ behave like $(x - iy)$ and $(x + iy)$ respectively.

Under boosts along the z-direction, $\alpha_1\beta_2^*$ and $\beta_1\alpha_2^*$ remain invariant just like $(x \pm iy)$. If we boost the system along the z-axis with the parameter ξ, then

$$\alpha_1\alpha_2^* \rightarrow e^{\xi}\alpha_1\alpha_2^* \qquad \text{and} \qquad \beta_1\beta_2^* \rightarrow e^{-\xi}\beta_1\beta_2^* , \tag{3.129}$$

just like $(t + z)$ and $(t - z)$ respectively. $\alpha_1\beta_2^*$ and $\beta_1\alpha_2^*$ remain invariant under this boost. If we boost the spinors along the x-direction, the combination $(\alpha_1\alpha_2^* - \beta_1\beta_2^*)$ which corresponds to z remain invariant. The combination $\frac{1}{2i}(\beta_1\alpha_2^* - \alpha_1\beta_2^*)$ which corresponds to y also remains invariant. Under the same boost, the combinations

$\frac{1}{2}(\alpha_1\alpha_2^* + \beta_1\beta_2^*)$ and $\frac{1}{2}(\alpha_1\beta_2^* + \beta_1\alpha_2^*)$, behave like t and x respectively, transform like t and x respectively.

Exercise 2. Show that the most general form of the $SL(2, c)$ matrix given in Eq. (3.25) can be written as a multiplication of a unimodular diagonal matrix, an upper and lower triangular matrices with unit diagonal elements [16].

We can write

$$W = \begin{pmatrix} a & b \\ c & d \end{pmatrix} = \begin{pmatrix} 1 & b' \\ 0 & 1 \end{pmatrix} \begin{pmatrix} 1 & 0 \\ c' & 0 \end{pmatrix} \begin{pmatrix} e^{a'/2} & 0 \\ 0 & e^{-a'/2} \end{pmatrix}, \tag{3.130}$$

which can be written as

$$[\exp(b'\sigma_+)]\,[\exp(c'\sigma_-)]\,[\exp(a'\sigma_3)]\,, \tag{3.131}$$

where the parameters b', c', and a' are complex numbers. Since the expression reduced the number of parameters into three complex numbers, using the fact that the determinant of the matrix is one. This is quite consistent with the form given in Eq. (3.6).

The result of the matrix multiplications in Eq. (3.130) is

$$W = \begin{pmatrix} (1 + b'c')e^{a'/2} & b'e^{-a'/2} \\ -c'e^{a'/2} & e^{-a'/2} \end{pmatrix}. \tag{3.132}$$

We can now identify $d = e^{-a'/2}$, $b = b'd$, $c = -c'/d$. a is determined from the condition that the determinant of W is one.

Exercise 3. In Eq. (3.80), and also in Sect. A.4 of Appendix A, we noted that $SL(2, c)$ is homomorphic to the conformal transformation

$$z' = \frac{az + b}{cz + d}. \tag{3.133}$$

Since the above expression is invariant under the sign change of a, b, c, and d, the correspondence is two-to-one. When $a = d = 1$ and $b = c = 0$, this transformation is an identity transformation. When $a = d = 1$, and b and c are small, we can define an infinitesimal transformation. Is it possible to formulate a Lie group and its Lie algebra for the function $f(z)$?

Let us go back to the expression of W given in Eq. (3.6). If the group parameters are allowed to be complex, it is sufficient to use only three generators for $SL(2, c)$. It is convenient to choose three rotation generators: J_1, J_2, and J_3, satisfying the commutation relations:

$$[J_3, J_\pm] = \pm J_\pm\,, \qquad [J_+, J_-] = 2J_+\,, \tag{3.134}$$

where

$$J_\pm = J_1 \pm iJ_2\,. \tag{3.135}$$

We are quite familiar with the differential forms of the J operators depending on three variables. In Eq. (3.67), we introduced those depending on two variables. This time, we have to introduce the J operators depending on only one variable. The one-variable forms are available in the literature [10, 16], and they are:

$$J_3 = i\left(-n + z\frac{d}{dz}\right), \qquad J_+ = i\left(2nz + z^2\frac{d}{dz}\right), \qquad J_- = -i\frac{d}{dz}. \qquad (3.136)$$

These operators satisfy the commutation relations of Eq. (3.134). For functions of z, we should use the operator:

$$T = (\exp(-ia'J_3))\,(\exp(-ic'J_-))\,(\exp(-ib'J_+)) \qquad (3.137)$$

in which the relation between the active and passive transformation is used [see Exercise 2 in Sect. B.6 of Appendix B]. The result of this operation on $f(z)$ is [16].

$$Tf(z) = (bz + d)^{2n} f\left(\frac{az + b}{bz + d}\right). \qquad (3.138)$$

In addition to the change in the z variable according to Eq. (3.133), there is a multiplication factor $(bz + d)^{2n}$. For this reason, this type of representation is called a multiplier representation [10].

Problem 1. In Eqs. (2.82) and (2.83) of Chap. 2, we introduced the Weyl representation of the Dirac matrices. However, in addition, there is another representation commonly used in the literature, in which γ_0 and γ_5 become interchanged:

$$\gamma_0 = \begin{pmatrix} I & 0 \\ 0 & -I \end{pmatrix}, \qquad \gamma_5 = \begin{pmatrix} 0 & I \\ I & 0 \end{pmatrix} \qquad (3.139)$$

but the matrices γ_i remain unchanged. This is called the Dirac representation. Show that the Dirac representation is unitarily equivalent to the Weyl representation [17]. What form do the generators of rotations and boosts take in the Dirac representation?

Problem 2. A more conventional way of solving the Dirac equation is to solve the eigenvalue equation for the Hamiltonian in the Dirac representation. The Hamiltonian in the Dirac representation takes the form:

$$H = \begin{bmatrix} \boldsymbol{\sigma} \cdot \mathbf{p} & M \\ M & -\boldsymbol{\sigma} \cdot \mathbf{p} \end{bmatrix}. \qquad (3.140)$$

Solve the eigenvalue equation in the Dirac representation, and compare the result with the expressions given in Sect. 3.6. Show that the helicity operator commutes with the Hamiltonian.

Problem 3. Show that the helicity and γ_5 can be diagonalized simultaneously, and show that, in this case, the Hamiltonian is not diagonal. Show that, if the helicity operator commutes with the Hamiltonian, they can be simultaneously diagonalized.

Show in this case that γ_5 is not diagonal. The Hamiltonian is diagonal in the Dirac representation, but is not in the Weyl representation.

Problem 4. The Dirac representation of the Dirac equation is commonly used for solving the hydrogen atom problem in the non-relativistic limit. Use the Weyl representation to do the same.

Problem 5. The Hamiltoninan is not diagonal in the Dirac or Weyl representation of the Dirac equation. Is it possible to obtain a representation in which the Hamiltonian is diagonal by making a unitary transformation of the Dirac or Weyl representation? The answer to this question is *yes*, and the transformation in question is called the Foldy-Wouthuysen transformation [18]. Solve the hydrogen atom problem in the Foldy-Wouthuysen representation. For applications of the Foldy-Wouthuysen transformation, see Streater and Wightman [19], Bogoliubov et al. [20], Barut and Raczka [21], and Ali in [22].

Problem 6. Construct the three-by-three rotation matrix from the two-by-two $SU(2)$ matrix. This problem is discussed in many textbooks [4, 5].

Problem 7. Let us go to Eq. (3.54). Starting from the rotation matrix for each spinor, show that the rotation matrix for the states $|1, 1 >, |1, 0 >$, and $|1, -1 >$ is the same as the rotation matrix for $Y_1^m(\theta, \phi)$.

Problem 8. Calculate the rotation matrix for the degenerate states of $j = 3/2$ given in Eq. (3.56). Compare the result with the matrix which can be computed from the general Eq. (3.70).

Problem 9. Show that the rotation matrices for the states given in Eq. (3.57) and in Eq. (3.58) are identical and are the same as the rotation matrix for the spinors α and β.

Problem 10. Starting from the most general form of the two-by-two matrix representing $SL(2, c)$ given in Eq. (3.25) with six independent real parameters, calculate the four-by-four Lorentz transformation matrix in terms of a, b, c, and d given in Eq. (3.78). Use the equation given in Eq. (3.77) [11, 23].

Problem 11. The four-by-four matrix of Eq. (3.78) is a complicated expression. Check whether this expression becomes that of each little group. Check in particular whether the expression reduces to that of Eq. (3.45) for the $SU(2)$ subgroup. Check also whether it becomes the $T(u, v)$ matrix of Eq. (2.39) in Chap. 2 when the $SL(2, c)$ transformation takes the form:

$$\begin{pmatrix} 1 & u - iv \\ 0 & 1 \end{pmatrix}. \tag{3.141}$$

Problem 12. Wigner's paper [12] contains a proof that the group of proper Lorentz transformations is simple. Use Cartan's criterion to show that $SL(2, c)$ is a simple group. Show also that $SU(2)$ and $SU(1, 1)$ are simple groups, but $W(2)$ [$E(2)$-like subgroup of $SL(2, c)$] is not.

Problem 13. Let us consider the conformal representation:

$$z' = \frac{az + b}{cz + d}.$$

(3.142)

For the $E(2)$-like little group, let $c = 0$ and the modulus of a be unity. The transformation is only a linear transformation. Discuss the property of transformations for $SL(2, c)$ and $SU(1, 1)$ in the complex z-plane.

Problem 14. There are two equivalent ways of representing the $O(2, 1)$-like subgroup of $SL(2, c)$. One is $SU(1, 1)$, and the other is $SL(2, r)$. Translate this equivalence into the language of the four-by-four matrix.

Problem 15. Calculate the boost matrices along the x-direction for the spinor combinations given in Eq. (3.73).

Problem 16. The operators J_3, N_1 and N_2 of Eq. (3.37) satisfy the Lie algebra of the $E(2)$ group. If we add K_3, the four generators satisfy the closed algebra of Eqs. (3.37) and (3.38). What are the transformations generated by these operators? See Janner and Janssen [24].

References

1. E. Cartan, *The Theory of SPINORS*, nachdr. edn. (Dover Publ, Mineola, NY, 2009). ISBN 9780486640709. (Reprinted unabridged from the original French version published in 1966.)
2. S. Başkal, Y. Kim, M. Noz, *Mathematical Devices for Optical Sciences*. (IOP Publishing, Bristol, UK, 2019). ISBN 978-0-7503-1612-5. URL https://dx.doi.org/10.1088/2053-2563/aafe78. (OCLC: 1034620988.)
3. H. Weyl, *The classical groups: their invariants and representations*, 2nd edn. Princeton landmarks in mathematics and physics Mathematics (Princeton University Press, Princeton, N.J. USA, 1997). ISBN 978-0-691-07923-3;978-0-691-05756-9. (Originally published 1947.)
4. M.E. Rose, *Elementary theory of angular momentum*, dover edn. (Dover, New York, 1995). ISBN 9780486684802. (Reprinted unabridged from the orginal version published by John Wiley and Sons, New Yotk, NY USA, 1957.)
5. A.R. Edmonds, *Angular momentum in quantum mechanics*. Princeton landmarks in physics (Princeton University Press, Princeton, 1996). ISBN 9780691025896. (Originally published 1957, Second Edition 1960, Revised printing 1968, Third printing with corrections 1974, Fourth printing 1996.)
6. E.P. Wigner, *Group Theory: And its Application to the Quantum Mechanics of Atomic Spectra* (Academic Press, New York, NY, USA, 1959). ISBN 978-0127505503. (Originally published as: Gruppentheorie und ihre Anwendung auf die Quantenmechanik der Atomspektren, Springer Verlag, Braunscheig, Germany 1931.)
7. M. Hamermesh, *Group theory and its application to physical problems*. Dover books on physics and chemistry (Dover Publications, New York, 1989). ISBN 978-0-486-66181-0. (Originally publushed 1962, Addison-Wesley, Reading MA, USA.)
8. V.B. Beresteckij, E.M. Lifshitz, L.P. Pitaevskij, *Quantum electrodynamics*, 2nd edn. Course of Theoretical Physics, Volume 4 (Butterworth-Heinemann Elsevier Ltd, Oxford, UK, 2008). ISBN 978-0-7506-3371-0. (Originally published 1980; second edition reprinted 2008; OCLC: 254580834; original authors: L. D. Landau and E. M. Lifshitz.)
9. I.V. Novozhilov, *Introduction to elementary particle theory*. No. v. 78 in International series of monographs in natural philosophy (Pergamon Press, Oxford, New York, 1975). ISBN 9780080179544

10. V. Bargmann, Irreducible Unitary Representations of the Lorentz Group, The Annals of Mathematics **48**(3), 568–640 (1947). DOI 10.2307/1969129. URL http://www.jstor.org/stable/1969129?origin=crossref

11. M. Naimark, Linear Representation of the Lorentz Group, Usp.Mat. Nauk **9**, 19–93 (1954). ISBN 9781483169170. (Naimark M A 1957 Linear Representation of the Lorentz Group Am. Math. Soc. Transl. Ser. 2 6 379–458 Engl. transl.; Naimark M A Linear Representations of the Lorentz Group International Series of Monographs in Pure and Appliead Mathematics vol 63, First Edition 1964, reprinted 2014, series editor: Farahat, H. K., Oxford UK: Pergamon, Engl. transl.)

12. E.P. Wigner, On Unitary Representations of the Inhomogeneous Lorentz Group, The Annals of Mathematics **40**(1), 149–204 (1939). DOI 10.2307/1968551. URL http://www.jstor.org/stable/1968551?origin=crossref

13. M. Jacob, G.C. Wick, On the general theory of collisions for particles with spin, Annals of Physics **7**(4), 404–428 (1959). DOI 10.1016/0003-4916(59)90051-X. URL http://linkinghub.elsevier.com/retrieve/pii/000349165990051X

14. M. Goldhaber, L. Grodzins, A.W. Sunyar, Helicity of Neutrinos, Physical Review **109**(3), 1015–1017 (1958). DOI 10.1103/PhysRev.109.1015. URL https://link.aps.org/doi/10.1103/PhysRev.109.1015

15. D. Han, Y.S. Kim, D. Son, E(2)–like little group for massless particles and neutrino polarization as a consequence of gauge invariance, Physical Review D **26**(12), 3717–3725 (1982). DOI 10.1103/PhysRevD.26.3717. URL https://link.aps.org/doi/10.1103/PhysRevD.26.3717

16. W.J. Miller, *Lie Theory and Special Functions* (Academic Press Elsevier Science, New York, NY, USA, 1968). ISBN 978-0-08-095551-3. URL https://books.google.se/books/about/Lie_Theory_and_Special_Functions.html?id=XJ9RiMLiOigC&redir_esc=y

17. J.D. Bjorken, S.D. Drell, *Relativistic quantum fields*. International series in pure and applied physics (McGraw-Hill, New York, NY, USA, 1965). ISBN 9780070054943. URL https://www.abebooks.com/9780486485881/Relativistic-Quantum-Fields-Dover-Books-0486485889/plp. (Reprinted in a Dover Edition 2014, ISBN 13: 9780486485881.)

18. L.L. Foldy, S.A. Wouthuysen, On the Dirac Theory of Spin 1/2 Particles and Its Non–Relativistic Limit, Physical Review **78**(1), 29–36 (1950). DOI 10.1103/PhysRev.78.29. URL https://link.aps.org/doi/10.1103/PhysRev.78.29

19. R.F. Streater, A.S. Wightman, *PCT, spin and statistics, and all that*, 1st edn. Princeton landmarks in physics (Princeton University Press, Princeton, NJ, USA, 2000). ISBN 9780691070629. (Originally published W. A. Benjamin Publishing LTD, Boston, MA USA, 1964.)

20. N.N. Bogolubov, A.A. Logunov, I.T. Todorov, *Introduction to Axiomatic Quantum Field Theory*. No. 18 in Mathematical Physics Monograph Series (W.A. Benjamin, Inc., Boston, MA, 1975). URL https://www.osti.gov/biblio/4139916

21. R. Raczka, A. Barut, *Theory Of Group Representations And Applications* (World Scientific Publishing Company, Singapore; Hackensack, NJ, USA, 1986). ISBN 978-981-310-387-0. URL https://books.google.com.tr/books?id=DAU8DQAAQBAJ

22. H.D. Doebner (ed.), *Differential geometric methods in mathematical physics. 1978: proceedings of the international conference held at the Techn. University of Clausthal, Germany, July 1978*. No. 139 in Lecture notes in physics (Springer, 1978). ISBN 9783540105787

23. V.I. Smirnov, *Linear algebra and group theory*. Dover books on mathematics (Dover Publications, Mineola, N.Y, 2011). ISBN 9780486482224. (Originally published 1961 by McGraw-Hill Book Co., New York, NY USA, OCLC: ocn706965516.)

24. A. Janner, T. Janssen, A charged particle in the field of a transverse electromagnetic plane wave. A group-theoretical analysis, Physica **60**(2), 292–321 (1972). DOI 10.1016/0031-8914(72)90107-3. URL https://linkinghub.elsevier.com/retrieve/pii/0031891472901073

Chapter 4
Group Contractions

Abstract Although the mathematics of group contraction is not widely discussed in the established physics curriculum, we very often think of $E(2)$ as a limiting case of $O(3)$. For instance, when we commute from home to school, we are making transformations within a two-dimensional Euclidean space. However, when we travel on the surface of the earth, we know that we are performing rotations. We then wonder how the motions on the $E(2)$ plane can be reconciled with those on the spherical surface. We explain first how the squeeze transformation can be used and that it is possible a straight line can be produced from a function on the xy-plane. We give a summary of Wigner's $E(2)$ little group and the cylindrical group, and contract the $O(3)$ into both the $E(2)$ and cylindrical groups. We then consider group contractions and unitary representations of $E(2)$. The Lorentz and the Galilei transformations are discussed as well as the $O(2, 1)$ Lorentz group. Additionally, we show how we can use the squeeze transformation to contract $O(2, 1)$ into the $E(2)$ and cylindrical groups and the Lorentz group $O(3, 1)$ into the Galilean group.

When we study the Lorentz group, it is often true that it would be convenient to transform a given group into a simpler group. For instance, take the surface of the earth, which has a radius large enough that the surface appears flat. It is possible to consider the two-dimensional Euclidean group as applying to a flat surface and that of the rotation group as applying to spheres. The technique of transforming one group to another is called group contraction.

Although the mathematics of group contraction is not widely discussed in the established physics curriculum, we very often think of $E(2)$ as a limiting case of $O(3)$. For instance, when we commute from home to school, we are making transformations within a two-dimensional Euclidean space. However, when we travel on the surface of the earth, we know that we are performing rotations. We then wonder how the motions on the $E(2)$ plane can be reconciled with those on the

S. Başkal et al., *Theory and Applications of the Poincaré Group*, Fundamental Theories of Physics 217, https://doi.org/10.1007/978-3-031-64376-7_4

spherical surface. Another example of group contraction is our belief that a Lorentz transformation becomes a Galilean transformation in the limit of small velocity compared with that of light. We shall discuss also these topics in this Chapter.

The method for group contraction was first introduced as a limiting process by Inönü and Wigner [1]. In this chapter, we first use group contraction to illustrate squeeze transformations, which are relatively new in the physics literature. It is rewarding then to formulate the group contraction as a limiting process in terms of this new mathematical technique.

Section 4.1, illustrates squeeze transformations, and shows for example that it is possible a straight line can be produced from a function on the xy-plane. In Sect 4.2, we study Wigner's $E(2)$-like little group in terms of transformations on the two-dimensional Euclidean plane by summarizing what we have already discussed in Appendices A and B. It is shown also that the translation-like degrees of freedom in the $E(2)$-like little group are gauge degrees of freedom. Additionally, we define and discuss the cylindrical group. In Sect. 4.3, a plane and a cylinder tangent to a sphere are considered. We then formulate how the rotation group can be contracted into the two-dimensional Euclidean group and to the cylindrical group.

In Sect. 4.4 we discuss group contractions and unitary representations of $E(2)$. Here we use the group contraction procedure to convert the spherical harmonics into the $E(2)$ harmonics which are essentially solutions of Laplace's equation.

Section 4.5, discusses the Lorentz and the Galilei transformations and presents in some detail the $O(2, 1)$ Lorentz group. We then use the methods already discussed to contract the $O(2, 1)$ group into the $E(2)$ group and the cylindrical group. We note that this is similar to the contraction of the $O(3)$ group to these two groups. Additionally, we contract the Lorentz group into the Galilean group. In Sect. 4.6, we discuss other related examples in the form of exercises and problems.

4.1 Contraction with Squeeze Transformations

With respect to the xy-plane, we are familiar with rotations and translations. However, the concept of squeeze transformations is relatively new. The point here is that when the x-axis is expanded the y-axis is contracted so that the product xy remains the same. This means that the squeeze is an area-preserving transformation.

The squeeze transformation matrix is in the form of a boost transformation that we have seen previously:

$$B(\eta) = \begin{pmatrix} e^{\eta/2} & 0 \\ 0 & e^{-\eta/2} \end{pmatrix} . \tag{4.1}$$

This squeeze matrix has an inverse given by

$$B(\eta)^{-1} = \begin{pmatrix} e^{-\eta/2} & 0 \\ 0 & e^{\eta/2} \end{pmatrix} , \tag{4.2}$$

with $B^{-1}B = I$ where I is, as usual, the unit matrix.

When the squeeze matrix is applied to the coordinate matrix

$$X = \begin{pmatrix} t+z & x+iy \\ x-iy & t-z \end{pmatrix} \tag{4.3}$$

that we saw in Eq. (3.8) of Chap. 3, which transforms as in Eq. (3.10) with $X' = B(\eta)^{\dagger} X B(\eta)$ and can be written explicitly as

$$\begin{aligned}
&\begin{pmatrix} e^{\eta/2} & 0 \\ 0 & e^{-\eta/2} \end{pmatrix} \begin{pmatrix} t+z & x+iy \\ x-iy & t-z \end{pmatrix} \begin{pmatrix} e^{\eta/2} & 0 \\ 0 & e^{-\eta/2} \end{pmatrix} \\
&= \begin{pmatrix} e^{\eta}(t+z) & x+iy \\ x-iy & e^{-\eta}(t-z) \end{pmatrix}.
\end{aligned} \tag{4.4}$$

In the limit of large η, this matrix becomes

$$X'_c = \begin{pmatrix} e^{\eta}(t+z) & x+iy \\ x-iy & 0 \end{pmatrix}. \tag{4.5}$$

Then we make the inverse transformation: $[B(\eta)^{\dagger}]^{-1} X'_c B(\eta)^{-1}$ which yields to

$$\begin{pmatrix} t+z & x+iy \\ x-iy & 0 \end{pmatrix}, \tag{4.6}$$

meaning $t-z = 0$. Since we need a space-like hypersurface, we choose $t = 1$, then $z = 1$. Writing the X matrix as a column vector, we have

$$\begin{pmatrix} 1 \\ 1 \\ x \\ y \end{pmatrix}. \tag{4.7}$$

This changed vector is called the *contracted vector*.

Suppose now $y = f(x)$, we can write the 2-dimensional column vector as

$$\begin{pmatrix} x \\ f(x) \end{pmatrix}. \tag{4.8}$$

If the squeeze matrix $B(\eta)$ is applied to this column vector, we obtain:

$$\begin{pmatrix} e^{\eta/2} & 0 \\ 0 & e^{-\eta/2} \end{pmatrix} \begin{pmatrix} x \\ f(x) \end{pmatrix} = \begin{pmatrix} e^{\eta/2} x \\ e^{-\eta/2} f(x) \end{pmatrix}. \tag{4.9}$$

Now in the limit of large η, the lower element of this matrix becomes very small, say ϵ. The column matrix then has the form

$$\begin{pmatrix} e^{\eta/2} x \\ \epsilon \end{pmatrix}. \tag{4.10}$$

Then if we apply the inverse of the squeeze matrix to this column vector we obtain:

$$\begin{pmatrix} e^{-\eta/2} & 0 \\ 0 & e^{\eta/2} \end{pmatrix} \begin{pmatrix} e^{\eta/2}\,x \\ \epsilon \end{pmatrix}. \tag{4.11}$$

The column matrix that results is

$$\begin{pmatrix} x \\ e^{\eta/2}\epsilon \end{pmatrix}. \tag{4.12}$$

Since ϵ is very small we can write this as:

$$\begin{pmatrix} x \\ 1 \end{pmatrix}. \tag{4.13}$$

For example, suppose $f(x)$ is $\sin(x)$, then $y = \sin(x)$. We illustrate the contraction procedure above in Fig. 4.1. The expression given in Eq. (4.13) is then the contracted $(x\ y)$ column vector.

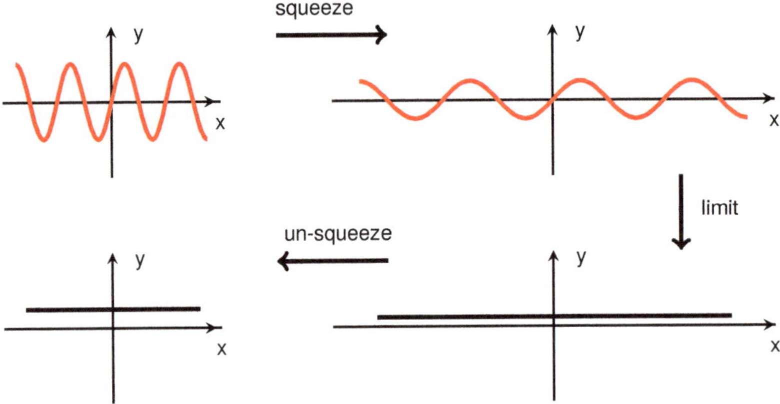

Fig. 4.1: Transforming the *sine* curve to $y = $ constant. As can be seen, the right side expands along the x-direction while it is contracted in the y-direction until it becomes almost a horizontal line with $y = \epsilon$, where ϵ is a very small constant. We can then un-squeeze this function. Finally, the sine curve becomes a horizontal line [2].

As another example we shall apply the squeeze transformation on the rotation matrix. In this two-dimensional space, the rotation is generated by σ_2. The rotation matrix is

$$\exp\left(-i\alpha\,\sigma_2\right) = \begin{pmatrix} \cos\alpha & -\sin\alpha \\ \sin\alpha & \cos\alpha \end{pmatrix}. \tag{4.14}$$

If we then apply the squeeze matrix $B(\eta)$ of Eq. (4.1) to σ_2 we obtain,

$$B(\eta)\sigma_2 B(\eta)^{-1} = \begin{pmatrix} 0 & -i\,e^{\eta} \\ i\,e^{-\eta} & 0 \end{pmatrix} . \tag{4.15}$$

In the limit of large η, we can write this expression as

$$\begin{pmatrix} 0 & -i\,e^{\eta} \\ 0 & 0 \end{pmatrix} . \tag{4.16}$$

Now we apply the inverse of the squeeze transformation with the result

$$\begin{pmatrix} 0 & -i \\ 0 & 0 \end{pmatrix} . \tag{4.17}$$

The resulting transformation matrix is

$$\exp\left\{ -i\alpha \begin{pmatrix} 0 & -i \\ 0 & 0 \end{pmatrix} \right\} = \begin{pmatrix} 1 & -\alpha \\ 0 & 1 \end{pmatrix} . \tag{4.18}$$

This matrix is indeed, the contracted rotation matrix.

If we now apply this contracted matrix to the $(x\,y)$ column vector in Eq. (4.8),

$$\begin{pmatrix} 1 & -\alpha \\ 0 & 1 \end{pmatrix} \begin{pmatrix} x \\ y \end{pmatrix} = \begin{pmatrix} x - \alpha y \\ y \end{pmatrix} . \tag{4.19}$$

We can see that the y variable remains invariant, but the translation αy along the x-direction now depends on the y variable.

We can also freely choose the contracted column vector to be the same form as Eq. (4.13). The result will be

$$\begin{pmatrix} 1 & -\alpha \\ 0 & 1 \end{pmatrix} \begin{pmatrix} x \\ 1 \end{pmatrix} = \begin{pmatrix} x - \alpha \\ 1 \end{pmatrix} . \tag{4.20}$$

Now the y variable remains invariant at 1, while the x variable has a transition solely depending on the α parameter.

4.2 Wigner's E(2)-Like Little Group

It is not possible to find a reference frame in which a massless particle is at rest. On the other hand Wigner showed that the little group for massless particles has three degrees of freedom, resembles the $E(2)$, which also is known as the two-dimensional Euclidean group. Therefore, the same Lie algebra satisfied by the generators of this little group should be shared with the two-dimensional Euclidean group. Each has one rotational and two translational degrees of freedom. One can easily associate the $E(2)$ rotational degree of freedom with the massless particle helicity. The question then is how can we interpret physically the two translational degrees of freedom?

Unfortunately, in his original paper [3], there was no answer to this question provided by Wigner.

For a massless particle, the energy-momentum four-vector is proportional to

$$p^{\mu} = (\omega, \omega, 0, 0). \tag{4.21}$$

Wigner [3] observed that this expression, while being invariant under rotations around the z-axis, is also invariant under the transformation

$$W(\gamma, \alpha) = \exp\left[-i\gamma(N_1 \cos\alpha + N_2 \sin\alpha)\right] \tag{4.22}$$

with

$$N_1 = K_1 - J_2 = \begin{pmatrix} 0 & 0 & i & 0 \\ 0 & 0 & i & 0 \\ i & -i & 0 & 0 \\ 0 & 0 & 0 & 0 \end{pmatrix} \quad \text{and} \quad N_2 = J_1 + K_2 = \begin{pmatrix} 0 & 0 & 0 & i \\ 0 & 0 & 0 & i \\ 0 & 0 & 0 & 0 \\ i & -i & 0 & 0 \end{pmatrix}. \tag{4.23}$$

These expressions for N_i were given in Eq. (2.44) of Chap. 2. Hence,

$$W(\gamma, \alpha) = \begin{pmatrix} 1 + \gamma^2/2 & -\gamma^2/2 & \gamma\cos\alpha & \gamma\sin\alpha \\ \gamma^2/2 & 1 - \gamma^2/2 & \gamma\cos\alpha & \gamma\sin\alpha \\ \gamma\cos\alpha & -\gamma\cos\alpha & 1 & 0 \\ \gamma\sin\alpha & -\gamma\sin\alpha & 0 & 1 \end{pmatrix}. \tag{4.24}$$

This little group has the generators N_1, N_2, and J_3. These generators satisfy the set of commutation relations:

$$[N_1, N_2] = 0, \quad [J_3, N_1] = iN_2, \quad \text{and} \quad [J_3, N_2] = -iN_1. \tag{4.25}$$

Wigner noted this Lie algebra and that of the two-dimensional Euclidean group are the same where

$$[P_1, P_2] = 0, \quad [J_3, P_1] = iP_2, \quad \text{and} \quad [J_3, P_2] = -iP_1. \tag{4.26}$$

Here, translations along the x- and y-directions are generated by P_1 and P_2, respectively. Thus the little group, generated by J_3, N_1, and N_2 was named by Wigner as the $E(2)$-like little group.

Later, the following operators were considered by Kim and Wigner [4, 5]:

$$Z_1 = -ix\frac{\partial}{\partial z} \quad \text{and} \quad Z_2 = -iy\frac{\partial}{\partial z}, \tag{4.27}$$

where $(x^2 + y^2)$ is constant. Translations along the z-direction are generated by the Z_i. These act on the surface of a circular cylinder which is illustrated in Fig. 4.2. Consistent with the P_i generators, the commutation relations satisfied by the Z_i are:

$$[Z_1, Z_2] = 0, \quad [J_3, Z_1] = iZ_2, \quad \text{and} \quad [J_3, Z_2] = -iZ_1. \tag{4.28}$$

This is, indeed, the Lie algebra for the *cylindrical group.*

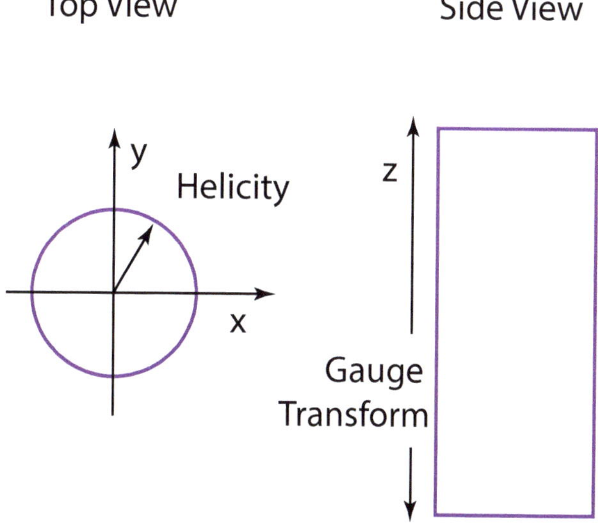

Fig. 4.2: This cylindrical picture shows the internal space-time structure of photons. The side view of the cylinder shows the up-down translation which illustrates a gauge transformation. The top view, a circle, illustrates that the rotational degree of freedom corresponds to the helicity of the photon [5].

Next, we consider a photon with momentum along the z-direction, having the four-potential

$$\left(A_0 , A_z , A_x , A_y \right) . \tag{4.29}$$

The Lorentz condition of $A_0 = A_z$, can be applied, and thus the four-potential is

$$\left(A_0 , A_0 , A_x , A_y \right) . \tag{4.30}$$

When the $W(\gamma, \alpha)$ of Eq. (4.24) is applied to this four-potential we obtain:

$$\begin{pmatrix} 1 + \gamma^2/2 & -\gamma^2/2 & \gamma \cos\alpha & \gamma \sin\alpha \\ \gamma^2/2 & 1 - \gamma^2/2 & \gamma \cos\alpha & \gamma \sin\alpha \\ \gamma \cos\alpha & -\gamma \cos\alpha & 1 & 0 \\ \gamma \sin\alpha & -\gamma \sin\alpha & 0 & 1 \end{pmatrix} \begin{pmatrix} A_0 \\ A_0 \\ A_x \\ A_y \end{pmatrix} , \tag{4.31}$$

resulting in

$$\begin{pmatrix} A_0 + \gamma \left(A_x \cos\alpha + A_y \sin\alpha \right) \\ A_0 + \gamma \left(A_x \cos\alpha + A_y \sin\alpha \right) \\ A_x \\ A_y \end{pmatrix} . \tag{4.32}$$

This little group transformation $W(\gamma, \alpha)$ provides an addition to A_0 but leaves the transverse components A_x and A_y invariant. This is in the form of a cylindrical transformation. This is a gauge transformation in the language of physics. Table 4.1 summarizes these results.

When boosted along the z-direction, the energy-momentum four-vector of Eq. (4.21) becomes:

$$p'^{\mu} = e^{\eta}(\omega, \omega, 0, 0), \tag{4.33}$$

and N_1 and N_2 become $e^{\eta} N_1$ and $e^{\eta} N_2$, respectively. J_3 remains invariant.

Table 4.1: The covariance of the energy-momentum relation is shown here, along with the covariance of the internal space-time symmetry. When a Lorentz boost along the z-direction is performed, the J_3 component of the angular momentum remains invariant. It is known as the helicity. The transverse N_1 and N_2 components are seen to collapse into a gauge transformation. Earlier papers [6, 7, 5], which used the four-by-four matrix formulation, have studied the γ parameter for the massless particle.

	Massive Slow	Lorentz Covariance	Massless Fast
Energy- Momentum	$E = p^2/2m$	Einstein's $E = [p^2 + m^2]^{1/2}$	$E = cp$
Spin, Helicity Gauge Transformation	J_3 N_1, N_2	Wigner's Little Group	Helicity Gauge Transformation

4.2.1 Geometric Interpretation of the E(2)-Like Little Group

The Euclidean group which consists of translations and a rotation in a two-dimensional Euclidean plane, is often referred to as $E(2)$. From Appendices A and B we are already familiar with the coordinate transformations applicable to this plane. They have been defined as:

$$x' = x \cos \alpha - y \sin \alpha + u$$

$$y' = x \sin \alpha + y \cos \alpha + v. \tag{4.34}$$

Translations in the xy-plane are represented by u and v. Using our four-dimensional coordinate system, we can write this in matrix form:

$$\begin{pmatrix} 1 \\ 1 \\ x' \\ y' \end{pmatrix} = \begin{pmatrix} 1 & 0 & 0 & 0 \\ 0 & 1 & 0 & 0 \\ 0 & u & \cos\alpha & -\sin\alpha \\ 0 & v & \sin\alpha & \cos\alpha \end{pmatrix} \begin{pmatrix} 1 \\ 1 \\ x \\ y \end{pmatrix}. \tag{4.35}$$

If we now consider a small xy-plane which is located at the north pole of a very large sphere, then x and y are very small since they are translations on a very small xy-plane. The above matrix can be written in exponential form as

$$\exp[\,-i(\,uP_1 + vP_2\,)\,]\exp(\,-i\alpha J_3\,). \tag{4.36}$$

where P_1 and P_2 represent translations in the xy-plane and J_3 is the usual rotation generator around the z-axis. We can write the translation generators in matrix form as:

$$P_1 = \begin{pmatrix} 0 & 0 & 0 & 0 \\ 0 & 0 & 0 & 0 \\ 0 & i & 0 & 0 \\ 0 & 0 & 0 & 0 \end{pmatrix} \quad \text{and} \quad P_2 = \begin{pmatrix} 0 & 0 & 0 & 0 \\ 0 & 0 & 0 & 0 \\ 0 & 0 & 0 & 0 \\ 0 & i & 0 & 0 \end{pmatrix}. \tag{4.37}$$

We can also write them as:

$$P_1 = -i\frac{\partial}{\partial x} \quad \text{and} \quad P_2 = -i\frac{\partial}{\partial y}. \tag{4.38}$$

The rotation generator J_3 takes the form

$$J_3 = -i\left(x\frac{\partial}{\partial y} - y\frac{\partial}{\partial x}\right). \tag{4.39}$$

These generators also satisfy the commutation relations of Eq. (4.25).

4.2.2 Cylindrical Aspects of the E(2)-Like Little Group

The $E(2)$-like little group has been discussed previously in terms of the cylindrical group. In cylindrical coordinates the transformation equations have the form

$$x' = x\cos\alpha - y\sin\alpha,$$

$$y' = x\sin\alpha + y\cos\alpha,$$

$$z' = z + ux + vy. \tag{4.40}$$

We see here that $(x^2 + y^2)$ is left invariant under this transformation, but that z can vary from $-\infty$ to $+\infty$. Hence the cylindrical group is represented by this linear transformation. This was illustrated in Fig. 4.2. Here z represents up and down translations on the z-axis of the cylinder and x and y are transformations on the circular part of the cylinder. If we use the four-dimensional coordinate system, these equations can be written in matrix can form as:

$$\begin{pmatrix} 1 \\ z' \\ x' \\ y' \end{pmatrix} = \begin{pmatrix} 1 & 0 & 0 & 0 \\ 0 & 1 & u & v \\ 0 & 0 & \cos\alpha & -\sin\alpha \\ 0 & 0 & \sin\alpha & \cos\alpha \end{pmatrix} \begin{pmatrix} 1 \\ z \\ x \\ y \end{pmatrix}. \tag{4.41}$$

Since z can be much larger than x and y, the above matrix can be written in exponential form as

$$\exp[-i(uZ_1 + vZ_2)] \exp(-i\alpha J_3). \tag{4.42}$$

Of course, J_3 represents rotations in the xy-plane around z, while the Z_i take the matrix form:

$$Z_1 = \begin{pmatrix} 0 & 0 & 0 & 0 \\ 0 & 0 & i & 0 \\ 0 & 0 & 0 & 0 \\ 0 & 0 & 0 & 0 \end{pmatrix} \quad \text{and} \quad Z_2 = \begin{pmatrix} 0 & 0 & 0 & 0 \\ 0 & 0 & 0 & i \\ 0 & 0 & 0 & 0 \\ 0 & 0 & 0 & 0 \end{pmatrix}. \tag{4.43}$$

The Z_i again satisfy the same commutation relations given in Eq. (4.28). The commutation relations in Eq. (4.26) and Eq. (4.28), it is worth noting, remain invariant when there is a sign change in P_i or Z_i. Hermitian conjugation also leaves them invariant. As J_3 is Hermitian, the relationship between P_i or Z_i can be written as:

$$Z_i = -(P_i)^\dagger \quad \text{and} \quad P_i = -(Z_i)^\dagger. \tag{4.44}$$

We illustrate in Fig. 4.3 these two geometric interpretations.

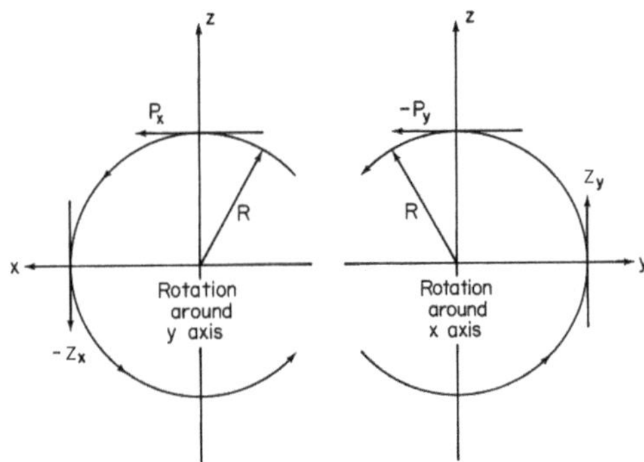

Fig. 4.3: Geometric aspects of the Euclidean and cylindrical groups. We can have, for a given sphere, both a plane tangential at the north pole, and a cylinder tangential to the equatorial belt. The x and y translations are on a plane. Up-down z translations along the circular axis of a cylindrical surface correspond to gauge transformations [4].

4.3 Contractions of the O(3) to E(2) and the Cylindrical Group

We are familiar with the three-dimensional rotation group. Here we start with a three-dimensional sphere where the rotation group is well known. We are interested now in both a plane tangent at the north pole as well as a cylindrical surface touching the equatorial belt. These are illustrated in Fig. 4.4. The transformation group on these tangential surfaces will be obtained using the squeeze procedure introduced in Sect. 4.1.

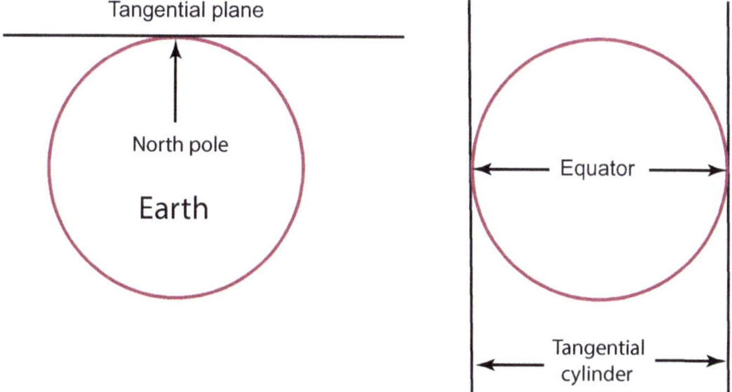

Fig. 4.4: Tangential surfaces attached to a spherical surface. Although we could consider a flat plane tangent at any surface point on the sphere, we choose for convenience, the north pole. We also can consider a cylindrical surface tangent to the equatorial belt [4]. By constructing symmetry properties on these surfaces, it is possible to contract the $O(3)$ rotation group [2].

Generated by $J_1, J_2,$ and J_3, the rotation group has the Lie algebra :

$$\left[J_i, J_j\right] = i\varepsilon_{ijk}J_k.\tag{4.45}$$

We defined the rotation generators in Chap. 2 as

$$J_1 = \begin{pmatrix} 0 & 0 & 0 & 0 \\ 0 & 0 & 0 & i \\ 0 & 0 & 0 & 0 \\ 0 & -i & 0 & 0 \end{pmatrix}, \quad J_2 = \begin{pmatrix} 0 & 0 & 0 & 0 \\ 0 & 0 & -i & 0 \\ 0 & i & 0 & 0 \\ 0 & 0 & 0 & 0 \end{pmatrix}, \quad \text{and} \quad J_3 = \begin{pmatrix} 0 & 0 & 0 & 0 \\ 0 & 0 & 0 & 0 \\ 0 & 0 & 0 & -i \\ 0 & 0 & i & 0 \end{pmatrix}.\tag{4.46}$$

In terms of the differential operators, the rotation generators can be written as

$$J_1 = -i\left(y\frac{\partial}{\partial z} - z\frac{\partial}{\partial y}\right), \quad J_2 = -i\left(z\frac{\partial}{\partial x} - x\frac{\partial}{\partial z}\right), \quad \text{and} \quad J_3 = -i\left(x\frac{\partial}{\partial y} - y\frac{\partial}{\partial x}\right).\tag{4.47}$$

These generators satisfy the commutation relations given in (4.45).

Using Fig. 4.5 let us consider the spherical surface of the earth. We first consider a flat plane tangent at the north pole. We also can consider a cylinder tangent at the equatorial belt. We achieve this by considering the squeeze process which will shorten the z-axis, while it expands the xy-plane. Using this process we are led to the plane tangent at the north pole. Using the squeeze process again in the opposite direction, namely expanding the z-axis while shrinking the xy-plane, will lead us to the single line along the z-axis.

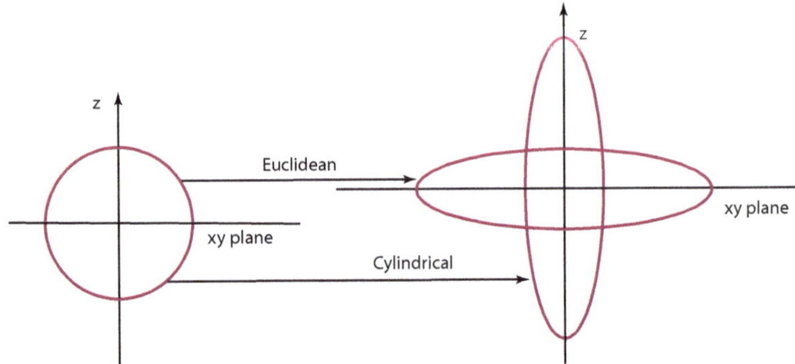

Fig. 4.5: Squeezing the sphere. We use the squeezing process to obtain the flat plane. Then, contracting the z-axis expands the xy-plane. This allows us to obtain the tangential plane at the pole. We can also obtain the tangential cylindrical surface when we expand the z-axis while at the same time contracting the radius of the xy-plane [2].

On the tangential plane at the north pole, we generate transformations by rotations in the z-direction. Translations are then generated along the x- and y-directions. The rotation generator, J_3, is given in Eq. (4.46). In Eq. (4.38) a set of translation generators which represented translations in the x- and y-axis were defined as:

$$P_1 = -i\frac{\partial}{\partial x} \qquad \text{and} \qquad P_2 = -i\frac{\partial}{\partial y} \,. \tag{4.48}$$

Together with J_3, they satisfy the following closed set of commutation relations

$$[\,P_1, P_2\,] = 0\,, \quad [\,J_3, P_1\,] = iP_2\,, \quad \text{and} \quad [\,J_3, P_2\,] = -iP_1\,. \tag{4.49}$$

These commutation relations do indeed form the Lie algebra for the two-dimensional Euclidean group, $E(2)$.

We can now use the squeeze matrix, similar to the one in Sect. 4.1, as

$$B(\eta) = \begin{pmatrix} e^{-\eta/2} & 0 & 0 & \\ 0 & e^{-\eta/2} & 0 & 0 \\ 0 & 0 & e^{\eta/2} & 0 \\ 0 & 0 & 0 & e^{\eta/2} \end{pmatrix}\,. \tag{4.50}$$

Using the same technique as in Eqs. (4.15), (4.16), (4.17), and (4.18), J_3 remains the same, but $J_2 \to P_1$ and $-J_1 \to P_2$. The generator J_3 as well as the P_i generators as given in Eq. (4.37) are:

$$J_3 = \begin{pmatrix} 0 & 0 & 0 & 0 \\ 0 & 0 & 0 & 0 \\ 0 & 0 & 0 & -i \\ 0 & 0 & i & 0 \end{pmatrix}, \quad P_1 = \begin{pmatrix} 0 & 0 & 0 & 0 \\ 0 & 0 & 0 & 0 \\ 0 & i & 0 & 0 \\ 0 & 0 & 0 & 0 \end{pmatrix}, \quad \text{and} \quad P_2 = \begin{pmatrix} 0 & 0 & 0 & 0 \\ 0 & 0 & 0 & 0 \\ 0 & 0 & 0 & 0 \\ 0 & i & 0 & 0 \end{pmatrix}. \tag{4.51}$$

The form of the transformation matrix which performs translations after rotation has the form:

$$\exp\left(-iuP_x - ivP_y\right)\exp\left(-i\alpha J_3\right) = \begin{pmatrix} 1 & 0 & 0 & 0 \\ 0 & 1 & 0 & 0 \\ 0 & u & \cos\alpha & -\sin\alpha \\ 0 & v & \sin\alpha & \cos\alpha \end{pmatrix}. \tag{4.52}$$

When this matrix is applied to the contracted vector $(1, 1, x, y)$, we have

$$\begin{pmatrix} 1 & 0 & 0 & 0 \\ 0 & 1 & 0 & 0 \\ 0 & u & \cos\alpha & -\sin\alpha \\ 0 & v & \sin\alpha & \cos\alpha \end{pmatrix}\begin{pmatrix} 1 \\ 1 \\ x \\ y \end{pmatrix} = \begin{pmatrix} 1 \\ 1 \\ u + x\ \cos\alpha - y\ \sin\alpha \\ v + x\ \sin\alpha + y\ \cos\alpha \end{pmatrix}. \tag{4.53}$$

Let us now take another limiting process by consider a ring around the equator of the sphere, as seen in Fig. 4.5. Here we see that z is much larger than the radius $R = \sqrt{x^2 + y^2}$. Once again we use the inverse of the above procedure to obtain the explicit form for Z_1 and Z_2 as:

$$B(\eta)^{-1}J_1 B(\eta) \to J_1' \to B(\eta)J_1'B(\eta)^{-1} \to Z_2,$$

$$B(\eta)^{-1}J_2 B(\eta) \to J_2' \to B(\eta)J_2'B(\eta)^{-1} \to -Z_1. \tag{4.54}$$

This results in J_3 remaining invariant while $-J_2 \to Z_1$ and $J_1 \to Z_2$:

$$J_3 = \begin{pmatrix} 0 & 0 & 0 & 0 \\ 0 & 0 & 0 & 0 \\ 0 & 0 & 0 & -i \\ 0 & 0 & i & 0 \end{pmatrix}, \quad Z_1 = \begin{pmatrix} 0 & 0 & 0 & 0 \\ 0 & 0 & i & 0 \\ 0 & 0 & 0 & 0 \\ 0 & 0 & 0 & 0 \end{pmatrix}, \quad \text{and} \quad Z_2 = \begin{pmatrix} 0 & 0 & 0 & 0 \\ 0 & 0 & 0 & i \\ 0 & 0 & 0 & 0 \\ 0 & 0 & 0 & 0 \end{pmatrix}. \tag{4.55}$$

This algebra is again the same as that for $E(2)$, the two-dimensional Euclidean group as given in Eq. (4.49). The cylindrical group is, therefore, locally isomorphic to the $E(2)$ group. We illustrate transformations in these two groups in Fig. 4.6.

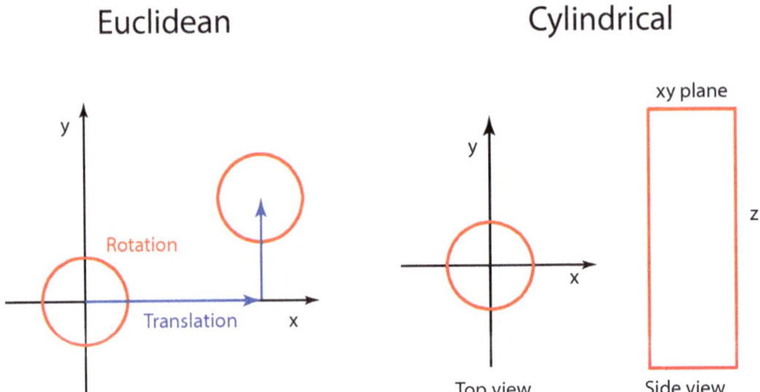

Fig. 4.6: Euclidean and cylindrical groups. In the Euclidean group, as we can see, translations along the x- and y-axis and rotations around the z-axis can be made. Whereas for the cylindrical group, rotations in the xy-plane and translations up and down on the z-axis can be made [2].

4.4 Group Contractions and Unitary Representations of E(2)

As was noted previously the unitary representations of the $E(2)$ group are infinite dimensional, and we shall see their representation space consists of solutions of the Bessel's differential equation. If we look at finite-dimensional representations which are contractions of the spherical harmonics, the dimension increases as the value ℓ increases. For this reason, we expect that the unitary representations of the $E(2)$ group can be achieved through the contraction of the spherical harmonics with an infinite value of ℓ [1].

Let us start with the spherical harmonics of the form

$$Y_\ell^m(\theta, \phi) = e^{im\phi} P_\ell^m(\cos \theta), \tag{4.56}$$

where we do not worry about the normalization constant. The associated Legendre function $P_\ell^m(\cos \theta)$ satisfies the differential equation:

$$\frac{d}{dz}\left((1 - z^2)\frac{d}{dz}P_\ell^m(z)\right) + \left(\ell(\ell + 1) - \frac{m^2}{(1 - z^2)}\right) P_\ell^m(z) = 0, \tag{4.57}$$

where

$$z = \cos \theta. \tag{4.58}$$

In the limit of small θ, the above differential equation becomes

$$\theta\left(\frac{d}{d\theta}\theta\frac{d}{d\theta}P_\ell^m\right) + \left(\ell(\ell + 1)\theta^2 - m^2\right) P_\ell^m = 0. \tag{4.59}$$

This is Bessel's differential equation, and its solutions are of course Bessel functions. In the small-θ limit, the spherical harmonics of Eq. (4.56) becomes

$$Y_\ell^m(\theta, \phi) \rightarrow e^{-im\phi} J_m(k\theta) , \tag{4.60}$$

where

$$k^2 = \ell(\ell + 1) . \tag{4.61}$$

The above form is just a representation space and the parameter k or ℓ has to be sufficiently large to make the argument of the Bessel function non-zero in the small-θ limit. When we make the approximation of keeping only the lowest-order term in $k\theta$, the representation becomes finite-dimensional and non-unitary.

The basic difference between the infinite-dimensional unitary representations and finite-dimensional non-unitary representations is that the representation space consisting of Bessel functions is square integrable and all the generators of the group are Hermitian. On the other hand, if the representation is finite-dimensional, the functions are not normalizable. They are only tensor representations of the coordinate variables.

Let us discuss one physical application of this limiting process. It is also a high-momentum limit. In scattering processes in which two incoming particles collide with each other resulting in two particles moving in different directions, we commonly use the Legendre polynomials $P_\ell(\cos \theta)$ to describe the dependence on the scattering angle. The quantum number ℓ is the angular momentum around the scattering center, and can be regarded as a measure of the incoming momentum multiplied by the impact parameter.

When particles move slowly, it is sufficient to consider only two or three lowest values ℓ. On the other hand, when the particles move with speed very close to that of light, the scattering becomes predominantly forward and becomes like the Fraunhofer diffraction. In this case, we have to deal with large values of ℓ. One way approach to this problem is to start from the operator

$$(J^2) = J_1^2 + J_2^2 + J_3^2 , \tag{4.62}$$

with the eigenvalue $\ell(\ell+1)$. For large values of ℓ, we can ignore J_3 whose eigenvalue is usually not larger than 1, and let

$$\ell(\ell + 1) = \left(\ell + \frac{1}{2}\right)^2 . \tag{4.63}$$

It is interesting to note [8] that, when the scattering angle is very small, we can replace $(J_1^2 + J_2^2)$ by

$$R^2(P_1^2 + P_2^2) = \left(\ell + \frac{1}{2}\right)^2 , \tag{4.64}$$

with suitable redefinitions for R and P_i. We can readily let R be (P_0/M). For a given value of P_0, the eigenvalue of

$$(P_1^2 + P_2^2) \tag{4.65}$$

will give a measure of ℓ. This new parameter will be that of the Bessel function. Thus, for large values of ℓ, the Legendre polynomial becomes the Bessel function:

$$P_\ell(\cos\theta) \to J_0(\alpha q\theta), \tag{4.66}$$

where

$$\alpha = P_0/M. \tag{4.67}$$

The parameter α now measures ℓ, and becomes continuous for large values of P_0 [9, 10, 11].

The above Bessel-function form is commonly used for studying high-energy data. It is interesting to note that the transition from the use of the Legendre polynomials for low-energy processes to that of the Bessel functions in high-energy scattering is a group contraction of $O(3)$ to $E(2)$.

4.5 Lorentz and Galilei Transformations

The purpose of this section is to indicate that there are many other interesting applications of group contractions. In Sect. 4.3, we studied the contraction of $O(3)$ to $E(2)$ using the notion of a plane tangent to a spherical surface. The concept of this tangent plane plays a very important role in many branches of science dealing with curved surfaces. As is illustrated in Figure (4.7), all the curved surfaces pictured contract to the tangent plane.

In Sect. 2.2 of Chap. 2 we considered the little group $O(2, 1)$ which is represented by the three-parameters which leave the four-momentum $(0, q, 0, 0)$ invariant. The rotation around the z-axis satisfies the requirement. Boosts along the x- and y-axes will also leave the representative four-vector invariant. These boosts together with the rotation around the z-axis form a three-dimensional Lorentz group operating on two spatial dimensions and one time-like dimension.

Since the rotation around the z-axis commutes with the boost along the same direction, it is sufficient to consider the transformations in the zx-plane. It is sufficient therefore to discuss the Lorentz group applicable to two space dimensions and one time dimension, namely to the (t, x, y)-coordinates. However, the rotation around the z-axis will also extend the representation to the full (t, z, x, y) space. Furthermore, physical applications seldom go beyond transformations in the (t, x, y)-coordinates. We call this smaller group the $O(2, 1)$-like little group. High-energy physics calculations involving Lorentz transformations, mostly are based on the $O(2, 1)$-like little group [2]. Furthermore, this group forms one of the basic languages in classical and modern optics [12, 13].

With this introduction let us consider the surface of the hyperbola in a three-dimensional space spanned by t, x, and y:

$$(kt)^2 - x^2 - y^2 = \text{Constant}, \tag{4.68}$$

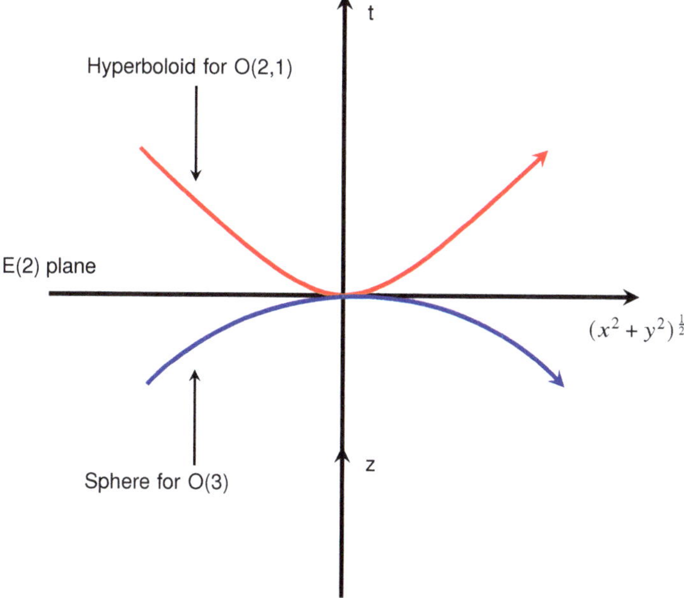

Fig. 4.7: Group contraction and tangent plane. The contraction of $O(3)$ or $O(2, 1)$ to $E(2)$ is essentially a tangent-plane approximation.

where k is a constant and may become very large. This is a description of the $O(2, 1)$ group consisting of Lorentz boosts along the x- and y-directions and rotations on the xy-plane.

We can now consider a plane tangent to the surface at $x = y = 0$. As the constant k becomes very large, the portion of surface in which $(x^2 + y^2)$ is finite becomes flat and coincides with the tangent plane. It is then not difficult to imagine that transformations on this tangent plane are Galilei transformations.

In order to see this point, let us start with coordinate transformations of the group $O(2, 1)$ generated by J_3, K_1, and K_2, where

$$J_3 = \begin{pmatrix} 0 & 0 & 0 & 0 \\ 0 & 0 & 0 & 0 \\ 0 & 0 & 0 & -i \\ 0 & 0 & i & 0 \end{pmatrix}, \quad K_1 = \begin{pmatrix} 0 & 0 & i & 0 \\ 0 & 0 & 0 & 0 \\ i & 0 & 0 & 0 \\ 0 & 0 & 0 & 0 \end{pmatrix}, \quad \text{and} \quad K_2 = \begin{pmatrix} 0 & 0 & 0 & i \\ 0 & 0 & 0 & 0 \\ 0 & 0 & 0 & 0 \\ i & 0 & 0 & 0 \end{pmatrix} \tag{4.69}$$

are applicable to the column vector $(kt, 1, x, y)$. K_1 and K_2 are the generators of Lorentz boosts along the x- and y-directions respectively.

The column vector $(kt, 1, x, y)$ can be written as

$$
\begin{pmatrix} kt \\ 1 \\ x \\ y \end{pmatrix} = \begin{pmatrix} k & 0 & 0 & 0 \\ 0 & 1 & 0 & 0 \\ 0 & 0 & 1 & 0 \\ 0 & 0 & 0 & 1 \end{pmatrix} \begin{pmatrix} t \\ 1 \\ x \\ y \end{pmatrix}. \tag{4.70}
$$

Then, as k becomes very large, the circumstance is identical to that of Eq. (4.53) and also Eq. (4.111). The resulting transformation matrix becomes

$$
\begin{pmatrix} t' \\ 1 \\ x' \\ y' \end{pmatrix} = \begin{pmatrix} 1 & 0 & 0 & 0 \\ 0 & 1 & 0 & 0 \\ u & 0 & \cos\alpha & -\sin\alpha \\ v & 0 & \sin\alpha & \cos\alpha \end{pmatrix} \begin{pmatrix} t \\ 1 \\ x \\ y \end{pmatrix}, \tag{4.71}
$$

This form is a rotation on the xy-plane followed by Galilei boosts along the x- and y-directions:

$$
t' = t,
$$

$$
x' = ut + x\cos\alpha - y\sin\alpha,
$$

$$
y' = vt + x\sin\alpha + y\cos\alpha. \tag{4.72}
$$

The generalization of the above procedure for three-dimensional space is straightforward [Problem 4 in Sect. 4.6]. Indeed, special relativity becomes Galilean relativity in the limit of large k [1].

Let us use $F(\alpha, u, v)$ for this matrix. This matrix can be decomposed into

$$
F(\alpha, u, v) = B(u, v)R(\alpha). \tag{4.73}
$$

The mathematical expressions for these matrices have already been given many times in connection with the $E(2)$ group. $B(u, v)$ is clearly an Abelian invariant subgroup of the homogeneous Galilean group represented by $F(\alpha, u, v)$.

We can write the inhomogeneous Galilean transformation as

$$
\begin{pmatrix} t' \\ 1 \\ x' \\ y' \end{pmatrix} = \begin{pmatrix} 1 & t_0 & 0 & 0 \\ 0 & 1 & 0 & 0 \\ u & x_0 & \cos\alpha & -\sin\alpha \\ v & y_0 & \sin\alpha & \cos\alpha \end{pmatrix} \begin{pmatrix} t \\ 1 \\ x \\ y \end{pmatrix}, \tag{4.74}
$$

which produces a translation on each of the coordinate variables in addition to the homogeneous Galilean transformation given in Eq. (4.72). Let us call the above four-by-four matrix G. Then this matrix can be decomposed to

$$
G = T(t_0, x_0, y_0)B(u, v)R(\alpha) \tag{4.75}
$$

where $T(t_0, x_0, y_0)$ is the translation matrix, and its form is obvious from the four-by-four matrix in Eq. (4.74) and from our experience with the $E(2)$ group.

The group of translation matrices is clearly an Abelian invariant subgroup of G. The boost matrix B is an Abelian invariant subgroup of the homogeneous linear transformation of Eq. (4.72). However, B is not an invariant subgroup of G. The group G has the five-parameter subgroup TB whose form is

$$TB = \begin{pmatrix} 1 & t_0 & 0 & 0 \\ 0 & 1 & 0 & 0 \\ u & x_0 & 1 & 0 \\ v & y_0 & 0 & 1 \end{pmatrix}. \tag{4.76}$$

The above TB matrix is a semidirect product of T and B, and T is an Abelian invariant subgroup. The inhomogeneous Galilean group is a semidirect product of R and TB. The group TB is not Abelian, but still is an invariant subgroup of G. Therefore, TB is a non-Abelian invariant subgroup of G [see Problem 5 in Sect. 4.6].

4.5.1 Contraction of O(2, 1) to E(2)

Previously we have seen that the $O(2, 1)$ group consists of rotations around the z-direction and Lorentz boosts along the x- and y-directions. We have seen that the column matrix of (t, z, x, y) is applied to this group to produce transformations. Here the rotation is around z-axis, therefore, takes the usual form:

$$\begin{pmatrix} 1 & 0 & 0 & 0 \\ 0 & 1 & 0 & 0 \\ 0 & 0 & \cos\phi & -\sin\phi \\ 0 & 0 & \sin\phi & \cos\phi \end{pmatrix} \begin{pmatrix} t \\ z \\ x \\ y \end{pmatrix} = \begin{pmatrix} t \\ z \\ x\cos\phi - y\sin\phi \\ x\sin\phi + y\sin\phi \end{pmatrix}. \tag{4.77}$$

The Lorentz boost along the x-direction is:

$$\begin{pmatrix} \cosh\lambda & 0 & \sinh\lambda & 0 \\ 0 & 1 & 0 & 0 \\ \sinh\lambda & 0 & \cosh\lambda & 0 \\ 0 & 0 & 0 & 1 \end{pmatrix} \begin{pmatrix} t \\ z \\ x \\ y \end{pmatrix} = \begin{pmatrix} t\cosh\lambda + x\sinh\lambda \\ z \\ t\sinh\lambda + x\cosh\lambda \\ y \end{pmatrix}. \tag{4.78}$$

The generators of this group, as given explicitly in Eq. (4.69), consist of J_3, K_1, and K_2. As usual J_3 generates rotations around the z-axis, while K_1 and K_2 generate Lorentz boosts along the x- and y-directions, respectively.

Let us now use, as we did in Sect. 4.3, the squeeze or contraction matrix given in Eq. (4.50). In this case, the z-coordinate does not participate in the transformations. Using this matrix, the (t, z, x, y) column matrix can be contracted as

$$\begin{pmatrix} t \\ z \\ x \\ y \end{pmatrix} \rightarrow \begin{pmatrix} 1 \\ 1 \\ x \\ y \end{pmatrix}. \tag{4.79}$$

As is now clear the Euclidean contraction procedure will leave J_3 invariant. Using the coordinate transform matrix in Eq. (4.71) and applying the same technique as in Eq. (4.54), we obtain:

$$K_1 \rightarrow P_1 = \begin{pmatrix} 0 & 0 & 0 & 0 \\ 0 & 0 & 0 & 0 \\ i & 0 & 0 & 0 \\ 0 & 0 & 0 & 0 \end{pmatrix} \quad \text{and} \quad K_2 \rightarrow P_2 = \begin{pmatrix} 0 & 0 & 0 & 0 \\ 0 & 0 & 0 & 0 \\ 0 & 0 & 0 & 0 \\ i & 0 & 0 & 0 \end{pmatrix}. \tag{4.80}$$

We can see that the form of this set of generators is the same as that given in Eq. (4.51) when contracting the $O(3)$ rotation group to $E(2)$.

For the cylindrical contraction we have:

$$K_1 \rightarrow Z_2 = \begin{pmatrix} 0 & 0 & 0 & i \\ 0 & 0 & 0 & 0 \\ 0 & 0 & 0 & 0 \\ 0 & 0 & 0 & 0 \end{pmatrix} \quad \text{and} \quad K_2 \rightarrow Z_1 = \begin{pmatrix} 0 & 0 & i & 0 \\ 0 & 0 & 0 & 0 \\ 0 & 0 & 0 & 0 \\ 0 & 0 & 0 & 0 \end{pmatrix}. \tag{4.81}$$

This set of generators also has the same form as that given in Sect. 4.3 for contracting the $O(3)$ rotation group to the cylindrical group.

4.5.2 Contraction of the Lorentz Group to Galilean Group

In Chap. 2, the proper Lorentz group, was defined as the group of Lorentz transformations, operating in the four-dimensional Minkowski space. The proper Lorentz group is a six-parameter group which is generated by four-by-four matrices applicable to the column vector of (t, z, x, y). In this space, the six generators were given in Eq. (2.27) of Chap. 2. The three rotation generators were defined as:

$$J_1 = \begin{pmatrix} 0 & 0 & 0 & 0 \\ 0 & 0 & 0 & i \\ 0 & 0 & 0 & 0 \\ 0 & -i & 0 & 0 \end{pmatrix}, \quad J_2 = \begin{pmatrix} 0 & 0 & 0 & 0 \\ 0 & 0 & -i & 0 \\ 0 & i & 0 & 0 \\ 0 & 0 & 0 & 0 \end{pmatrix}, \quad \text{and} \quad J_3 = \begin{pmatrix} 0 & 0 & 0 & 0 \\ 0 & 0 & 0 & 0 \\ 0 & 0 & 0 & -i \\ 0 & 0 & i & 0 \end{pmatrix}. \tag{4.82}$$

The three boost generators take the form:

$$K_1 = \begin{pmatrix} 0 & 0 & i & 0 \\ 0 & 0 & 0 & 0 \\ i & 0 & 0 & 0 \\ 0 & 0 & 0 & 0 \end{pmatrix}, \quad K_2 = \begin{pmatrix} 0 & 0 & 0 & i \\ 0 & 0 & 0 & 0 \\ 0 & 0 & 0 & 0 \\ i & 0 & 0 & 0 \end{pmatrix}, \quad \text{and} \quad K_3 = \begin{pmatrix} 0 & i & 0 & 0 \\ i & 0 & 0 & 0 \\ 0 & 0 & 0 & 0 \\ 0 & 0 & 0 & 0 \end{pmatrix}. \tag{4.83}$$

Let us now perform the contraction procedures by using a four-by-four squeeze matrix which takes the form:

$$B(\eta) = \begin{pmatrix} e^{-\eta/2} & 0 & 0 & 0 \\ 0 & e^{\eta/2} & 0 & 0 \\ 0 & 0 & e^{\eta/2} & 0 \\ 0 & 0 & 0 & e^{\eta/2} \end{pmatrix}. \tag{4.84}$$

This process leaves the the rotation generators J_i unchanged, but the boost generators K_1, K_2, and K_3, take the following form:

$$K_1 \rightarrow P_1 = \begin{pmatrix} 0 & 0 & 0 & 0 \\ 0 & 0 & 0 & 0 \\ i & 0 & 0 & 0 \\ 0 & 0 & 0 & 0 \end{pmatrix}, \quad K_2 \rightarrow P_2 = \begin{pmatrix} 0 & 0 & 0 & 0 \\ 0 & 0 & 0 & 0 \\ 0 & 0 & 0 & 0 \\ i & 0 & 0 & 0 \end{pmatrix},$$

$$K_3 \rightarrow P_3 = \begin{pmatrix} 0 & 0 & 0 & 0 \\ i & 0 & 0 & 0 \\ 0 & 0 & 0 & 0 \\ 0 & 0 & 0 & 0 \end{pmatrix}, \tag{4.85}$$

respectively. The transformation to which these matrices lead takes the form:

$$\exp\left(-i\left[v_1 P_1 + v_2 P_2 + v_3 P_3\right]\right) = \begin{pmatrix} 1 & 0 & 0 & 0 \\ v_3 & 1 & 0 & 0 \\ v_1 & 0 & 1 & 0 \\ v_2 & 0 & 0 & 1 \end{pmatrix}. \tag{4.86}$$

When we apply this matrix to the column vector (t, z, x, y),

$$\begin{pmatrix} 1 & 0 & 0 & 0 \\ v_3 & 1 & 0 & 0 \\ v_1 & 0 & 1 & 0 \\ v_2 & 0 & 0 & 1 \end{pmatrix} \begin{pmatrix} t \\ z \\ x \\ y \end{pmatrix} = \begin{pmatrix} t \\ z + v_3 t \\ x + v_1 t \\ y + v_2 t \end{pmatrix}. \tag{4.87}$$

If we apply the rotation matrix to this four-dimensional space we obtain:

$$\begin{pmatrix} 1 & 0 & 0 & 0 \\ 0 & & \text{3-by-3} & \\ 0 & & \text{rotation} & \\ 0 & & \text{matrix} & \end{pmatrix}. \tag{4.88}$$

If this matrix is then applied to the four-vector, we have

$$\begin{pmatrix} 1 & 0 & 0 & 0 \\ 0 & & \text{3-by-3} & \\ 0 & & \text{rotation} & \\ 0 & & \text{matrix} & \end{pmatrix} \begin{pmatrix} t \\ z \\ x \\ y \end{pmatrix} = \begin{pmatrix} t \\ z' \\ x' \\ y' \end{pmatrix}. \tag{4.89}$$

Here z', x', and y', the results of the three-dimensional rotation, take the usual familiar forms.

Thus the rotation followed by the Galilean boost matrix is

$$
\begin{pmatrix} 1 & 0 & 0 & 0 \\ v_3 & 1 & 0 & 0 \\ v_1 & 0 & 1 & 0 \\ v_2 & 0 & 0 & 1 \end{pmatrix}
\begin{pmatrix} 1 & 0 & 0 & 0 \\ 0 & & \text{3-by-3} & \\ 0 & & \text{rotation} & \\ 0 & & \text{matrix} & \end{pmatrix}
=
\begin{pmatrix} 1 & 0 & 0 & 0 \\ v_3 & & \text{3-by-3} & \\ v_1 & & \text{rotation} & \\ v_1 & & \text{matrix} & \end{pmatrix} .
\tag{4.90}
$$

This transformation thus has the net effect of

$$
\begin{pmatrix} 1 & 0 & 0 & 0 \\ v_3 & & \text{3-by-3} & \\ v_1 & & \text{rotation} & \\ v_1 & & \text{matrix} & \end{pmatrix}
\begin{pmatrix} t \\ z \\ x \\ y \end{pmatrix}
=
\begin{pmatrix} t \\ z' + v_3 t \\ x' + v_1 t \\ y' + v_2 t \end{pmatrix} .
\tag{4.91}
$$

Applying this transformation matrix to the four-vector with the elements $(1, z, x, y)$, the transformation results in:

$$
\begin{pmatrix} 1 & 0 & 0 & 0 \\ v_3 & & \text{3-by-3} & \\ v_1 & & \text{rotation} & \\ v_2 & & \text{matrix} & \end{pmatrix}
\begin{pmatrix} 1 \\ z \\ x \\ y \end{pmatrix}
=
\begin{pmatrix} 1 \\ z' + v_3 \\ x' + v_1 \\ y' + v_2 \end{pmatrix} .
\tag{4.92}
$$

This is clearly a three-dimensional transformation in Euclidean space.

4.6 Exercises and Problems

Exercise 1. We discussed in Chap. 3 the $SU(2)$ group and that it is homomorphic (two to one correspondence) with the O(3) group which becomes $E(2)$ during the contraction process. According to Cartan's criterion, the Cartan determinant given in Sect. B.5 of Appendix B vanishes for $E(2)$ while it is non-zero for $SU(2)$. Show how the Cartan determinant vanishes during this contraction procedure.

Let us start from the Lie algebra of $SU(2)$:

$$
[J_i, J_j] = i C_{ij}^{k} J_k,
\tag{4.93}
$$

where

$$
C_{ij}^{k} = \varepsilon_{ij}{}^{k} .
\tag{4.94}
$$

Cartan's determinant is the determinant of the three-by-three matrix:

$$
g_{ij} = \varepsilon_{ik}^{\ell} \varepsilon_{j\ell}^{k} .
\tag{4.95}
$$

The explicit form of this matrix is

$$g_{ij} = \begin{pmatrix} -2 & 0 & 0 \\ 0 & -2 & 0 \\ 0 & 0 & -2 \end{pmatrix}. \tag{4.96}$$

The determinant of this matrix does not vanish. According to the group contraction procedure discussed in Sects. 4.3, 10.1, and 10.8, we are led to consider the algebra of J_3, N_1, and N_2 given in Eq. (10.125). Then

$$C_{12}^3 = -\left(\frac{M}{E}\right)^2, \qquad C_{23}^1 = C_{31}^2 = 1, \tag{4.97}$$

and the Cartan matrix becomes

$$\begin{pmatrix} 2(M/E)^2 & 0 & 0 \\ 0 & -2 & 0 \\ 0 & 0 & -2 \end{pmatrix}. \tag{4.98}$$

where M and E are, as before, the mass and energy of the particle respectively. The determinant of this matrix becomes zero in the high-energy limit.

Exercise 2. The group contraction we studied in this Chapter is essentially a flat-surface approximation of curved surfaces. This procedure can prove to be very useful in many branches of physics and engineering. Let us consider here an application in geometric optics. The surface of lens is approximately spherical. However, it would be more convenient if the light rays can be expressed as linear functions. Find matrix representations of light rays going through a thin lens.

Let us consider a straight line in the xy-coordinate system. We can write the equation for this line as

$$y = ax + b. \tag{4.99}$$

The first derivative of this function or the slope of the line is independent of the x variable. Thus for a given x variable, the y variable and the slope completely determine the line. Therefore, we can use the column vector

$$\begin{pmatrix} y \\ a \end{pmatrix}_x, \tag{4.100}$$

to specify the line. If we wish to move from the coordinate point (x_0, y_0) to (x_1, y_1), we can represent this transformation as

$$\begin{pmatrix} y_1 \\ a \end{pmatrix}_{x_1} = \begin{pmatrix} 1 & (x_1 - x_0) \\ 0 & 1 \end{pmatrix} \begin{pmatrix} y_0 \\ a \end{pmatrix}_{x_0}. \tag{4.101}$$

The matrix on the right-hand side of the above expression is called the translation matrix, and the algebraic property of this matrix has been studied in Chap. 3 in connection with the $E(2)$-like little group for massless particles.

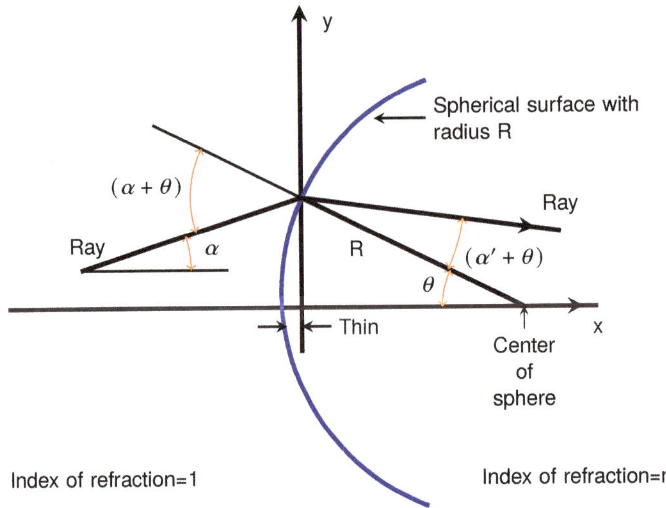

Fig. 4.8: Incident ray on a spherical lens surface. Matrix optics results from the linear approximation of the relation between the slope of the ray and the distance of the ray from the axis. The lens surface is spherical or composed of any other curved surface which can be approximated as a spherical surface near the lens axis.

Let us next consider a light ray going through a convex spherical surface of a transparent material where the index of refraction is n and radius is R, as is specified in Figure 4.8. We assume that the angle between the ray and the axis of the surface is very small and that the distance between the ray and the axis is much smaller than radius of the spherical surface. Then from Snell's law,

$$\sin(\alpha + \theta) = n \sin(\alpha' + \theta) \,, \tag{4.102}$$

where the index of refraction of air is assumed to be 1. Since all the angles in the above expression are small,

$$\alpha = a \quad \text{and} \quad \alpha' = a' \,, \quad \theta = y/R \,, \tag{4.103}$$

and

$$a + \frac{y}{R} = na' + n\frac{y}{R} \,. \tag{4.104}$$

This means that the slope undergoes the change through

$$a' = \frac{1}{n}a + \left(\frac{1-n}{nR}\right) y \,, \tag{4.105}$$

while there is no change in y. The slopes, a and a', of the rays are in the air and the refractive material respectively. We can ignore the variation in the x-coordinate, which is of order $(y/R)^2$, in the spirit of the group contraction discussed in this

Chapter. Thus the relation between a and a' can be represented by the matrix relation

$$\begin{pmatrix} y \\ a' \end{pmatrix}_x = \begin{pmatrix} 1 & 0 \\ (1-n)/nR & 1/n \end{pmatrix} \begin{pmatrix} y \\ a \end{pmatrix}_x . \tag{4.106}$$

If the ray comes out from the refractive material through another convex surface with radius of curvature R' facing in the opposite direction, then the applicable matrix is

$$\begin{pmatrix} 1 & 0 \\ (1-n)/R' & n \end{pmatrix} . \tag{4.107}$$

If the distance between the two convex surface is small, they constitute a thin lens. The overall effect is the multiplication of the matrix of Eq. (4.106) from left by the matrix of Eq. (4.107). The net effect is the matrix

$$\begin{pmatrix} 1 & 0 \\ -1/f & 1 \end{pmatrix} , \tag{4.108}$$

where

$$\frac{1}{f} = (n-1) \left(\frac{1}{R} + \frac{1}{R'} \right) . \tag{4.109}$$

f is called the focal length of the thin lens. The above expression is called the lens makers formula and was discovered first experimentally. The matrix of Eq. (4.108) is called the lens matrix. This matrix, together with the translation matrix of Eq. (4.101), forms the basic building blocks for a group theoretical approach to modern geometrical optics [14]. See Chap. 15 as well as Blaker [15] for a more detailed treatment of the matrix approach to geometrical optics.

Problem 1. Throughout this Chapter, we used the small area on the north pole of the spherical Earth as a model for group contraction. It is also possible to make approximations on other areas. For example, it is possible to choose a narrow belt around the equator as a basis for approximation. Show in this case that the Legendre function becomes a harmonic oscillator wave function. See Talman [16] and Gilmore [17].

Problem 2. Consider the following operations on the rotation generators of Eq. (4.46):

$$P_1 = \lim_{R \to \infty} \frac{1}{R} \left[B(R)^{-1} J_2 B(R) \right] \quad \text{and} \quad P_2 = \lim_{R \to \infty} \frac{-1}{R} \left[B(R)^{-1} J_1 B(R) \right] , \tag{4.110}$$

where

$$B(R) = \begin{pmatrix} 1 & 0 & 0 & 0 \\ 0 & R & 0 & 0 \\ 0 & 0 & 1 & 0 \\ 0 & 0 & 0 & 1 \end{pmatrix} \tag{4.111}$$

with

$$J_3 = B(R)^{-1} J_3 B(R) . \tag{4.112}$$

Here, R can be interpreted as the radius of a sphere in three dimensions, which becomes very large during the process. Then show that P_1, P_2, and J_3 take the form of Eq. (4.51), and thus satisfy the same commutation relations of Eq. (4.49).

Problem 3. Construct the Lie algebra for the three-dimensional (homogeneous) Galilean group and that for the inhomogeneous Galilean group.

Problem 4. It is clear from Eq. (4.71) that the two-dimensional Galilean group is isomorphic to $E(2)$. Is the three-dimensional Galilean group isomorphic to $E(3)$?

Problem 5. Show that the Poincaré group can be obtained from the $(3 + 2)$ de Sitter group or $O(3, 2)$ through group contraction. See Chaps. 7 and 11.

Problem 6. Show that the three-dimensional inhomogeneous Galilean group can also be obtained as a contraction of $O(4, 1)$ or $O(3, 2)$. See Brooke [18, 19].

Problem 7. We discussed in Sect. 4.5 how the Galilean transformation in two-dimensional space can be obtained from $O(2, 1)$. Show that the Galilean transformation in three-dimensional space can be obtained from $O(3, 1)$ through group contraction. Show also that we can get the same result by contracting $O(4)$. See Holman [20].

Problem 8. Show that the subgroup consisting of Galilei boosts and translations in three-dimensional space is a non-Abelian invariant subgroup of the inhomogeneous Galilean group.

Problem 9. Work out the irreducible unitary representations of the inhomogeneous Galilean group applicable to the time-independent Schrödinger wave functions. See Inönü and Wigner [21] and Bargmann [22].

Problem 10. Discuss the contraction of $SU(2)$ to $E(2)$ using the conformal mapping for $SL(2, c)$:

$$z' = \frac{az + b}{cz + d}. \tag{4.113}$$

Problem 11. Show that the harmonic oscillator group mentioned in Sect. 5.7 (Problem 6) of Chap. 5 is a contraction of $O(3, 2)$. See Roman and Haavisto [23].

Problem 12. The role of the Pauli spin matrices in non-relativistic quantum mechanics is well known, and they are identical to those used for the generators of rotations in $SL(2, c)$. What happens to the boost operators of $SL(2, c)$ when we contract $O(3, 1)$ to the three-dimensional Galilean group?

Problem 13. A spherical surface can always be approximated as a parabolic surface. Is it possible to modify the group contraction procedure discussed in this Chapter to obtain a parabolic approximation? This procedure is the calculation of aberration.

Problem 14. From the γ-matrices and their products fifteen independent four-by-four matrices can be formed. They serve to generate the fifteen parameter $O(3, 3)$ Lorentz group. (See Sects. 7.6 and 7.7 of Chap. 7). Show that there are eighteen different ways to express $E(2)$-like algebra by using the fifteen independent γ-matrices in the

Weyl representation and by employing a similar contraction procedure of Problem 2, where in this case the boost matrices take the role Eq. (4.111). See [24].

References

1. E. Inönü, E.P. Wigner, On the Contraction of Groups and Their Representations, Proceedings of the National Academy of Sciences **39**(6), 510–524 (1953). DOI 10.1073/pnas.39.6.510. URL http://www.pnas.org/cgi/doi/10.1073/pnas.39.6.510
2. S. Başkal, Y.S. Kim, M.E. Noz, *Physics of the Lorentz Group (Second Edition): Beyond high-energy physics and optics* (IOP Publishing, Bristol, UK, 2021). DOI 10.1088/978-0-7503-3607-9. ISBN 978-0-7503-3607-9. URL https://iopscience.iop.org/book/978-0-7503-3607-9
3. E.P. Wigner, On Unitary Representations of the Inhomogeneous Lorentz Group, The Annals of Mathematics **40**(1), 149–204 (1939). DOI 10.2307/1968551. URL http://www.jstor.org/stable/1968551?origin=crossref
4. Y.S. Kim, E.P. Wigner, Cylindrical group and massless particles, Journal of Mathematical Physics **28**(5), 1175–1179 (1987). DOI 10.1063/1.527824. URL http://aip.scitation.org/doi/10.1063/1.527824
5. Y.S. Kim, E.P. Wigner, Space–time geometry of relativistic particles, Journal of Mathematical Physics **31**(1), 55–60 (1990). DOI 10.1063/1.528827. URL http://aip.scitation.org/doi/10.1063/1.528827
6. D. Han, Y.S. Kim, D. Son, E(2)–like little group for massless particles and neutrino polarization as a consequence of gauge invariance, Physical Review D **26**(12), 3717–3725 (1982). DOI 10.1103/PhysRevD.26.3717. URL https://link.aps.org/doi/10.1103/PhysRevD.26.3717
7. D. Han, Y.S. Kim, D. Son, Eulerian parametrization of Wigner's little groups and gauge transformations in terms of rotations in two–component spinors, Journal of Mathematical Physics **27**(9), 2228–2235 (1986). DOI 10.1063/1.526994. URL http://aip.scitation.org/doi/10.1063/1.526994
8. S.P. Misra, J. Maharana, Diffraction scattering and group contraction, Physical Review D **14**(1), 133–139 (1976). DOI 10.1103/PhysRevD.14.133. URL https://link.aps.org/doi/10.1103/PhysRevD.14.133
9. R. Blankenbecler, M.L. Goldberger, Behavior of Scattering Amplitudes at High Energies, Bound States, and Resonances, Physical Review **126**(2), 766–786 (1962). DOI 10.1103/PhysRev.126.766. URL https://link.aps.org/doi/10.1103/PhysRev.126.766
10. S.J. Wallace, High-Energy Expansions of Scattering Amplitudes, Physical Review D **8**(6), 1846–1863 (1973). DOI 10.1103/PhysRevD.8.1846. URL https://link.aps.org/doi/10.1103/PhysRevD.8.1846
11. S.J. Wallace, Correspondence of partial-wave and impact-parameter representations, Physical Review D **9**(2), 406–407 (1974). DOI 10.1103/PhysRevD.9.406. URL https://link.aps.org/doi/10.1103/PhysRevD.9.406
12. Y.S. Kim, M.E. Noz, *Phase space picture of quantum mechanics: group theoretical approach*. No. 40 in Lecture notes in physics series (World Scientific Publishing Co., Singapore; Hackensack, NJ, USA, 1991). ISBN 978-981-02-0360-3,978-981-02-0361-0. URL https://doi.org/10.1142/1197
13. S. Başkal, Y.S. Kim, Lorentz group in ray and polarization optics, in *Mathematical Optics: Classical, Quantum and Computational Methods*, ed. by V. Lakshminarayanan, M. L. Calvo, T. Alieva (Taylor and Francis, Boca Raton, FL, USA, 2013), 303–349. ISBN 978-1-4398-6961-1. URL https://www.taylorfrancis.com/books/9781439869611
14. H. Bacry, M. Cadilhac, Metaplectic group and Fourier optics, Physical Review A **23**(5), 2533–2536 (1981). DOI 10.1103/PhysRevA.23.2533. URL https://link.aps.org/doi/10.1103/PhysRevA.23.2533

15. J.W. Blaker, *Geometric Optics - The Matrix Theory* (Marcel Decker, New York, NY, USA, 1971). ISBN 978-0824710460

16. J.D. Talman, *Special Functions: a Group Theoretical Approach Based on Lectures By Eugene P Wigner* (W.A. Benjamin, Inc. (Benjamin Cummings), San Francisco, CA, USA, 1968). URL https://books.google.se/books?id=m-IbtAEACAAJ. 260 pp., lccn=68-54038.

17. R. Gilmore, *Lie groups, Lie algebras, and some of their applications* (Dover Publications, Mineola, NY, USA, 2005). ISBN 978-0-486-44529-8. (Originally published: 1974, John Wiley and Sons, New York, NY, USA.)

18. J.A. Brooke, A Galileian formulation of spin. I. Clifford algebras and spin groups, Journal of Mathematical Physics **19**(5), 952–959 (1978). DOI 10.1063/1.523798. URL http://aip.scitation.org/doi/10.1063/1.523798

19. J.A. Brooke, A Galileian formulation of spin. II. Explicit realizations, Journal of Mathematical Physics **21**(4), 617–621 (1980). DOI 10.1063/1.524506. URL http://aip.scitation.org/doi/10.1063/1.524506

20. W.J. Holman, The asymptotic forms of the fano function: The representation functions and wigner coefficients of SO(4) and E(3), Annals of Physics **52**(1), 176–191 (1969). DOI 10.1016/0003-4916(69)90323-6. URL https://linkinghub.elsevier.com/retrieve/pii/0003491669903236

21. E. Inönü, E.P. Wigner, Representations of the Galilei group, Il Nuovo Cimento **9**(8), 705–718 (1952). DOI 10.1007/BF02782239. URL http://link.springer.com/10.1007/BF02782239

22. V. Bargmann, On Unitary Ray Representations of Continuous Groups, The Annals of Mathematics **59**(1), 1–46 (1954). DOI 10.2307/1969831. URL https://www.jstor.org/stable/1969831?origin=crossref

23. P. Roman, J. Haavisto, Relativistic quantum dynamical group for hadrons, Journal of Mathematical Physics **22**(2), 403–411 (1981). DOI 10.1063/1.524906. URL http://aip.scitation.org/doi/10.1063/1.524906

24. S. Başkal, Dirac gamma matrices as representations of the algebra of little groups, Modern Physics Letters A **15**(13), 833–839 (2000). DOI 10.1142/S0217732300000827. URL https://www.worldscientific.com/doi/abs/10.1142/S0217732300000827

Chapter 5
Covariant Harmonic Oscillator Formalism

Abstract The relativistic wave function quest originated with Schrödinger's attempt to formulate his wave mechanics using the relativistic wave equation. The resulting Klein-Gordon equation has a negative energy problem and difficulties with a probability interpretation. The first problem was later solved by the second quantization procedure, but not the probability interpretation. Dirac's equation for electrons, successful for electron properties in the static limit, has the same relativistic feature as the Klein-Gordon equation. In 1945, Dirac suggested the use of normalizable relativistic harmonic oscillator wave functions. Yukawa in 1953 constructed relativistic harmonic oscillator wave functions which led to infinite-component wave functions, but suggested a subsidiary condition involving the four-momentum of the particle was needed. Feynman et al. advocated the use of relativistic harmonic oscillators instead of Feynman diagrams for studying hadron structures and interactions. The basic problem facing any relativistic harmonic oscillator equation is the negative-energy spectrum due to time-like excitations. Eliminating time-like excitations was thought to lead to a violation of probability conservation. Harmonic oscillator wave functions without time-like excitations, but with a probability interpretation can be constructed. These form the vector spaces for unitary irreducible representations of the Poincaré group. We discuss this in detail here.

Because wave functions play a central role in non-relativistic quantum mechanics, one method of combining quantum mechanics and special relativity takes the form of efforts to construct relativistic wave functions with an appropriate probability interpretation. The story of relativistic wave functions begins with Schrödinger's original attempt to formulate his wave mechanics using the relativistic wave equation which is known today as the Klein-Gordon equation [1]. Unlike the Schrödinger wave function, the solutions of the Klein-Gordon equation have the well-known negative-energy problem. The Klein-Gordon wave function also has difficulties with

S. Başkal et al., *Theory and Applications of the Poincaré Group*, Fundamental Theories of Physics 217, https://doi.org/10.1007/978-3-031-64376-7_5

a probability interpretation. The negative-energy problem was later *solved* by the second quantization procedure, but the difficulty with a probability interpretation still persists.

We can next mention Dirac's equation for electrons. This equation has been strikingly successful in explaining the properties of the electron in the static limit. However, its relativistic feature is not different from that of the Klein-Gordon equation.

In spite of continued efforts to construct relativistic wave functions, this approach was overshadowed by the successes of quantum field theory in quantum electrodynamics. It is quite fair to say that we are still living under this shadow.

Because of its mathematical simplicity, the harmonic oscillator has served as the first concrete solution to many new physical theories. It played a key role in the developing stages of non-relativistic quantum mechanics, statistical mechanics, theory of specific heat, molecular theory, quantum field theory, theory of superconductivity, theory of coherent light, and many others. It is, therefore, quite natural to expect that the first nontrivial relativistic wave function would be a relativistic harmonic oscillator wave function.

As early as in 1945, Dirac suggested the use of normalizable relativistic harmonic oscillator wave functions for studying relativistic Fock space needed in quantum electrodynamics [2]. In connection with relativistic particles with internal space-time structure, Yukawa attempted to construct relativistic harmonic oscillator wave functions in 1953 [3]. Yukawa observed that an attempt to solve a relativistic harmonic oscillator wave equation in general leads to infinite-component wave functions, and that finite-component wave functions may be chosen if a subsidiary condition involving the four-momentum of the particle is considered. This proposal of Yukawa was further developed by others [4, 5, 6, 7, 8].

The effectiveness of Yukawa's harmonic oscillator wave function in the relativistic quark model was first demonstrated by Fujimura et al. [9] who showed that the Yukawa wave function leads to the correct high-energy asymptotic behavior of the nucleon form factor. The harmonic oscillator wave function was also rediscovered by Feynman et al. [10] who advocated the use of relativistic harmonic oscillators instead of Feynman diagrams for studying hadron structures and interactions. The paper of Feynman et al. contains all the troubles expected from relativistic wave equations, and the authors of this paper did not make any attempt to hide those troubles.

The basic problem facing any relativistic harmonic oscillator equation is the negative-energy spectrum due to time-like excitations. It had once been widely believed that any attempt to obtain finite-component wave functions by eliminating time-like excitations would lead to a violation of probability conservation. This belief did not turn out to be true. It is now possible to construct harmonic oscillator wave functions without time-like wave functions which form the vector spaces for unitary irreducible representations of the Poincaré group.

In Sect. 5.1, we formulate the problem by writing down the relativistically invariant differential equation which leads to the covariant harmonic oscillator formalism. In Sect. 5.2, we study solutions of the harmonic oscillator differential equation which are normalizable in the four-dimensional (t, z, x, y) space. In Sect. 5.3, representations of the Poincaré group for massive hadrons are constructed from the normalizable harmonic oscillator wave functions. It is shown that they form the basis for unitary irreducible representations of the Poincaré group, as well as that for the $O(3)$-like little group for massive particles.

In Sect. 5.4, Lorentz transformation properties of the harmonic oscillator wave functions are studied. The linear unitary representation of Lorentz transformation is provided for the harmonic oscillator wave functions. In Sect. 5.5, we study the relativistic harmonic oscillators using the language of the four-dimensional Euclidean coordinate system. We then study in Sect. 5.6 how Lorentz transformations are different from rotations in the $O(4)$-coordinate system, and investigate the nature of the Wick rotation commonly used in physics. Section 5.7 consists of exercises and problems.

5.1 Covariant Harmonic Oscillator Differential Equations

Developing a new physical theory usually requires a new set of mathematical formulae. There are in general two different approaches to this problem. According to Eddington, we have to understand all the physical principles before writing down the first mathematical formula. According to Dirac, however, it is more profitable to construct plausible mathematical devices which can describe quantitatively the real world, and then add physical interpretations to the mathematical formalism [11], (this paper contains an interesting discussion on Eddington's versus Dirac's view). Both special relativity and quantum mechanics were developed in Dirac's way, and most of the new physical models these days are developed in this way.

The Klein-Gordon equation for a single particle is quite familiar to us. This equation takes the form:

$$\left[\left(\frac{\partial}{\partial x}\right)^2 + m^2\right]\phi(x) = 0, \tag{5.1}$$

where

$$\left(\frac{\partial}{\partial x}\right)^2 = \frac{\partial^2}{\partial t^2} - \frac{\partial^2}{\partial z^2} - \frac{\partial^2}{\partial x^2} - \frac{\partial^2}{\partial y^2}. \tag{5.2}$$

In physics this equation as well as its solutions are well known. For two independent particles with the coordinates x_a and x_b, respectively, the equation is

$$\left[\left(\frac{\partial}{\partial x_a}\right)^2 + \left(\frac{\partial}{\partial x_b}\right)^2 + m_a^2 + m_b^2\right]\phi(x_a)\phi(x_b) = 0. \tag{5.3}$$

The coordinates x_a and x_b are four-vectors, and the notation

$$x_\mu^2 \equiv t^2 - z^2 - x^2 - y^2 \tag{5.4}$$

will be used for both of them. The physics of this system for two free particles is also well known. The question then becomes what happens when we replace $\left(m_a^2 + m_b^2\right)$ with a term containing $(x_a - x_b)^2$.

With this point in mind, let us consider the differential equation of Feynman et al. [10] for a hadron consisting of two quarks bound together by a harmonic oscillator potential of unit strength:

$$\left\{2\left[\left(\frac{\partial}{\partial x_a^\mu}\right)^2 + \left(\frac{\partial}{\partial x_b^\mu}\right)^2\right] + \left(-\frac{1}{16}\right)(x_{a\mu} - x_{b\mu})^2 + m_0^2\right\}\phi(x_a, x_b) = 0, \qquad (5.5)$$

where x_a and x_b are space-time coordinates for the first and second quarks respectively. This partial differential equation has many different solutions depending on the choice of variables and boundary conditions.

In order to simplify the above differential equation, let us introduce new coordinate variables:

$$X = \frac{(x_a + x_b)}{2}, \qquad x = \frac{(x_a - x_b)}{2\sqrt{(2)}}. \qquad (5.6)$$

The four-vector X specifies where the hadron is located in space-time, while the variable x measures the space-time separation between the quarks. In terms of these variables, Eq. (5.5) can be written as

$$\left[\frac{\partial^2}{\partial X_\mu^2} + m_0^2 + \frac{1}{2}\left(\frac{\partial^2}{\partial x_\mu^2} - x^{\mu 2}\right)\right]\phi(X, x) = 0. \qquad (5.7)$$

This equation is separable in the X and x variables. Thus

$$\phi(X, x) = f(X)\psi(x), \qquad (5.8)$$

and $f(X)$ and $\psi(x)$ satisfy the following differential equations respectively:

$$\left[\frac{\partial^2}{\partial X_\mu^2} + m_0^2 + (\lambda + 1)\right]f(X) = 0, \qquad (5.9)$$

$$\frac{1}{2}\left(\frac{\partial^2}{\partial x_\mu^2} - x^{\mu 2}\right)\psi(x) = (\lambda + 1)\psi(x). \qquad (5.10)$$

Eq. (5.9) is a Klein-Gordon equation, and its solution takes the form

$$f(X) = \exp[\pm i p_\mu X^\mu], \qquad (5.11)$$

with

$$P^2 = P_\mu P^\mu = M^2 = m_0^2 + (\lambda + 1), \qquad (5.12)$$

where M and P are the mass and four-momentum of the hadron respectively. The eigenvalue λ is determined from the solution of Eq. (5.10). We are using the same notation for the operator and eigenvalue for the hadron four-momentum. This should not cause any confusion since we are dealing only with free hadron states with a definite four-momentum.

As for the four-momenta of the quarks p_a and p_b, we can combine them into the total four-momentum and momentum-energy separation between the quarks:

$$P = p_a + p_b, \quad q = \sqrt{2}(p_a - p_b), \tag{5.13}$$

P is the hadron four-momentum conjugate to X. The internal momentum-energy separation q is conjugate to x provided that there exist wave functions which can be Fourier-transformed. If the momentum-energy wave functions can be obtained from the Fourier transformation of the space-time wave function, the differential equation in the q space is identical to the harmonic oscillator equation for the x space given in Eq. (5.10).

We now have a set of equations to study. The first question is whether the above set of equations generates a mathematics which has enough aesthetic values. The second question is whether this mathematics can serve as a device which can describe the real world. The third question is whether the mathematical device we use is consistent with the existing rules of quantum mechanics and special relativity. We shall study the first question in this Chapter and discuss the second question in Chaps. 12 and 13. We shall examine the the third question in Chap. 6.

5.2 Normalizable Solutions of the Relativistic Harmonic Oscillator

We are quite familiar with the fact that the three-dimensional harmonic oscillator equation is separable in several coordinate systems. Indeed, the four-dimensional differential equation of Eq. (5.10) is separable in at least thirty-four different coordinate systems [12]. However, we are only interested in those solutions which are useful for constructing physically relevant representations of the Poincaré group.

There are solutions which are normalizable and those which are not. In this section, we discuss one of the solutions in four-dimensional space-time. This solution will form a basis for unitary irreducible representations of the Poincaré group. There are also interesting solutions which are not normalizable. We shall discuss these solutions in Chaps. 9 and 10.

Since we are quite familiar with the three-dimensional harmonic oscillator equation from non-relativistic quantum mechanics, we are naturally led to consider the separation of the space and time variables and write the four-dimensional harmonic oscillator equation of Eq. (5.10) as

$$\frac{1}{2}\left[-\nabla^2 +\frac{\partial^2}{\partial t^2} - \left(t^2 - x_i^2\right)\right]\psi(x) = (\lambda + 1)\psi(x) . \tag{5.14}$$

However, the xt system is not the only coordinate system in which the differential equation takes the above form.

If the hadron moves along the Z-direction which is also the z-direction, then the hadron factor $f(X)$ of Eq. (5.11) is Lorentz-transformed in the same manner as the scalar particles are transformed, as is discussed in Sect. 2.5 of Chap. 2. The Lorentz transformation of the internal coordinates from the laboratory frame to the hadron rest frame takes the form

$$t' = (t - \beta z)/(1 - \beta^2)^{1/2} ,$$

$$z' = (z - \beta t)/(1 - \beta^2)^{1/2} ,$$

$$x' = x, \quad y' = y , \tag{5.15}$$

where β is the velocity of the hadron moving along the z-direction. The primed quantities are the coordinate variables in the hadron rest frame. In terms of the primed variables, the harmonic oscillator differential equation is

$$\frac{1}{2}\left[-\nabla'^2 +\frac{\partial^2}{\partial t'^2} - \left(t'^2 - x_i'^2\right)\right]\psi(x') = (\lambda + 1)\psi(x') . \tag{5.16}$$

This form is identical to that of Eq. (5.14), due to the fact that the harmonic oscillator differential equation is Lorentz-invariant [13].

Among many possible solutions of the above differential equation, let us consider the form

$$\psi_\eta(x) = \left(\frac{1}{\pi}\right)\left(\frac{1}{2}\right)^{(k+n+a+b)/2}\left(\frac{1}{k!n!a!b!}\right)^{1/2} H_k(t'^2)H_n(z'^2)$$

$$\times H_a(x'^2)H_b(y'^2)\exp\left[-\frac{1}{2}\left(t'^2 + z'^2 + x'^2 + y'^2\right)\right] , \tag{5.17}$$

where $k, n, a,$ and b are integers, $H_k(t'), H_n(z'), \ldots$ are the Hermite polynomials, and η represents a Lorentz boost along the z-axis resulting in a hadron velocity of β. This wave function is normalizable, but the eigenvalue takes the values

$$\lambda = (k - n - a - b) . \tag{5.18}$$

Thus for a given finite value of λ, there are infinitely many possible combinations of $k, n, a,$ and b. The most general solution of the harmonic oscillator differential equation is infinitely degenerate [3].

Because the wave functions are normalizable, all the generators of the Lorentz transformations given in Eq. (2.28) of Chap. 2 are Hermitian operators. The Lorentz transformation applicable to this function space is therefore a *unitary* transformation.

Indeed, we can write any function of the coordinate variables t, z, x, and y as a linear combination of the above solutions. In particular, a solution of the harmonic oscillator equation with a given set of quantum numbers in the hadron rest frame can be written as a linear sum of infinitely many solutions in the hadron rest frame as we shall see in Sect. 5.4.

It is very difficult, if not impossible to give physical interpretations to infinite-component wave functions. For this reason, it is quite natural to seek a finite set from the infinite number of wave functions at least in one Lorentz frame. Fortunately, there is a solution.

From atomic spectroscopy, the transition time and line broadening enabled the formulation of a time-energy uncertainty relation even before 1927 [14]. As we shall see in Chap. 6, in 1927, Dirac studied this same uncertainty relation between the time and energy variables [15]. Since, in the time-like direction there are no excitations, Dirac noted that the time variable must be a *c-number*. From Heisenberg's position-momentum uncertainty it is clear that there are excitations along the space-like longitudinal direction. We know that when the observer is moving, the space and time coordinates become mixed up as is illustrated in Fig. 5.1.

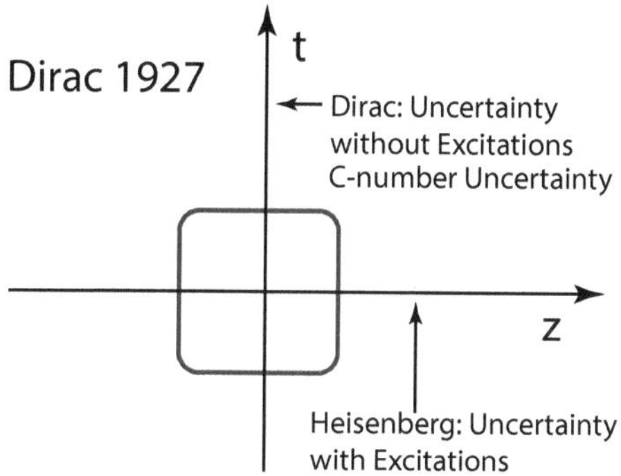

Fig. 5.1: Space-time picture of quantum mechanics. Dirac noted that because of the uncertainty in the time and energy variables, along the time axis there are no excitations [16]. This is what Dirac meant by the *c-number time-energy uncertainty relation*. Therefore, only the ground state is allowed along the time axis [17, 18].

With this point in mind, the simplest way to obtain such a finite set of wave functions is to invoke the restriction that there be no time-like oscillations in the Lorentz frame in which the hadron is at rest and that the integer k in Eqs. (5.17) and (5.18) be zero. In doing so, we are led to the following two fundamental questions:

(a) Is it possible to give physical interpretations to the wave functions belonging to the resulting finite set?
(b) Is it still possible to maintain Lorentz covariance with this condition?

We shall study the first question in Chap. 6. Let us examine the second question closely here.

When the hadron moves along the z-axis, the $k = 0$ condition is equivalent to

$$\left(t' + \frac{\partial}{\partial t'} \right) \psi_\eta(x) = 0 . \tag{5.19}$$

The most general form of the above condition is

$$p_\mu \left(x^\mu + \frac{\partial}{\partial x_\mu} \right) \psi_\eta(x) = 0 . \tag{5.20}$$

Thus the $k = 0$ condition is covariant. Once this condition is set, we can write the wave function belonging to this finite set as

$$\psi_\eta(x) = \left(\frac{1}{\pi} \right) \left[\frac{1}{(2^a 2^b 2^n a! b! n!)} \right]^{\frac{1}{2}} H_n(z') H_a(x') H_b(y')$$

$$\times \exp\left[-\frac{1}{2} \left(t'^2 + z'^2 + x'^2 + y'^2 \right) \right] . \tag{5.21}$$

Except for the Gaussian factor in the t' variable, the above expression is the wave function for the three-dimensional isotropic harmonic oscillator. This means that we can use the spherical coordinate system for the z', x', and y' variables. We shall see in Sect. 5.3 how these ideas form the basis for constructing representations of the Poincaré group.

Since the above harmonic oscillator wave functions are separable in the Cartesian coordinate system, and since the transverse coordinate variables are not affected by the boost along the z-direction, we can omit the factors depending on the x and y variables when studying their Lorentz transformation properties. The most general form of the wave function given in Eq. (5.17) becomes

$$\psi_\eta^{n,k}(t', z') = \left[\frac{1}{(\pi 2^n 2^k n! k!)} \right]^{\frac{1}{2}} H_k(t') H_n(z') \exp\left[-\frac{1}{2} \left(t'^2 + z'^2 \right) \right] , \tag{5.22}$$

with

$$\lambda = (n - k) . \tag{5.23}$$

The wave functions satisfying the subsidiary condition of Eq. (5.20) take the simple form

$$\psi_\eta^n(t, z) = \left[\frac{1}{(\pi 2^n n!)} \right]^{\frac{1}{2}} H_n(z') \exp\left[-\frac{1}{2} \left(t'^2 + z'^2 \right) \right] , \tag{5.24}$$

with

$$\lambda = n . \tag{5.25}$$

As we shall see in Chaps. 6 and 10, this normalizable wave function without excitations along the t'-axis describes the internal space-time structure of the hadron moving along the z-direction with the velocity parameter β. If $\beta = 0$, then the wave function becomes

$$\psi_0^n(t, z) = \left[\frac{1}{(\pi 2^n n!)}\right]^{\frac{1}{2}} H_n(z) \exp\left[-\frac{1}{2}(t^2 + z^2)\right].$$ (5.26)

Thus

$$\psi_\eta^n(t, z) = \psi_0^n(t', z').$$ (5.27)

We have therefore obtained the Lorentz-boosted wave function by making a passive coordinate transformation on the t- and z-coordinate variables.

Let us next study the orthogonality relations of the wave functions. Since the volume element is Lorentz-invariant:

$$dz\,dt = dz'\,dt',$$ (5.28)

there is no difficulty in understanding the orthogonality relation:

$$\int \psi_\eta^{n'}(t, z)\psi_\eta^n(t, z)dz\,dt = \int \psi_0^{n'}(t, z)\psi_0^n(t, z)dz\,dt = \delta_{n'n}.$$ (5.29)

However, a more interesting problem is the inner product of two wave functions belonging to different Lorentz frames. As is shown in Exercise 4 in Sect. 5.7, the inner product becomes

$$\int \psi_0^{n'}(t, z)\psi_\eta^n(t, z)dz\,dt = \delta_{n'n}[1 - \beta^2]^{\frac{(n+1)}{2}}.$$ (5.30)

The remarkable fact is that the orthogonality in the quantum number n is still preserved, because of the Lorentz invariance of the harmonic oscillator differential equation. The harmonic oscillator equation does not depend on the velocity parameter β.

Creation or step-up operators increase the number of particles in a given state by one, while annihilation or step-down operators decrease the number of particles by one. As for the factor $[1 - \beta^2]^{\frac{(n+1)}{2}}$ in Eq. (5.30), we note first that, when the harmonic oscillator is in the ground state, it becomes like a Lorentz contraction of a rigid rod by $[1 - \beta^2]^{\frac{1}{2}}$. Excited-state wave functions are obtained from the ground state wave function through repeated applications of the step-up operator:

$$|n, \eta\rangle = \sqrt{\frac{1}{n!}}\left(z' - \frac{\partial}{\partial z'}\right)^n |0, \eta\rangle.$$ (5.31)

The transformation property of each step-up operator is like that of z. Therefore, if the ground-state wave function is like a rigid rod along the z-direction, the nth excited state should behave like a multiplication of $(n + 1)$ rigid rods [19]. The additional contraction factor $[1 - \beta^2]^{\frac{(n+1)}{2}}$ comes from the step-up operator. This contraction factor is summarized in Fig. 5.2.

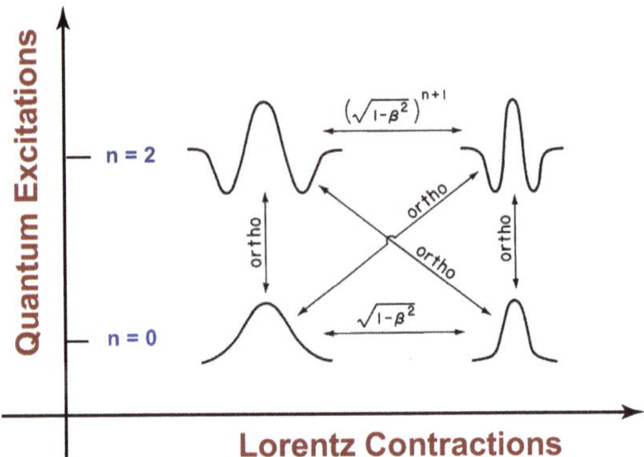

Fig. 5.2: Orthogonality and Lorentz contraction properties of the normalizable harmonic oscillator wave functions. The ground-state wave function is contracted like a rigid rod. The nth excited state is contracted like a product of $(n + 1)$ rigid rods [19, 20].

Let us recall Eq. (5.26) which had the form:

$$\psi_0^n(t, z) = \left[\frac{1}{(\pi 2^n n!)}\right]^{1/2} H_n(z) \exp[-(1/2)(t^2 + z^2)] . \qquad (5.32)$$

Here the $H_n(z)$ is the Hermite polynomial of order n. The fact that this formula is important in the formulation of the squeezed states of light [21], has been systematically discussed [22].

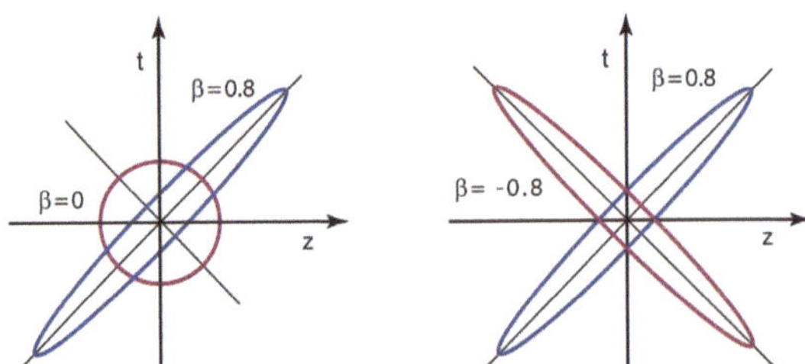

Fig. 5.3: Two overlapping wave functions. On the left we see two wave functions with $\beta = 0$ and $\beta = 0.8$ which overlap. The figure on the right shows that not only can they overlap but additionally, the wave functions can move in the opposite directions with $\beta = \pm 0.8$ [13, 18].

Let us now consider the overlap of the wave function $\psi_\eta^n(t, z)$ with $\psi_0^n(t, z)$ as indicated in Fig. 5.3. We are interested in the integral

$$\int \left(\psi_{\eta}^{n'}(t,z)\right)^{*} \psi_{0}^{n}(t,z)dz\,dt, \tag{5.33}$$

where $\psi_{0}^{n}(t,z)$, given in Eq. (5.32), can be written as

$$\psi_{0}^{n}(t,z) = \chi_{n}(z)\chi_{0}(t). \tag{5.34}$$

Then the evaluation of the integral in Eq. (5.33) is straight-forward and the result is [23]

$$\left(\psi_{\eta}^{n'},\psi_{0}^{n}\right) = \left(1-\beta^{2}\right)^{\frac{(n+1)}{2}} \delta_{nn'}, \tag{5.35}$$

as we saw previously and means that the orthogonality relation is preserved between two wave functions in two different Lorentz frames.

Now let us replace the value of β in one of the wave functions with nonzero β'. Of particular interest is the case with $\beta' = -\beta$, as shown in Fig. 5.3. Then there is an overlap of two wave functions moving in the opposite directions.

These orthogonality and contraction properties give us a basis to give the covariant harmonic oscillator, in the Lorentz-covariant world, a quantum probability interpretation. This was shown in Fig. 5.2. However, the time-separation variable between the quarks is not accounted for here. The meaning of this variable, in the present form of quantum mechanics, is completely hidden [24]. For clarification of this issue we need Feynman's concept of the rest of the universe [18, 25, 26].

5.3 Irreducible Unitary Representations of the Poincaré Group

As is explained in Chap. 2, the Poincaré group consists of space-time translations and Lorentz transformations. Let us go back to the quark coordinates x_{a} and x_{b} in Eq. (5.6), and consider performing Poincaré transformations on the quarks. The same Lorentz transformation matrix is applicable to x_{a}, x_{b}, x, as well as X. However, under the space-time translation which changes x_{a} and x_{b} to $x_{a} + a$ and $x_{b} + a$ respectively,

$$X \to X + a, \quad x \to x. \tag{5.36}$$

The quark separation coordinate x is not affected by translations. For this reason, the generators of translations for this system are

$$P_{\mu} = i\frac{\partial}{\partial X^{\mu}}, \tag{5.37}$$

while the generators of Lorentz transformations are

$$M_{\mu\nu} = L_{\mu\nu}^{*} + L_{\mu\nu}, \tag{5.38}$$

where

$$L^*_{\mu\nu} = i\left(X_\mu \frac{\partial}{\partial X^\nu} - X_\nu \frac{\partial}{\partial X^\mu}\right), \tag{5.39}$$

$$L_{\mu\nu} = i\left(x_\mu \frac{\partial}{\partial x^\nu} - x_\nu \frac{\partial}{\partial x^\mu}\right). \tag{5.40}$$

It is straight-forward to check that the ten generators defined in Eqs. (5.37) and (5.38) satisfy the commutation relations of the Poincaré group given in Eq. (2.56) of Chap. 2. We are interested in constructing normalizable wave functions which are diagonal in the Casimir operators P^2 and W^2:

$$P^2 = \frac{1}{2}\left(\frac{\partial^2}{\partial X_\mu^2} - (X^\mu)^2\right) - m_0^2, \tag{5.41}$$

$$W^2 = M^2(\mathbf{L'})^2 \tag{5.42}$$

where

$$L'_i = i\varepsilon_{ijk}x'^{\,j}\frac{\partial}{\partial x'^{\,k}}. \tag{5.43}$$

The eigenvalue of P^2 is $M^2 = m_0^2 + (\lambda + 1)$, and that for W^2 is $M^2\ell(\ell + 1)$. M is the hadron mass, and ℓ is the total intrinsic angular momentum of the hadron due to internal motion of the spinless quarks [27].

In addition, we can choose the solutions to be diagonal in the component of the intrinsic angular momentum along the direction of the motion. This component is often called the helicity. If the hadron moves along the Z-direction, the helicity operator is J_3.

Because the spatial part of the harmonic oscillator equation in Eq. (5.16) is separable also in the spherical coordinate system, we can write its solution using spherical variables in the hadron rest frame space spanned by z', x', and y'. The most general form of the solution is

$$\psi^{k,m}_{\eta\lambda\ell}(x) = R^\ell_\mu(r')Y^m_\ell(\theta', \phi')\left(\frac{1}{\sqrt{\pi}2^k k!}\right)^{1/2} H_k(t')e^{-t'^2/2}, \tag{5.44}$$

where

$$r' = \left(x'^2 + y'^2 + z'^2\right)^{1/2}, \tag{5.45}$$

$$\theta' = \cos^{-1}\frac{z'}{r'}, \qquad \phi' = \tan^{-1}\frac{y'}{x'},$$

and

$$\lambda = 2\mu + \ell - k. \tag{5.46}$$

$R^\ell_\mu(r')$ is the normalized radial wave function for the three-dimensional harmonic oscillator:

$$R_\mu^\ell(r) = \left(\frac{2(\mu!)}{[\Gamma(\mu + \ell + 3/2)]^3}\right)^{1/2} r^\ell L_\mu^{\ell+1/2}(r^2)e^{-r^2/2}, \tag{5.47}$$

where $L_\mu^{\ell+1/2}(r^2)$ is the associated Laguerre function [28]. The above radial wave function satisfies the orthonormality condition [29]:

$$\int_0^\infty r^2 R_\mu^\ell(r) R_\nu^\ell(r)dr = \delta_{\mu\nu}. \tag{5.48}$$

The spherical form given in Eq. (5.44) can of course be expressed as a linear combination of the wave functions in the Cartesian coordinate system given in Eq. (5.21).

The wave function of Eq. (5.44) is diagonal in the Casimir operators of Eqs. (5.41) and (5.42), as well as in J_3. It indeed forms a vector space for the $O(3)$-like little group [30, 31]. However, the system is infinitely degenerate due to excitations along the t'-axis. As we did in Sect. 5.2, we can suppress the time-like oscillation by imposing the subsidiary condition of Eq. (5.20), or by restricting k to be zero in Eq. (5.46). The solution then takes the form

$$\psi_{\eta\lambda\ell}^m(x) = R_\mu^\ell(r')Y_\ell^m(\theta', \phi')\left[(1/\pi)^{1/4}\exp(-t'^2/2)\right], \tag{5.49}$$

with

$$\lambda = 2\mu + \ell. \tag{5.50}$$

Thus for a given λ, there are only a finite number of solutions. The above spherical form can be expressed as a linear combination of the solutions without time-like excitations in the Cartesian coordinate system given in Eq. (5.21).

We can now write the solution of the differential equation of Eq. (5.7) as

$$\phi(X, x) = \exp(\pm iP_\mu X^\mu)\psi_{\eta\lambda\ell}^m(x). \tag{5.51}$$

This wave function describes a free hadron with a definite four-momentum having an internal space-time structure which can be described by an irreducible unitary representation of the Poincaré group. The representation is unitary because the portion of the wave function depending on the internal variable x is square-integrable, and all the generators of Lorentz transformations are Hermitian operators. We shall study in Sect. 5.4 how these wave functions are Lorentz transformed.

5.4 Transformation Properties of Harmonic Oscillator Wave Functions

If the hadron velocity is zero, then its rest frame coincides with the laboratory frame. The wave function then is

$$\psi_0(x) = R_\mu^\ell(r)Y_\ell^m(\theta, \phi)\left[(1/\pi)^{1/4}\exp(-t^2/2)\right].\tag{5.52}$$

The simplest way to obtain the wave function for the moving hadron is to replace the t, r, θ, ϕ variables in the above expression by their primed counterparts. This produces Eq. (5.44). However, we are interested in obtaining the wave function for a moving hadron as a linear combination of the wave functions for the rest frame. If we apply the boost operator to the wave function for the hadron at rest,

$$\psi_{\eta\lambda}^{\ell m}(x) = \left[e^{-i\eta K_3}\right]\psi_{0\lambda}^{\ell m}(x),\tag{5.53}$$

where K_3 is the boost generator along the z-axis. Its form is

$$K_3 = -i\left(t\frac{\partial}{\partial z} + z\frac{\partial}{\partial t}\right),\tag{5.54}$$

and η is related to velocity parameter β by

$$\sinh\eta = \frac{\beta}{(1-\beta^2)^{1/2}}.\tag{5.55}$$

Both the rest-frame and moving-frame wave functions have the same set of eigenvalues for the Casimir operators P^2 and W^2 of the Poincaré group.

These eigenstate wave functions are linear combinations of the Cartesian forms in their respective coordinate systems. If the hadron moves along the z-direction, the x and y variables remain invariant. Therefore, we use the wave function of Eq. (5.24) with $\beta = 0$:

$$\psi_0^{n,0}(t,z) = \left[\frac{1}{(\pi 2^n n!)}\right]^{1/2}H_n(z)\exp[-(1/2)(t^2 + z^2)].\tag{5.56}$$

The superscript 0 indicates that there are no time-like excitations: $k = 0$. We are now led to consider the transformation

$$\psi_\eta^{n,0}(t,z) = [\exp(-i\eta K_3)]\,\psi_0^{n,0}(t,z)$$

$$= \psi_0^{n,0}(t',z'),\tag{5.57}$$

and ask what the boost operator $\exp(-i\eta K_3)$ does to $\psi_0^{n,0}(t,z)$.

This boost operator of course changes z and t to z' and t' respectively as is indicated above. However, we are interested in whether the transformation can take the linear form

$$\psi_\eta^{n,0}(t,z) = \sum_{n',k'} A_{n',k'}^n(\beta)\psi_0^{n',k'}(t,z).\tag{5.58}$$

Because the harmonic oscillator differential equation is Lorentz invariant, the eigenvalue λ of Eq. (5.22) remains invariant, and only the terms which satisfy the condition

$$n = (n' - k')$$

(5.59)

make non-zero contributions in the sum. Thus the above expression can be simplified to

$$\psi_\eta^{n,0}(t,z) = \sum_{k=0}^{\infty} A_k^n(\beta)\psi_0^{n+k,k}(t,z) .$$

(5.60)

This is indeed a *linear unitary representation of the Lorentz group*. The representation is infinite-dimensional because the sum over k is extended from zero to infinity [32].

The remaining problem is to determine the coefficient $A_k^n(\beta)$. Using the orthogonality relation, we can write

$$A_k^n(\beta) = \int \psi_0^{n+k,k}(t,z)\psi_\eta^{n,0}(t,z)\,dzdt$$

$$= \frac{1}{\pi}\left(\frac{1}{2}\right)^n\left(\frac{1}{2}\right)^{1/2}\left(\frac{1}{n!(n+k)!}\right)^{1/2}$$

$$\times \int H_{n+k}(z)H_k(t)H_n(z')$$

$$\times \exp\left[-\frac{1}{2}(t^2 + t'^2 + z^2 + z'^2)\right]dzdt .$$

(5.61)

In this integral, the Hermite polynomials and the Gaussian form are mixed with the kinematics of Lorentz transformation. However, if we use the generating function for the Hermite polynomial, the evaluation of the integral is straightforward [Exercise 4 in Sect. 5.7], and the result is

$$A_k^n(\beta) = (1 - \beta^2)^{(1+n)/2}\beta^k\left(\frac{(n+k)!}{n!k!}\right)^{1/2} .$$

(5.62)

Thus the linear expansion given in Eq. (5.58) can be written as

$$\psi_\eta^{n,0}(t,z) = \left[\frac{1}{(2^n\pi)}\right]^{1/2}(1 - \beta^2)^{(1+n)/2}\exp[-(t^2 + z^2)/2]$$

$$\times \sum_{k=0}^{\infty}\left(\frac{1}{2}\right)^k\left(\frac{\beta}{2}\right)^k H_{n+k}(z)H_k(t) .$$

(5.63)

As was indicated with respect to Eq. (5.26), this linear transformation has to be unitary. Let us check this by calculating the sum

$$S = \sum_{k=0}^{\infty}|A_k^n(\eta)|^2 .$$

(5.64)

According to Eq. (5.62), this sum is

$$S = (1 - \beta^2)^{(n+1)} \sum_{k=0}^{\infty} \frac{(n+k)!}{n!k!} (\beta^2)^k \, . \tag{5.65}$$

On the other hand, the binomial expansion of $[1 - \beta^2]^{-(n+1)}$ takes the form

$$[1 - \beta^2]^{-(n+1)} = \sum_{k=0}^{\infty} \frac{(n+k)!}{n!k!} \beta^{2k} \, . \tag{5.66}$$

Therefore the sum S is equal to one. The linear transformation of Eq. (5.60) is indeed a unitary transformation.

We have discussed above the transformation of the physical wave function with no time-like excitations. In general, for a given value of λ, the infinite-by-infinite transformation matrix can be written as

$$\begin{pmatrix} \psi_\eta^{n,0} \\ \psi_\eta^{n+1,1} \\ \psi_\eta^{n+2,2} \\ \cdots \\ \cdots \\ \cdots \end{pmatrix} = \begin{pmatrix} b_{00} & b_{01} & b_{02} & \cdots & \cdots & \cdots \\ b_{10} & b_{11} & b_{12} & \cdots & \cdots & \cdots \\ b_{20} & b_{21} & b_{22} & \cdots & \cdots & \cdots \\ \cdots & \cdots & \cdots & \cdots & \cdots \\ \cdots & \cdots & \cdots & \cdots & \cdots \\ \cdots & \cdots & \cdots & \cdots & \cdots \end{pmatrix} \begin{pmatrix} \psi_0^{n,0} \\ \psi_0^{n+1,1} \\ \psi_0^{n+2,2} \\ \cdots \\ \cdots \\ \cdots \end{pmatrix} \, . \tag{5.67}$$

This is the most general form of the unitary irreducible representation of the Lorentz group applicable to the covariant harmonic oscillator wave functions.

The expansion coefficients we calculated from Eq. (5.60) to Eq. (5.62) constitute the first row of above matrix. The remaining coefficients can also be calculated [Exercise 5 in Sect. 5.7]. When β is zero, the above matrix becomes an infinite-by-infinite unitary matrix. From the orthogonality of the wave functions:

$$\left(\psi_\eta^{n+k',k'}, \psi_\eta^{n+k,k} \right) = \delta_{k',k} \, , \tag{5.68}$$

it is clear that this matrix has to be unitary and orthogonal:

$$\sum_{k=0}^{\infty} b_{ik}, b_{jk} = \delta_{ij} \, . \tag{5.69}$$

Eqs. (5.64) to (5.66) constitute an explicit calculation of

$$\sum_{k} |b_{0k}|^2 = 1 \, . \tag{5.70}$$

We have so far carried out the study of the unitarity of the unitary matrix representation using the solutions in the Cartesian coordinate system. Although the representations diagonal in the Casimir operators of the Poincaré group can be obtained as linear combinations of these solutions, it is still of interest to see how the transformation is achieved directly in terms of the solutions which are eigenstates of the Casimir operators. For this purpose let us note first that these solutions are constructed in terms of the spherical coordinate variables for the three-dimensional (z, x, y) space and treat t separately. If the hadron is at rest,

$$\psi_{0\lambda\ell}^{k,m}(x) = R_\mu^\ell(r)Y_\ell^m(\theta, \phi)\left[\frac{1}{\sqrt{(\pi)}2^k k!}\right]^{1/2} H_k(t)e^{-t^2/2}. \tag{5.71}$$

Thus we have to write the generators of Lorentz transformations in terms of these variables. The three rotation generators can be written as [28]:

$$J_3 = -i\frac{\partial}{\partial\phi},$$

$$J_\pm = J_1 \pm J_2$$

$$= \pm e^{\pm i\phi}\left(\frac{\partial}{\partial\theta} \pm i\cot\theta\frac{\partial}{\partial\phi}\right). \tag{5.72}$$

It is not difficult to calculate the three boost generators. They take the form

$$iK_3 = \cos\theta\left(r\frac{\partial}{\partial t} + t\frac{\partial}{\partial r}\right) - \frac{t}{r}\sin\theta\frac{\partial}{\partial\theta},$$

$$iK_\pm = K_1 \pm iK_2$$

$$= e^{\pm i\phi}\left(r\frac{\partial}{\partial t} + t\sin\theta\frac{\partial}{\partial r} - \frac{t}{r}\cos\theta\frac{\partial}{\partial\theta} \pm \frac{t}{r\sin\theta}\frac{\partial}{\partial\phi}\right). \tag{5.73}$$

The rotation generators affect only the spherical harmonics in the wave function of Eq. (5.71). Thus

$$J_3\psi_{0\lambda\ell}^{k,m} = m\psi_{0\lambda\ell}^{k,m},$$

$$J_\pm\psi_{0\lambda\ell}^{k,m} = \sqrt{(\ell \mp m)(\ell \mp m + 1)}\psi_{0\lambda\ell}^{k,m\pm1}. \tag{5.74}$$

The above relations mean that rotations do not change the quantum numbers λ, ℓ, and k. They only change m. Eq. (5.74) indeed corresponds to the fact that the little group for massive hadrons is like $O(3)$.

On the other hand, if we apply the boost generators, we end up with somewhat complicated formulae:

$$iK_3\psi_{\lambda\ell}^{k,m} = \left[\frac{(\ell+m+1)(\ell-m+1)}{(2\ell+1)(2\ell+3)}\right]^{1/2} Y_{\ell+1}^m(\theta,\phi)Q_{-\ell}F_{\lambda\ell}^k(r,t)$$

$$+\left[\frac{(\ell+1)(\ell-m)}{(2\ell+1)(2\ell-1)}\right]^{1/2} Y_{\ell-1}^m(\theta,\phi)Q_{\ell+1}F_{\lambda\ell}^k(r,t),$$

$$iK_\pm\psi_{\lambda\ell}^{k,m} = \left[\frac{(\ell\pm m+1)(\ell\pm m+2)}{(2\ell+1)(2\ell+3)}\right]^{1/2} Y_{\ell+1}^{m\pm1}(\theta,\phi)Q_{\pm\ell}F_{\lambda\ell}^k(r,t)$$

$$+\left[\frac{(\ell\mp m)(\ell\mp m-1)}{(2\ell+1)(2\ell-1)}\right]^{1/2} Y_{\ell-1}^{m\pm1}(\theta,\phi)Q_{(\pm\ell+l)}F_{\lambda\ell}^k(r,t), \quad (5.75)$$

where

$$Q_\ell = \left(t\frac{\partial}{\partial r} + r\frac{\partial}{\partial t} + \ell\frac{t}{r}\right),$$

$$F_{\lambda\ell}^k(r,t) = R_\mu^\ell(r)\left[\frac{1}{\sqrt{(\pi)}2^k k!}\right] H_k(t)\exp(-t^2/2). \quad (5.76)$$

K_3 does not change the value of m, while K_+ and K_- change m by $+1$ and -1 respectively. In addition, unlike the rotation operators, the boost generators change λ, ℓ, and k. This is a manifestation of the fact that the unitary representation is infinite-dimensional as is indicated in Eq. (5.60) [Problem 1 in Sect. 5.7].

It is possible to finish the calculation by explicitly carrying out the differentiations contained in the Q_ℓ operators [Problem 4 in Sect. 5.7]. However, this does not appear necessary, because we already know what the answer should be from our experience with the Cartesian coordinate system.

5.5 Harmonic Oscillators in the Four-Dimensional Euclidean Space

Since Euclidean space is easier for us to visualize, it is not uncommon to use this space when we study physics and mathematics formulated in Minkowski space. There are many formalisms and mathematical theorems concerning the relationship between the four-dimensional Minkowski and Euclidean spaces. However, there are not many concrete examples which illustrate the relation between these two important coordinate systems.

The harmonic oscillator wave functions discussed in Sects. 5.2 to 5.4 are normalizable in the four-dimensional Euclidean space of t, z, x, and y. The solutions of the Lorentz-invariant harmonic oscillator equation can be written as linear combinations of the solutions of the differential equation:

$$\left(\frac{1}{2}\right)\left[\nabla^2 + \left(\frac{\partial}{\partial t}\right)^2 - (t^2 + \mathbf{x}^2)\right] u(x) = (\sigma + 2)u(x).$$ (5.77)

This differential equation is also separable in many different coordinate systems. It is separable in the Cartesian coordinate system, and the procedure for constructing wave functions is identical to that for the wave functions of Sect. 5.4 with $\beta = 0$. Indeed the solution takes the form

$$u(x) = f_a(x)f_b(y)f_n(z)f_k(t),$$ (5.78)

where $f_a, f_b, \ldots$ are normalized one-dimensional harmonic oscillator wave functions, and can be written as

$$f_n(z) = \left[\frac{1}{(\sqrt{\pi}2^n n!)}\right]^{1/2} H_n(z)\exp(-z^2/2).$$ (5.79)

The above solutions form a complete orthonormal set. The eigenvalue σ takes the integer values

$$\sigma = a + b + n + k.$$ (5.80)

Thus, for a given value of σ, there are only a finite number of possible combinations of the quantum numbers a, b, n, and k.

Let us next consider the effect of Lorentz transformations. The differential equation is not Lorentz-invariant. For this reason, the harmonic oscillator wave functions of Sects. 5.2 to 5.4 with non-zero β are no longer solutions of the differential equation of Eq. (5.77). However, it is still possible to write wave functions as linear combinations of the solutions given in Eq. (5.78), because they form a complete orthonormal set. The linear forms in Eq. (5.60) and Eq. (5.67) are precisely those expansions. Indeed, all of the normalizable solutions discussed in Sects. 5.2 to 5.4 are linear combinations of the solutions in the four-dimensional Euclidean space [33].

The differential equation of Eq. (5.77) is not Lorentz-invariant, but is invariant under rotations in the four-dimensional Euclidean space. In this space, it is convenient to use the coordinate variables ρ, α, θ, and ϕ related to the Cartesian variables by

$$t = \rho \cos \alpha,$$

$$z = \rho \sin \alpha \cos \theta,$$

$$x = \rho \sin \alpha \sin \theta \cos \phi,$$

$$y = \rho \sin \alpha \sin \theta \sin \phi.$$ (5.81)

In terms of the new variables, we can write the solution of the differential equation given in Eq. (5.77) as [Problem 2 in Sect. 5.7]

$$u_{\mu n}^{\ell m}(x) = S_\mu^n(\rho)Z_{n+1}^{\ell m}(\alpha, \theta, \phi),$$ (5.82)

where

$$Z_{n+1}^{\ell m}(\alpha, \theta, \phi) = (\sin \alpha)^{1/2} P_{n+3/2}^{-n-1/2}(\cos \alpha) Y_{\ell}^{m}(\theta, \phi) . \tag{5.83}$$

$S_{\mu}^{n}(\rho)$ satisfies the radial differential equation:

$$\frac{1}{2}\left[-\frac{\partial^2}{\partial \rho^2} - \frac{3}{\rho}\frac{d}{d\rho} + \frac{n(n+2)}{\rho^2} + \rho^2 \right] S_{\mu}^{n}(\rho) = \sigma S_{\mu}^{n}(\rho) . \tag{5.84}$$

$P_{n+3/2}^{-n-1/2}(\cos \alpha)$ is an associated Legendre function commonly known as the Gegenbauer polynomial [29]. The index n in the above expression should not be confused with the one used in Eqs. (5.78) and (5.79), and is the maximum value the quantum number ℓ can take:

$$\ell = 0, 1, 2, ..., n . \tag{5.85}$$

The solution of the radial equation is

$$S_{\mu}^{n}(\rho) = \left(\frac{2(\mu!)}{[\Gamma(\mu + n + 2)]^3} \right)^{1/2} \rho^n L_{\mu}^{n+1}(\rho^2) e^{(-\rho^2/2)} , \tag{5.86}$$

satisfying the orthonormality condition:

$$\int_0^\infty \rho^3 S_{\mu}^{n}(\rho) S_{\nu}^{n}(\rho) d\rho = \delta_{\mu\nu} . \tag{5.87}$$

$L_{\mu}^{n+1}(\rho^2)$ is the associated Laguerre polynomial [28]. The eigenvalue σ in this case is

$$\sigma = (2\mu + n) . \tag{5.88}$$

While the radial wave functions for the three-dimensional harmonic oscillator are commonly available in various textbooks, it is not easy to find solutions for the four-dimensional problem. Let us write down explicitly the wave functions in the $O(4)$-coordinate system for some low eigenvalues of σ. If $\sigma = 0$, the system is in the ground state, and the wave function is

$$u_{00}^{00}(x) = \left(\frac{1}{\pi} \right) \exp(-\rho^2/2) . \tag{5.89}$$

In terms of the Cartesian variables, it is easy to construct the first four excited-state wave functions:

$$\left(\frac{\sqrt{2}}{\pi} \right) x_i \exp(-\rho^2/2), \tag{5.90}$$

where $x_i = t, z, x$, or y. In terms of the $O(4)$-coordinate variables, when $\sigma = n = 1$, the wave functions are written as

$$u_{01}^{\ell m}(x) = \rho \exp(-\rho^2) Z_2^{\ell m}(\alpha, \theta, \phi), \tag{5.91}$$

with

$$Z_2^{00}(\alpha, \theta, \phi) = \frac{\sqrt{2}}{\pi} \cos \alpha,$$

$$Z_2^{1m}(\alpha, \theta, \phi) = \left(\frac{8}{3\pi}\right)^{1/2} \sin \alpha\, Y_1^m(\theta, \phi), \qquad m = 1, 0, -1. \tag{5.92}$$

Here again the wave function is four-fold degenerate. The above wave-functions can be written as linear combinations of the Cartesian forms given in Eq. (5.79).

For $\sigma = 2$, there are four Cartesian wave functions of the form

$$\left(\frac{1}{\pi\sqrt{2}}\right)(2x_i^2 - 1)\exp(-\rho^2/2), \tag{5.93}$$

and six wave functions of the form

$$\left(\frac{2}{\pi}\right)x_i x_j \exp(-\rho^2/2), \qquad \text{with} \qquad i \neq j. \tag{5.94}$$

If we use the $O(4)$-coordinate system, we have to consider two different values of the radial quantum number μ:

$$[a] \qquad \mu = 1 \qquad \text{with} \qquad n = 0,$$

$$[b] \qquad \mu = 0 \qquad \text{with} \qquad n = 2.$$

If $\mu = 1$ and $n = 0$, the wave function becomes

$$u_{10}^{00}(x) = \left(\frac{1}{\sqrt{2\pi}}\right)(2 - \rho^2)\exp(-\rho^2/2). \tag{5.95}$$

If, on the other hand, $\mu = 0$ and $n = 2$, the wave function is

$$u_{02}^{\ell m}(x) = \left(\frac{1}{\sqrt{3}}\right)\rho^2 \exp(-\rho^2/2) Z_3^{\ell m}(\alpha, \theta, \phi). \tag{5.96}$$

For $\ell = 0$, there is one wave function with

$$Z_3^{00} = \frac{\sqrt{2}}{\pi}(4\cos^2 \alpha - 1). \tag{5.97}$$

For $\ell = 1$, there are three wave functions with

$$Z_3^{1m} = \left(\frac{16}{\pi}\right)^{1/2}\cos \alpha \sin \alpha\, Y_\ell^m(\theta, \phi). \tag{5.98}$$

For $\ell = 2$, there are five wave functions with

$$Z_3^{2m} = \left(\frac{16}{5\pi}\right)^{1/2}\sin^2 \alpha\, Y_2^m(\theta, \phi). \tag{5.99}$$

We have given above enough explicit calculations to show how to construct the harmonic oscillator wave functions using the *O(4) harmonics*, and how they can be converted to the Cartesian forms. The above calculation also illustrates that the solutions diagonal in the Casimir operators of the $O(4)$ group are not diagonal in those of the Poincaré group discussed in Sect. 5.4. Since, however, the $O(4)$ solutions form a complete set, we can construct the wave functions needed for the representations of the Poincaré group by making suitable linear combinations of the $O(4)$ solutions [Problem 5 in Sect. 5.7].

5.6 Moving O(4)-Coordinate System

We considered in Sect. 5.5 the solutions of the differential equation of Eq. (5.77). It was noted that this differential equation is separable in both the Cartesian and $O(4)$-coordinate systems. We observed that the harmonic oscillator formalism can serve as an illustrative example of how rotations in the $O(4)$-coordinate system are different from Lorentz transformations, while the Lorentz transformation of the normalizable harmonic oscillator wave functions can be regarded as a unitary transformation in the $O(4)$-space. The difference is of course due to the fact that the differential equation of Eq. (5.77) is not Lorentz-invariant. The linear expansion of Eq. (5.63), although realizable in the $O(4)$ space, is not a transformation which preserves the eigenvalue σ which remains invariant under the $O(4)$ rotations.

Mathematically, the difference between the above-mentioned two differential equations is very simple. We can obtain one from the other by replacing the variable x by (ix). This is in fact one of the most commonly-used procedures in physics, especially in connection with the Bethe-Salpeter equation [34, 35]. However, the physics of this procedure requires further explanation.

In order to complete the discussion of this problem, let us go back to the Lorentz-invariant differential equation of Eq. (5.14) written in terms of the x variables for the hadron at rest and that of Eq. (5.16) written in the Lorentz-boosted variables x' for the hadron moving along the z-direction. Because of the Lorentz invariance, both Eq. (5.14) and Eq. (5.16) are the same differential equation. In Sect. 5.5, we discussed only the $O(4)$ counterpart of Eq. (5.14). The $O(4)$ counterpart of Eq. (5.16) can be written simply as

$$\frac{1}{2}\left[\nabla'^2 + \left(\frac{\partial}{\partial t'}\right)^2 - \left(t'^2 + \mathbf{x}'^2\right)\right] u(x') = (\sigma' + 2)u(x'). \qquad (5.100)$$

It is possible to repeat the calculation of Sect. 5.5 and write all the solutions in terms of the primed variables. It is even possible to write the $O(4)$ solution in one Lorentz frame as a linear expansion of those in another Lorentz frame, although the mathematics is rather cumbersome.

However, the important point is that the above differential equation is only a Wick-rotated form of Eq. (5.16), and is different from the $O(4)$ counterpart of Eq. (5.14) given in Eq. (5.77). The only common connection the moving $O(4)$ system has with the rest-frame $O(4)$ system is that the moving system becomes that of the rest frame when the hadron velocity is zero. Indeed, we can see why the Wick-rotation is not a Lorentz-invariant procedure by carrying out the above-mentioned mathematics [33].

In addition to the above-mentioned problem of Lorentz invariance, the procedure of replacing x by (ix) has the following complication. Although the differential equation can be obtained by a simple replacement of the variable, the solutions of the differential equation discussed in Sect. 5.5 are not an analytic continuation of those given in Sect. 5.2. Both solutions are normalizable. However, the analytic continuation from x to (ix) or vice versa leads to non-normalizable solutions. This means that the boundary condition on the solutions is an independent condition, and cannot be regarded as a part of the Wick rotation [Problem 10 in Sect. 5.7].

In Chap. 9, we shall discuss the connection between the solutions in the $O(4)$-coordinate system with those which are diagonal in the Casimir operators of the Lorentz group.

5.7 Exercises and Problems

Exercise 1. It is widely believed that we can obtain momentum wave functions in the harmonic oscillator regime by replacing the coordinate variables by the their conjugate momentum variables. For this reason, there is a tendency to overlook the details. For instance, what happens to phases? Consider the one-dimensional harmonic oscillator, and use $f_n(x)$ as the wave function for the nth excited state. Consider next an arbitrary function $F(x)$ which can be expanded as

$$F(x) = \sum_n A_n f_n(x). \tag{5.101}$$

Then, is it possible to obtain the momentum wave function by simply replacing x by p?

If we write the momentum wave function as a Fourier transform of the spatial wave function it takes the form:

$$g_n(p) = \left(\frac{1}{2\pi}\right)^{1/2} \int f_n(x) e^{(-ipx)} dx. \tag{5.102}$$

Then, for the ground state, we can show

$$g_0(p) = f_0(x). \tag{5.103}$$

However, since the nth excited-state wave functions are obtained by repeated application of step-up operators on the ground-state wave function, we can show that

$$g_n(p) = (-i)^n f_n(p). \tag{5.104}$$

Thus, the Fourier transformation of F(x) is

$$G(p) = \sum_n (-1)^n A_n f(p) \,. \tag{5.105}$$

For a more detailed treatment of this problem, see K. B. Wolf [36]. This point is important when we analyze the high-energy data using models based on mixed harmonic oscillator states. See Le Yaouanc et al. [37, 38].

Exercise 2. Throughout this Chapter, we have been using the harmonic oscillator equations and wave functions in configuration space and time. Discuss the harmonic oscillator formalism using momentum-energy coordinates.

The mathematics of the harmonic oscillator is the same for both configuration space and momentum space. The Lorentz transformation of the momentum-energy four-vector is also identical to that for the space-time four-vector. For this reason, we expect that the analysis in momentum-energy space be identical to that for space-time configuration space.

However, in view of the phase factors of the momentum wave function discussed in Exercise 1, the momentum wave functions are either real or imaginary. For this reason, we have to be careful when we take the inner product of two wave functions as in the case of Eq. (5.30). In addition, we have to modify the expansion formula given in Eq. (5.60). The easiest way to approach this problem would be to note that $(i)(-i) = 1$, and make the momentum wave function absorb an appropriate power of i, while the coefficient $A_k^n(\beta)$ can absorb the same power of $(-i)$.

The momentum-energy wave function can be obtained from the Fourier transformation of the space-time wave function only when the Fourier integral exists.

Exercise 3. While mesons are believed to be two-body bound states of a quark and antiquark, baryons such as protons and neutrons are bound states of three quarks. Assuming that the interaction between each pair of quarks is like that of the harmonic oscillator, write down the covariant harmonic oscillator equation for a hadron consisting of three quarks, and carry out the separation of variable suitable for the description of massive hadrons.

If we denote the space-time coordinates of three quarks by x_a, x_b, and x_c, the three-particle harmonic oscillator equation can be written as

$$\left\{ 3\left[\left(\frac{\partial}{\partial x_a}\right)^2 + \left(\frac{\partial}{\partial x_b}\right)^2 + \left(\frac{\partial}{\partial x_c}\right)^2 \right] + \left(-\frac{1}{36}\right)\left[(x_a - x_b)^2 + (x_b - x_c)^2 + (x_c - x_a)^2 \right] + m_0^2 \right\} \psi(x_a, x_b, x_c) = 0 \,. \tag{5.106}$$

This equation can be separated if we use the variables X, r, and s. Following Feynman et al. [10] define the following three variables:

$$X = \frac{x_a + x_b + x_c}{3} \,, \qquad r = \frac{x_b + x_c - 2x_a}{6} \,, \qquad s = \frac{x_b - x_c}{2\sqrt{3}} \,, \tag{5.107}$$

and

$$x_a = X - 2r , \quad x_b = X + r + \sqrt{3}s , \quad x_c = X + r - \sqrt{3}s . \tag{5.108}$$

In terms of the r and s variables, the quadratic form becomes

$$18\left(r^2 + s^2\right) , \tag{5.109}$$

and does not depend on the X variable, which specifies the space-time coordinate of the proton.

In terms of these new variables, the wave function in Eq. (5.106) can be written as

$$\psi(x_a, x_b, x_c) = f(X)R(r)S(s) , \tag{5.110}$$

where $f(X)$, $R(r)$ and $S(s)$ satisfy the following equations:

$$\left[\left(\frac{\partial}{\partial X}\right)^2 + m_0^2 + \lambda_r + \lambda_s\right] f(X) = 0 ,$$

$$\frac{1}{2}\left[\left(\frac{\partial}{\partial r^\mu}\right)^2 - r_\mu^2\right] R(r) = \lambda_r R(r) ,$$

$$\frac{1}{2}\left[\left(\frac{\partial}{\partial s^\mu}\right)^2 - s_\mu^2\right] S(s) = \lambda_s S(s) . \tag{5.111}$$

Here again, $f(X)$ is a solution of the free-particle Klein-Gordon equation with four-momentum P satisfying

$$-P^2 = m_0^2 + \lambda_r + \lambda_s . \tag{5.112}$$

We can eliminate time-like excitations in both the r- and s-coordinate systems by imposing the conditions:

$$P^\mu\left(r_\mu + \frac{\partial}{\partial r^\mu}\right) R(r) = 0$$

$$P^\mu\left(s_\mu + \frac{\partial}{\partial s^\mu}\right) S(s) = 0 . \tag{5.113}$$

If we use p_1, p_2, and p_3 for the four-momenta for the first, second, and third quarks respectively, and P, q, and k for the momentum variables conjugate to X, r, and s respectively, then we can introduce, for the energy-momentum four-vectors the following three variables:

$$P = p_a + p_b + p_c , \quad q = p_b + p_c - 2p_a , \quad k = \sqrt{3}\left(p_b - p_c\right) . \tag{5.114}$$

Then

$$p_a = \frac{1}{3}P - \frac{1}{3}q\,, \quad p_b = \frac{1}{3}P + \frac{1}{6}q + \frac{1}{2\sqrt{3}}k\,, \quad p_c = \frac{1}{3}P + \frac{1}{6}q - \frac{1}{2\sqrt{3}}k\,. \tag{5.115}$$

In terms of these variables, we are led to consider the quadratic form of

$$18\left(q^2 + k^2\right). \tag{5.116}$$

This form does not depend on the variable P, which measures the momentum and energy of the proton. The harmonic oscillator wave functions in the momentum space are identical to those in the space-time coordinates, except the phase factors discussed in Exercise 2. See [10, 39, 18].

Exercise 4. Derive the result of Eq. (5.62) by evaluating the integral of Eq. (5.61).

Let us replace the Hermite polynomials in the integrand by their respective generating functions:

$$G(r, z) = \exp(-r^2 + 2rz)$$

$$= \sum_{m=0}^{\infty} \frac{r^m}{m!} H_m(z)\,. \tag{5.117}$$

Then the calculation is reduced to the evaluation of the integral:

$$I = \int G(r, z)G(s, t)G(r', z')$$

$$\times \exp[-(t^2 + z^2 + t'^2 + z'^2)/2]\, dz dt\,. \tag{5.118}$$

The integrand in the above expression becomes one exponential function whose argument is quadratic in z and t. This quadratic form can be diagonalized, and the result is

$$I = \pi(1 - \beta^2)^{1/2} \exp(2\beta rs) \exp[2rr'(1 - \beta^2)^{1/2}]\,. \tag{5.119}$$

In Eq. (5.119) above, I, recalling that the boost parameter is η, can also be expanded as [32, 22]:

$$I = \left[\frac{\pi}{\cosh\eta}\right] \exp\left(2rs\tanh\eta\right) \exp\left(\frac{2rr'}{\cosh\eta}\right), \tag{5.120}$$

where from Eq. (5.55)

$$\frac{1}{\cosh\eta} = (1 - \beta^2)^{1/2} \quad \text{and} \quad \tanh(\eta) = \beta\,. \tag{5.121}$$

By expanding the above exponential factors and choosing the term containing the r^{n+k}, s^k, and $r'^{n'}$ for $H_{n+k}(z)$, $H_k(t)$, and $H_{n'}(z')$ respectively, we arrive at the result of Eq. (5.62). If we set $k = 0$, and $n = n'$, the result is Eq. (5.63) [23, 33]. In terms of Eq. (5.120), Eq. (5.63) can be expanded as [23, 32, 22]

$$\psi_\eta^{n,0}(t,z) = \left(\frac{1}{\cosh\eta}\right)^{(n+1)} \frac{1}{\sqrt{\pi}2^n n!} \exp[-(t^2+z^2)/2] \tag{5.122}$$

$$\times \sum_k \left[\frac{(n+k)!}{n!k!}\right]^{1/2} (\tanh\eta)^k H_{k+n}(z)H_k(t).$$

Exercise 5. Calculate the elements of the infinite-by-infinite matrix of Eq. (5.67) for Lorentz transformation of the covariant harmonic oscillators.

This exercise is an extension of Exercise 3 to include excitations in the t' variable. Thus we have to consider the transformation of the function

$$\psi_{n.m}(t,z) = f_n(z)f_m(t), \tag{5.123}$$

where

$$f_n(z) = \left[\frac{1}{(\sqrt{\pi}2^n n!)}\right]^{1/2} H_n(z)e^{-z^2/2}. \tag{5.124}$$

This means that we have to compute the coefficients in the following linear expansion:

$$f_n(z')f_m(t') = \sum_k \sum_j A_{nm}^{kj} f_k(z)f_j(t). \tag{5.125}$$

Thus

$$A_{nm}^{kj} = \int f_k(z)f_j(t)f_n(z')f_m(t')\, dzdt. \tag{5.126}$$

Again we can use the generating function of Hermite polynomials to evaluate the above integral. This time, we have to evaluate the integrand involving four generating functions:

$$\int G(r,z)G(s,t)G(r',z')G(s',t')$$

$$\times \exp\left[-(t^2+z^2+t'^2+z'^2)/2\right]\, dzdt. \tag{5.127}$$

The procedure for evaluation of this integral is similar to that for Exercise 4. However, the calculation is far more complicated. Rotbart [40] carried out this calculation and obtained explicit expressions for all the elements of the infinite-by-infinite matrix given in Eq. (5.67). The result of Rotbart's calculation is

$$A_{nm}^{kj} = \beta^{k-n}[n!m!k!j!]^{1/2}\delta_{j-m,k-n}$$

$$\times \sum_{s=0}^{s_0} \frac{\left[\frac{-\beta^2}{(1-\beta^2)}\right]^s}{s!(m-s)!(n-s)!(k-n+1)!}, \tag{5.128}$$

where s_0 is the smaller of n or m.

Problem 1. Derive the relations given in Eq. (5.74) using Eq. (5.71), Eq. (5.72), and the recurrence relation for the spherical harmonics.

Problem 2. Write down the Casimir operators of the $O(4)$ group and the differential equation given in Eq. (5.77) in terms of the ρ, α, θ, and ϕ variables. The $O(4)$ group has been extensively discussed in connection with the symmetry of the hydrogen atom. See L. C. Biedenharn [41] for a complete and thorough discussion of this problem. See G. Domokos and P. Suranyi [42] for a discussion of the Klein-Gordon equation in the $O(4)$-coordinate system.

Problem 3. The degeneracies of the $O(4)$ harmonics given in Sect. 5.5 are like those of the Rydberg energy levels of the hydrogen atom. Starting from the Hamiltonian of the hydrogen atom, show that the energy levels should exhibit the $O(4)$ symmetry. This problem is extensively discussed in the literature. See, for instance, the above-mentioned paper of Biedenharn [41]. See also Gilmore [43].

Problem 4. From the transformation matrix of Eq. (5.67) based on the Cartesian form of normalizable wave functions, we know that the derivatives of the right-hand sides of Eq. (5.75) should be replaced in terms of the Laguerre functions in r and the Hermite polynomials in t. In order to solve this problem, we have to know the recurrence relations for the associated Laguerre functions and those for the Hermite polynomials. The recurrence relations for the Hermite polynomials are readily available, while those for the Laguerre functions needed for solving this problem are not yet available in standard mathematical tables. However, it is possible to write the associated Laguerre polynomials in finite power series. From these power series, derive the recurrence relation for the covariant harmonic oscillator wave functions starting from Eq. (5.75).

Problem 5. Discuss degeneracies of the harmonic oscillator wave functions in the four-dimensional isotropic space using the Cartesian coordinate system. Find the unitary matrix which will transform the solutions given in Sect. 5.5 into the Cartesian forms.

Problem 6. The non-relativistic harmonic oscillator serves as an illustrative example for many branches of physics [44]. It is therefore quite natural for physicists to have developed a group theory of harmonic oscillators. Let us consider a one-dimensional harmonic oscillator. We are quite familiar with the step-up and step-down operators:

$$a^{\dagger} = \frac{1}{\sqrt{2}} \left(x - \frac{\partial}{\partial x} \right) ,$$

$$a = \frac{1}{\sqrt{2}} \left(x + \frac{\partial}{\partial x} \right) , \qquad (5.129)$$

and the number operator

$$N = a^{\dagger} a . \qquad (5.130)$$

Write down the commutation relations for $(ia^\dagger)$, (ia), and (iN). Show that the (iI) has to be added to form a closed Lie algebra. The group generated by these four operators is called the harmonic oscillator group. Show that the following four-parameter matrix forms one of the representations of the harmonic oscillator group [45] :

$$
\begin{pmatrix}
1 & ce^d & a & d \\
0 & e^d & b & 0 \\
0 & 0 & 1 & 0 \\
0 & 0 & 0 & 1
\end{pmatrix}.
\tag{5.131}
$$

Problem 7. The problem becomes more complicated in the case of non-relativistic three-dimensional isotropic harmonic oscillators. The additional symmetry we have to consider is that of three-dimensional rotation. What is the resulting Lie algebra? See Gilmore [43].

Problem 8. If we introduce special relativity to the three-dimensional harmonic oscillator, what is the resulting Lie algebra? Translate the treatment of Sect. 5.3 into the language of this algebra. See Shapiro [46].

Problem 9. Discuss the Lie algebra for the four-dimensional isotropic harmonic oscillator using the appropriate step-up and step-down operators. Discuss the connection between this algebra and that of Problem 8. This could be a future research problem.

Problem 10. Compare the Wick rotation applicable to the Bethe-Salpeter equation and wave function to those of the harmonic oscillator equations and wave functions discussed in this Chapter.

Problem 11. In this Chapter we have been primarily concerned with the use of harmonic oscillators in relativistic physics. Another area of physics in which the harmonic oscillator plays the central role is the coherent state representation. This representation is useful in many branches of modern physics, including optics [47, 48, 49], quantum mechanics [50, 51, 52, 53], and high-energy scattering theory [54, 55]. In order to understand the basic mathematics of the coherent state representation, let us consider the following problems starting from the one-dimensional harmonic oscillator.

(i) In Dirac's notation, the nth excited harmonic oscillator is written as $|n\rangle$. The step-up operator $a^\dagger$ and the step-down operator a act on the Fock space with the property

$$
a^\dagger |n\rangle = \sqrt{n+1}\,|n+1\rangle \qquad \text{and} \qquad a|n\rangle = \sqrt{n}\,|n-1\rangle
\tag{5.132}
$$

and are given in Eq. (5.129). The coherent state is defined as

$$
|\alpha\rangle = e^{-|\alpha|^2/2} \sum_n \frac{\alpha^n}{\sqrt{n!}}\,|n\rangle
\tag{5.133}
$$

where α is a complex number. Show that the above form is an eigenstate of the step-down operator a with eigenvalue α.

(ii) From above, show that $|\alpha\rangle$ can also be defined as displaced vacuum:

$$|\alpha\rangle = \hat{D}(\alpha)\,|0\rangle \tag{5.134}$$

where $\hat{D}(\alpha)$ is the displacement operator

$$\hat{D}(\alpha) = \exp(\alpha a^{\dagger} - \alpha^{*} a)\,. \tag{5.135}$$

(iii) Use the generating function for the Hermite polynomials to show that the above form is a Gaussian function whose origin is $x = \alpha$, and therefore that it satisfies the minimum uncertainty relation.

(iv) Show that the inner product of the two coherent states satisfies the relation [56]:

$$|\langle \alpha \mid \alpha' \rangle|^{2} = e^{-|\alpha - \alpha'|^{2}}\,. \tag{5.136}$$

(v) Replace α by αe^{-iwt}. Show then that the probability density of the coherent state describes a simple harmonic oscillation of the Gaussian distribution around $x = 0$. See Goldin [49].

Problem 12. Coherent states are not the only states that possess the minimum uncertainty property. Squeezed states offer such an example, when the variances of the quadrature distributions are not required to be equal [22]. The simplest single mode squeezed state is generated by applying the squeeze operator

$$\hat{S}(\xi) = \exp\left\{\frac{1}{4}(\xi\, a^{\dagger} a^{\dagger} - \xi^{*}\, aa)\right\} \tag{5.137}$$

to the vacuum $|0\rangle$, where the squeeze parameter ξ is a complex number with $\xi = \lambda e^{i\phi}$. This can be expanded as [57, 58]:

$$\hat{S}(\xi) = \exp\left\{\frac{1}{2}\left(e^{i\phi}\tanh\frac{\lambda}{2}\right) a^{\dagger} a^{\dagger}\right\} \exp\left\{-\ln\left(\cosh\frac{\lambda}{2}\right)\left(a^{\dagger} a + \frac{1}{2}\right)\right\}$$

$$\times \exp\left\{-\frac{1}{2}\left(e^{-i\phi}\tanh\frac{\lambda}{2}\right) aa\right\}\,. \tag{5.138}$$

Show that the squeezed vacuum

$$\hat{S}(\xi)\,|0\rangle = \left(\cosh\frac{\lambda}{2}\right)^{\frac{1}{2}} \sum_{n} \frac{\sqrt{(2n)!}}{2^{n} n!}\left(e^{i\phi}\tanh\frac{\lambda}{2}\right)^{n} |2n\rangle \tag{5.139}$$

can be expressed as a superposition even number of photon states. Accordingly, it is occasionally called a two-photon coherent state [21]. Single mode states can naturally be generalized to two-mode states. We shall dwell on further mathematical properties of two-mode squeezed states in the context of their symmetries in Sect. 7.4 and in Sect. 7.5 of Chap. 7. Among many other benefits of practical applications of squeezed light as in optical communications and quantum imaging [59], most prominently it is used in the detection of gravitational waves to increase the sensitivity of the interferometer. This latter has been successfully realized at GEO 600 [60] and at LIGO [61, 62].

References

1. P.A.M. Dirac, *The development of quantum theory; J. Robert Oppenheimer memorial prize acceptance speech* (Gordon and Breach Science Publishers, New York, 1971). ISBN 978-0-677-02975-7. (OCLC: 146520.)
2. P.A.M. Dirac, Unitary Representations of the Lorentz Group, Proceedings of the Royal Society A: Mathematical, Physical, and Engineering Sciences **183**(994), 284–295 (1945). DOI 10.1098/rspa.1945.0003. URL http://rspa.royalsocietypublishing.org/cgi/doi/10.1098/rspa.1945.0003
3. H. Yukawa, Structure and Mass Spectrum of Elementary Particles. I. General Considerations, Physical Review **91**(2), 415–416 (1953). DOI 10.1103/PhysRev.91.415.2. URL https://link.aps.org/doi/10.1103/PhysRev.91.415.2
4. M. Markov, On dynamically deformable form factors in the theory of elementary particles, Il Nuovo Cimento **3**(S4), 760–772 (1956). DOI 10.1007/BF02746074. URL http://link.springer.com/10.1007/BF02746074
5. T. Takabayasi, Oscillator model for particles underlying unitary symmetry, Il Nuovo Cimento **33**(2), 668–672 (1964). DOI 10.1007/BF02750221. URL http://link.springer.com/10.1007/BF02750221
6. T. Takabayasi, Relativistic Mechanics of Confined Particles as Extended Model of Hadrons: The Bilocal Case, Progress of Theoretical Physics Supplement **67**, 1–68 (1979). DOI 10.1143/PTPS.67.1. URL https://academic.oup.com/ptps/article-lookup/doi/10.1143/PTPS.67.1
7. I. Sogami, The Non-Local Field Theory of the Quark Model, Progress of Theoretical Physics **41**(5), 1352–1365 (1969). DOI 10.1143/PTP.41.1352. URL https://academic.oup.com/ptp/article-lookup/doi/10.1143/PTP.41.1352
8. S. Ishida, "Ur-citon": An Attempt for Unified Theory of Hadrons, Progress of Theoretical Physics **46**(6), 1905–1923 (1971). DOI 10.1143/PTP.46.1905. URL https://academic.oup.com/ptp/article-lookup/doi/10.1143/PTP.46.1905
9. K. Fujimura, T. Kobayashi, M. Namiki, Nucleon Electromagnetic Form Factors at High Momentum Transfers in an Extended Particle Model Based on the Quark Model, Progress of Theoretical Physics **43**(1), 73–79 (1970). DOI 10.1143/PTP.43.73. URL https://academic.oup.com/ptp/article-lookup/doi/10.1143/PTP.43.73
10. R.P. Feynman, M. Kislinger, F. Ravndal, Current Matrix Elements from a Relativistic Quark Model, Physical Review D **3**(11), 2706–2732 (1971). DOI 10.1103/PhysRevD.3.2706. URL https://link.aps.org/doi/10.1103/PhysRevD.3.2706
11. P.A.M. Dirac, Quantum electrodynamics, Commun. Dublin Inst. Adv. Stud., A **1**, 36 (1943). URL http://cds.cern.ch/record/230780
12. E.G. Kalnins, W. Miller, Lie theory and the wave equation in space–time. I. The Lorentz group, Journal of Mathematical Physics **18**(1), 1–16 (1977). DOI 10.1063/1.523130. URL http://aip.scitation.org/doi/10.1063/1.523130
13. Y.S. Kim, M.E. Noz, Covariant Harmonic Oscillators and the Quark Model, Physical Review D **8**(10), 3521–3527 (1973). DOI 10.1103/PhysRevD.8.3521. URL https://link.aps.org/doi/10.1103/PhysRevD.8.3521
14. E.P. Wigner, On time-energy uncertainty relation, in *Aspects of Quantum Theory, in Honour of P.A.M. Dirac's 70th Birthday*, ed. by A. Salam, E.P. Wigner (Cambridge University Press, London UK, 1972), 237–248. ISBN 978-0521131032
15. P.A.M. Dirac, The Quantum Theory of the Emission and Absorption of Radiation, Proceedings of the Royal Society A: Mathematical, Physical, and Engineering Sciences **114**(767), 243–265

(1927). DOI 10.1098/rspa.1927.0039. URL http://rspa.royalsocietypublishing.org/cgi/doi/10.1098/rspa.1927.0039

16. P.A.M. Dirac, The Quantum Theory of Dispersion, Proceedings of the Royal Society A: Mathematical, Physical, and Engineering Sciences **114**(769), 710–728 (1927). DOI 10.1098/rspa.1927.0071. URL http://rspa.royalsocietypublishing.org/cgi/doi/10.1098/rspa.1927.0071

17. Y.S. Kim, M.E. Noz, Integration of Dirac's Efforts to Construct a Quantum Mechanics Which is Lorentz-Covariant, Symmetry **12**(8), 1270–1–30 (2020). DOI 10.3390/sym12081270. URL https://www.mdpi.com/2073-8994/12/8/1270

18. S. Başkal, Y.S. Kim, M.E. Noz, *Physics of the Lorentz Group (Second Edition): Beyond high-energy physics and optics* (IOP Publishing, Bristol, UK, 2021). DOI 10.1088/978-0-7503-3607-9. ISBN 978-0-7503-3607-9. URL https://iopscience.iop.org/book/978-0-7503-3607-9

19. Y.S. Kim, M.E. Noz, Group theory of covariant harmonic oscillators, American Journal of Physics **46**(5), 480–483 (1978). DOI 10.1119/1.11239. URL http://aapt.scitation.org/doi/10.1119/1.11239

20. Y.S. Kim, M.E. Noz, Lorentz Harmonics, Squeeze Harmonics, and Their Physical Applications, Symmetry **3**(4), 16–36 (2011). DOI 10.3390/sym3010016. URL http://www.mdpi.com/2073-8994/3/1/16/

21. H.P. Yuen, Two–photon coherent states of the radiation field, Physical Review A **13**(6), 2226–2243 (1976). DOI 10.1103/PhysRevA.13.2226. URL https://link.aps.org/doi/10.1103/PhysRevA.13.2226

22. S. Başkal, Y. Kim, M. Noz, *Mathematical Devices for Optical Sciences.* (IOP Publishing, Bristol, UK, 2019). ISBN 978-0-7503-1612-5. URL https://dx.doi.org/10.1088/2053-2563/aafe78. (OCLC: 1034620988.)

23. M.J. Ruiz, Orthogonality relation for covariant harmonic-oscillator wave functions, Physical Review D **10**(12), 4306–4307 (1974). DOI 10.1103/PhysRevD.10.4306. URL https://link.aps.org/doi/10.1103/PhysRevD.10.4306

24. Y.S. Kim, M.E. Noz, Feynman's Decoherence, Optics and Spectroscopy **94**(5), 733–740 (2003). DOI 10.1134/1.1576844. URL http://link.springer.com/10.1134/1.1576844

25. D. Han, Y.S. Kim, M.E. Noz, Illustrative example of Feynman's rest of the universe, American Journal of Physics **67**(1), 61–66 (1999). DOI 10.1119/1.19192. URL http://aapt.scitation.org/doi/10.1119/1.19192

26. Y.S. Kim, M.E. Noz, Dirac Matrices and Feynman's Rest of the Universe, Symmetry **4**(4), 626–643 (2012). DOI 10.3390/sym4040626. URL http://www.mdpi.com/2073-8994/4/4/626/

27. Y.S. Kim, M.E. Noz, S.H. Oh, Representations of the Poincaré group for relativistic extended hadrons, Journal of Mathematical Physics **20**(7), 1341–1344 (1979). DOI 10.1063/1.524237. URL http://aip.scitation.org/doi/10.1063/1.524237. See also, Physics Auxiliary Publication Service Document No. PAPS JMAPA-20-1336-12

28. G.B. Arfken, H.J. Weber, F.E. Harris, *Mathematical methods for physicists: a comprehensive guide*, 7th edn. (Elsevier, Amsterdam NL; Boston, MA, USA, 2013). ISBN 978-0-12-384654-9. (Originally published in 1966.)

29. P.M. Morse, H. Feshbach, *Methods of theoretical physics.* International series in pure and applied physics (McGraw-Hill, Boston, Mass, 1999). ISBN 978-0-07-043316-8. (Orginally published 1953.)

30. D. Han, M.E. Noz, Y.S. Kim, D. Son, Space-time symmetries of confined quarks, Physical Review D **25**(6), 1740–1743 (1982). DOI 10.1103/PhysRevD.25.1740. URL https://link.aps.org/doi/10.1103/PhysRevD.25.1740

31. D. Han, M.E. Noz, Y.S. Kim, D. Son, c -number time-energy uncertainty relation in the quark model, Physical Review D **27**(12), 3032–3035 (1983). DOI 10.1103/PhysRevD.27.3032. URL https://link.aps.org/doi/10.1103/PhysRevD.27.3032

32. Y.S. Kim, M.E. Noz, S.H. Oh, A simple method for illustrating the difference between the homogeneous and inhomogeneous Lorentz groups, American Journal of Physics **47**(10), 892–897 (1979). DOI 10.1119/1.11622. URL http://aapt.scitation.org/doi/10.1119/1.11622

33. Y.S. Kim, M.E. Noz, S.H. Oh, Lorentz deformation in the O(4) and light-cone coordinate systems, Journal of Mathematical Physics **21**(5), 1224–1228 (1980). DOI 10.1063/1.524513. URL http://aip.scitation.org/doi/10.1063/1.524513

34. G.C. Wick, Properties of Bethe-Salpeter Wave Functions, Physical Review **96**(4), 1124–1134 (1954). DOI 10.1103/PhysRev.96.1124. URL https://link.aps.org/doi/10.1103/PhysRev.96.1124

35. N. Nakanishi, Indefinite-Metric Quantum Field Theory, Progress of Theoretical Physics Supplement **51**, 1–95 (1972). DOI 10.1143/PTPS.51.1. URL https://academic.oup.com/ptps/article/doi/10.1143/PTPS.51.1/2946860

36. K.B. Wolf, *Integral Transforms in Science and Engineering* (Springer US : Imprint : Springer, Boston, MA, 1979). ISBN 978-1-4757-0872-1. URL https://doi.org/10.1007/978-1-4757-0872-1. (Originally published 1979, by Plenum Publishing Company Limited, NY, NY, USA; OCLC: 1113652314.)

37. A.L. Le Yaouanc, L. Oliver, O. Pène, J.C. Raynal, SU(6) strong breaking: structure functions and static properties of the nucleon, Physical Review D **12**(7), 2137–2156 (1975). DOI 10.1103/PhysRevD.12.2137. URL https://link.aps.org/doi/10.1103/PhysRevD.12.2137

38. A.L. Le Yaouanc, L. Oliver, O. Pène, J.C. Raynal, Combined effects of internal quark motion and SU(6) breaking on the properties of the baryon ground state, Physical Review D **15**(3), 844–853 (1977). DOI 10.1103/PhysRevD.15.844. URL https://link.aps.org/doi/10.1103/PhysRevD.15.844

39. Y.S. Kim, M.E. Noz, Covariant Harmonic Oscillators and Excited Baryon Decays, Progress of Theoretical Physics **57**(4), 1373–1386 (1977). DOI 10.1143/PTP.57.1373. URL https://academic.oup.com/ptp/article-lookup/doi/10.1143/PTP.57.1373

40. F.C. Rotbart, Complete orthogonality relations for the covariant harmonic oscillator, Physical Review D **23**(12), 3078–3080 (1981). DOI 10.1103/PhysRevD.23.3078. URL https://link.aps.org/doi/10.1103/PhysRevD.23.3078

41. L.C. Biedenharn, Wigner Coefficients for the R_4 Group and Some Applications, Journal of Mathematical Physics **2**(3), 433–441 (1961). DOI 10.1063/1.1703728. URL http://aip.scitation.org/doi/10.1063/1.1703728

42. G. Domokos, P. Suranyi, Bound states and analytic properties in angular momentum, Nuclear Physics **54**, 529–548 (1964). DOI 10.1016/0029-5582(64)90432-8. URL https://linkinghub.elsevier.com/retrieve/pii/0029558264904328

43. R. Gilmore, *Lie groups, Lie algebras, and some of their applications* (Dover Publications, Mineola, NY, USA, 2005). ISBN 978-0-486-44529-8. (Originally published: 1974, John Wiley and Sons, New York, NY, USA.)

44. M. Moshinsky, *Harmonic oscillator in modern physics: from atoms to quarks* (Gorden and Breach, New York, NY, 1969). ISBN 978-0-677-02450-9. (OCLC: 233550291.)

45. W. Miller, *Symmetry groups and their applications*. No. 50 in Pure and applied mathematics; a series of monographs and textbooks (Academic Press, New York, NY, USA, 1972). ISBN 978-0-12-497460-9

46. J.A. Shapiro, A quarks-on-springs model of the baryons, Annals of Physics **47**(3), 439–467 (1968). DOI 10.1016/0003-4916(68)90209-1. URL https://linkinghub.elsevier.com/retrieve/pii/0003491668902091

47. R.J. Glauber, Coherent and Incoherent States of the Radiation Field, Physical Review **131**(6), 2766–2788 (1963). DOI 10.1103/PhysRev.131.2766. URL https://link.aps.org/doi/10.1103/PhysRev.131.2766

48. J.R. Klauder, E.C.G. Sudarshan, *Fundamentals of quantum optics* (Dover Publications, Mineola, N.Y, 2006). ISBN 978-0-486-45008-7. (Originally published: New York, NY, USA : W.A. Benjamin, 1968.)

49. E. Goldin, *Waves and photons: an introduction to quantum optics*. Wiley series in pure and applied optics (John Wiley and Sons, New York, NY, USA, 1982). ISBN 978-0471085928

50. P. Carruthers, M.M. Nieto, Coherent States and the Number-Phase Uncertainty Relation, Physical Review Letters **14**(11), 387–389 (1965). DOI 10.1103/PhysRevLett.14.387. URL https://link.aps.org/doi/10.1103/PhysRevLett.14.387

51. R. Glauber, Classical behavior of systems of quantum oscillators, Physics Letters **21**(6), 650–652 (1966). DOI 10.1016/0031-9163(66)90111-9. URL https://linkinghub.elsevier.com/retrieve/pii/0031916366901119

52. V. Moncrief, Coherent states and quantum nonperturbing measurements, Annals of Physics **114**(1-2), 201–214 (1978). DOI 10.1016/0003-4916(78)90266-X. URL https://linkinghub.elsevier.com/retrieve/pii/000349167890266X

53. J.R. Klauder, Path integrals and stationary-phase approximations, Physical Review D **19**(8), 2349–2356 (1979). DOI 10.1103/PhysRevD.19.2349. URL https://link.aps.org/doi/10.1103/PhysRevD.19.2349

54. P. Carruthers, C.C. Shih, Correlations and fluctuations in hardonic multiciplicity distribution: The meaning of KNO scaling, Physics Letters B **127**(3-4), 242–250 (1983). DOI 10.1016/0370-2693(83)90884-5. URL http://linkinghub.elsevier.com/retrieve/pii/0370269383908845

55. P. Carruthers, F. Zachariasen, Quantum collision theory with phase-space distributions, Reviews of Modern Physics **55**(1), 245–285 (1983). DOI 10.1103/RevModPhys.55.245. URL https://link.aps.org/doi/10.1103/RevModPhys.55.245

56. S.M. Barnett, P.M. Radmore, *Methods in theoretical quantum optics*, reprint edn. No. 15 in Oxford series in optical and imaging sciences (Clarendon Press, Oxford, UK, 2005). ISBN 978-0-19-856361-7. (Originially published 1997; OCLC: 316132663.)

57. C.M. Caves, B.L. Schumaker, New formalism for two–photon quantum optics. I. Quadrature phases and squeezed states, Physical Review A **31**(5), 3068–3092 (1985). DOI 10.1103/PhysRevA.31.3068. URL https://link.aps.org/doi/10.1103/PhysRevA.31.3068

58. C.M. Caves, B.L. Schumaker, New formalism for two-photon quantum optics. II. Mathematical foundation and compact notation, Physical Review A **31**(5), 3093–3111 (1985). DOI 10.1103/PhysRevA.31.3093. URL https://link.aps.org/doi/10.1103/PhysRevA.31.3093

59. H. Bachor, T.C. Ralph, Applications of Squeezed Light, in *A Guide to Experiments in Quantum Optics*, ed. by H. Bachor, T.C. Ralph, 1st edn. (Wiley, 2004), 310–342. DOI 10.1002/9783527619238.ch10. ISBN 9783527403936,9783527619238. URL https://onlinelibrary.wiley.com/doi/10.1002/9783527619238.ch10

60. R. Demkowicz-Dobrzański, K. Banaszek, R. Schnabel, Fundamental quantum interferometry bound for the squeezed-light-enhanced gravitational wave detector GEO 600, Physical Review A **88**(4), 041,802 (2013). DOI 10.1103/PhysRevA.88.041802. URL https://link.aps.org/doi/10.1103/PhysRevA.88.041802

61. LIGO Scientific Collaboration, A gravitational wave observatory operating beyond the quantum shot-noise limit, Nature Physics **7**(12), 962–965 (2011). DOI 10.1038/nphys2083. URL https://www.nature.com/articles/nphys2083

62. L. Barsotti, J. Harms, R. Schnabel, Squeezed vacuum states of light for gravitational wave detectors, Reports on Progress in Physics **82**(1), 016,905 (2019). DOI 10.1088/1361-6633/aab906. URL https://iopscience.iop.org/article/10.1088/1361-6633/aab906

Chapter 6
Dirac's Form of Relativistic Quantum Mechanics

Abstract We give a physical interpretation to the covariant harmonic oscillator formalism by incorporating Dirac's ideas on relativistic quantum mechanics. His papers from 1927 discuss the time-energy uncertainty relation and how it differs from Heisenberg's. Dirac observed that time is a c-number. In 1945, Dirac attempted to use the four-dimensional harmonic oscillator with normalizable time-like wave functions. This use of the covariant oscillator formalism is consistent with Dirac's overall plan to construct a relativistic quantum mechanics. In his 1949 paper, Dirac emphasizes that the task of constructing a relativistic dynamics is equivalent to constructing a representation of the Poincaré group. We discuss the front, instant, and point forms of quantum mechanics which Dirac proposes along with the problems Dirac had. Dirac's light-cone coordinate system in which the longitudinal and time-like coordinate variables undergo only scale changes under Lorentz boosts is discussed. This can be helpful in describing Lorentz deformation of relativistic extended hadrons. We show that the constraint applicable to the point form is useful in describing the conservation of probability under Lorentz transformations. This allows us to construct Lorentz-transformed relativistic bound-state wave functions, needed in understanding relativistic hadrons as bound states of quarks and/or antiquarks.

The purpose of this Chapter is to give a physical interpretation to the covariant harmonic oscillator formalism by incorporating Dirac's ideas on relativistic quantum mechanics. Dirac's form of relativistic quantum mechanics serves a very useful purpose in understanding Lorentz deformation properties of relativistic extended hadrons, such as hadrons in the quark model.

As early as in 1927 [1, 2], Dirac observed that the uncertainty relation applicable to the time and energy variables is different from Heisenberg's uncertainty relation applicable to the position and momentum variables, and that this space-time asymmetry is one of the problems we have to face in the process of making quantum

S. Başkal et al., *Theory and Applications of the Poincaré Group*, Fundamental Theories of Physics 217, https://doi.org/10.1007/978-3-031-64376-7_6

mechanics relativistic. In 1945 [3], Dirac considered the possibility of using the four-dimensional harmonic oscillator with normalizable time-like wave functions in connection with relativistic Fock space. The covariant harmonic oscillator wave functions discussed in Chap. 5 are of Dirac's type. Indeed, the use of the covariant oscillator formalism is perfectly consistent with Dirac's overall plan to construct a relativistic quantum mechanics.

Dirac's plan to construct a *relativistic dynamics of atom* using *Poisson brackets* is contained in his 1949 [4] paper entitled *Forms of Relativistic Dynamics*. Here Dirac emphasizes that the task of constructing a relativistic dynamics is equivalent to constructing a representation of the Poincaré or inhomogeneous Lorentz group.

In an attempt to find a three-dimensional space in which non-relativistic quantum mechanics is valid, Dirac in his 1949 paper considered three constraint conditions, each of which reduces the four-dimensional Minkowski space-time into a three-dimensional Euclidean space. Dirac called the forms of relativistic quantum mechanics with these constraints *instant form*, *front form*, and *point form*. The purpose of this Chapter is to show that the covariant harmonic oscillator formalism for massive hadrons based on the $O(3)$-like little group constitutes Dirac's *instant form* quantum mechanics, while displaying the space-time asymmetry in uncertainty relations.

In addition, we shall discuss in detail the kinematics of the front and point forms. In his *front form* quantum mechanics, Dirac introduced the light-cone coordinate system. The beauty of the light-cone coordinate system is that the longitudinal and time-like coordinate variables undergo only scale changes under Lorentz boosts. We shall study in this Chapter how Dirac's light-cone coordinate system can serve useful purposes in describing Lorentz deformation of relativistic extended hadrons. Furthermore, we shall show that the constraint applicable to Dirac's *point form* quantum mechanics is very useful in describing the conservation of probability under Lorentz transformations.

Dirac's approach to relativistic quantum mechanics allows us to construct relativistic bound-state wave functions which can be Lorentz-transformed. These wave functions are needed in understanding relativistic hadrons which are believed to be bound states of quarks and/or antiquarks. We shall study in this Chapter Lorentz deformation properties of relativistic hadrons. In Chap. 13, we shall discuss some observable consequences of the Lorentz deformation, particularly the peculiarities observed in Feynman's parton picture [5, 6].

The ultimate purpose of studying the models of relativistic quantum mechanics is to study how the time-energy uncertainty relation can be combined with the position-momentum uncertainty, in a manner consistent with special relativity and with what we observe in the experimental world. We shall study this question using extensively space-time and momentum-energy diagrams. Indeed, the light-cone coordinate system is an effective language for this pictorial approach.

In Sect. 6.1, we study the present form of the uncertainty relation applicable to the time and energy variables. It is noted that there exists an uncertainty relation between the time and energy variables. However, since the time variable is a c-number [1, 2] there is no Hilbert space associated with this variable [7]. We shall call this the *c-number time-energy uncertainty relation*. We then study how this form of the uncertainty relation is applicable to the time separation variable in the quark model, using the harmonic oscillator formalism.

In Sect. 6.2, we examine closely how the covariant harmonic oscillator model can serve as a solution of the Dirac's *Poisson brackets* in his *instant form* quantum mechanics. It is noted that the use of the c-number time-energy relation and the concept of off-mass-shell particles allow us to avoid the difficulties mentioned in Dirac's 1949 paper. We study in detail the manner in which the time-energy uncertainty relation is combined covariantly with the usual position-momentum uncertainty relation.

Also in his 1949 paper, Dirac introduced the light-cone coordinate system in order to simplify the mathematics of the Lorentz transformation. In Sect. 6.3, the geometry of the light-cone coordinate system is discussed in detail. The light-cone coordinate system allows us to regard the Lorentz transformation as a simple scale transformation preserving the area in the two-dimensional coordinate system of longtitudinal and time-like variables.

In Sect. 6.4, we discuss the covariant harmonic oscillator formalism using the light-cone coordinate system. In Sect. 6.5, it is shown that, in spite of the Lorentz deformation of hadron distribution, it is still possible to define the uncertainty relations in a Lorentz-invariant manner.

Section 6.6 contains exercises and problems which illustrate some immediate applications of Dirac's form of relativistic quantum mechanics. Chap. 13 will deal with observable consequences of the Lorentz-Dirac deformation.

6.1 C-Number Time-Energy Uncertainty Relation

From atomic spectroscopy, the transition time and line broadening enabled the formulation of a time-energy uncertainty relation even before 1927 [8]. In 1927, however, Dirac studied this same uncertainty relation between the time and energy variables [1, 2]. Later in 1927, Heisenberg formulated his uncertainty relation. Then Dirac considered whether the two uncertainty relations could be combined into a Lorentz-covariant uncertainty relation [1, 2].

As we noted in Chap. 5 in the time-like direction there are no excitations, Dirac noted that the time variable must be a c-number. From the position-momentum uncertainty we know there are excitations along the space-like longitudinal direction. We also know that when the observer is moving, the space and time coordinates become mixed up. How then, can this space-time asymmetry be made consistent with Lorentz covariance?

The time-energy uncertainty relation in the form of $(\Delta t)(\Delta E) = 1$ was known to exist even before the present form of quantum mechanics was formulated. However, the treatment of this subject in the existing quantum mechanics textbooks is not adequate. Students, as well as physicists, are likely to be confused on the following three issues:

(a) Is the time-energy uncertainty relation a consequence of the time-dependent Schrödinger equation? Or, is this relation expected to hold even in systems which cannot be described by the Schrödinger equation?
(b) While there exists the time-energy uncertainty relation in the real world, possibly with the form $[t, H] = -i$ [9], this commutator is zero in the case of Schrödinger quantum mechanics. As was noted by Dirac in 1927, the time variable is a c-number. Then, is the c-number time-energy uncertainty relation universal, or true only in non-relativistic quantum mechanics?
(c) If the time variable is a c-number and the position variables are q-numbers, then the coordinate variables in different Lorentz frames are mixtures of c and q numbers. This cannot be consistent with quantum mechanics and special relativity, as was also pointed out by Dirac [1].

The reason why we are not able to get satisfactory answers to these questions from the existing literature is very simple. While the uncertainty relation is to be formulated from experimental observations [10, 11], there are not many experimental phenomena which can be regarded as direct manifestations of the time-energy uncertainty relation. In fact, the connection between the life-time and the energy-width of unstable states is the only direct application of this important relation [8].

It is widely believed that off-mass-shell intermediate particles in quantum field theory are due to the time-energy uncertainty relation [12, 13]. Since, however, those off-mass-shell particles are not observable, the relation between those two concepts is not directly observable and requires further investigation [14].

We study in this section the relativistic quark model as a physical example in which the time-energy uncertainty relation leads to a directly observable effect. In this model, hadrons are bound states of quarks. For a hadron consisting of two quarks whose space-time coordinates are x_a and x_b, we can define new variables:

$$X = \frac{(x_a + x_b)}{2},$$

$$x = \frac{(x_a - x_b)}{2\sqrt{2}}, \tag{6.1}$$

as we did in Chap. 5. As we noted there, X and x correspond respectively to the overall hadron coordinate and space-time separation between the quarks. The spatial component of the four-vector X specifies where the hadron is, and its time component tells how old the hadron and the quarks become. The spatial components of the four-vector x specify the relative spatial separation between the quarks. Its time component is the time interval or separation between the quarks [15, 16, 17, 18, 19, 20, 21].

Because the time-separation variable is not contained in non-relativistic quantum mechanics, the quark model provides an excellent testing ground to examine whether there exists the time-energy uncertainty relation which does not come from the Schrödinger equation. We are interested in whether there exists an uncertainty relation along the time separation coordinate. We would like then to know whether this time-energy uncertainty relation is the same as the currently accepted form largely based on non-relativistic quantum mechanics.

According to the currently accepted version [7], the time variable is a c-number or

$$[t, H] = 0, \qquad (6.2)$$

where H is the Hamiltonian or the energy operator. In non-relativistic quantum mechanics, this Hamiltonian does not depend on time and commutes with the time variable. Because the commutator vanishes, the Robertson procedure [22] applicable to Heisenberg's position-momentum uncertainty relation does not work here. Classically, this corresponds to the fact that t and H are not canonically conjugate variables. In quantum mechanics, the above commutator means that there is no Hilbert space in which t and $i\frac{\partial}{\partial t}$ act as operators. However, it is important to note that there still exists a *Fourier relation* between time and energy which limits the precision to $(\Delta t)(\Delta E) = 1$ [1, 12, 7].

In order to discuss the uncertainty relations, we need momentum-energy variables in addition to the space-time coordinates. Let us use the four-momenta:

$$P = p_a + p_b, \qquad q = \sqrt{2}(p_a - p_b), \qquad (6.3)$$

which were introduced in Chap. 5. p_a and p_b are the four-momenta of the first and second quarks respectively. The spatial and time-like components of the four-vector P are the sum of the momenta and the energy of the two quarks respectively. We assume here that the system is in an eigenstate of this four-momentum, and that every component of P is a sharply defined number. The spatial component of q measures the momentum difference between the quarks. The time-like component of q is the energy difference between the quarks. The concept of this energy separation does not exist in non-relativistic quantum mechanics. If the wave functions can be Fourier-transformed, P and q are conjugate to X and x respectively. We are particularly interested in the uncertainty relation applicable between the time-like components of the four-vectors x and q.

If the hadron has a definite four-momentum and moves along the z-direction with velocity parameter β, it is possible to find the Lorentz frame in which the hadron is at rest. We shall use t', z', x', and y' to denote the space-time separations in this frame, and q'_0, q'_z, q'_x, q'_y for momentum-energy separations. In the Lorentz frame where the hadron is at rest, the uncertainty principle applicable to the space-time separation of quarks is expected to be the same as the presently accepted form largely based on non-relativistic quantum mechanics. The usual Heisenberg uncertainty relation holds for each of the three spatial coordinates:

$$[z', q'_z] = i,$$

$$[x', q'_x] = i,$$

$$[y', q'_y] = i. \qquad (6.4)$$

On the other hand, the time-separation variable is a c-number and therefore does not cause quantum excitations. The commutator of Eq. (6.2) in this case takes the form

$$[t', q'_0] = 0. \tag{6.5}$$

However, according to the presently accepted form of time-energy uncertainty relation, this commutator allows the existence of the Fourier relation between time and energy variables [7] resulting in

$$(\Delta t')(\Delta q'_0) = 1. \tag{6.6}$$

The crucial question is how these uncertainty relations appear to an observer in the laboratory frame with the space-time separation variables t, z, x, and y.

We assume here that the hadron moves along the z-axis. Thus the x- and y-coordinates are not affected by boosts, and the second two commutation relations of Eq. (6.4) remain invariant:

$$[x, q_x] = i,$$

$$[y, q_y] = i. \tag{6.7}$$

If we consider only the ground state, the first, longitudinal commutator of Eq. (6.4) can be quantified as

$$(\Delta z')(\Delta q'_z) = 1. \tag{6.8}$$

Then the uncertainty relations associated with both the longitudinal and time-separation variables will lead to a distribution centered around the origin in the $t'z'$ coordinate system, as is illustrated in Fig. 6.1. It is clear from Fig. 6.1 that the distribution, with its simpler mathematical form exhibit all qualitative properties associated with the uncertainty relations.

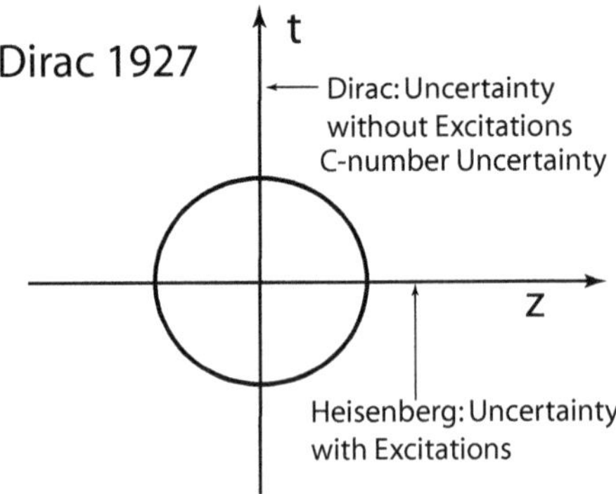

Fig. 6.1: The hadron distribution dictated by the uncertainty relations. t and z are the time-like and longitudinal separations between two quarks inside a hadron. The distribution is localized around the origin. There can be excited states along the longitudinal axis, while excitations are forbidden along the time-like direction. The circular distribution, with its simple mathematics, is expected to share all qualitative features with the arbitrary distribution [23].

Next, we should examine this localization region in the laboratory frame where the hadron moves along the z-direction with the velocity parameter β. The Lorentz transformation from the hadron rest frame to the lab frame takes the form

$$t = \frac{t' + \beta z'}{(1 - \beta^2)^{1/2}} ,$$

$$z = \frac{z' + \beta t'}{(1 - \beta^2)^{1/2}} . \tag{6.9}$$

The above form is the inverse of the transformation given in Eq. (5.15).

In studying the connection between the localized region and the commutation relations, the harmonic oscillator is the standard language [22]. In order to study the localization region in the two different Lorentz frames, we use the normalizable covariant harmonic oscillator wave functions which we discussed in Chap. 5. While the exact form for the hadron wave function diagonal in the Casimir operators of the Poincaré group is somewhat complicated, the essential element in the wave function using the boost parameter η, takes the form

$$\psi_\eta^{n,k}(x') = \left[\frac{1}{(\pi 2^{n+k} n! k!)} \right]^{\frac{1}{2}} H_k(t') H_n(z') \exp\left[-\frac{1}{2} \left(t'^2 + z'^2 \right) \right] , \tag{6.10}$$

where we have suppressed all the factors which are not affected by the Lorentz transformation along the z-axis. This is possible because the oscillator wave functions are separable in both the Cartesian and spherical coordinate systems. t' and z' are the time-like and longitudinal coordinate variables respectively in the hadron rest frame.

In terms of the standard step-up and step-down operators:

$$a_\mu^\dagger = \frac{1}{\sqrt{2}} \left(x_\mu - \frac{\partial}{\partial x^\mu} \right) ,$$

$$a_\mu = \frac{1}{\sqrt{2}} \left(x_\mu + \frac{\partial}{\partial x^\mu} \right) , \tag{6.11}$$

the harmonic oscillator wave function of Eq. (6.10) satisfies the differential equation:

$$a_\mu^\dagger a^\mu \psi(x) = (\lambda + 1)\psi(x) , \tag{6.12}$$

where the eigenvalue λ, together with transverse excitations, determines the $(mass)^2$ of the hadron.

The operators defined in Eq. (6.11) satisfy the algebraic relation

$$[a_\mu, a_\nu^\dagger] = \eta_{\mu\nu} , \tag{6.13}$$

defined in Chap. 2. This commutation relation is Lorentz-invariant [24, 21]. The time-like component of the above commutator is 1 in every Lorentz frame. This allows time-like excitations. Indeed, Rotbart [25] discussed the covariant Hilbert space of harmonic oscillator wave functions in which time-like excitations are allowed in all Lorentz frames. The above commutation relation allows the construction of representations diagonal in the Casimir operators of the $O(3, 1)$ group. These issues have been discussed in Chap. 5.

On the other hand, there is no evidence to indicate the existence of such time-like excitations in the real world. This is perfectly consistent with the fact that the basic space-time symmetry of confined quarks is that of the $O(3)$-like little group of the Poincaré group. We can suppress time-like excitations in the hadron rest frame by imposing the subsidiary condition:

$$P^\mu a_\mu^\dagger \psi_{nk}(x) = 0, \tag{6.14}$$

where P_μ is the hadron four-momentum. Then only the solutions with $k = 0$ are allowed, and the commutator given in Eq. (6.13) is not consistent with the above subsidiary condition.

How can we then construct a covariant commutator consistent with Eq. (6.14)? In order to attack this problem, let us divide the four-dimensional Minkowski space-time into the one-dimensional time-like space parallel to the hadron four-momentum and the three-dimensional space-like hyperplane perpendicular to the four-momentum [26, 27]. This hyperplane accommodates the internal space-time symmetry dictated by the $O(3)$-like little group. This leads us to consider the operator

$$b_\mu = a_\mu - \left(\frac{P_\mu P^\nu}{M^2}\right) a_\nu . \tag{6.15}$$

Then b_μ satisfies the constraint condition:

$$P^\mu b_\mu = P^\mu b_\mu^\dagger = 0, \tag{6.16}$$

and has only three independent components. Thus, for b_μ and $b_\mu^\dagger$, we can write the covariant commutation relation:

$$[b_\mu, b_\nu^\dagger] = -\eta_{\mu\nu} + \frac{P_\mu P_\nu}{M^2} . \tag{6.17}$$

The right hand side of the above expression, symmetric in μ and ν, satisfies the relation:

$$P^\mu \left(-\eta_{\mu\nu} + \frac{P_\mu P_\nu}{M^2}\right) = 0 . \tag{6.18}$$

Therefore, the covariant commutation relation given in Eq. (6.17) is consistent with the subsidiary condition of Eq. (6.14).

The covariant form of Eq. (6.17) represents the usual Heisenberg uncertainty relations on the three-dimensional space-like hypersurface perpendicular to the hadron four-momentum. This form enables us to treat separately the uncertainty relation applicable to the time-like direction, without destroying covariance. The existence of the t' distribution due to the ground-state wave function in Eq. (6.10) restricted by Eq. (6.14) allows us to write the time-energy uncertainty relation in the form

$$(\Delta t')(\Delta E') = 1, \tag{6.19}$$

without postulating the commutation relation. E' in this case is the energy separation between the quarks in the Lorentz frame in which the hadron is at rest. As we shall see in Chap. 13, the parton phenomenon, together with other high-energy features in the quark model, indicates clearly the existence of this uncertainty relation.

As has been expected, the uncertainty relation exists between the time and energy separation variables in the quark model. The remarkable fact is that this uncertainty relation is just like the one expected in all other physical phenomena [28, 29, 30, 31, 32, 33, 34, 35, 36, 37, 38, 39, 40, 41, 42, 43, 44, 14, 7].

6.2 Dirac's Form of Relativistic Theory of Atom

As the Lorentz group is the language of special relativity, Dirac, during the second war world, considered using harmonic oscillators as these provide a staring point for the present form of quantum mechanics. His idea was to use harmonic oscillator wave functions to obtain representations of the Lorentz group [3]. By using harmonic oscillator wave functions to construct representations of the Lorentz group, Dirac suggested that it might be possible to make quantum mechanics Lorentz-covariant. Therefore, Dirac studied the following Gaussian form in his 1945 paper:

$$\exp\left[-\frac{1}{2}(t^2 + z^2 + x^2 + y^2)\right]. \tag{6.20}$$

If we consider only Lorentz boosts along the z, the x-and y-coordinates can be dropped and the above expression can be written as:

$$\exp\left[-\frac{1}{2}(t^2 + z^2)\right]. \tag{6.21}$$

However, this expression is strange because $(t^2 - z^2)$ is the usually accepted Lorentz invariant quantity. The expression in Eq. 6.21 is, however, consistent with Dirac's 1927 paper [1] where he proposed the time-energy uncertainty relation. There he observed there are no excitations along the time axis as time is a c-number. This is therefore, consistent with is his earlier paper.

Let us look at Fig. 6.1 carefully. This can be seen as a pictorial representation of Eq. 6.20. Since there is localization in both the space and time coordinates, the fundamental of question of Dirac would then be: is it possible to make this picture Lorentz-covariant This is illustrated in Fig. 6.2. Dirac, in his 1945 paper, stops there. This is, however, not the end of Dirac's story.

In 1949 [4], Dirac wrote a further paper in which he proposed using the ten generators of the Poincaré group. With them he constructed three forms of relativistic

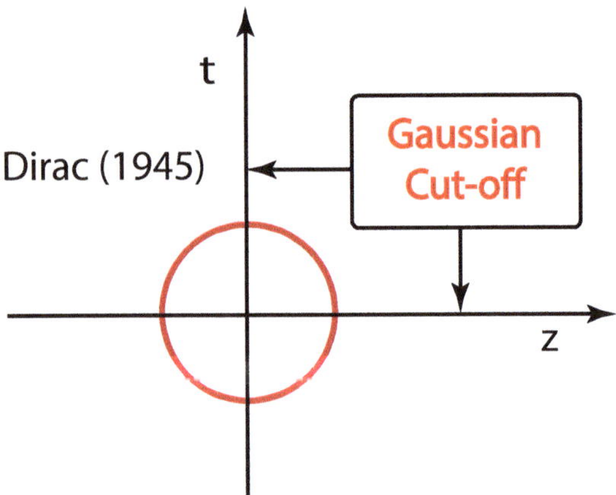

Fig. 6.2: Dirac's four-dimensional oscillators localized in a closed space-time region. This concept is not Lorentz-invariant. The question then is how to make this Lorentz-covariant [45].

dynamics of the atom. The wave functions that resulted were apparently not Lorentz-covariant. Even after subsidiary conditions necessary because of current form of quantum mechanics, Dirac found inconsistencies in each of these forms.

Dirac's atom in modern language is a hadron that is a bound state of quarks and/or antiquarks. We have defined in Sect. 6.1 the kinematical variables of the quarks and hadrons. In order to embed quantum mechanics into relativistic space-time, Dirac in his 1949 paper was interested in constructing a three-dimensional subspace of the four-dimensional Minkowski space in which non-relativistic quantum mechanics is valid.

Dirac considered three possible constraint conditions which are called *instant form*, *front form*, and *point form*. Each of these forms imposes one constraint condition which reduces the four-dimensional Minkowski space into a three-dimensional Euclidean space. In his instant form, Dirac considered the condition

$$x_0 \sim 0, \tag{6.22}$$

whose covariant form is

$$P_\mu x^\mu \sim 0, \tag{6.23}$$

where P is the total momentum of the hadron. Eq. (6.23) becomes Eq. (6.22) when the hadron is at rest. Dirac avoided using the exact numerical equality in writing down the above constraint in order to allow further physical interpretations consistent with quantum mechanics and special relativity. In particular, Dirac had in mind the possibility of the left-hand side becoming an operator acting on state vectors.

After introducing constraints, Dirac emphasizes that the relativistic dynamical equations should consist of transformation operators which generate space-time translations, rotations, and Lorentz boosts. He points out further that those transformation operators should be generators of the Poincaré group. Dirac then writes down the *Poisson bracket* relations these generators should satisfy:

$$[P_\mu, P_\nu] = 0,$$

$$[M_{\mu\nu}, M_{\rho\sigma}] = i(\eta_{\mu\rho}P_\nu - \eta_{\nu\rho}P_\mu),$$

$$[M_{\mu\nu}, M_{\rho\sigma}] = i(\eta_{\mu\rho}M_{\nu\sigma} - \eta_{\nu\rho}M_{\mu\sigma} + \eta_{\mu\sigma}M_{\rho\nu} - \eta_{\nu\sigma}M_{\rho\mu}) \quad (6.24)$$

where P_μ and $M_{\mu\nu}$ are the generators of space-time translations and Lorentz transformations respectively. Here again, we are using the same notation for the operator and eigenvalue of the hadron four-momentum. If we translate the above brackets into the language of quantum mechanics, they become the commutators for the generators of the Poincaré group discussed in Chap. 2.

We are thus led to Sect. 5.3 of Chap. 5 where we constructed representations of the Poincaré group using the covariant harmonic oscillator. This procedure has been thoroughly discussed in Chap. 5.

The only remaining step in constructing Dirac's dynamical system is therefore to make the constraint condition consistent with the generators of the Poincaré group. Dirac noted in particular that the constraint condition of Eq. (6.23) can be an operator equation, and that its Poisson brackets with other dynamical variables should be zero or become zero in the manner in which the right-hand side of Eq. (6.23) vanishes.

In terms of the step-up and step-down operators introduced in Sect. 6.1, we can rewrite the harmonic oscillator equation given in Eq. (5.7) as

$$\left[\left(\frac{\partial}{\partial X_\mu}\right)^2 + \frac{1}{2}a_\mu^\dagger a^\mu + m_0^2\right]\phi(X,x) = 0, \quad (6.25)$$

with the subsidiary condition given in Eq. (6.14). The $(mass)^2$ of the hadron P^2 is constructed by the eigenvalue of the oscillator differential equation of

$$P^2 = \lambda + m_0^2$$

$$= m_0^2 + \frac{1}{2}a_\mu^\dagger a^\mu . \quad (6.26)$$

As was pointed out before, Dirac was interested in a solution of his Poisson bracket equations consistent with the instant form constraint of Eq. (6.23). The key question is whether the subsidiary condition of Eq. (6.14) meets this requirement.

What Dirac wanted from his conditional equality of Eq. (6.23) was to freeze the motion along the time separation variable in a manner consistent with quantum mechanics and relativity. This means that we can allow a time-energy uncertainty along this time-like axis without excitations, in accordance with the c-number time-energy uncertainty relation discussed in Sect. 6.1. The c-number in matrix language is a one-by-one matrix, and is the ground state with no excitations in the harmonic oscillator system.

We have observed in Sect. 5.2 of Chap. 5, the subsidiary condition of Eq. (6.14) becomes

$$(a'_0)^\dagger \psi(x) = 0, \tag{6.27}$$

which forbids time-like excitations in the hadron rest frame. We can therefore conclude that the subsidiary condition of Eq. (6.14) is a quantum mechanical form of Dirac's instant form constraint given in Eq. (6.23)

In order that the dynamical system be completely consistent, the subsidiary condition should commute with the generators of the Poincaré group:

$$[P_\sigma, P^\mu a_\mu^\dagger] = 0,$$

$$[M_{\sigma\beta}, P^\mu a_\mu^\dagger] = 0. \tag{6.28}$$

The above equations follow immediately from the fact that the operator $P^\mu a_\mu^\dagger$ is invariant under translations and Lorentz transformations.

Since the Casimir operators are constructed from the generators of the Poincaré group, we are tempted to say that the constraint operator commutes with the invariant Casimir operators. However, we have to note that the operator P^2 also takes the form of Eq. (6.26). Therefore, in order that the dynamical system be completely consistent, the operator $P^\mu a_\mu^\dagger$ should also commute with the form of P^2 given in Eq. (6.26). Let us see whether this is the case. If we compute the commutator using the expression of Eq. (6.26) for P^2,

$$[P^2, P^\mu a_\mu^\dagger] = P^\mu a_\mu^\dagger. \tag{6.29}$$

The right-hand side of this commutator does not vanish. This, however, should not alarm us. Because of the subsidiary condition of Eq. (6.14), the right-hand side vanishes in the applicable Hilbert space.

The constraint condition of Eq. (6.14) and its commutator with other operators produce zero either identically or in the manner in which the Eq. (6.14) is zero. Therefore, the subsidiary condition of Eq. (6.14) satisfies all the requirements of the instant form constraint of Eq. (6.23). Indeed, the covariant harmonic oscillator formalism, while serving as the basis for the representations of the Poincaré group, is a solution of Dirac's Poisson bracket equations consistent with the instant form constraint [46, 47].

In his 1949 paper, Dirac notes a *real difficulty* associated with making his dynamical models consistent with the energy-momentum relation for each particle participating in the interacting system. Let us examine why we do not have this difficulty in the harmonic oscillator formalism. The basic difference between Dirac's original paper and the present case is that, for each quark in the dynamics, Dirac uses the free-particle energy-momentum relation:

$$E_a = [(p_a)^2 + m_a^2]^{1/2}. \tag{6.30}$$

In the harmonic oscillator formalism, the Casimir operators of the Poincaré group indicate clearly that the mass of the hadron is a Poincaré-invariant constant, but they do not tell us anything about the masses of the constituent particles. In fact, the

$(mass)^2$ operator for an individual quark $(p_a)^2$ does not commute with P^2 of Eq. (6.26) [Problem 7 in Sect. 6.6]:

$$[p_a^2, P^2] \neq 0 ,$$

$$[p_b^2, P^2] \neq 0 . \tag{6.31}$$

The above non-vanishing commutators caused very serious difficulties in 1949. It is interesting to note however that Dirac's 1949 difficulty can be resolved by the c-number time-energy uncertainty relation which Dirac discussed in 1927 [48, 47].

6.3 Dirac's Light-Cone Coordinate System

The most uncomfortable feature in Lorentz transformation is to deal with any skew symmetric coordinate system. Many attempts have been made in the past to rectify this situation. Among them, the most common practice has been to use a four-dimensional Euclidean geometry with three real spatial coordinates and one imaginary time variable. This practice has not been helpful in understanding the fundamental aspects of the subject.

In an attempt to formulate a *relativistic dynamics of atom*, Dirac introduced a coordinate system which allows us to use a rectangular coordinate system commonly known as the light-cone coordinate system. In 1949 [4], Dirac observed that it would be more convenient to use the variables:

$$z_+ = (t + z)/\sqrt{2} ,$$

$$z_- = (t - z)/\sqrt{2} , \tag{6.32}$$

for the system being boosted along the z-direction. The above combinations of t and z are commonly called the light-cone variables.

In order to see how these light-cone variables are Lorentz transformed, let us write down the light-cone variables in the hadron rest frame z'_+ and z'_- defined as

$$z'_+ = (t' + z')/\sqrt{2} ,$$

$$z'_- = (t' - z')/\sqrt{2} . \tag{6.33}$$

Then the transformation from z'_+ and z'_- to z_+ and z_- can be written as

$$z_+ = \left[\frac{1 + \beta}{1 - \beta}\right]^{1/2} z'_+ ,$$

$$z_- = \left[\frac{1 - \beta}{1 + \beta}\right]^{1/2} z'_- . \tag{6.34}$$

Under the Lorentz transformation, one axis becomes elongated while the other goes through a contraction so that the product z_+z_- will stay constant [49]:

$$z_+z_- = z'_+z'_-$$
$$= (t^2 - z^2)/2 = [(t')^2 - (z')^2]/2. \tag{6.35}$$

This transformation property is illustrated in Fig. 6.3. If the stationary space-time region is a square, then the moving region will appear like a rectangle with the same area.

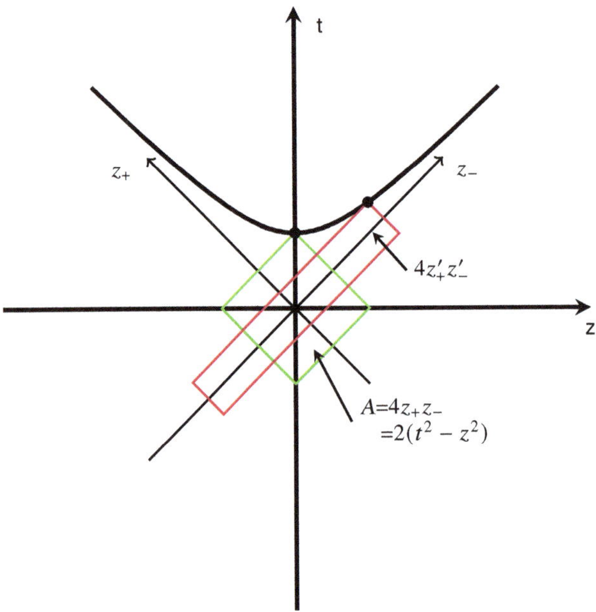

Fig. 6.3: Space-time diagram of a Lorentz boost in the light-cone coordinate system. The coordinate system remains orthogonal. The boost traces a point along the hyperbola and squeezes the square into a rectangle. The area of both the square and the rectangle is $2(t^2 - z^2)$ and thus remains invariant under the Lorentz boost.

In his 1949 paper [4], Dirac also discussed his *point form* constraint, which imposes the restriction

$$x^\mu x_\mu = \text{constant}. \tag{6.36}$$

In terms of the light-cone variables, Eq. (6.35) and Fig. 6.3 illustrate the content of point form constraint.

The Lorentz deformation property in the $q_z q_0$-plane is the same as that for the zt-plane, and it is possible to use the light-cone coordinate system for these variables:

$$q_+ = (q_0 + q_z)/\sqrt{2},$$

$$q_- = (q_0 - q_z)/\sqrt{2}, \tag{6.37}$$

with the transformation property:

$$q_+ = \left[\frac{1+\beta}{1-\beta}\right]^{1/2} q'_+,$$

$$q_- = \left[\frac{1-\beta}{1+\beta}\right]^{1/2} q'_- . \tag{6.38}$$

The Lorentz deformation property in the momentum-energy plane can also be seen in Fig. 6.3.

The basic advantage of using the light-cone variables is that the coordinate system remains orthogonal. It is not difficult to visualize the deformation of the regions given in Fig. 6.4 due to the boost. The circular region in the hadron rest frame will appear as an ellipse to the lab-frame observer.

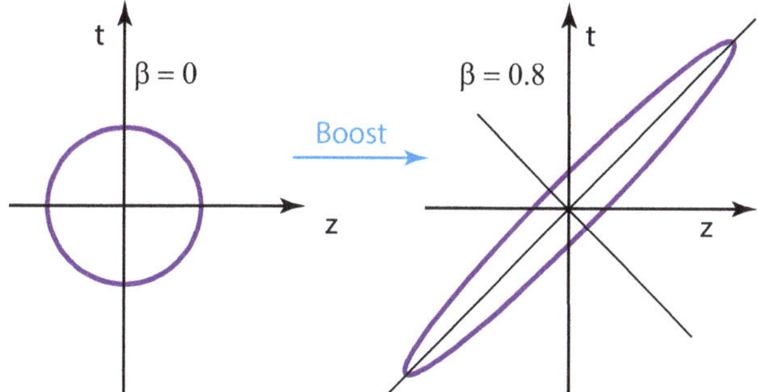

Fig. 6.4: Lorentz deformation in the longitudinal and time-like coordinate system. The left-hand side of this figure is identical to Fig. 6.1. It is clear that the circular region in the zt-plane will undergo an elliptic deformation [23].

With this understanding, let us use the Gaussian wave function which has the circular uncertainty distribution in the hadron rest frame:

$$\psi_\eta(t', z') = (1/\sqrt{\pi}) \exp(-(1/2)[(t')^2 + (z')^2]) . \tag{6.39}$$

This is the ground-state space-time wave function in the covariant oscillator formalism discussed in Chap. 5. Since

$$(t')^2 + (z')^2 = (z'_+)^2 + (z'_-)^2$$

$$= \left[\frac{1-\beta}{1+\beta}\right]^{1/2} z_+^2 + \left[\frac{1+\beta}{1-\beta}\right]^{1/2} z_-^2 . \tag{6.40}$$

The Gaussian form of Eq. (6.39) can be written as

$$\psi_\eta(t, z) = (1/\sqrt{\pi}) \exp\left(-\frac{1}{4}\left[\frac{1-\beta}{1+\beta}(t+z)^2 \right.\right.$$

$$\left.\left. +\frac{1+\beta}{1-\beta}(t-z)^2\right]\right). \tag{6.41}$$

The momentum-energy wave function becomes:

$$\phi(q_0, q_z) = \left(\frac{1}{2\pi}\right)\int \exp[i(q_t t - q_z z)\psi(t,z)]dzdt$$

$$= \left(\frac{1}{\sqrt{\pi}}\right)\exp\left(-\frac{1}{2}\right)\left[i(q_0')^2 - (q_z')^2\right]. \tag{6.42}$$

This wave function can also be written as

$$\phi(q_0, q_z) = \left(\frac{1}{\sqrt{\pi}}\right)\exp\left(-\frac{1}{4}\left[\frac{1-\beta}{1+\beta}(q_0+q_z)^2 \right.\right.$$

$$\left.\left. +\frac{1+\beta}{1-\beta}(q_0-q_z)^2\right]\right). \tag{6.43}$$

Because $q^\mu x_\mu$ takes the form

$$q^\mu x_\mu = q_0 t - q_z z - q_x x - q_y y$$

$$= -q_x x - q_y y + \frac{1}{2}\left[(q_0 - q_z)(t+z) + (q_0+q_z)(t-z)\right]. \tag{6.44}$$

The Fourier relations between the space-time and momentum-energy coordinates are [48]:

$$q_{z+} = q_- = -i\frac{\partial}{\partial z_+}, \quad q_{z-} = q_+ = -i\frac{\partial}{\partial z_-}. \tag{6.45}$$

This means that the major and minor axes of the momentum-energy coordinates are the *Fourier conjugates* of the minor and major axes of the space-time coordinates respectively. Thus we have the following Lorentz invariant relations:

$$(\Delta z_+)(\Delta q_-) = (\Delta z_+')(\Delta q_-') = 1,$$

$$(\Delta z_-)(\Delta q_+) = (\Delta z_-')(\Delta q_+') = 1. \tag{6.46}$$

This is indeed a Lorentz-invariant statement of the c-number time-energy uncertainty relation combined with Heisenberg's position-momentum uncertainty relation.

The above statement was based on the uncertainty relation along the t'-direction and that associated with the ground state along the longitudinal direction. However, we have to take into account the fact that there are excitations along the z'-direction. We discuss this problem in Sect. 6.5.

6.4 Harmonic Oscillators in the Light-Cone Coordinate System

We have so far been interested in Lorentz deformation properties of the ground-state harmonic oscillator wave function using the light-cone coordinate system. In this section, we shall study the harmonic oscillator formalism including all excited states.

It was noted in Sects. 5.5 and 5.6 of Chap. 5 that the covariant harmonic oscillator wave functions, forming the basis for the $O(3)$-like little group, can be written as a linear sum of the orthonormal solutions in the $O(4)$ coordinate system. It was noted also that, although not Lorentz-invariant, the $O(4)$-invariant differential equation is covariant in the sense that its $O(3, 1)$ counterpart is invariant under Lorentz transformations.

Because the light-cone coordinate system is only a $45°$ rotation of the $O(4)$-coordinate system in the subspace of longitudinal and time-like coordinates, we can write the $O(4)$-invariant oscillator equation in terms of the light-cone variables as

$$\frac{1}{2}\left[\left(\frac{\partial^2}{\partial x^2} + \frac{\partial^2}{\partial y^2} + \frac{\partial^2}{\partial z_+^2} + \frac{\partial^2}{\partial z_-^2}\right) - (x^2 + y^2 + z_+^2 + z_-^2)\right] g(x) = (\lambda + 2)g(x). \quad (6.47)$$

When the system is boosted along the z-direction, z_+ and z_- are to be replaced by z'_+ and z'_- respectively. Since we are not interested in the transverse components which remain invariant under Lorentz boosts along the z-direction, the above differential equation can now be written as [50]:

$$\frac{1}{2}\left[\left(\frac{\partial^2}{\partial z_+'^2} + \frac{\partial^2}{\partial z_-'^2}\right) - (z_+'^2 + z_-'^2)\right] g'(z'_+, z'_-) = (\lambda + 1)g'(z'_+, z'_-) . \quad (6.48)$$

The above differential equation is separable in the z_+ and z_- variables, and remains separable under the boost because the Lorentz boost does not mix z_+ and z_-. It is therefore easy to construct the normalizable solutions of the above differential equation. For a given integer value of $\lambda = n$, we can write

$$g_n(x') = \sum_{m=0}^{n} C_n^m f_m(z'_+) f_{n-m}(z'_-) , \quad (6.49)$$

where $f_m(z_+)$ is the normalized one-dimensional harmonic oscillator wave function defined in Sect. 5.2 of Chap. 5.

If we insist on the subsidiary condition of Eq. (6.14) forbidding time-like oscillations in the hadron rest frame, the oscillator wave function should have excitations only along the z'-direction, or along $(z'_+ - z'_-)/\sqrt{2}$. In terms of the z'_+ and z'_- variables, the Hermite polynomial $H_n(z')$ can be written as [51]

$$H_n(z') = H_n\left(\frac{[z'_+ - z'_-]}{\sqrt{2}}\right)\left(\frac{1}{2}\right)^{n/2}\sum_{m=0}^{n}\binom{n}{m}H_{n-m}(z'_+)H_m(z'_-).\qquad(6.50)$$

Thus the explicit form of the physical wave function becomes [50]:

$$\psi_\eta^n(z,t) = \left(\frac{1}{2}\right)^n\left(\frac{1}{\pi n!}\right)^{1/2}\left[\sum_{m=0}^{n}\binom{n}{m}H_{n-m}(z'_+)H_m(z'_-)\right]$$

$$\times\exp\left[-\frac{1}{2}(z'^2_+ - z'^2_-)\right].\qquad(6.51)$$

In order to express this in terms of the oscillator functions of the z_+ and z_- variables, we still have to make an expansion in an infinite series. The power of Hermite polynomials remains the same. However, the Gaussian factor requires an infinite sum of the wave functions in terms of the z_+ and z_- variables. This calculation remains as a future research problem.

6.5 Lorentz-Invariant Uncertainty Relations

As was pointed out in Sect. 6.1, the uncertainty principle applicable to the space-time separation of quarks in the hadron rest frame is the same as the currently accepted form based on existing theories and observations. The usual Heisenberg uncertainty relation holds for each of the three spatial coordinates. The time-separation variable is a c-number and therefore does not cause quantum excitations. However, this does not forbid the existence of the *Fourier relation* between time and energy variables resulting in the time-energy uncertainty relation given in Eq. (6.19).

Also in Sect. 6.1, it was noted that this peculiar time-energy uncertainty can be combined with Heisenberg's position-momentum uncertainty relation to give a covariant form for the internal space-time separation variables in the quark model within the framework of the covariant harmonic oscillator formalism. It was shown then that the uncertainty relations describing the oscillator system can be generalized into a *covariant commutator* form. This means that the uncertainty principle which holds in the harmonic oscillator formalism is applicable to all other models designed to explain hadron structures in a covariant manner.

Furthermore, as is shown in Sect. 6.4, the covariant form of the uncertainty relations given in Eq. (6.17) and Eq. (6.19) takes a *Lorentz-invariant form* in the light-cone coordinate system, as is shown in Eq. (6.46). However, the crucial question is how these uncertainty relations appear to an observer in the laboratory frame with the space-time separation variables t and z. Another important question is whether the uncertainty relations in the observer's frame can be stated in a Lorentz-invariant manner.

If we consider only the ground-state wave function, then the localization dictated by the uncertainty relations associated with both space and time separation variables, as is illustrated in Fig. 6.2, will lead to a distribution centered around the origin in the $t'z'$ coordinate system.

The question then is how this circularly-shaped region appears to the lab-frame observer, while the coordinates of the two different frames are related by the Lorentz transformation of Eq. (6.9). It is very easy to describe the Lorentz deformation of this region in the light-cone coordinate, in which the Lorentz transformation is simply a scale transformation along the light cones. If we use the Lorentz transformation parameter η where

$$\sinh \eta = \frac{\beta}{\left(1 - \beta^2\right)^{\frac{1}{2}}} , \tag{6.52}$$

then the transformation of Eq. (6.34) can be written as:

$$z_+ = [\exp(\eta)]z'_+ , \quad z_- = [\exp(-\eta)]z'_- . \tag{6.53}$$

The Lorentz transformation in the light-cone coordinate system is illustrated in Fig. 6.3, where the area of the rectangle remains the same as the area of the square.

The basic advantage of using the light-cone variables is that the coordinate system remains orthogonal. It is not difficult to visualize the deformation of the regions given in Fig. 6.4 due to the boost. The circular region in the hadron rest frame will appear as an ellipse to the an observer in the laboratory frame. It is therefore sufficient to study the circular region in Fig. 6.2 in order to study the transformation properties of more complicated regions. With this understanding, let us go back to the oscillator formalism which has the circular uncertainty distribution in the hadron rest frame. We can use the space-time wave function of Eqs. (6.39) and (6.41) and the momentum-energy wave function of Eqs. (6.42) and (6.43).

Again, in terms of the light-cone variables, we write the transformation of the momentum-energy variables as:

$$q_+ = [\exp(\eta)]q'_+ , \quad q_- = [\exp(-\eta)]q'_- . \tag{6.54}$$

Within the frame work of the harmonic oscillator formalism, the momentum-energy wave function has the same Lorentz-transformation property as the space-time wave function.

It is then not difficult to arrive at the Lorentz-invariant uncertainty products given in Eq. (6.46). However, we are interested in what implications these invariant relations have on the t- and z-coordinates. We started with the uncertainty relations:

$$(\Delta z')(\Delta q'_z) = 1 \qquad \text{and} \qquad (\Delta t')(\Delta q'_0) = 1 , \tag{6.55}$$

in the hadron rest frame. The question is whether or not we are allowed to write

$$(\Delta z)(\Delta q_z) = 1 \qquad \text{and} \qquad (\Delta t)(\Delta q_0) = 1 , \tag{6.56}$$

in the laboratory frame. In order to answer this question, let us look carefully at the deformation properties described in Fig. 6.4.

If we measure the z distribution along the z-axis where $t = 0$, the distribution is contracted by

$$\langle \Delta z \rangle_c = \langle \Delta z' \rangle \left[\cosh(2\eta) \right]^{-1/2} . \tag{6.57}$$

On the other hand, if we look at the distribution along the light-cone axis and its projection on the z-direction, then the distribution is elongated by

$$\langle \Delta z \rangle_e = \langle \Delta z' \rangle \left[\cosh(2\eta) \right]^{1/2} . \tag{6.58}$$

The geometry is the same along the t-axis, and also for the momentum-energy coordinate system. We can therefore write

$$(\Delta z)_c (\Delta q_z)_e = 1 \quad \text{and} \quad (\Delta z)_e (\Delta q_z)_c = 1 ,$$

$$(\Delta t)_c (\Delta q_0)_e = 1 \quad \text{and} \quad (\Delta t)_e (\Delta q_0)_c = 1 . \tag{6.59}$$

The uncertainty relations of Eq. (6.56) are basically correct if they are interpreted in the above manner.

As is illustrated in Fig. 6.4, the above interpretation requires the concept that the hadron is contracted if looked at in one way and elongated if looked at in another way along the longitudinal and time-like directions. This should be true also in the momentum-energy coordinate system. The important question is whether this dual nature of Lorentz deformation is consistent with the real world. The concept of *Lorentz contraction* of hadron matter is not new, and there are several physical models based on the idea that the longitudinal dimension of the hadron becomes shortened [52, 53, 54].

The concept of elongation is also familiar to us, because it is built into the parton model. First of all, it has been firmly established that the momentum distribution becomes wide-spread as the hadron moves very fast. This is precisely the elongation in the momentum-energy coordinate system. Since the Lorentz deformation property in the space-time coordinate system is identical to that in the momentum-energy coordinate system, there should also be an elongation in the space-time coordinate system. Then, does this effect manifest itself in high-energy laboratories? We shall discuss this and other related questions in Chap. 13.

6.6 Exercises and Problems

Exercise 1. In Sect. 6.3, we introduced the light-cone variables z_+ and z_- and discussed their Lorentz transformation properties. How are their derivatives transformed? How does the wave equation appear in terms of the light-cone variables?

Because the z_+ and z_- variables do not mix with each other when boosted along the z-direction, their derivatives undergo simple scale transformations. From Eq. (6.53),

$$\frac{\partial}{\partial z'_+} = e^{\eta}\frac{\partial}{\partial z_+} \qquad \text{and} \qquad \frac{\partial}{\partial z'_-} = e^{-\eta}\frac{\partial}{\partial z_-}. \tag{6.60}$$

The above form is consistent with the definition of the conjugate variables in Eq. (6.45) and the transformation property of the q_+ and q_- variables. It is clear from the above expression that the second derivative $\partial^2/\partial z_+ \partial z_-$ is a Lorentz-invariant operator. In terms of z_+ and z_-, the differential operators in the conventional z and t variables can be written as

$$\frac{\partial}{\partial z} = \left(\frac{1}{\sqrt{2}}\right)\left(\frac{\partial}{\partial z_+} + \frac{\partial}{\partial z_-}\right) \qquad \text{and} \qquad \frac{\partial}{\partial t} = \left(\frac{1}{\sqrt{2}}\right)\left(\frac{\partial}{\partial z_+} - \frac{\partial}{\partial z_-}\right). \tag{6.61}$$

Then, after a simple calculation,

$$\frac{\partial^2}{\partial z_+ \partial z_-} = \left(\frac{\partial}{\partial z}\right)^2 - \left(\frac{\partial}{\partial t}\right)^2. \tag{6.62}$$

Both sides of the equation are Lorentz-invariant. The immediate application of the above relation is the wave equation, which can now be written as

$$\frac{\partial^2}{\partial z_+ \partial z_-}\psi(z_+, z_-) = 0. \tag{6.63}$$

This equation tells us that solutions of the wave equation depend either on z_+ or z_-, not on both. Thus the most general form of the solution takes the form

$$\psi(z_+, z_-) = f(z_+) + g(z_-), \tag{6.64}$$

$f(z_+)$ and $g(z_-)$ depend only on $(t + z)$ and $(t - z)$ respectively [55, 56].

Exercise 2. In the Lorentz frame where the hadron is at rest, the uncertainty principle applicable to the space-time separation of quarks is expected to be the same as the presently accepted form largely based on non-relativistic quantum mechanics. The usual Heisenberg uncertainty relation holds for each of the three spatial coordinates:

$$[x', q'_x] = i, \qquad [y', q'_y] = i, \qquad [z', q'_z] = i. \tag{6.65}$$

On the other hand, the time-separation variable is a c-number and therefore does not cause quantum excitations. The commutator in this case takes the form

$$[t', q'_0] = 0, \tag{6.66}$$

accompanied by the *Fourier relation*:

$$(\Delta t')(\Delta q'_0) = 1. \tag{6.67}$$

How would these uncertainty relations appear to an observer in the laboratory frame?

Since the hadron moves along the z-axis, the x- and y-coordinates are not affected by boosts, and the first two commutation relations of Eq. (6.65) remain invariant:

$$[x, q_x] = i, \qquad [y, q_y] = i. \tag{6.68}$$

As for the third commutation relation of Eq. (6.65) for the longitudinal direction, we have to consider the Lorentz transformation of the coordinate system:

$$z = \left(\frac{z' + \beta t'}{(1 - \beta^2)^{1/2}} \right), \qquad t = \left(\frac{t' + \beta z'}{(1 - \beta^2)^{1/2}} \right). \tag{6.69}$$

Likewise, the transformation equations for the momentum-energy variables can be written as

$$q_z = \left(\frac{q'_z + \beta q'_0}{(1 - \beta^2)^{1/2}} \right), \qquad q_0 = \left(\frac{q'_0 + \beta q'_z}{(1 - \beta^2)^{1/2}} \right). \tag{6.70}$$

In terms of variables in the laboratory frame, the uncertainty relation $[z', q'_z] = i$ of Eq. (6.65) and the time-energy relation of Eq. (6.66) can be written as

$$[z, q_z] = \frac{i}{(1 - \beta^2)}, \qquad [t, q_0] = \frac{i\beta^2}{(1 - \beta^2)}. \tag{6.71}$$

In addition, because the Lorentz transformation is not an orthogonal transformation, the commutation relations between z and q_0 and between t and q_z do not vanish:

$$[z, q_0] = \frac{i\beta}{(1 - \beta^2)}, \qquad [t, q_z] = \frac{i\beta}{(1 - \beta^2)}. \tag{6.72}$$

The commutation relations of Eqs. (6.68), (6.71), and (6.72) can now be combined into a covariant form:

$$[x_\mu, q_\nu] = -\eta_{\mu\nu} + \frac{P_\mu P_\nu}{M^2}, \tag{6.73}$$

with the covariant form of Eq. (6.67):

$$(\Delta(P_\mu x^\mu / M)\Delta(P_\nu q^\nu / M)) = 1, \tag{6.74}$$

where M is the hadron mass. The above form of the uncertainty relations have two troublesome features. According to Eq. (6.72), Planck's constant appears frame-dependent. According to Eq. (6.72), there is a lack of the orthogonality relation among independent variables. If we use the light-cone coordinate system, we can see how these two troubles cancel each other out, and how the concept of the uncertainty relations becomes Lorentz-invariant [57].

Problem 1. Discuss time dilation, space contraction and clock synchronization using the light-cone coordinate system [58].

Problem 2. Write down Maxwell's equations and derive the wave equations using the light-cone coordinate system [58].

Problem 3. Discuss the Lorentz transformation properties of electromagnetic potentials and fields in terms of the light-cone variables. See Chap. 10.

Problem 4. Show that the time-energy uncertainty relation in quantum field theory is the same as the c-number time-energy uncertainty relation we used in this Chapter. See [14].

Problem 5. It was observed in Chap. 5 that the $(mass)^2$ spectrum has no lower bound if excitations along the time-like axis are allowed. Show that there is no lower energy bound in non-relativistic quantum mechanics if the time variable becomes a q number. See [44].

Problem 6. Since it is believed that the time-energy uncertainty relation is independent of the Schrödinger equation, there has been a tendency to formulate the problem in terms of other quantities. Show that the time-energy uncertainty can be stated in terms of the density operators [36, 7].

Problem 7. Show that the individual quark mass cannot be simultaneously diagonalized with the hadron mass. Prove Eq. (6.31). See [46].

Problem 8. The solution of the harmonic oscillator equation given in Eq. (6.49) is essentially a harmonic oscillator wave function in a two-dimensional space. One-dimensional and three-dimensional harmonic oscillators are discussed extensively in the literature. However, two-dimensional oscillators are seldom discussed. Show that the Lorentz deformation is essentially a problem of two-dimensional harmonic oscillators.

Problem 9. Describe the localization of light waves using the light-cone coordinate system.

Problem 10. Consider a blinking red light. For a stationary observer, the uncertainty relation is $(\Delta t)(\Delta q_0) = 1$. To an observer in an automobile moving toward the light, the color would be green if his/her vehicle moves sufficiently fast. This observer would insist on $(\Delta t')(\Delta q_0') = 1$ in his/her moving frame. Since neither t nor q_0 is a Lorentz-invariant quantity, the two observers will disagree on whose relation is more fundamental. Use the light-cone coordinate system to resolve this disagreement.

References

1. P.A.M. Dirac, The Quantum Theory of the Emission and Absorption of Radiation, Proceedings of the Royal Society A: Mathematical, Physical, and Engineering Sciences **114**(767), 243–265 (1927). DOI 10.1098/rspa.1927.0039. URL http://rspa.royalsocietypublishing.org/cgi/doi/10.1098/rspa.1927.0039
2. P.A.M. Dirac, The Quantum Theory of Dispersion, Proceedings of the Royal Society A: Mathematical, Physical, and Engineering Sciences **114**(769), 710–728 (1927). DOI 10.1098/rspa.1927.0071. URL http://rspa.royalsocietypublishing.org/cgi/doi/10.1098/rspa.1927.0071
3. P.A.M. Dirac, Unitary Representations of the Lorentz Group, Proceedings of the Royal Society A: Mathematical, Physical, and Engineering Sciences **183**(994), 284–295 (1945). DOI 10.1098/rspa.1945.0003. URL http://rspa.royalsocietypublishing.org/cgi/doi/10.1098/rspa.1945.0003

4. P.A.M. Dirac, Forms of Relativistic Dynamics, Reviews of Modern Physics **21**(3), 392–399 (1949). DOI 10.1103/RevModPhys.21.392. URL https://link.aps.org/doi/10.1103/RevModPhys.21.392

5. R.P. Feynman, Very High–Energy Collisions of Hadrons, Physical Review Letters **23**(24), 1415–1417 (1969). DOI 10.1103/PhysRevLett.23.1415. URL https://link.aps.org/doi/10.1103/PhysRevLett.23.1415

6. R.P. Feynman, The Behavior of Hadron Collisions at Extreme Energies, in *Proceedings of the 3rd International Conference on High Energy Collisions*, ed. by C. Yang, et al. (Gordon and Breach, New York, NY, USA, 1969), 237–249. (Stony Brook, New York, USA, 5-6-September.)

7. C.H. Blanchard, Density matrix and energy–time uncertainty, American Journal of Physics **50**(7), 642–645 (1982). DOI 10.1119/1.12772. URL http://aapt.scitation.org/doi/10.1119/1.12772

8. E.P. Wigner, On time-energy uncertainty relation, in *Aspects of Quantum Theory, in Honour of P.A.M. Dirac's 70th Birthday*, ed. by A. Salam, E.P. Wigner (Cambridge University Press, London UK, 1972), 237–248. ISBN 978-0521131032

9. W. Heisenberg, Multi–body problem and resonance in quantum mechanics, Z. Phys **43**, 172–198 (1927). DOI http://dx.doi.org/10.1007/BF01397280

10. N. Bohr, *Atomic physics and human knowledge*. Dover books on physics (Dover Publications, Mineola, NY, USA, 2010). ISBN 978-0-486-47928-6. (Originally publshed 1958 by John Wiley and Sons, New York, NY, USA; OCLC: ocn642624678.)

11. W. Heisenberg, Development of concepts in the history of quantum theory, American Journal of Physics **43**(5), 389–394 (1975). DOI 10.1119/1.9833. URL http://aapt.scitation.org/doi/10.1119/1.9833

12. W. Heitler, *The quantum theory of radiation*, 3rd edn. (Dover Publications, New York, NY, USA, 1984). ISBN 978-0-486-64558-2. (Originally published: Oxford : Clarendon Press, 1954. IN: The International series of monographs on physics.)

13. S.S. Schweber, *An Introduction to Relativistic Quantum Field Theory* (Dover Books on Physics, Dover Publications, Inc, New York, NY, USA, 2005). ISBN 978-0-486-44228-0. (Originally published 1961, Harper & Row, Publishers, New York, NY, USA.)

14. D. Han, Y.S. Kim, M.E. Noz, Physical principles in quantum field theory and in covariant harmonic oscillator formalism, Foundations of Physics **11**(11-12), 895–905 (1981). DOI 10.1007/BF00727106. URL http://link.springer.com/10.1007/BF00727106

15. G.C. Wick, Properties of Bethe-Salpeter Wave Functions, Physical Review **96**(4), 1124–1134 (1954). DOI 10.1103/PhysRev.96.1124. URL https://link.aps.org/doi/10.1103/PhysRev.96.1124

16. Y.S. Kim, M.E. Noz, Covariant Harmonic Oscillators and the Quark Model, Physical Review D **8**(10), 3521–3527 (1973). DOI 10.1103/PhysRevD.8.3521. URL https://link.aps.org/doi/10.1103/PhysRevD.8.3521

17. T. Goto, The New Wave Equation of the Bi-Local Field and Its Mechanical Model, Progress of Theoretical Physics **58**(5), 1635–1644 (1977). DOI 10.1143/PTP.58.1635. URL https://academic.oup.com/ptp/article-lookup/doi/10.1143/PTP.58.1635

18. D. Dominici, G. Longhi, Covariant harmonic oscillator with half-integer spin, Il Nuovo Cimento A **42**(2), 235–258 (1977). DOI 10.1007/BF02724585. URL http://link.springer.com/10.1007/BF02724585

19. J. Lukierski, M. Oziewicz, Relative time dependence as gauge freedom and bilocal models of hadrons, Physics Letters B **69**(3), 339–342 (1977). DOI 10.1016/0370-2693(77)90561-5. URL https://linkinghub.elsevier.com/retrieve/pii/0370269377905615

20. J. Jersák, D. Rein, Spin forces in harmonic model of confinement, Zeitschrift für Physik C Particles and Fields **3**(4), 339–344 (1980). DOI 10.1007/BF01414186. URL http://link.springer.com/10.1007/BF01414186

21. I. Sogami, H. Yabuki, Relativistic wave equation for the harmonically confined quark-antiquark system, Physics Letters B **94**(2), 157–160 (1980). DOI 10.1016/0370-2693(80)90847-3. URL https://linkinghub.elsevier.com/retrieve/pii/0370269380908473

22. H.P. Robertson, The Uncertainty Principle, Physical Review **34**(1), 163–164 (1929). DOI 10.1103/PhysRev.34.163. URL https://link.aps.org/doi/10.1103/PhysRev.34.163

23. S. Başkal, Y.S. Kim, M.E. Noz, *Physics of the Lorentz Group (Second Edition): Beyond high-energy physics and optics* (IOP Publishing, Bristol, UK, 2021). DOI 10.1088/978-0-7503-3607-9. ISBN 978-0-7503-3607-9. URL https://iopscience.iop.org/book/978-0-7503-3607-9

24. L.P. Horwitz, C. Piron, Relativistic dynamics, Helvetica Physica Acta **46**(3), 316–326 (1973)

25. F.C. Rotbart, Complete orthogonality relations for the covariant harmonic oscillator, Physical Review D **23**(12), 3078–3080 (1981). DOI 10.1103/PhysRevD.23.3078. URL https://link.aps.org/doi/10.1103/PhysRevD.23.3078

26. G.N. Fleming, Covariant Position Operators, Spin, and Locality, Physical Review **137**(1B), B188–B197 (1965). DOI 10.1103/PhysRev.137.B188. URL https://link.aps.org/doi/10.1103/PhysRev.137.B188

27. G.N. Fleming, A Manifestly Covariant Description of Arbitrary Dynamical Variables in Relativistic Quantum Mechanics, Journal of Mathematical Physics **7**(11), 1959–1981 (1966). DOI 10.1063/1.1704880. URL http://aip.scitation.org/doi/10.1063/1.1704880

28. V. Weisskopf, E.P. Wigner, Calculation of the natural brightness of spectral lines on the basis of Dirac's theory, Zeitschrift für Physik **63**(1-2), 54–73 (1930). DOI 10.1007/BF01336768. URL http://link.springer.com/10.1007/BF01336768

29. M. Moshinsky, Quantum Mechanics in Fock Space, Physical Review **84**(3), 533–540 (1951). DOI 10.1103/PhysRev.84.533. URL https://link.aps.org/doi/10.1103/PhysRev.84.533

30. M. Moshinsky, Boundary Conditions for the Description of Nuclear Reactions, Physical Review **81**(3), 347–352 (1951). DOI 10.1103/PhysRev.81.347. URL https://link.aps.org/doi/10.1103/PhysRev.81.347

31. M. Moshinsky, Diffraction in Time, Physical Review **88**(3), 625–631 (1952). DOI 10.1103/PhysRev.88.625. URL https://link.aps.org/doi/10.1103/PhysRev.88.625

32. L.D. Landau, E.M. Lifshiftz, *Quantum mechanics: non-relativistic theory*, 3rd edn. No. Vol. 3 in Course of theoretical physics / by L. D. Landau and E. M. Lifshitz (Elsevier [u.a.], Singapore, 2007). ISBN 978-0-7506-3539-4, 978-981-272-088-7, 978-7-5062-4257-8. (Second Edition published 1958 by Pergamon Press: New York, NY, USA)

33. Y. Aharonov, D. Bohm, Time in the Quantum Theory and the Uncertainty Relation for Time and Energy, Phys. Rev. **122**(5), 1649–1658 (1961). DOI 10.1103/PhysRev.122.1649. URL https://link.aps.org/doi/10.1103/PhysRev.122.1649

34. Y. Aharonov, D. Bohm, Answer to Fock Concerning the Time Energy Indeterminacy Relation, Phys. Rev. **134**(6B), B1417–B1418 (1964). DOI 10.1103/PhysRev.134.B1417. URL https://link.aps.org/doi/10.1103/PhysRev.134.B1417

35. V.A. Fock, Criticism of an attempt to disprove the uncertainty relation between time and energy, Sov. Phys. JETP **15**(4), 784–786 (1962). URL http://jetp.ras.ru/cgi-bin/dn/e_015_04_0784

36. J.H. Eberly, L.P.S. Singh, Time Operators, Partial Stationarity, and the Energy-Time Uncertainty Relation, Physical Review D **7**(2), 359–362 (1973). DOI 10.1103/PhysRevD.7.359. URL https://link.aps.org/doi/10.1103/PhysRevD.7.359

37. M. Bauer, P.A. Mello, On the lifetime-width relation for a decaying state and the uncertainty principle, Proceedings of the National Academy of Sciences **73**(2), 283–285 (1976). DOI 10.1073/pnas.73.2.283. URL https://pnas.org/doi/full/10.1073/pnas.73.2.283

38. M. Bauer, P. Mello, The time-energy uncertainty relation, Annals of Physics **111**(1), 38–60 (1978). URL https://linkinghub.elsevier.com/retrieve/pii/0003491678902233

39. E.W.R. Papp, Quantum theory of the natural space-time units, in *The Uncertainty Principle and Quantum Mechanics*, ed. by W.C. Price, S.S. Chissick (John Wiley and Sons, New York, NY, USA, 1977), 29–50. ISBN 978-0471994145

40. R. Recami, A time operator and the time-energy uncertainty relation, in *The Uncertainty Principle and Quantum Mechanics*, ed. by W.C. Price, S.S. Chissick (John Wiley and Sons, New York, NY, USA, 1977), 21–28. ISBN 978-0471994145

41. J. Rayski, J.M. Rayski, On the meaning of the time-energy uncertainty relation, in *The Uncertainty Principle and Quantum Mechanics*, ed. by W.C. Price, S.S. Chissick (John Wiley and Sons, New York, NY, USA, 1977), 13–20. ISBN 978-0471994145

42. I. Fujiwara, K. Wakita, H. Yoro, Explicit Construction of Time-Energy Uncertainty Relationship in Quantum Mechanics, Progress of Theoretical Physics **64**(2), 363–379 (1980). DOI 10.1143/PTP.64.363. URL https://academic.oup.com/ptp/article-lookup/doi/10.1143/PTP.64.363

43. W. Pauri, Canonical (possibly Lagrangian) realizations of the Poincaré group with increasing mass-spin trajectories, in *Group Theorectical Methods in Physics, ninth international Colloquium, Cocoyoc, Mexico, June 23-27, 1980*, ed. by K.B. Wolf (Springer-Verlag, Berlin, Germany, 1980), 615–622. ISBN 978-3540102717

44. E. Prugovečki, *Quantum mechanics in Hilbert space*, 2nd edn. (Dover Publications, Mineola, N.Y, 2007). ISBN 978-0-486-45327-9. (Originally published by Academic Press, New York, NY 1981.)

45. Y.S. Kim, M.E. Noz, Integration of Dirac's Efforts to Construct a Quantum Mechanics Which is Lorentz-Covariant, Symmetry **12**(8), 1270-1–30 (2020). DOI 10.3390/sym12081270. URL https://www.mdpi.com/2073-8994/12/8/1270

46. D. Han, Y.S. Kim, Yukawa's Approach and Dirac's Approach to Relativistic Quantum Mechanics: Relativistic Harmonic Oscillator Model, Progress of Theoretical Physics **64**(5), 1852–1860 (1980). DOI 10.1143/PTP.64.1852. URL https://academic.oup.com/ptp/article-lookup/doi/10.1143/PTP.64.1852

47. D. Han, Y.S. Kim, Dirac's form of relativistic quantum mechanics, American Journal of Physics **49**(12), 1157–1161 (1981). DOI 10.1119/1.12563. URL http://aapt.scitation.org/doi/10.1119/1.12563

48. Y.S. Kim, M.E. Noz, Physical basis for minimal time-energy uncertainty relation, Foundations of Physics **9**(5-6), 375–387 (1979). DOI 10.1007/bf00708529. URL https://link.springer.com/article/10.1007/BF00708529

49. Y.S. Kim, M.E. Noz, Dirac's light-cone coordinate system, American Journal of Physics **50**(8), 721–724 (1982). DOI 10.1119/1.12737. URL http://aapt.scitation.org/doi/10.1119/1.12737

50. Y.S. Kim, M.E. Noz, S.H. Oh, Lorentz deformation in the O(4) and light-cone coordinate systems, Journal of Mathematical Physics **21**(5), 1224–1228 (1980). DOI 10.1063/1.524513. URL http://aip.scitation.org/doi/10.1063/1.524513

51. W. Magnus, F. Oberhettinger, R.P. Soni, *Formulas and theorems for the special functions of mathematical physics* (Springer-Verlag, Berlin, Germany; New York, NY, USA, 1966). ISBN 978-3-662-11761-3978-3-662-11763-7. URL http://books.google.com/books?id=KoJQAAAAMAAJ. (Originally published 1949; OCLC: 557712575.)

52. N. Byers, C.N. Yang, πp Charge-Exchange Scattering and a "Coherent Droplet" Model of High-Energy Exchange Processes, Physical Review **142**(4), 976–981 (1966). DOI 10.1103/PhysRev.142.976. URL https://link.aps.org/doi/10.1103/PhysRev.142.976

53. T.T. Chou, C.N. Yang, Model of Elastic High-Energy Scattering, Physical Review **170**(5), 1591–1596 (1968). DOI 10.1103/PhysRev.170.1591. URL https://link.aps.org/doi/10.1103/PhysRev.170.1591

54. F. Gürsey, S. Orfanidis, Extended hadrons, scaling variables and the poincaré group, Il Nuovo Cimento A **11**(2), 225–278 (1972). DOI 10.1007/BF02728874. URL http://link.springer.com/10.1007/BF02728874

55. H. Bacry, M. Cadilhac, Metaplectic group and Fourier optics, Physical Review A **23**(5), 2533–2536 (1981). DOI 10.1103/PhysRevA.23.2533. URL https://link.aps.org/doi/10.1103/PhysRevA.23.2533

56. Y.S. Kim, M.E. Noz, Symplectic formulation of relativistic quantum mechanics, Journal of Mathematical Physics **22**(10), 2289–2293 (1981). DOI 10.1063/1.524763. URL http://aip.scitation.org/doi/10.1063/1.524763

57. P.E. Hussar, Y.S. Kim, M.E. Noz, Time–energy uncertainty relation and Lorentz covariance, American Journal of Physics **53**(2), 142–147 (1985). DOI 10.1119/1.14099. URL http://aapt.scitation.org/doi/10.1119/1.14099

58. L. Parker, G.M. Schmieg, A Useful Form of the Minkowski Diagram, American Journal of Physics **38**(11), 1298–1302 (1970). DOI 10.1119/1.1976076. URL http://aapt.scitation.org/doi/10.1119/1.1976076

Chapter 7
Symmetries of Dirac's Coupled Oscillators and Dirac's Matrices

Abstract In his 1963 paper, Dirac started from two annihilation and creation operators, and constructed four operators which resulted in sixteen linear independent operators, organized to represent two coupled harmonic oscillators. Therefore, we discuss the mathematics of two coupled harmonic oscillators as these are used in many physics models. We show the Lorentz covariance of this system. From the Schrödinger equation for two harmonic oscillators, Dirac observed that unitary transformations applicable to the ground-state function could be generated by ten combinations of his sixteen bilinear forms. He then observed that these ten combinations form the Lie algebra for the $(3+2)$ de Sitter, or $O(3, 2)$ Lorentz group. In 1976, Yuen used these bilinear forms to generate the two-photon coherent state and in 1986, Yurke et al., used them to develop the symmetries of the two photon problem or the squeezed states of light. We give a discussion of this and note that only a subset of Dirac's operators was used. A discussion of the symmetries of the Dirac matrices and of the fifteen generators derivable from them is included. These fifteen generators form the $O(3, 3)$ Lorentz group. Additionally, we discuss the symmetries associated with the $O(3, 2)$ Lorentz group and detail the procedure for contracting this group to the Poincaré group.

We saw in Chap. 6, that in 1927 [1, 2], Paul A. M. Dirac started his discussion of how quantum mechanics and special relativity could be combined. Indeed, Dirac made this his life long ambition, publishing in 1945 [3] and 1949 [4] papers where he pointed out that the task of constructing a relativistic dynamics is to construct a representation of the Poincaré group. In 1963 [5], Dirac studied carefully the symmetries generated by two harmonic oscillators. Using creation or step-up and annihilation or step-down operators, he was able to construct sixteen independent linear forms. These operators were chosen to represent two coupled harmonic oscillators.

S. Başkal et al., *Theory and Applications of the Poincaré Group*, Fundamental Theories of Physics 217, https://doi.org/10.1007/978-3-031-64376-7_7

The mathematical basis provided by the physics of two coupled harmonic oscillators [6] is fundamental for many solvable models in physics. Included in these are the Lee model in quantum field theory [7, 8], the Bogoliubov transformation in superconductivity [9, 10, 11], relativistic models of elementary particles [12, 13], and squeezed states of light [5, 14, 15, 16] also known as the two-mode coherent photon state. Another issue has been finding a coordinate transformation to decouple the oscillators. This lead to the study of the symmetry of two-mode squeezed states of light and all that has been in optical sciences since 1976 [14, 17]. This is commonly done with two coupled harmonic oscillators. From the Schrödinger equation for two harmonic oscillators, Dirac observed that unitary transformations applicable to the ground-state function could be generated by ten combinations of his sixteen bilinear forms.

Dirac then wrote down the ten generators of this group and their closed set of commutation relations. This set is known as the Lie algebra of the Poincaré group. However, as the decoupling of the harmonic oscillators was done using Lorentz transformations, it has a much broader mathematics basis.

Indeed, in 1976, Yuen used Dirac's bilinear forms to generate the two-photon coherent state and in 1986, Yurke et al., used them to develop the symmetries of the two photon problem or the squeezed states of light. We give a discussion of this and note only a subset of Dirac's sixteen linear operators was used. From his set of sixteen linear operators, Dirac chose ten operators to represent coupled harmonic oscillators. These ten operators satisfied a closed set of commutation relations.

Dirac also noted, in the same 1963 paper [5], that the commutation relations of these ten operators formed the Lie algebra for the $(3 + 2)$ de Sitter group which is isomorphic to the $O(3, 2)$ group. This latter group is the Lorentz group applicable to three space and two time dimensions.

We discuss the symmetries associated with the $O(3, 2)$ Lorentz group. The ten generators of $O(3, 2)$ also form the Lie algebra of the $Sp(4)$ group. The generators of this group are ten canonical four-by-four traceless matrices with imaginary elements.

Additionally, we are quite familiar with the four Dirac matrices which Dirac constructed for his relativistic electron equation. They are γ_1, γ_2, γ_3, and γ_0. With these matrices, it is possible construct fifteen traceless matrices. In the Majorana representation [18], all the elements in these matrices are imaginary, leading to four-by-four transformation matrices with real elements. Therefore, there are fifteen possible four-by-four matrices that can be constructed from the Dirac matrices. Only ten of them lead to canonical transformations. This means that there are five matrices which do not lead to canonical transformations. These five non-canonical matrices cannot be interpreted in terms of the present form of quantum mechanics.

Indeed, the system of fifteen matrices constructed from the Dirac matrices leads to the Lie algebra for the $O(3, 3)$ Lorentz group. It is known that these four-by-four matrices can also serve as the generators of the $SL(4, r)$ group [19, 20], which is locally isomorphic to the Lorentz group $O(3, 3)$ applicable to three space and three time dimensions [19, 20]. A discussion of the symmetries of the Dirac matrices and of the generators derivable from them is included.

There are now two sets of four-by-four matrices constructed by Dirac. The first set consists of his ten harmonic oscillator matrices, and the second of the fifteen generators formed from the γ matrices that come from Dirac's relativistic electron equation. The question is then whether this difference can be explained within the framework of the harmonic oscillator formalism with tangible physics. We discuss in this Chapter the physics of this difference. Dirac's ten matrices for coupled harmonic oscillators explain the quantum world for both oscillators. The set of Dirac's fifteen γ matrices contains his ten harmonic oscillator matrices as a subset.

Lastly, we discuss the symmetries associated with the $O(3,2)$ Lorentz group and detail the procedure for contracting this group to the Poincaré group. From the mathematical point of view, it is straightforward to contract one of the two time-like dimensions of $O(3,2)$ to construct the Poincaré group from $O(3,2)$. This is what we present in this Chapter. However, from the physics point of view, we are deriving the Poincaré symmetry for the Lorentz-covariant quantum world purely from the symmetries of two coupled harmonic oscillators.

In Sect. 7.1, we study the mathematics of two coupled harmonic oscillators as they are used in many physics models. In Sect. 7.2, we consider the Lorentz covariance of these two coupled harmonic oscillators. Dirac's algebra for two coupled harmonic oscillators, later became the fundamental mathematical language for two-mode squeezed states in quantum optics [14, 21, 12, 22].

Section 7.3 presents Dirac's two oscillator system. Indeed Dirac's ten oscillator matrices play a fundamental role in modern physics. In Sect. 7.4 the symmetries of two photons are discussed and in Sect. 7.5, those of Dirac's two oscillator system.

Section 7.6 discusses in detail the symmetries of the Dirac matrices. Section 7.7 expands the $O(3,2)$ group to $O(3,3)$ and provides the five additional generators. Finally, in Sect. 7.8 the contraction procedure is carried out to obtain the Poincaré group from $O(3,2)$.

7.1 Coupled Harmonic Oscillators

The Hamiltonian for two harmonic oscillators is given by:

$$H_1 = \frac{1}{2}\left(\frac{p_1^2}{m_1} + k_1 x_1^2\right), \qquad H_2 = \frac{1}{2}\left(\frac{p_2^2}{m_2} + k_2 x_2^2\right). \tag{7.1}$$

When the oscillators are coupled, the Hamiltonian can be written as:

$$H = \frac{1}{2}\left(\frac{p_1^2}{m_1} + \frac{p_2^2}{m_2} + k_1 x_1^2 + k_2 x_2^2\right) + K\left(x_1 - x_2\right)^2. \tag{7.2}$$

By making the following scale changes

$$x_1 \to \frac{x_1}{\sqrt{k_1}}, \quad x_2 \to \frac{x_2}{\sqrt{k_2}} \quad \text{and} \quad p_1 \to p_1 \sqrt{m_1}, \quad p_2 \to p_2 \sqrt{m_2} \tag{7.3}$$

the Hamiltonian becomes

$$H = \frac{1}{2}\left(p_1^2 + p_2^2 + x_1^2 + x_2^2\right) + K\left(\frac{x_1}{\sqrt{k_1}} - \frac{x_2}{\sqrt{k_2}}\right)^2 . \tag{7.4}$$

The space of (x_1, x_2) and (p_1, p_2) is rotated

$$x_1 \to x_1 \cos\theta - x_2 \sin\theta, \qquad x_2 \to x_1 \sin\theta + x_2 \cos\theta,$$

$$p_1 \to p_1 \cos\theta - p_2 \sin\theta, \qquad p_2 \to p_1 \sin\theta + p_2 \cos\theta, \tag{7.5}$$

so that the Hamiltonian becomes

$$H = \frac{1}{2}\left(p_1^2 + p_2^2 + x_1^2 + x_2^2\right) + K\left\{\left(\frac{x_1 \cos\theta - x_2 \sin\theta}{\sqrt{k_1}}\right) - \left(\frac{x_1 \sin\theta + x_2 \cos\theta}{\sqrt{k_2}}\right)\right\}^2 . \tag{7.6}$$

This can then be rewritten as:

$$H = \frac{1}{2}\left(p_1^2 + p_2^2 + x_1^2 + x_2^2\right) + K\left\{A(x_1 - x_2) + B(x_1 + x_2)\right\}^2 , \tag{7.7}$$

with

$$A = \frac{1}{2}\left\{\left(\frac{1}{\sqrt{k_1}} + \frac{1}{\sqrt{k_2}}\right)\cos\theta + \left(\frac{1}{\sqrt{k_1}} - \frac{1}{\sqrt{k_2}}\right)\sin\theta\right\},$$

$$B = \frac{1}{2}\left\{\left(\frac{1}{\sqrt{k_1}} - \frac{1}{\sqrt{k_2}}\right)\cos\theta - \left(\frac{1}{\sqrt{k_1}} + \frac{1}{\sqrt{k_2}}\right)\sin\theta\right\} . \tag{7.8}$$

If $B = 0$ then

$$\tan\theta = \frac{\sqrt{k_2} - \sqrt{k_1}}{\sqrt{k_2} + \sqrt{k_1}} . \tag{7.9}$$

This will then lead us to

$$A = \sqrt{\frac{1}{2}\left(\frac{1}{k_1} + \frac{1}{k_2}\right)} . \tag{7.10}$$

The Hamiltonian is then

$$H = \frac{1}{2}\left(p_1^2 + p_2^2\right) + \frac{1}{2}\left\{x_1^2 + x_2^2 + \left[\frac{K(k_1 + k_2)}{k_1 k_2}\right](x_1 - x_2)^2\right\} . \tag{7.11}$$

We make another coordinate transformation

$$x_+ = \frac{x_1 + x_2}{\sqrt{2}}, \quad x_- = \frac{x_1 - x_2}{\sqrt{2}} \quad \text{and} \quad p_+ = \frac{p_1 + p_2}{\sqrt{2}}, \quad p_- = \frac{p_1 - p_2}{\sqrt{2}} . \tag{7.12}$$

Now the Hamiltonian takes the form

$$H = \frac{1}{2}\left[\left(p_+^2 + p_-^2\right) + \left(x_+^2 + e^{4\eta}x_-^2\right)\right],\tag{7.13}$$

where

$$e^{4\eta} = 1 + \frac{K(k_1 + k_2)}{k_1 k_2}.\tag{7.14}$$

When we examine whether the coupling produces a canonical transformation this parametrization becomes convenient. This is because if $\eta = 0$, the coupling with K does not exist as $K = 0$.

We can renormalise this Hamiltonian by multiplying by $e^{-\eta}$. This results in:

$$H = \frac{1}{2}\left[\left(x_+^2 + p_+^2\right)e^{-\eta} + \left(x_-^2 e^{2\eta} + p_-^2 e^{-2\eta}\right)e^{\eta}\right].\tag{7.15}$$

The following transformations

$$x_+ \to e^{\eta/2}x_+, \quad p_+ \to e^{\eta/2}p_+ \quad \text{and} \quad x_- \to e^{-\eta/2}x_-, \quad p_- \to e^{-\eta/2}p_-,\tag{7.16}$$

renders the Hamiltonian in the form

$$H = \frac{1}{2}\left(e^{2\eta}x_-^2 + e^{-2\eta}p_-^2\right) + \frac{1}{2}\left(x_+^2 + p_+^2\right).\tag{7.17}$$

Within only the system of x_+ and p_+, we now make a scale change of

$$x_+ \to e^{-\eta}x_+ \quad \text{and} \quad p_+ \to e^{\eta}p_+,\tag{7.18}$$

then the Hamiltonian becomes

$$H = \frac{1}{2}\left(e^{-2\eta}x_+^2 + e^{2\eta}p_+^2\right) + \frac{1}{2}\left(e^{2\eta}x_-^2 + e^{-2\eta}p_-^2\right).\tag{7.19}$$

For this Hamiltonian the ground-state wave function then is

$$\psi_\eta(x_1, x_2) = \frac{1}{\sqrt{\pi}}\exp\left\{-\frac{1}{4}\left[(x_1 + x_2)^2 e^{-2\eta} + (x_1 - x_2)^2 e^{2\eta}\right]\right\}.\tag{7.20}$$

Then we have a momentum wave function of the form

$$\phi_\eta(p_1, p_2) = \frac{1}{2\pi}\int \exp\left[i(p_1 x_1 + p_2 x_2)\right]\psi_\eta(x_1, x_2)\,dx_1 dx_2,\tag{7.21}$$

which leads to

$$\phi_\eta(p_1, p_2) = \frac{1}{\sqrt{\pi}}\exp\left\{-\frac{1}{4}\left[(p_1 + p_2)^2 e^{2\eta} + (p_1 - p_2)^2 e^{-2\eta}\right]\right\}.\tag{7.22}$$

These form the wave functions for the coupled harmonic oscillators.

7.2 Lorentz-Covariant Oscillators and Entangled States

In the previous section we studied the details of scale transformations and rotations of the Hamiltonian and its associated wave function, for coupled two oscillators. In this section we deal with their Lorentz transformation properties and discuss the difference between coupled and covariant oscillators. In addition to the Hamiltonian approach, it is also worthwhile to investigate the properties of the Lorentz invariant differential equation of this system, whose solutions will lead to entangled excited states.

7.2.1 Lorentz-Covariant Oscillators

As we saw in Chap. 5, the differential equation for the harmonic oscillator can be separated in the space and time variables. When the system is boosted in the z-direction, the transverse components are not changed. Consequently, only the variables t and z need to be considered when a Lorentz boost is applied to the system. We now use x_1 and x_2 for the z and t variables respectively, so that the Hamiltonian for the harmonic oscillator function becomes

$$H_- = \frac{1}{2}\left(x_1^2 + p_1^2\right) - \frac{1}{2}\left(x_2^2 + p_2^2\right) , \tag{7.23}$$

which can then be reformulated as:

$$H_- = \frac{1}{2}\left(x_1^2 - x_2^2\right) + \frac{1}{2}\left(p_1^2 - p_2^2\right) . \tag{7.24}$$

H_- is invariant under the squeeze transformations:

$$\begin{pmatrix} x_1' \\ x_2' \end{pmatrix} = \begin{pmatrix} \cosh\eta & \sinh\eta \\ \sinh\eta & \cosh\eta \end{pmatrix} \begin{pmatrix} x_1 \\ x_2 \end{pmatrix} , \tag{7.25}$$

$$\begin{pmatrix} p_1' \\ p_2' \end{pmatrix} = \begin{pmatrix} \cosh\eta & \pm\sinh\eta \\ \pm\sinh\eta & \cosh\eta \end{pmatrix} \begin{pmatrix} p_1 \\ p_2 \end{pmatrix} . \tag{7.26}$$

If we use a Lorentz transformation of the form

$$x_1 \to x_1\cosh\eta + x_2\sinh\eta , \qquad x_2 \to x_1\sinh\eta + x_2\cosh\eta ,$$

$$p_1 \to p_1\cosh\eta + p_2\sinh\eta , \qquad p_2 \to p_1\sinh\eta + p_2\cosh\eta , \tag{7.27}$$

then (7.24) is left invariant. We rewrite this transformation in terms of $x_\pm$ and $p_\pm$, to obtain

$$x_\pm \to e^{\pm\eta}x_\pm \qquad \text{and} \qquad p_\pm \to e^{\pm\eta}p_\pm . \tag{7.28}$$

For the space coordinates, the wave function has the form

$$f_\eta (x_1, x_2) = \frac{1}{\sqrt{\pi}} \exp \left\{ -\frac{1}{4} \left[(x_1 + x_2)^2 \, e^{-2\eta} + (x_1 - x_2)^2 \, e^{2\eta} \right] \right\} . \qquad (7.29)$$

This is the same wave function as given in (7.20) for the coupled oscillators. However, the momentum wave function for the covariant harmonic oscillator is

$$g_\eta (p_1, p_2) = \frac{1}{2\pi} \int \exp \left[i \, (p_1 x_1 - p_2 x_2) \right] \psi_\eta (x_1, x_2) \, dx_1 dx_2 , \qquad (7.30)$$

which results in

$$g_\eta (p_1, p_2) = \frac{1}{\sqrt{\pi}} \exp \left\{ -\frac{1}{4} \left[(p_1 + p_2)^2 \, e^{-2\eta} + (p_1 - p_2)^2 \, e^{2\eta} \right] \right\} . \qquad (7.31)$$

The difference between the coupled and covariant harmonic oscillators is illustrated in Fig. 7.1 .

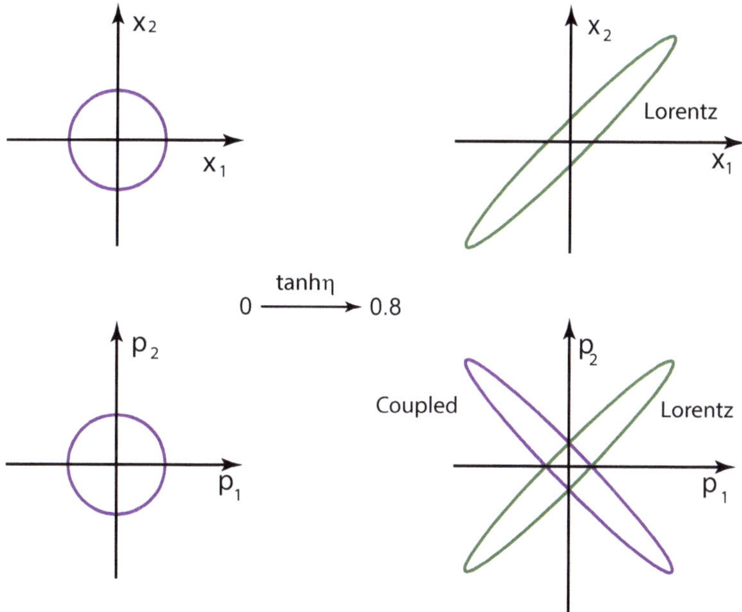

Fig. 7.1: Coupled and covariant oscillators. As can be seen from the illustration, the momentum wave functions are transformed differently while the space coordinates transform in the same way [23].

7.2.2 Entangled Excited States

The coupled harmonic oscillator equation

$$\frac{1}{2}\left\{\left(x_1^2 - \frac{\partial^2}{\partial x_1^2}\right) - \left(x_2^2 - \frac{\partial^2}{\partial x_2^2}\right)\right\} \chi_n(x_1)\chi_m(x_2) = (n - m)\chi_n(x_1)\chi_m(x_2) \quad (7.32)$$

is a quadratic equation with two independent variables x_1 and x_2, where they behave as space-like and time-like variables, respectively. While this equation is Lorentz invariant, its solution $\chi_n(x_1)\chi_m(x_2)$ is subject to the conditions of Lorentz covariance. For a single excitation state, where no time-like excitations are allowed, the wave function can be expressed as

$$\chi_n(x_1)\chi_0(x_2) = \left[\frac{1}{\pi 2^n n!}\right]^{\frac{1}{2}} H_n(x_1) \exp\left\{-\frac{1}{2}\left(x_1^2 + x_2^2\right)\right\}. \quad (7.33)$$

In the Lorentzian regime, this function is squeezed by replacing x_1 with x_1' and x_2 with x_2', using the inverse transformation of Eq. (7.27). Then this squeezed function can be expanded in form of a series

$$\chi_n(x_1')\chi_0(x_2') = \sum_{k'k} A_{k'k}(n)\chi_{k'}(x_1)\chi_k(x_2) \quad (7.34)$$

with $k' - k = n$. In view of Exercise 4 of Chap. 5 and also of [24], the coefficient A_{nk} can be calculated. Therefore we attain the wave function

$$\chi_n(x_1')\chi_0(x_2') = \left[\frac{1}{\cosh\eta}\right]^{n+1} \sum_{k}\left[\frac{(n+k)}{n!k!}\right]^{\frac{1}{2}} (\tanh\eta)^k \chi_{k+n}(x_1)\chi_k(x_2). \quad (7.35)$$

For the ground state, $n = 0$, this expression should be equal to Eq. (7.29) (or equally well to Eq. (7.51)), leading to

$$\frac{1}{\cosh\eta} \sum_{k} (\tanh\eta)^k \chi_k(x_1)\chi_k(x_2)$$

$$= \frac{1}{\sqrt{\pi}} \exp\left\{-\frac{1}{4}\left[(x_1 + x_2)^2 e^{-2\eta} + (x_1 - x_2)^2 e^{2\eta}\right]\right\}. \quad (7.36)$$

Here, we have the series expansion of squeezed Gaussian form which is not separable in the x_1 and x_2 variables. Thus the wave function is *entangled* in terms of its continuous variables. The importance of this expression in quantum information is discussed in the context of quantum optics, where it accounts for the fact that entanglement arises from the superposition of infinitely many ways in which equal number of photons can be distributed in each mode [25]. Gaussian entanglement is also known to be applicable to many other dynamical systems [26, 27, 28].

7.3 Dirac's Two Oscillator System

Annihilation and creation operators are, in both quantum mechanics and quantum field theory, viewed as being an essential component of their respective programs. In particular, an annihilation (or step-down) operator, decreases the number of particles by one, while its adjoint, the creation (or step-up) operator increases the number of particles in a given state by one. The space in which they act is called the Fock space [29]. These operators satisfy the following commutation relations:

$$\left[a, a^{\dagger}\right] = 1, \qquad [a, a] = 0, \qquad \text{and} \qquad \left[a^{\dagger}, a^{\dagger}\right] = 0. \tag{7.37}$$

They can be expressed in the form

$$a = \frac{1}{\sqrt{2}}\,(x + ip) = \frac{1}{\sqrt{2}}\left(x + \frac{\partial}{\partial x}\right),$$

$$a^{\dagger} = \frac{1}{\sqrt{2}}\,(x - ip) = \frac{1}{\sqrt{2}}\left(x - \frac{\partial}{\partial x}\right). \tag{7.38}$$

If there is more than one mode then Eq. (7.37) is rewritten as

$$\left[a_i, a_j^{\dagger}\right] = \delta_{ij}, \qquad [a_i, a_j] = 0, \qquad \text{and} \qquad \left[a_i^{\dagger}, a_j^{\dagger}\right] = 0 \tag{7.39}$$

where subscripts i, j denote each individual mode. Consequently, the operators are redefined

$$a_i = \frac{1}{\sqrt{2}}\,(x_i + ip_i) = \frac{1}{\sqrt{2}}\left(x_i + \frac{\partial}{\partial x_i}\right),$$

$$a_i^{\dagger} = \frac{1}{\sqrt{2}}\,(x_i - ip_i) = \frac{1}{\sqrt{2}}\left(x_i - \frac{\partial}{\partial x_i}\right). \tag{7.40}$$

We can use the one-dimensional harmonic oscillator as an example of using these operators. We'll therefore, drop the subscripts i and j. Hence, the operators of Eq. (7.40) can be expressed as in Eq. (7.38).

In his 1963 paper [5] Dirac considered two harmonic oscillators with bilinear forms in terms of annihilation and creation operators

$$a_i a_j, \quad a_i^{\dagger} a_j^{\dagger}, \quad a_i^{\dagger} a_j, \quad \text{and} \quad a_i a_j^{\dagger}. \tag{7.41}$$

Further, these were combined to obtain sixteen bilinear forms. The importance of this two harmonic oscillator system is that from it, using these bilinear forms we can generate the two-photon coherent state [14].

In addition, by starting with the two harmonic oscillator Schrödinger equation Dirac, in this same paper, observed that, by using ten combinations of these bilinear forms, unitary transformations could be generated which were applicable to the ground-state harmonic oscillator wave function. He noted as well that these ten combinations formed the (3 + 2) de Sitter group which is isomorphic to the $O(3,2)$ Lorentz group. Since this latter group is applicable to three space-like and two time-like coordinates, this group can be thought of as two coupled Lorentz groups. Later [19] it was recognized that the Lie algebra of the $O(3,3)$ group could be formed from fifteen combinations of these bilinear forms. This Lorentz group is applicable to three space-like and three time-like coordinates. Later in this Chapter we discuss in detail the relationship between this group and the Dirac matrices.

7.4 Two Photons and Their Symmetries

For a given quantum state, we can use the mathematics of harmonic oscillators to describe the number of particles in that state. Then the nth excited state can mean that in that quantum state there are n particles. Here the state of two photons is what is of interest. We call this coherent state of two photons the *two-mode squeezed state*, or simply the *squeezed state*. Formal definitions and further properties are given in Problem 12 of Sect. 5.7 in Chap. 5 and in [23].

We can write the Hamiltonians of Eq. (7.1) for two harmonic oscillators as:

$$H_1 = \frac{1}{2}\left(p_1^2 + x_1^2\right), \qquad H_2 = \frac{1}{2}\left(p_2^2 + x_2^2\right). \tag{7.42}$$

Then writing the Schrödinger equation, the corresponding solution becomes

$$\chi_n(x) = \left(\frac{1}{2^n n! \sqrt{\pi}}\right)^{1/2} H_n(x) \exp\left(-\frac{1}{2}x^2\right), \tag{7.43}$$

for each mode. This well known form represents the harmonic oscillator in the nth excited state using the Hermite polynomial $H_n(x)$.

Now we shall use the operators a and $a^\dagger$ defined in Eq. (7.38) as

$$a = \frac{1}{\sqrt{2}}\left(x + \frac{\partial}{\partial x}\right) = \frac{1}{\sqrt{2}}\left(x + ip\right),$$

$$a^\dagger = \frac{1}{\sqrt{2}}\left(x - \frac{\partial}{\partial x}\right) = \frac{1}{\sqrt{2}}\left(x - ip\right), \tag{7.44}$$

for each photon. If we then apply these operators to the wave function, the result is:

$$a\,\chi_n(x) = \sqrt{n}\,\chi_{n-1}(x) \quad \text{and} \quad a^\dagger\,\chi_n(x) = \sqrt{n+1}\,\chi_{n+1}(x). \tag{7.45}$$

Hence we can use $\chi_n(x)$ for the state of n photons. Now a and $a^\dagger$ can serve as the annihilation and creation operators, respectively. Therefore, using the notation

$\chi_n(x) = |n\rangle$, we can write

$$a^\dagger |0\rangle = |1\rangle , \quad \left(a^\dagger\right)^2 |0\rangle = \sqrt{2}\,|2\rangle , \quad \text{and} \quad \left(a^\dagger\right)^n |0\rangle = \sqrt{n!}\,|n\rangle , \qquad (7.46)$$

to deduce

$$|n\rangle = \frac{1}{\sqrt{n!}} \left(a^\dagger\right)^n |0\rangle . \qquad (7.47)$$

Next we shall consider two different photons, and write $|n_1, n_2\rangle$ where n_1 is the state of the first kind of photons, and n_2 is the state of the photons of the second kind [14]. Then we can write the exponential form

$$|\beta\rangle = \left(1 - \beta^2\right)^{1/2} \exp\left\{\beta a_1^\dagger a_2^\dagger\right\} |0,0\rangle . \qquad (7.48)$$

Here the real number β is smaller than one. If we apply a Taylor expansion to this formula we obtain:

$$|\beta\rangle = \left(1 - \beta^2\right)^{1/2} \sum_k \beta^k |k, k\rangle . \qquad (7.49)$$

This is the two-photon coherent state [14].

Although this state is normalized with $\langle\beta|\beta\rangle = 1$, the operator $\left(a_1^\dagger a_2^\dagger\right)$ is not Hermitian, and the operator

$$\exp\left(\beta a_1^\dagger a_2^\dagger\right) \qquad (7.50)$$

is not unitary.

To obtain a unitary operator to change $|0\rangle$ to $|\beta\rangle$, the wave function for the Lorentz-covariant oscillators which was given in Eq. (7.29) can be written as:

$$f_\eta(x_1, x_2) = \frac{1}{\sqrt{\pi}} \exp\left\{-\frac{1}{4}\left[(x_1 + x_2)^2 e^{-2\eta} + (x_1 - x_2)^2 e^{2\eta}\right]\right\} , \qquad (7.51)$$

where η is the boost parameter. We can see that this is the same as the coupled oscillator wave function which was given in Eq. (7.20) where η is the coupling parameter.

The question of symmetries associated with the two-photon problem goes beyond the coupling through the η parameter. Yurke $et\ al.$ [21], looked into the possibility that interferometers could be applied to the two-mode photon problem. They first picked these three Hermitian operators:

$$J_1 = \frac{1}{2}\left(a_1^\dagger a_2 + a_2^\dagger a_1\right),$$

$$J_2 = \frac{1}{2i}\left(a_1^\dagger a_2 - a_2^\dagger a_1\right),$$

$$J_3 = \frac{1}{2}\left(a_1^\dagger a_1 - a_2^\dagger a_2\right), \qquad (7.52)$$

and added to them the number operator

$$N = a_1^\dagger a_1 + a_2^\dagger a_2.$$ (7.53)

The J_i operators, of course, have the following commutation relations

$$\left[J_i, J_j \right] = i\epsilon_{ijk} J_j ,$$ (7.54)

which means it is possible to use the three-dimensional rotation group to study the symmetry properties of the two-photon problem.

They also introduced another additional set of operators which could effect the two-mode states which were

$$Q_3 = \frac{i}{2} \left(a_1^\dagger a_2^\dagger - a_1 a_2 \right),$$

$$K_3 = \frac{1}{2} \left(a_1^\dagger a_2^\dagger + a_1 a_2 \right),$$

$$S_3 = \frac{1}{2} \left(a_1^\dagger a_1 + a_2 a_2^\dagger \right),$$ (7.55)

with the closed set of commutation relations:

$$[K_3, Q_3] = -iS_3, \qquad [Q_3, S_3] = iK_3, \qquad [S_3, K_3] = iQ_3.$$ (7.56)

These commutation relations are similar to those for the $SU(1, 1)$ group, isomorphic to $O(2, 1)$, the Lorentz group composed of two space dimensions and one time dimension.

The question then becomes whether we can construct one set of closed commutation relations with the six operators given in Eq. (7.52) and Eq. (7.56). Dirac provided the answer to this question in 1963 [5].

It is interesting indeed that the same set of formulae is applicable to three different physical cases, namely two coupled harmonic oscillators, Lorentz-covariant harmonic oscillators, and the two-mode squeezed state.

7.5 Symmetries of Dirac's Two Oscillator System

In 1963 [5], Dirac discussed the ground state for two harmonic oscillators, and considered the following relations

$$\left[a_i , a_j^\dagger \right] = \delta_{ij} ,$$ (7.57)

where i and j represent 1 or 2. Using Eq. (7.44) we have:

$$a_i = \frac{1}{\sqrt{2}}\left(x_i + ip_i\right), \qquad a_i^\dagger = \frac{1}{\sqrt{2}}\left(x_i - ip_i\right). \tag{7.58}$$

From these, Dirac then constructed following ten operators. He started with the four rotation-like generators:

$$J_1 = \frac{1}{2}\left(a_1^\dagger a_2 + a_2^\dagger a_1\right), \qquad J_2 = \frac{1}{2i}\left(a_1^\dagger a_2 - a_2^\dagger a_1\right),$$

$$J_3 = \frac{1}{2}\left(a_1^\dagger a_1 - a_2^\dagger a_2\right), \qquad S_3 = \frac{1}{2}\left(a_1^\dagger a_1 + a_2 a_2^\dagger\right). \tag{7.59}$$

He then constructed six squeeze-like generators:

$$K_1 = -\frac{1}{4}\left(a_1^\dagger a_1^\dagger + a_1 a_1 - a_2^\dagger a_2^\dagger - a_2 a_2\right),$$

$$K_2 = +\frac{i}{4}\left(a_1^\dagger a_1^\dagger - a_1 a_1 + a_2^\dagger a_2^\dagger - a_2 a_2\right),$$

$$K_3 = +\frac{1}{2}\left(a_1^\dagger a_2^\dagger + a_1 a_2\right), \tag{7.60}$$

together with

$$Q_1 = -\frac{i}{4}\left(a_1^\dagger a_1^\dagger - a_1 a_1 - a_2^\dagger a_2^\dagger + a_2 a_2\right),$$

$$Q_2 = -\frac{1}{4}\left(a_1^\dagger a_1^\dagger + a_1 a_1 + a_2^\dagger a_2^\dagger + a_2 a_2\right),$$

$$Q_3 = +\frac{i}{2}\left(a_1^\dagger a_2^\dagger - a_1 a_2\right). \tag{7.61}$$

These generators are Hermitian and generate unitary transformations. From these ten generators, Dirac formed the following set of commutation relations:

$$[J_i, J_j] = i\epsilon_{ijk}J_k, \quad [J_i, K_j] = i\epsilon_{ijk}K_k,$$

$$[J_i, Q_j] = i\epsilon_{ijk}Q_k, \quad [K_i, K_j] = [Q_i, Q_j] = -i\epsilon_{ijk}J_k,$$

$$[K_i, Q_j] = -i\delta_{ij}S_3, \quad [J_i, S_3] = 0, \quad [K_i, S_3] = -iQ_i, \quad [Q_i, S_3] = iK_i. \tag{7.62}$$

Table 7.1: Generators for the two coupled harmonic oscillator system written in Dirac's two oscillator form as well as in five-by-five matrix form.

Generators	Two-oscillators	5x5 Matrix form
J_1	$\frac{1}{2}\left(a_1^\dagger a_2 + a_2^\dagger a_1\right)$	$\begin{pmatrix} 0 & 0 & 0 & 0 & 0 \\ 0 & 0 & 0 & 0 & 0 \\ 0 & 0 & 0 & 0 & i \\ 0 & 0 & 0 & 0 & 0 \\ 0 & 0 & -i & 0 & 0 \end{pmatrix}$
J_2	$\frac{1}{2}\left(a_1^\dagger a_2 - a_2^\dagger a_1\right)$	$\begin{pmatrix} 0 & 0 & 0 & 0 & 0 \\ 0 & 0 & 0 & 0 & 0 \\ 0 & 0 & 0 & -i & 0 \\ 0 & 0 & i & 0 & 0 \\ 0 & 0 & 0 & 0 & 0 \end{pmatrix}$
J_3	$\frac{1}{2}\left(a_1^\dagger a_1 - a_2^\dagger a_2\right),$	$\begin{pmatrix} 0 & 0 & 0 & 0 & 0 \\ 0 & 0 & 0 & 0 & 0 \\ 0 & 0 & 0 & 0 & 0 \\ 0 & 0 & 0 & 0 & -i \\ 0 & 0 & 0 & i & 0 \end{pmatrix}$
S_3	$\frac{1}{2}\left(a_1^\dagger a_1 + a_2 a_2^\dagger\right),$	$\begin{pmatrix} 0 & -i & 0 & 0 & 0 \\ i & 0 & 0 & 0 & 0 \\ 0 & 0 & 0 & 0 & 0 \\ 0 & 0 & 0 & 0 & 0 \\ 0 & 0 & 0 & 0 & 0 \end{pmatrix}$
K_1	$-\frac{1}{4}\left(a_1^\dagger a_1^\dagger + a_1 a_1 - a_2^\dagger a_2^\dagger - a_2 a_2\right)$	$\begin{pmatrix} 0 & 0 & 0 & 0 & 0 \\ 0 & 0 & 0 & i & 0 \\ 0 & 0 & 0 & 0 & 0 \\ 0 & i & 0 & 0 & 0 \\ 0 & 0 & 0 & 0 & 0 \end{pmatrix}$
K_2	$+\frac{i}{4}\left(a_1^\dagger a_1^\dagger - a_1 a_1 + a_2^\dagger a_2^\dagger - a_2 a_2\right)$	$\begin{pmatrix} 0 & 0 & 0 & 0 & 0 \\ 0 & 0 & 0 & 0 & i \\ 0 & 0 & 0 & 0 & 0 \\ 0 & 0 & 0 & 0 & 0 \\ 0 & i & 0 & 0 & 0 \end{pmatrix}$
K_3	$\frac{1}{2}\left(a_1^\dagger a_2^\dagger + a_1 a_2\right)$	$\begin{pmatrix} 0 & 0 & 0 & 0 & 0 \\ 0 & 0 & i & 0 & 0 \\ 0 & i & 0 & 0 & 0 \\ 0 & 0 & 0 & 0 & 0 \\ 0 & 0 & 0 & 0 & 0 \end{pmatrix}$
Q_1	$-\frac{i}{4}\left(a_1^\dagger a_1^\dagger - a_1 a_1 - a_2^\dagger a_2^\dagger + a_2 a_2\right)$	$\begin{pmatrix} 0 & 0 & 0 & i & 0 \\ 0 & 0 & 0 & 0 & 0 \\ 0 & 0 & 0 & 0 & 0 \\ i & 0 & 0 & 0 & 0 \\ 0 & 0 & 0 & 0 & 0 \end{pmatrix}$
Q_2	$-\frac{1}{4}\left(a_1^\dagger a_1^\dagger + a_1 a_1 + a_2^\dagger a_2^\dagger + a_2 a_2\right)$	$\begin{pmatrix} 0 & 0 & 0 & 0 & i \\ 0 & 0 & 0 & 0 & 0 \\ 0 & 0 & 0 & 0 & 0 \\ 0 & 0 & 0 & 0 & 0 \\ i & 0 & 0 & 0 & 0 \end{pmatrix}$
Q_3	$\frac{i}{2}\left(a_1^\dagger a_2^\dagger - a_1 a_2\right)$	$\begin{pmatrix} 0 & 0 & i & 0 & 0 \\ 0 & 0 & 0 & 0 & 0 \\ i & 0 & 0 & 0 & 0 \\ 0 & 0 & 0 & 0 & 0 \\ 0 & 0 & 0 & 0 & 0 \end{pmatrix}$

These commutation relations satisfy the Lie algebra for the $(3 + 2)$ de Sitter group as well as the $O(3, 2)$ group which is the Lorentz group applicable to three-dimensional space with two time variables. This group plays an important role in space-time symmetries. These ten generators along with their five-by-five matrix form are summarized in Table 7.1.

Let us use the notation (s, t, z, x, y), with (z, x, y) as space coordinates and (s, t) as two time coordinates. The harmonic oscillator serves as the language of quantum mechanics. From the commutation relations given in Eq. (7.62) we see that the three rotations operators, J_i, form the generators of the rotation group. If we further consider the generators J_i together with the generators K_i, they form the Lie algebra for the Lorentz group in the four-dimensional Minkowski space. The same can be said for the generators J_i and Q_i. Therefore there are a rotation group and two Lorentz groups as subgroups contained in the $(3 + 2)$ de Sitter group.

Furthermore, in his 1963 paper [5], Dirac noted that this set of commutation relations can also serve as the Lie algebra for the four-dimensional symplectic group $Sp(4)$. Thus the $Sp(4)$ group also contains a rotation group as well as two Lorentz groups as subgroups. The group Sp(4) has an important role in squeezed state of light, polarization optics, and when special relativity is incorporated with quantum mechanics. It was shown in [23] that the $Sp(4)$ group was the symmetry group for two Wigner phase spaces. The generators of this group are all defined as four-by-four traceless matrices with only imaginary elements. Since $Sp(4)$ contains a rotation group and two Lorentz groups as subgroups, like $O(3, 2)$, the ten generators can be written like J_i, S_3, K_i, and Q_i where i equals 1, 2, 3. These generators can be defined in terms of four-by-four canonical variables [23].

7.6 Symmetries for Dirac's Matrices

In Sect. 7.3 we saw that in Dirac's two oscillator system there were sixteen linear combinations of two harmonic oscillators. Dirac, in his 1963 paper [5] used ten combinations of them to define the generators for the $O(3, 2)$ group. From Table 7.1, we can see that all the matrices associated with the generators are traceless and all the elements are imaginary. Hence we know that all the transformation variables of these generators are real.

Let us now turn to Dirac's four γ matrices which he defined for his relativistic electron equation. Dirac chose his matrices to correspond to the internal space-time symmetries of particles possessing spin. These γ matrices are four-by-four. In Chap. 2 and in Chap. 3 we used the Weyl representation of the these γ matrices. Here we use them in the Majorana representation [18, 30, 31, 20], in which all the matrix elements are imaginary and traceless. When considering four-by-four matrices all with imaginary elements, there are, like Dirac's coupled oscillators, sixteen independent matrices. By requiring these matrices to be traceless, there are only fifteen remaining independent matrices.

We now want to examine if it is possible to construct fifteen matrices from Dirac's γ matrices [20]. We start from Dirac's four γ matrices and write them in the Majorana representation [18]:

$$\gamma_1 = i \begin{pmatrix} I & 0 \\ 0 & I \end{pmatrix} \sigma_3 , \quad \gamma_2 = \begin{pmatrix} 0 & -I \\ I & 0 \end{pmatrix} \sigma_2 , \quad \gamma_3 = -i \begin{pmatrix} I & 0 \\ 0 & I \end{pmatrix} \sigma_1 , \quad \gamma_0 = \begin{pmatrix} 0 & I \\ I & 0 \end{pmatrix} \sigma_2 . \quad (7.63)$$

One pseudo-scalar matrix can be constructed with these four matrices:

$$\gamma_5 = i\gamma_0\gamma_1\gamma_2\gamma_3 = \begin{pmatrix} I & 0 \\ 0 & -I \end{pmatrix} \sigma_2 . \quad (7.64)$$

In addition one pseudo-vector $i\gamma_5\gamma_\mu$ which is composed of

$$i\gamma_5\gamma_1 = i \begin{pmatrix} -I & 0 \\ 0 & I \end{pmatrix} \sigma_1 , \qquad i\gamma_5\gamma_2 = -i \begin{pmatrix} 0 & I \\ I & 0 \end{pmatrix} ,$$

$$i\gamma_5\gamma_3 = -i \begin{pmatrix} I & 0 \\ 0 & -I \end{pmatrix} \sigma_3 , \qquad i\gamma_5\gamma_0 = i \begin{pmatrix} 0 & I \\ -I & 0 \end{pmatrix} \quad (7.65)$$

can be constructed. Furthermore, in view of Eq. (2.84)

$$L_{\mu\nu} = \frac{i}{4} \left[\gamma_\mu, \gamma_\nu \right] \quad (7.66)$$

(with $\mu \neq \nu$), the following six matrices can be constructed:

$$i\gamma_0\gamma_1 = -i \begin{pmatrix} 0 & I \\ I & 0 \end{pmatrix} \sigma_1 , \quad i\gamma_0\gamma_2 = i \begin{pmatrix} I & 0 \\ 0 & -I \end{pmatrix} , \qquad i\gamma_0\gamma_3 = -i \begin{pmatrix} 0 & I \\ I & 0 \end{pmatrix} \sigma_3 ,$$

$$i\gamma_1\gamma_2 = i \begin{pmatrix} 0 & -I \\ I & 0 \end{pmatrix} \sigma_1 , \quad i\gamma_2\gamma_3 = i \begin{pmatrix} 0 & I \\ -I & 0 \end{pmatrix} \sigma_3 , \quad i\gamma_3\gamma_1 = i \begin{pmatrix} I & 0 \\ 0 & I \end{pmatrix} \sigma_2 . \quad (7.67)$$

We now have fifteen traceless matrices with imaginary elements. In Sect. 7.4 we showed that Dirac's ten generators for two coupled harmonic oscillators formed the Lie algebra for the $(3 + 2)$ de Sitter group, the $O(3, 2)$ Lorentz group, as well as the $Sp(4)$ group. It was shown in [23] that the $Sp(4)$ group was the symmetry group for two Wigner phase spaces. The generators of this group are all defined as four-by-four traceless matrices with only imaginary elements. Since $Sp(4)$ contains a rotation group and two Lorentz groups as subgroups, like $O(3, 2)$, the ten generators can be written like J_i, S_3, K_i, and Q_i where i equals 1, 2, 3. These generators can be defined in terms of four-by-four canonical variables [23].

We have mentioned earlier in Sect. 3.3 of Chap. 3 that the symplectic group is governed by the symplectic condition:

$$M J \tilde{M} = J . \quad (7.68)$$

Here J is given by [19]:

$$J = \begin{pmatrix} 0 & 1 & 0 & 0 \\ -1 & 0 & 0 & 0 \\ 0 & 0 & 0 & 1 \\ 0 & 0 & -1 & 0 \end{pmatrix} \tag{7.69}$$

and

$$M_{ij} = \frac{\partial}{\partial \eta_j} \xi_i , \tag{7.70}$$

where M transforms

$$(x_1, p_1, x_2, p_2) = (\eta_1, \eta_2, \eta_3, \eta_4) \tag{7.71}$$

to $(\xi_1, \xi_2, \xi_3, \xi_4)$. The matrix J can also be written in a block diagonal form

$$J = i \begin{pmatrix} I & 0 \\ 0 & I \end{pmatrix} \sigma_2 . \tag{7.72}$$

This matrix is different from the traditional J matrix:

$$\begin{pmatrix} 0 & I \\ -I & 0 \end{pmatrix} , \tag{7.73}$$

where the canonical variables are taken as (x_1, x_2, p_1, p_2). Our choice of ordering as in Eq. (7.71) is convenient for studying the two uncoupled harmonic oscillator with (x_1, p_1) and (x_2, p_2), as well as expanding and shrinking the phase space.

We also recall that

$$M = e^{-i\alpha G} \tag{7.74}$$

where G represents the generators that are purely imaginary matrices. Then the symplectic condition of Eq. (7.68) dictates that G be symmetric and anti-commute with J, or anti-symmetric and commute with J. Therefore, we observe four anti-symmetric matrices that commute with J, which are J_i, S_3, and six symmetric generators that anti-commute with J, which are K_i, Q_i. So, they take the form, here given with their Dirac matrix equivalents as:

$$J_1 = -\frac{1}{2} \begin{pmatrix} 0 & I \\ I & 0 \end{pmatrix} \sigma_2 = -\frac{1}{2}\gamma_0 , \qquad J_2 = \frac{i}{2} \begin{pmatrix} 0 & -I \\ I & 0 \end{pmatrix} I = -\frac{i}{2}\gamma_5\gamma_0 ,$$

$$J_3 = \frac{1}{2} \begin{pmatrix} -I & 0 \\ 0 & I \end{pmatrix} \sigma_2 = -\frac{1}{2}\gamma_5 , \qquad S_3 = \frac{1}{2} \begin{pmatrix} I & 0 \\ 0 & I \end{pmatrix} \sigma_2 = -\frac{i}{2}\gamma_1\gamma_3 \tag{7.75}$$

and

$$K_1 = \frac{i}{2}\begin{pmatrix} I & 0 \\ 0 & -I \end{pmatrix}\sigma_1 = \frac{i}{2}\gamma_1\gamma_5 \,, \qquad K_2 = \frac{i}{2}\begin{pmatrix} I & 0 \\ 0 & I \end{pmatrix}\sigma_3 = \frac{1}{2}\gamma_1 \,,$$

$$K_3 = \frac{-i}{2}\begin{pmatrix} 0 & I \\ I & 0 \end{pmatrix}\sigma_1 = \frac{i}{2}\gamma_0\gamma_1 \,, \qquad Q_1 = \frac{i}{2}\begin{pmatrix} -I & 0 \\ 0 & I \end{pmatrix}\sigma_3 = \frac{i}{2}\gamma_5\gamma_3 \,,$$

$$Q_2 = \frac{i}{2}\begin{pmatrix} I & 0 \\ 0 & I \end{pmatrix}\sigma_1 = -\frac{1}{2}\gamma_3 \,, \qquad Q_3 = \frac{i}{2}\begin{pmatrix} 0 & I \\ I & 0 \end{pmatrix}\sigma_3 = -\frac{i}{2}\gamma_0\gamma_3 \,. \tag{7.76}$$

The set of ten generators from Sect. 7.5, shown to be the generators for $Sp(4)$, are now expressed in terms of Dirac matrices. Table 7.1 contains the five-by-five matrices of $O(3,2)$ acting on $(3+2)$-dimensional space-time. The generators of $Sp(4)$ given above terms of Dirac matrices satisfy the same commutation relations of Eq. (7.62) as those of $O(3,2)$. Table 7.2 give the generators for $Sp(4)$ in terms of Dirac matrices and differential forms.

Table 7.2: The generators for $Sp(4)$, in terms of Dirac matrices and differential forms [24]. They satisfy the same commutation relations as that of $O(3,2)$.

Generator	Dirac matrices	Differential form
J_1	$-\frac{1}{2}\gamma_0$	$\frac{i}{2}\left\{\left(x_1\frac{\partial}{\partial p_2} - p_2\frac{\partial}{\partial x_1}\right) + \left(x_2\frac{\partial}{\partial p_1} - p_1\frac{\partial}{\partial x_2}\right)\right\}$
J_2	$-\frac{i}{2}\gamma_5\gamma_0$	$\frac{i}{2}\left\{\left(x_1\frac{\partial}{\partial x_2} - x_2\frac{\partial}{\partial x_1}\right) + \left(p_1\frac{\partial}{\partial p_2} - p_2\frac{\partial}{\partial p_1}\right)\right\}$
J_3	$-\frac{1}{2}\gamma_5$	$-\frac{i}{2}\left\{\left(x_1\frac{\partial}{\partial p_1} - p_1\frac{\partial}{\partial x_1}\right) - \left(x_2\frac{\partial}{\partial p_2} - p_2\frac{\partial}{\partial x_2}\right)\right\}$
S_3	$-\frac{i}{2}\gamma_1\gamma_3$	$-\frac{i}{2}\left\{\left(x_1\frac{\partial}{\partial p_1} - p_1\frac{\partial}{\partial x_1}\right) + \left(x_2\frac{\partial}{\partial p_2} - p_2\frac{\partial}{\partial x_2}\right)\right\}$
K_1	$\frac{i}{2}\gamma_1\gamma_5$	$-\frac{i}{2}\left\{\left(x_1\frac{\partial}{\partial p_1} + p_1\frac{\partial}{\partial x_1}\right) - \left(x_2\frac{\partial}{\partial p_2} + p_2\frac{\partial}{\partial x_2}\right)\right\}$
K_2	$\frac{1}{2}\gamma_1$	$\frac{i}{2}\left\{\left(x_1\frac{\partial}{\partial x_1} + x_2\frac{\partial}{\partial x_2}\right) - \left(p_1\frac{\partial}{\partial p_1} + p_2\frac{\partial}{\partial p_2}\right)\right\}$
K_3	$\frac{i}{2}\gamma_0\gamma_1$	$-\frac{i}{2}\left\{\left(x_1\frac{\partial}{\partial p_2} + p_2\frac{\partial}{\partial x_1}\right) + \left(x_2\frac{\partial}{\partial p_1} + p_1\frac{\partial}{\partial x_2}\right)\right\}$
Q_1	$\frac{i}{2}\gamma_5\gamma_3$	$\frac{i}{2}\left\{\left(x_1\frac{\partial}{\partial x_1} - p_1\frac{\partial}{\partial p_1}\right) - \left(x_2\frac{\partial}{\partial x_2} - p_2\frac{\partial}{\partial p_2}\right)\right\}$
Q_2	$-\frac{1}{2}\gamma_3$	$-\frac{i}{2}\left\{\left(x_1\frac{\partial}{\partial p_1} + p_1\frac{\partial}{\partial x_1}\right) + \left(x_2\frac{\partial}{\partial p_2} + p_2\frac{\partial}{\partial x_2}\right)\right\}$
Q_3	$-\frac{i}{2}\gamma_0\gamma_3$	$-\frac{i}{2}\left\{\left(x_2\frac{\partial}{\partial x_1} + x_1\frac{\partial}{\partial x_2}\right) - \left(p_2\frac{\partial}{\partial p_1} + p_1\frac{\partial}{\partial p_2}\right)\right\}$

7.7 The O(3, 3) group and Dirac's Matrices

In the previous section, we have used up ten of the fifteen γ matrices. Let us now consider another generator which we call G_3 and compose as:

$$G_3 = \frac{i}{2}\begin{pmatrix} I & 0 \\ 0 & -I \end{pmatrix} I = \frac{i}{2}\gamma_0\gamma_2\,. \tag{7.77}$$

Next, we take the commutator of G_3 with the previous ten generators of $Sp(4)$. Although the G_3 matrix commutes with S_3, J_3, K_1, K_2, Q_1, and Q_2, when commuted with the remaining matrices four additional generators are produced:

$$[G_3, J_1] = iG_2\,, \qquad [G_3, J_2] = -iG_1\,,$$

$$[G_3, K_3] = iS_2\,, \qquad [G_3, Q_3] = -iS_1\,. \tag{7.78}$$

In terms of the γ matrices these generators have the form:

$$S_1 = \frac{i}{2}\begin{pmatrix} 0 & I \\ -I & 0 \end{pmatrix} \sigma_3 = \frac{i}{2}\gamma_2\gamma_3\,, \qquad S_2 = \frac{i}{2}\begin{pmatrix} 0 & -I \\ I & 0 \end{pmatrix} \sigma_1 = \frac{i}{2}\gamma_1\gamma_2\,,$$

$$G_1 = -\frac{i}{2}\begin{pmatrix} 0 & I \\ I & 0 \end{pmatrix} I = \frac{i}{2}\gamma_2\gamma_5\,, \qquad G_2 = \frac{1}{2}\begin{pmatrix} 0 & I \\ I & 0 \end{pmatrix} \sigma_2 = \frac{1}{2}\gamma_2\,. \tag{7.79}$$

These five generators together with the ten generators defined in Sect. 7.5 give us fifteen generators which satisfy the following set of commutation relations:

$$[J_i, J_j] = i\epsilon_{ijk}J_k\,, \qquad [S_i, S_j] = i\epsilon_{ijk}S_k\,, \quad [J_i, S_j] = 0\,,$$

$$[J_i, G_j] = i\epsilon_{ijk}G_k\,, \quad [J_i, K_j] = i\epsilon_{ijk}K_k\,,$$

$$[J_i, Q_j] = i\epsilon_{ijk}Q_k\,, \quad [K_i, K_j] = [Q_i, Q_j] = [G_i, G_j] = -i\epsilon_{ijk}J_k\,, \tag{7.80}$$

and

$$[K_i, Q_j] = -i\delta_{ij}S_3\,, \quad [Q_i, G_j] = -i\delta_{ij}S_1\,, \quad [G_i, K_j] = -i\delta_{ij}S_2\,,$$

$$[K_i, S_3] = -iQ_i\,, \qquad [Q_i, S_3] = iK_i\,, \qquad\qquad [K_i, S_1] = 0\,,$$

$$[G_i, S_3] = 0\,, \qquad\qquad [Q_i, S_2] = 0, \qquad\qquad [Q_i, S_1] = -iG_i\,,$$

$$[G_i, S_1] = iQ_i\,, \qquad\qquad [K_i, S_2] = iG_i\,, \qquad\qquad [G_i, S_2] = -iK_i\,. \tag{7.81}$$

These commutation relations form the Lie algebra for the group $SL(4, r)$ or the group of four-by-four unimodular matrices with real elements [20]. This set can also serve as the Lie algebra for the $O(3, 3)$ Lorentz group applicable to the six-dimensional space consisting of three space and three time coordinates. This six-dimensional space has three space coordinates denoted as (z, x, y) and three time coordinates (u, s, t). We order these variables as (u, s, t, z, x, y) [19]. We now have two sets of rotation generators and three sets of squeeze generators [19] in the $O(3,3)$ and $SL(4, r)$ regime. They take the form of six-by-six matrices. These matrices are given explicitly in [23]. The additional five generators together the corresponding Dirac matrices and differential forms are shown in Table 7.3.

Table 7.3: Additional generators for $O(3, 3)$. There are six generators in this table, where five of those are additional to complete the group. The set (S_1, S_2, S_3) forms a subgroup and is responsible for rotations among the time-like components (u, s, t).

Generator	Dirac matrices	Differential forms
S_1	$\frac{i}{2}\gamma_2\gamma_3$	$-\frac{i}{2}\left\{\left(x_1\frac{\partial}{\partial x_2} - x_2\frac{\partial}{\partial x_1}\right) - \left(p_1\frac{\partial}{\partial p_2} - p_2\frac{\partial}{\partial p_1}\right)\right\}$
S_2	$\frac{i}{2}\gamma_1\gamma_2$	$-\frac{i}{2}\left\{\left(x_1\frac{\partial}{\partial p_2} - p_2\frac{\partial}{\partial x_1}\right) + \left(x_2\frac{\partial}{\partial p_1} - p_1\frac{\partial}{\partial x_2}\right)\right\}$
S_3	$-\frac{i}{2}\gamma_1\gamma_3$	$-\frac{i}{2}\left\{\left(x_1\frac{\partial}{\partial p_1} - p_1\frac{\partial}{\partial x_1}\right) + \left(x_2\frac{\partial}{\partial p_2} - p_2\frac{\partial}{\partial x_2}\right)\right\}$
G_1	$\frac{i}{2}\gamma_2\gamma_5$	$-\frac{i}{2}\left\{\left(x_1\frac{\partial}{\partial x_2} + x_2\frac{\partial}{\partial x_1}\right) + \left(p_1\frac{\partial}{\partial p_2} + p_2\frac{\partial}{\partial p_1}\right)\right\}$
G_2	$\frac{1}{2}\gamma_2$	$\frac{i}{2}\left\{\left(x_1\frac{\partial}{\partial p_2} + p_2\frac{\partial}{\partial x_1}\right) - \left(x_2\frac{\partial}{\partial p_1} + p_1\frac{\partial}{\partial x_2}\right)\right\}$
G_3	$\frac{i}{2}\gamma_0\gamma_2$	$-\frac{i}{2}\left\{\left(x_1\frac{\partial}{\partial x_1} + p_1\frac{\partial}{\partial p_1}\right) + \left(x_2\frac{\partial}{\partial p_2} + p_2\frac{\partial}{\partial x_2}\right)\right\}$

It is possible to formulate $O(3, 3)$ in this form:

$$\begin{pmatrix} \text{3-by-3 rotation} & 0 \\ 0 & \text{3-by-3 rotation} \end{pmatrix} \begin{pmatrix} \text{3-dimensional space} \\ \text{3-dimensional time} \end{pmatrix}. \tag{7.82}$$

In this six-by-six representation, the Lorentz boost along x with respect to the time variable t becomes

$$\begin{pmatrix} 1 & 0 & 0 & 0 & 0 & 0 \\ 0 & 1 & 0 & 0 & 0 & 0 \\ 0 & 0 & \cosh\eta & 0 & \sinh\eta & 0 \\ 0 & 0 & 0 & 1 & 0 & 0 \\ 0 & 0 & \sinh\eta & 0 & \cosh\eta & 0 \\ 0 & 0 & 0 & 0 & 0 & 1 \end{pmatrix} \begin{pmatrix} u \\ s \\ t \\ z \\ x \\ y \end{pmatrix} = \begin{pmatrix} u \\ s \\ x\,\sinh\eta + t\,\cosh\eta \\ z \\ x\,\cosh\eta + t\,\sinh\eta \\ y \end{pmatrix}. \tag{7.83}$$

We are not able to see direct physical applications of this extensive space-time symmetry group. The Lorentz group $O(3, 1)$ and the $(3 + 2)$ de Sitter or $O(3, 2)$ group are subgroups of this $O(3, 3)$ group.

Let us borrow the language of harmonic oscillators. Each harmonic oscillator contains the symmetry of the $Sp(2)$ group with three generators. These generators are J_2, K_1, and K_3. Nine others are from the way they are coupled mathematically. We studied the mathematical coupling of the harmonic oscillators in detail in Sect. 7.1.

7.8 Contraction of O(3, 2) to the Poincaré Group

We have explicitly given the five-by-five matrices representing the ten generators of the $O(3, 2)$ group in Table 7.1. From this it can be seen that the six J_i and K_i matrices, generating rotations and boosts in the (t, z, x, y) space, contain only zero elements in their first rows and columns. Thus, with respect to them, the s-coordinate remains invariant. This indicates that the the generators J_i and K_i constitute the familiar four-by-four matrices of the Lorentz group $O(3, 1)$ operating in Minkowski space.

The remaining three boost generators, Q_i, produce boosts along the s variable. Additionally, the matrix S_3, generates rotations between the two time variables t and s. These four matrices have elements only in the first row and column. We are interested now in converting the S_3 and Q_i generators into the translation generators P_0 and P_i which can be applied to homogeneous Lorentz group (t, z, x, y) represented by the six generators J_i and K_i to enlarge it into Poincaré group.

We'll apply the group contraction procedure introduced first by Inönü and Wigner [32] and use the procedure formulated in Chap. 4, in terms of squeeze transformations. This will allow us to eliminate the elements in the first row of the S_3 and Q_i matrices. Essentially, this will result in the contraction of $O(3, 2)$ to the Poincaré group. To do this we introduce the five-by-five contraction matrix [33, 34]

$$C(\epsilon) = \begin{pmatrix} \epsilon & 0 & 0 & 0 & 0 \\ 0 & 1 & 0 & 0 & 0 \\ 0 & 0 & 1 & 0 & 0 \\ 0 & 0 & 0 & 1 & 0 \\ 0 & 0 & 0 & 0 & 1 \end{pmatrix}, \tag{7.84}$$

where ϵ can become very large. The inverse of this matrix is

$$C(\epsilon)^{-1} = \begin{pmatrix} 1/\epsilon & 0 & 0 & 0 & 0 \\ 0 & 1 & 0 & 0 & 0 \\ 0 & 0 & 1 & 0 & 0 \\ 0 & 0 & 0 & 1 & 0 \\ 0 & 0 & 0 & 0 & 1 \end{pmatrix}. \tag{7.85}$$

We note that this matrix and the inverse leave the last four columns and rows invariant. This results in the four-dimensional Minkowski sub-space of (t, z, x, y) staying invariant. For the generators S_3 and Q_i, this is not true. We illustrate in Figure 7.2, that these contraction matrices can flatten the hyperbola for Q_i and flatten the circle for S_3. As for the boost with respect to the s variable, according to the procedure

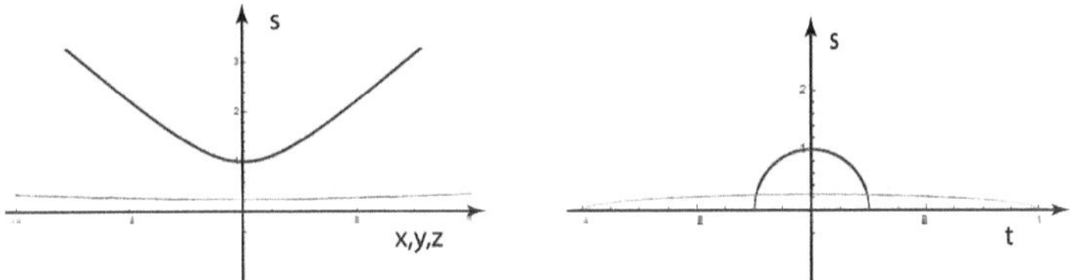

Fig. 7.2: Illustration of the Inönü-Wigner contraction procedure for $O(3, 2)$ [32]. By using a squeeze transformation, the hyperbola on the left is straightened to become a horizontal line on the axis. An inverse squeeze will make it another horizontal line with any s value used. Similarly the same process is applied to the circular curve on the right [23].

spelled out in Ref. [33, 34], the contracted boost generator becomes

$$Q_3^c = \lim_{\epsilon \to \infty} \frac{1}{\epsilon} \left[C^{-1}(\epsilon) \, Q_3 \, C(\epsilon) \right] = \begin{pmatrix} 0 & 0 & 0 & 0 & 0 \\ 0 & 0 & 0 & 0 & 0 \\ i & 0 & 0 & 0 & 0 \\ 0 & 0 & 0 & 0 & 0 \\ 0 & 0 & 0 & 0 & 0 \end{pmatrix}. \tag{7.86}$$

As this matrix now has only one element we can write this contraction procedure as

$$P_3 = \lim_{\epsilon \to \infty} \frac{1}{\epsilon} \left[C^{-1}(\epsilon) \, Q_3 \, C(\epsilon) \right]. \tag{7.87}$$

The explicit five-by-five matrix for P_3 is given in Eq. (7.89), and for the contraction procedure in Eq. (7.86). Likewise

$$P_1 = \lim_{\epsilon \to \infty} \frac{1}{\epsilon} \left[C^{-1} \, Q_1 \, C \right],$$

$$P_2 = \lim_{\epsilon \to \infty} \frac{1}{\epsilon} \left[C^{-1} \, Q_2 \, C \right],$$

$$P_0 = \lim_{\epsilon \to \infty} \frac{1}{\epsilon} \left[C^{-1} \, S_3 \, C \right]. \tag{7.88}$$

Therefore, we write the translation matrices obtained from the four matrices, S_3 and Q_i as:

$$
S_3 = \begin{pmatrix} 0 & -i & 0 & 0 & 0 \\ i & 0 & 0 & 0 & 0 \\ 0 & 0 & 0 & 0 & 0 \\ 0 & 0 & 0 & 0 & 0 \\ 0 & 0 & 0 & 0 & 0 \end{pmatrix} \quad \rightarrow \quad P_0 = \begin{pmatrix} 0 & 0 & 0 & 0 & 0 \\ i & 0 & 0 & 0 & 0 \\ 0 & 0 & 0 & 0 & 0 \\ 0 & 0 & 0 & 0 & 0 \\ 0 & 0 & 0 & 0 & 0 \end{pmatrix},
$$

$$
Q_3 = \begin{pmatrix} 0 & 0 & i & 0 & 0 \\ 0 & 0 & 0 & 0 & 0 \\ i & 0 & 0 & 0 & 0 \\ 0 & 0 & 0 & 0 & 0 \\ 0 & 0 & 0 & 0 & 0 \end{pmatrix} \quad \rightarrow \quad P_3 = \begin{pmatrix} 0 & 0 & 0 & 0 & 0 \\ 0 & 0 & 0 & 0 & 0 \\ i & 0 & 0 & 0 & 0 \\ 0 & 0 & 0 & 0 & 0 \\ 0 & 0 & 0 & 0 & 0 \end{pmatrix},
$$

$$
Q_1 = \begin{pmatrix} 0 & 0 & 0 & i & 0 \\ 0 & 0 & 0 & 0 & 0 \\ 0 & 0 & 0 & 0 & 0 \\ i & 0 & 0 & 0 & 0 \\ 0 & 0 & 0 & 0 & 0 \end{pmatrix} \quad \rightarrow \quad P_1 = \begin{pmatrix} 0 & 0 & 0 & 0 & 0 \\ 0 & 0 & 0 & 0 & 0 \\ 0 & 0 & 0 & 0 & 0 \\ i & 0 & 0 & 0 & 0 \\ 0 & 0 & 0 & 0 & 0 \end{pmatrix},
$$

$$
Q_2 = \begin{pmatrix} 0 & 0 & 0 & 0 & i \\ 0 & 0 & 0 & 0 & 0 \\ 0 & 0 & 0 & 0 & 0 \\ 0 & 0 & 0 & 0 & 0 \\ i & 0 & 0 & 0 & 0 \end{pmatrix} \quad \rightarrow \quad P_2 = \begin{pmatrix} 0 & 0 & 0 & 0 & 0 \\ 0 & 0 & 0 & 0 & 0 \\ 0 & 0 & 0 & 0 & 0 \\ 0 & 0 & 0 & 0 & 0 \\ i & 0 & 0 & 0 & 0 \end{pmatrix}. \tag{7.89}
$$

We notice that these translation generators have zero elements in the first rows.

These four contracted generators lead to the five-by-five transformation matrix

$$
\exp\left\{ -i\left(aP_0 + bP_3 + cP_1 + dP_2 \right) \right\} \tag{7.90}
$$

performing translations

$$
\begin{pmatrix} 1 & 0 & 0 & 0 & 0 \\ a & 1 & 0 & 0 & 0 \\ b & 0 & 1 & 0 & 0 \\ c & 0 & 0 & 1 & 0 \\ d & 0 & 0 & 0 & 1 \end{pmatrix} \begin{pmatrix} 1 \\ t \\ z \\ x \\ y \end{pmatrix} = \begin{pmatrix} 1 \\ t + a \\ z + b \\ x + c \\ y + d \end{pmatrix} \tag{7.91}
$$

in the four-dimensional Minkowski space. This means that the $O(3, 2)$ group becomes the inhomogeneous Lorentz group governing the Poincaré symmetry for quantum mechanics and quantum field theory.

Let us translate this into the language of differential operators with the translation operators

$$
P_0 = i\frac{\partial}{\partial t}, \qquad P_z = -i\frac{\partial}{\partial z}, \qquad P_x = -i\frac{\partial}{\partial x}, \qquad P_y = -i\frac{\partial}{\partial y}. \tag{7.92}
$$

Table 7.4: Among the ten coupled harmonic oscillator forms of Dirac, these are six generators that satisfy the Lie algebra of the Lorentz group applicable to the Minkowski space with $x_\mu = (t, \ -z, \ -x, \ -y)$. This table shows the generators in Dirac's two-oscillator form and in differential form.

Generators	Dirac's Two-oscillators	Differential
J_3	$\frac{1}{2}\left(a_1^\dagger a_1 - a_2^\dagger a_2\right)$	$-i\left(x\frac{\partial}{\partial y} - y\frac{\partial}{\partial x}\right)$
J_1	$\frac{1}{2}\left(a_1^\dagger a_2 + a_2^\dagger a_1\right)$	$-i\left(y\frac{\partial}{\partial z} - z\frac{\partial}{\partial y}\right)$
J_2	$\frac{1}{2i}\left(a_1^\dagger a_2 - a_2^\dagger a_1\right)$	$-i\left(z\frac{\partial}{\partial x} - x\frac{\partial}{\partial z}\right)$
K_3	$\frac{1}{2}\left(a_1^\dagger a_2^\dagger + a_1 a_2\right)$	$-i\left(z\frac{\partial}{\partial t} + t\frac{\partial}{\partial z}\right)$
K_1	$-\frac{1}{4}\left(a_1^\dagger a_1^\dagger + a_1 a_1 - a_2^\dagger a_2^\dagger - a_2 a_2\right)$	$-i\left(x\frac{\partial}{\partial t} + t\frac{\partial}{\partial x}\right)$
K_2	$+\frac{i}{4}\left(a_1^\dagger a_1^\dagger - a_1 a_1 + a_2^\dagger a_2^\dagger - a_2 a_2\right)$	$-i\left(y\frac{\partial}{\partial t} + t\frac{\partial}{\partial y}\right)$

Among the ten generators of Dirac's $O(3, 2)$, J_i and K_i are tabulated in Table 7.4. In addition to the six quadratic forms given in Table 7.4, Dirac constructed four additional quadratic forms. They correspond to the differential operators given in Table 7.5. The differential operators in Table 7.5 do not depend on the time variable t, but depend only on the second time variable s. We are now interested in converting these differential forms in this table into four translation generators, using the group contraction procedure outlined in this Section.

Table 7.5: The four additional coupled harmonic forms in Dirac's paper of 1963 [5]. These generators are those which operate on the second time variable, s. The generators are shown in differential and contracted differential form as well as in coupled harmonic oscillator form.

Generators	Dirac's Two-oscillators	Differential	Contracted to
S_3	$\frac{1}{2}\left(a_1^\dagger a_1 + a_2 a_2^\dagger\right)$	$-i\left(t\frac{\partial}{\partial s} - s\frac{\partial}{\partial t}\right)$	$i\frac{\partial}{\partial t}$
Q_3	$\frac{i}{2}\left(a_1^\dagger a_2^\dagger - a_1 a_2\right)$	$-i\left(z\frac{\partial}{\partial s} + s\frac{\partial}{\partial z}\right)$	$-i\frac{\partial}{\partial z}$
Q_1	$-\frac{i}{4}\left(a_1^\dagger a_1^\dagger - a_1 a_1 - a_2^\dagger a_2^\dagger + a_2 a_2\right)$	$-i\left(x\frac{\partial}{\partial s} + s\frac{\partial}{\partial x}\right)$	$-i\frac{\partial}{\partial x}$
Q_2	$-\frac{1}{4}\left(a_1^\dagger a_1^\dagger + a_1 a_1 + a_2^\dagger a_2^\dagger + a_2 a_2\right)$	$-i\left(y\frac{\partial}{\partial s} + s\frac{\partial}{\partial y}\right)$	$-i\frac{\partial}{\partial y}$

This contraction procedure tells us to fix the s time variable and set $s = 1$ for the generators given in Table 7.5 [35, 36]. They then become contracted to the translation generators given also in Table 7.5. Indeed, the six generators of the Lorentz group given in Table 7.4 together with the four translation generators constitute the ten generators of the Poincaré group or Einstein's Lorentzian system of space and time.

This contraction procedure produces four translation generators corresponding to the energy-momentum four-vector in the Lorentzian system. They lead to the on shell condition

$$p_0^2 - p_3^2 - p_1^2 - p_2^2 = \text{constant}. \tag{7.93}$$

This energy-momentum relation is widely known as Einstein's $E = mc^2$ [37, 35].

We can now compare the three different systems, namely, Galilean, Lorentzian, and Dirac's two-oscillator systems which leads to the system of the Poincaré group. They are compared in Table 7.6.

Table 7.6: Comparison of the three different systems: the traditional Galilean system, which has three rotation and three translation degrees of freedom; the Lorentzian system which has three possible boost operations and one additional translation, along the time direction; and the space-time symmetry from two harmonic oscillators which Dirac constructed in 1963 [5].

Systems	Galilean	Lorentzian	Two-oscillators
Rotations	J_z, J_x, J_y	J_z, J_x, J_y	J_z, J_x, J_y
Boosts	None	K_z, K_x, K_y	K_z, K_x, K_y
Translations	P_z, P_x, P_y	P_0, P_z, P_x, P_y	$\begin{pmatrix} S_3, Q_z, Q_x, Q_y \\ \text{contracted to} \\ P_0, P_z, P_x, P_y \end{pmatrix}$

References

1. P.A.M. Dirac, The Quantum Theory of the Emission and Absorption of Radiation, Proceedings of the Royal Society A: Mathematical, Physical, and Engineering Sciences **114**(767), 243–265 (1927). DOI 10.1098/rspa.1927.0039. URL http://rspa.royalsocietypublishing.org/cgi/doi/10.1098/rspa.1927.0039
2. P.A.M. Dirac, The Quantum Theory of Dispersion, Proceedings of the Royal Society A: Mathematical, Physical, and Engineering Sciences **114**(769), 710–728 (1927). DOI 10.1098/rspa.1927.0071. URL http://rspa.royalsocietypublishing.org/cgi/doi/10.1098/rspa.1927.0071
3. P.A.M. Dirac, Unitary Representations of the Lorentz Group, Proceedings of the Royal Society A: Mathematical, Physical, and Engineering Sciences **183**(994), 284–295 (1945). DOI 10.1098/rspa.1945.0003. URL http://rspa.royalsocietypublishing.org/cgi/doi/10.1098/rspa.1945.0003
4. P.A.M. Dirac, Forms of Relativistic Dynamics, Reviews of Modern Physics **21**(3), 392–399 (1949). DOI 10.1103/RevModPhys.21.392. URL https://link.aps.org/doi/10.1103/RevModPhys.21.392
5. P.A.M. Dirac, A Remarkable Representation of the 3 + 2 de Sitter Group, Journal of Mathematical Physics **4**(7), 901–909 (1963). DOI 10.1063/1.1704016. URL http://aip.scitation.org/doi/10.1063/1.1704016
6. H. Goldstein, *Classical Mechanics*, 2nd edn. Addison-Wesley series in physics (Addison-Wesley Pub. Co, Reading, MA, USA, 1980). ISBN 978-0-201-02918-5. (Originally published 1952.)
7. T.D. Lee, Some Special Examples in Renormalizable Field Theory, Physical Review **95**(5), 1329–1334 (1954). DOI 10.1103/PhysRev.95.1329. URL https://link.aps.org/doi/10.1103/PhysRev.95.1329
8. S.S. Schweber, *An Introduction to Relativistic Quantum Field Theory* (Dover Books on Physics, Dover Publications, Inc, New York, NY, USA, 2005). ISBN 978-0-486-44228-0. (Originally published 1961, Harper & Row, Publishers, New York, NY, USA.)
9. N.N. Bogoliubov, On a new method in the theory of superconductivity, Nuovo Cimento **7**, 794–804 (1958). DOI 10.1007/BF02745585

10. A.L. Fetter, J.D. Walecka, *Quantum theory of many-particle systems* (Dover Publications, Mineola, NY, USA, 2003). ISBN 978-0-486-42827-7. (Originally published: New York, NY, USA, McGraw-Hill, 1971.)

11. M. Tinkham, *Introduction to superconductivity*, 2nd edn. Dover books on physics (Dover Publ, Mineola, NY, USA, 2004). ISBN 978-0-486-43503-9. (Origiinally publlished: Krieger, Malabar; Florida USA; 1975; OCLC: 728146785.)

12. Y.S. Kim, M.E. Noz, *Phase space picture of quantum mechanics: group theoretical approach*. No. 40 in Lecture notes in physics series (World Scientific Publishing Co., Singapore; Hackensack, NJ, USA, 1991). ISBN 978-981-02-0360-3,978-981-02-0361-0. URL https://doi.org/10.1142/1197

13. H. Van Dam, Y.J. Ng, L.C. Biedenharn, A comment on fermionic tachyons and poincaré representations, Physics Letters B **158**(3), 227–230 (1985). DOI 10.1016/0370-2693(85) 90961-X. URL https://linkinghub.elsevier.com/retrieve/pii/037026938590961X

14. H.P. Yuen, Two–photon coherent states of the radiation field, Physical Review A **13**(6), 2226–2243 (1976). DOI 10.1103/PhysRevA.13.2226. URL https://link.aps.org/doi/10.1103/PhysRevA.13.2226

15. C.M. Caves, Quantum-mechanical noise in an interferometer, Physical Review D **23**(8), 1693–1708 (1981). DOI 10.1103/PhysRevD.23.1693. URL https://link.aps.org/doi/10.1103/PhysRevD.23.1693

16. R.F. Bishop, A. Vourdas, General two-mode squeezed states, Zeitschrift für Physik B Condensed Matter **71**(4), 527–529 (1988). DOI 10.1007/BF01313941. URL http://link.springer.com/10.1007/BF01313941

17. S. Başkal, Y. Kim, M. Noz, *Mathematical Devices for Optical Sciences.* (IOP Publishing, Bristol, UK, 2019). ISBN 978-0-7503-1612-5. URL https://dx.doi.org/10.1088/2053-2563/aafe78. (OCLC: 1034620988.)

18. E. Majorana, Teoria Relativistica di Particelle Con Momento Intrinseco Arbitrario Relativistic particles with arbitrary intrinsic angular momentum, Il Nuovo Cimento **9**(10), 335–344 (1932). DOI 10.1007/BF02959557. URL http://link.springer.com/10.1007/BF02959557

19. D. Han, Y.S. Kim, M.E. Noz, O(3,3)–like symmetries of coupled harmonic oscillators, Journal of Mathematical Physics **36**(8), 3940–3954 (1995). DOI 10.1063/1.530940. URL http://aip.scitation.org/doi/10.1063/1.530940

20. D.G. Lee, The Dirac gamma matrices as "relics" of a hidden symmetry?: As fundamental representations of the algebra sp(4,R), Journal of Mathematical Physics **36**(1), 524–530 (1995). DOI 10.1063/1.531320. URL http://aip.scitation.org/doi/10.1063/1.531320

21. B. Yurke, S.L. McCall, J.R. Klauder, SU(2) and SU(1,1) interferometers, Physical Review A **33**(6), 4033–4054 (1986). DOI 10.1103/PhysRevA.33.4033. URL https://link.aps.org/doi/10.1103/PhysRevA.33.4033

22. D. Han, Y.S. Kim, M.E. Noz, L. Yeh, Symmetries of two–mode squeezed states, Journal of Mathematical Physics **34**(12), 5493–5508 (1993). DOI 10.1063/1.530318. URL http://aip.scitation.org/doi/10.1063/1.530318

23. S. Başkal, Y.S. Kim, M.E. Noz, *Physics of the Lorentz Group (Second Edition): Beyond high-energy physics and optics* (IOP Publishing, Bristol, UK, 2021). DOI 10.1088/978-0-7503-3607-9. ISBN 978-0-7503-3607-9. URL https://iopscience.iop.org/book/978-0-7503-3607-9

24. S. Başkal, Y.S. Kim, M.E. Noz, Entangled Harmonic Oscillators and Space-Time Entanglement, Symmetry **8**(7), 55–80 (2016). DOI 10.3390/sym8070055. URL http://www.mdpi.com/2073-8994/8/7/55

25. D.F. Walls, G.J. Milburn, *Quantum optics*, 2nd edn. (Springer, Berlin, Germany, 2008). ISBN 978-3-540-28573-1

26. A. Ferraro, S. Olivares, M.G.A. Paris, *Gaussian States in Quantum Information*. Napoli Series on physics and Astrophysics (Bibliopolis, Napoles, Italy, 2005). ISBN 88-7088-483-X

27. G. Adesso, S. Ragy, A.R. Lee, Continuous Variable Quantum Information: Gaussian States and Beyond, Open Systems & Information Dynamics **21**(01n02), 1440,001 (2014). DOI 10.1142/S1230161214400010. URL https://www.worldscientific.com/doi/abs/10.1142/S1230161214400010

28. C. Weedbrook, S. Pirandola, R. García-Patrón, N.J. Cerf, T.C. Ralph, J.H. Shapiro, S. Lloyd, Gaussian quantum information, Reviews of Modern Physics **84**(2), 621–669 (2012). DOI 10.1103/RevModPhys.84.621. URL https://link.aps.org/doi/10.1103/RevModPhys.84.621

29. V. Fock, Quanten Elecktrodynamik, Physikalische Zeitschrift der Sovietunion **6**, 425–469 (1934)

30. Y.S. Kim, M.E. Noz, Dirac Matrices and Feynman's Rest of the Universe, Symmetry **4**(4), 626–643 (2012). DOI 10.3390/sym4040626. URL http://www.mdpi.com/2073-8994/4/4/626/

31. C. Itzykson, J.B. Zuber, *Quantum field theory*, dover edn. Dover books on physics (Dover Publications, Mineola, NY, USA, 2005). ISBN 978-0-486-44568-7. (With new preface and list of errata. Originally published: McGraw-Hill, New York, NY, USA 1980, in series: International series in pure and applied physics.)

32. E. Inönü, E.P. Wigner, On the Contraction of Groups and Their Representations, Proceedings of the National Academy of Sciences **39**(6), 510–524 (1953). DOI 10.1073/pnas.39.6.510. URL http://www.pnas.org/cgi/doi/10.1073/pnas.39.6.510

33. Y.S. Kim, E.P. Wigner, Cylindrical group and massless particles, Journal of Mathematical Physics **28**(5), 1175–1179 (1987). DOI 10.1063/1.527824. URL http://aip.scitation.org/doi/10.1063/1.527824

34. Y.S. Kim, E.P. Wigner, Space–time geometry of relativistic particles, Journal of Mathematical Physics **31**(1), 55–60 (1990). DOI 10.1063/1.528827. URL http://aip.scitation.org/doi/10.1063/1.528827

35. S. Başkal, Y.S. Kim, M.E. Noz, Einstein's $E = mc^2$ Derivable from Heisenberg's Uncertainty Relations, Quantum Reports **1**(2), 236–251 (2019). DOI 10.3390/quantum1020021. URL https://www.mdpi.com/2624-960X/1/2/21

36. Y.S. Kim, M.E. Noz, Integration of Dirac's Efforts to Construct a Quantum Mechanics Which is Lorentz-Covariant, Symmetry **12**(8), 1270–1–30 (2020). DOI 10.3390/sym12081270. URL https://www.mdpi.com/2073-8994/12/8/1270

37. S. Başkal, Y.S. Kim, M.E. Noz, Poincaré Symmetry from Heisenberg's Uncertainty Relations, Symmetry **11**(3), 409–1–9 (2019). DOI 10.3390/sym11030409. URL https://www.mdpi.com/2073-8994/11/3/409

Chapter 8
Representations and Applications of O(2, 1), SU(1, 1), and Sp(2)

Abstract The groups $O(2, 1)$, $SU(1, 1)$, $SL(2, r)$, and $Sp(2)$ are revisited. The importance of the group $Sp(2)$ is well-known in the Hamiltonian formulations of both classical and quantum mechanics. It has also found a myriad of applications in relatively new research areas such as quantum information and computing. The correspondence of $SU(1, 1)$ with $SL(2, r)$ or $Sp(2)$, consisting of real two-by-two matrices, is given. In this setting, the geometry of $SL(2, r)$ or $Sp(2)$ is made clear. Wigner, Bargmann, and Iwasawa decompositions are discussed within the context of little groups. It is shown that the study of $O(2, 1)$ is that of the Legendre functions. Complex angular momentum is introduced, and its physical applications are demonstrated. We also outline the steps in examining the unitary irreducible representations of $SU(1, 1)$.

After studying the little groups which preserve time-like and light-like four-momenta, we are led to consider the little group which leaves the space-like four-momentum invariant. Unlike the two previous cases, there is no clearly defined physical motivation to study this group as a little group, because we have to deal here with free particles travelling faster than light which are often called tachyons [1]. Although tachyons are intrinsically interesting from a group theoretical point of view [2, 3], there are many other applications of $O(2, 1)$ of immediate physical and mathematical interest.

First, $SU(1, 1)$ is isomorphic to the simplest symplectic group $Sp(2)$. The word *symplectic group* was introduced by Weyl in 1938 [4] and it plays a central role in the current development of physics, including classical mechanics [5, 6], plasma physics [7, 8, 9] nuclear physics [10, 11] statistical mechanics [12], general relativity and its alternative approaches [13, 14, 15, 16], optical sciences [17, 18, 19], quantum information [20, 21] and computation [22, 23], and high-energy physics [24, 25]. The reason is very simple. Symplectic transformations preserve skew symmetric

S. Başkal et al., *Theory and Applications of the Poincaré Group*, Fundamental Theories of Physics 217, https://doi.org/10.1007/978-3-031-64376-7_8

products, and there are many skew symmetric products in physics [26, 19]. Now that we recognize the importance of this subject matter, we shall discuss it in due detail.

It is not difficult to find discussions of the symplectic group in journals and textbooks. However, in most of the group theory textbooks, the symplectic group is examined in the context of abstract group theory. In most of the research articles, this group is introduced as a *new device* with which one can generate sophisticated research problems. While the symplectic group plays an increasingly important role in physics research, the existing literature gives the impression that one has to learn completely *new* material in order to understand this group. In this environment, it is gratifying that we can study $Sp(2)$ in conjunction with our effort to understand $O(2, 1)$ and $SU(1, 1)$.

Second, when we study Lorentz transformations, especially in physical applications, we often use transformations in the xyz-coordinate system with boosts along only one of the coordinate axis. We seldom discuss Lorentz transformations in the three-dimensional xyz-coordinate system. In spite of this simplification, the conventional method of computing velocity additions and successive Lorentz boosts is still complicated. Therefore, it is of practical interest to see whether there can be a further simplification. For instance, by restricting the problem from $O(3, 1)$ to $O(2, 1)$, we reduce the size of the matrix from four-by-four to three-by-three. We can reduce further the size of matrix to two-by-two by using $Sp(2)$.

Third, the study of $O(2, 1)$ is essentially that of the Legendre functions with parameters not necessarily confined to those of the rotation group. The Legendre function with complex angular momentum has an important physical application in scattering theory, commonly called the Regge-pole model [27, 28, 29].

Fourth, $SU(1, 1)$ and $O(2, 1)$ constitute a starting point for studying non-compact groups. These groups, like the rotation group, have three-parameters but are general enough to illustrate many of the properties of non-compact groups. Therefore, since the publication of Bargmann's original work [30], this group has been thoroughly and exhaustively discussed in the literature [31, 32, 33, 34, 35, 36, 37, 38, 39, 40, 41, 42, 43, 44, 45, 46, 47, 48]. As far as $SU(1, 1)$ and $O(2, 1)$ are concerned, the mathematics is far ahead of physics. At the present, we require more physical motivations.

In Sect. 8.1, we take advantage of the fact that $SU(1, 1)$ is unitarily equivalent to $SL(2, r)$ or $Sp(2)$ consisting of real two-by-two unimodular matrices. In this case, we can study transformation properties of the group using a two-dimensional graph. The two-to-one correspondence with $O(2, 1)$ is also studied.

In Sect. 8.2, finite-dimensional non-unitary representations of $O(2, 1)$ are studied as a preliminary step for later sections. It is illustrated in Sect. 8.3 that the study of $O(2, 1)$ is essentially that of the Legendre functions for the magnitude of its argument greater than one. In Sect. 8.4, we outline the first step which might be needed in studying unitary irreducible representations of $SU(1, 1)$.

Section 8.5 consists of exercises and problems. In view of the extensiveness of the coverage of this subject in the mathematical literature, and of the briefness of the discussion given in this Chapter, the list of mathematical exercises can become endless. We therefore choose only those problems which may have physical applications or may serve an illustrative purpose for the items discussed in earlier chapters.

8.1 Geometry of SL(2, r) and Sp(2)

Matrices representing $SU(1, 1)$ take the form

$$W = \begin{pmatrix} a & b \\ b^* & a^* \end{pmatrix}, \qquad \text{with} \qquad |a|^2 - |b|^2 = 1. \tag{8.1}$$

This group is generated by

$$J_3 = \frac{1}{2}\sigma_3, \qquad K_1 = \frac{i}{2}\sigma_1, \qquad K_2 = \frac{i}{2}\sigma_2. \tag{8.2}$$

If we perform a 90° rotation around the y-axis followed by another 90° rotation around the x-axis, then

$$x \to z, \qquad z \to y, \qquad y \to x. \tag{8.3}$$

The generators of Eq. (8.2) become pure imaginary and take the form

$$J_2 = \begin{pmatrix} 0 & -i/2 \\ i/2 & 0 \end{pmatrix}, \qquad K_3 = \begin{pmatrix} i/2 & 0 \\ 0 & -i/2 \end{pmatrix}, \qquad K_1 = \begin{pmatrix} 0 & i/2 \\ i/2 & 0 \end{pmatrix}. \tag{8.4}$$

The above generators satisfy the commutation relations:

$$[K_1, K_3] = iJ_2, \qquad [K_3, J_2] = -iK_1, \qquad [K_1, J_2] = iK_3 \tag{8.5}$$

and the matrix W becomes real.

The group of real W matrices generated by the above three imaginary matrices is called $SL(2, r)$ or alternatively, the two-dimensional symplectic group which is commonly called $Sp(2)$. Let M represent any group element obtained by the generators of $Sp(2)$. Then each element of $Sp(2)$ satisfy the condition:

$$\tilde{M}JM = J, \tag{8.6}$$

where

$$J = \begin{pmatrix} 0 & 1 \\ -1 & 0 \end{pmatrix}. \tag{8.7}$$

Here $\tilde{M}$ is the transpose of M. This is another definition of the group $Sp(2)$.

Because this group consists only of real two-by-two transformation matrices, it is possible to study this group using two-dimensional geometry, which can be sketched easily on a piece of paper. In fact, K_1, K_3, and J_2 generate the following transformation matrices in the two-dimensional xy-plane:

$$B(0,\xi) = \exp(-i\xi K_3) = \begin{pmatrix} e^{\xi/2} & 0 \\ 0 & e^{-\xi/2} \end{pmatrix}, \tag{8.8}$$

$$B\left(\frac{\pi}{2},\eta\right) = \exp(-i\eta K_1) = \begin{pmatrix} \cosh\frac{\eta}{2} & \sinh\frac{\eta}{2} \\ \sinh\frac{\eta}{2} & \cosh\frac{\eta}{2} \end{pmatrix}, \tag{8.9}$$

$$R(\theta) = \exp(-i\theta J_2) = \begin{pmatrix} \cos\frac{\theta}{2} & -\sin\frac{\theta}{2} \\ \sin\frac{\theta}{2} & \cos\frac{\theta}{2} \end{pmatrix}. \tag{8.10}$$

The R matrix in the above expression represents a rotation by angle θ. $B(0,\xi)$ is the elongation/contraction along the x/y-axes respectively. $B(\frac{\pi}{2},\eta)$ is the elongation/contraction along the $45°$-direction. The elongation matrix along the $(\theta/2)$-direction is

$$B(\theta,\eta) = R(\theta)B(0,\eta)R(-\theta)$$

$$= \begin{pmatrix} \cosh\frac{\eta}{2} + \left(\sinh\frac{\eta}{2}\right)\cos\theta & \left(\sinh\frac{\eta}{2}\right)\sin\theta \\ \left(\sinh\frac{\eta}{2}\right)\sin\theta & \cosh\frac{\eta}{2} + \left(\sinh\frac{\eta}{2}\right)\cos\theta \end{pmatrix}. \tag{8.11}$$

With this point in mind, let us start with a circle of unit radius centered around the origin in the Cartesian coordinate system with the coordinate variables x^* and y^*, as is illustrated in Fig. 8.1 . The equation for this circle is

$$(x^*)^2 + (y^*)^2 = 1. \tag{8.12}$$

This equation is invariant under rotation of the x^*y^*-coordinate system. The area of this circle is π.

We can next consider an ellipse, whose equation in the xy-coordinate system is

$$e^{-\eta}\left[x\cos\frac{\theta}{2} - y\sin\frac{\theta}{2}\right]^2 - e^{\eta}\left[x\sin\frac{\theta}{2} + y\cos\frac{\theta}{2}\right]^2 = 1. \tag{8.13}$$

If η is positive, the major and minor axes of the ellipse are $e^{\eta/2}$ and $e^{-\eta/2}$ respectively. The major axis is along the $(\theta/2)$-direction. The area of this ellipse is π and remains invariant as we change the values of η and θ. This is illustrated also in Fig. 8.1.

It is then not difficult to imagine that we can obtain the ellipse of Eq. (8.13) from the circle of Eq. (8.12) by making a coordinate transformation from the x^*y^* to the xy system. Since the area of the ellipse is conserved, we can consider the following two *area-conserving* coordinate transformations:

(a) We can rotate the coordinate system without changing the area.
(b) We can perform the following scale transformation without changing the area element on the xy-plane:

$$x \rightarrow x\exp(\eta/2), \qquad y \rightarrow y\exp(-\eta/2). \tag{8.14}$$

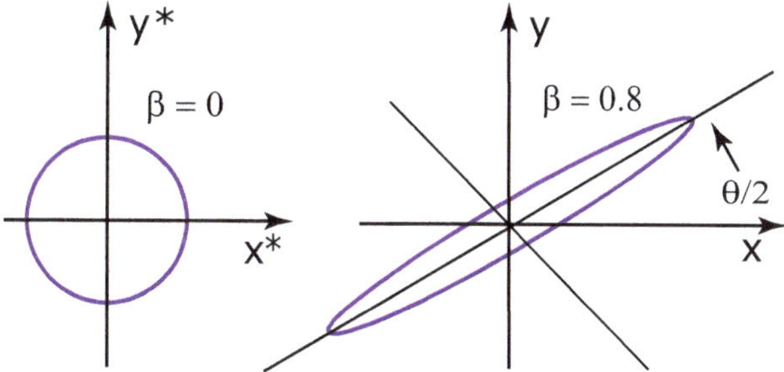

Fig. 8.1: Symplectic transformation of the circle in the x^*y^*-coordinate system to the ellipse in the xy-coordinate system. The area of the ellipse remains the same throughout the transformation process. In general, this elliptic deformation is preceded by a rotation in the x^*y^*-coordinate system [49]. Here, β is the velocity parameter v/c.

Option (a) is well known. The scale transformation of Eq. (8.14) is also very easy to visualize. However, not everybody is familiar with the combined effect of rotations and scale transformations. This is what the $Sp(2)$ group is about.

In fact, the coordinate transformation from the x^*y^* system for the circle of Eq. (8.12) to the xy system for the ellipse of Eq. (8.13) is achieved through the transformation matrix given in Eq. (8.11).

In order to see the properties of successive scale transformations, let us consider the $x'y'$ coordinate system which is related to the xy system through the transformation

$$x' = x \, \exp(\eta/2), \qquad y' = y \, \exp(-\eta/2).\tag{8.15}$$

Then the ellipse of Eq. (8.13) is transformed into a new ellipse with the same area whose equation is

$$e^{-\eta'}\,[x' \cos(\theta'/2) - y' \sin(\theta'/2)]^2 +$$

$$+ e^{\eta'}\,[x' \sin(\theta'/2) + y' \cos(\theta'/2)]^2 = 1,\tag{8.16}$$

where [50, 40]

$$\cosh\eta' = (\cosh\eta)\cosh\xi + (\sinh\eta)(\sinh\xi)\cos\theta,\tag{8.17}$$

and

$$\tan\theta' = \frac{(\sin\theta)[\sinh\xi + \tanh\eta(\cosh\xi - 1)\cos\theta]}{(\sinh\xi)\cos\theta + \tanh\eta[1 + (\cosh\xi - 1)\cos^2\theta]}.\tag{8.18}$$

The above result shows that the $x'y'$-coordinate system is obtained from the x^*y^* system through the transformation matrix $B(\theta', \eta')$.

It should be noted that, $B(\theta', \eta')$ is not a simple product of $B(\theta, \eta)$ and $B(0, \xi)$. An explicit calculation shows that

$$B(\theta, \eta)B(0, \xi) = B(\theta', \eta')R(\omega), \tag{8.19}$$

where the angle ω is determined from θ, ξ and η from the requirement that the $B(\theta', \eta')$ matrix be symmetric as indicated in Eq. (8.11) [51]. From this calculation we have

$$\tan\left(\frac{\omega}{2}\right) = \frac{\sin\theta}{\cos\theta + \coth(\xi/2)\coth(\eta/2)}. \tag{8.20}$$

The existence of this rotation matrix in Eq. (8.19) does not affect the geometry of circles and ellipses because the $R(\omega)$ matrix in Eq. (8.19) does not change the circle in the x^*y^*-coordinate system. This means that the two-parameter scale transformation matrices $B(\theta, \eta)$ alone cannot form a group, but they have to be supplemented by the rotation matrix $R(\omega)$. Indeed, the $Sp(2)$ group consists of both the B and R matrices, and is therefore a three-parameter group.

8.1.1 Sp(2) and O(2, 1)

If we use (t, x, y) to specify the coordinates in this $(2+1)$ space, then $O(2, 1)$ consists of Lorentz transformations along the x- and y-directions and of the rotation on the xy-plane around the axis perpendicular to this plane. The generators of this group are from Eq. (2.27) of Chap. 2:

$$K_1 = \begin{pmatrix} 0 & 0 & i & 0 \\ 0 & 0 & 0 & 0 \\ i & 0 & 0 & 0 \\ 0 & 0 & 0 & 0 \end{pmatrix}, \quad K_2 = \begin{pmatrix} 0 & 0 & 0 & i \\ 0 & 0 & 0 & 0 \\ 0 & 0 & 0 & 0 \\ i & 0 & 0 & 0 \end{pmatrix}, \quad \text{and} \quad J_3 = \begin{pmatrix} 0 & 0 & 0 & 0 \\ 0 & 0 & 0 & 0 \\ 0 & 0 & 0 & -i \\ 0 & 0 & i & 0 \end{pmatrix}. \tag{8.21}$$

They satisfy the following commutation relations:

$$[K_1, K_2] = -iJ_3, \qquad [K_2, J_3] = iK_1, \qquad [K_1, J_3] = -iK_2. \tag{8.22}$$

These commutation relations are like those for the generators of the $Sp(2)$ group given in Eq. (8.5). This is why we say that the $Sp(2)$ and $O(2, 1)$ groups are locally isomorphic to each other.

In order to study the content of this isomorphism, let us write the generators of the $O(2, 1)$ group in $SL(2, c)$ as we did in Eq. (8.2):

$$J_3 = \frac{1}{2}\sigma_3, \qquad K_1 = \frac{i}{2}\sigma_1, \qquad K_2 = \frac{i}{2}\sigma_2. \tag{8.23}$$

Let us then write the transformation matrices generated by the above matrices:

$$B(0, \xi) = \exp[-i\xi K_1] = \begin{pmatrix} \cosh \xi & \sinh \xi \\ \sinh \xi & \cosh \xi \end{pmatrix}, \tag{8.24}$$

$$B(\theta = \pi/2, \eta) = \exp[-i\eta K_2] = \begin{pmatrix} \cosh \eta & \sinh \eta \\ \sinh \eta & \cosh \eta \end{pmatrix}, \tag{8.25}$$

$$R(\theta) = \exp[-i\theta J_3] = \begin{pmatrix} \cos \theta & -\sin \theta \\ \sin \theta & \cos \theta \end{pmatrix}. \tag{8.26}$$

$B(0, \xi)$ and $B(\theta = \pi/2, \eta)$ are the Lorentz transformation or boost matrices along the x- and y-directions, respectively. The $R(\theta)$ matrix performs a rotation around the axis perpendicular to the xy-plane. The boost along the direction which makes an angle θ with the x-axis is therefore

$$B(\theta, \eta) = R(\theta)B(0, \eta)R(-\theta). \tag{8.27}$$

This relation is very similar to Eq. (8.11) for the $Sp(2)$ group. The above matrix is symmetric, and its determinant is unity.

We can next study successive Lorentz transformations. Let us consider a boost along the x-direction followed by another boost along θ-direction. Then we should end up with a boost in a new direction preceded by a rotation:

$$B(\theta, \eta)B(0, \xi) = B(\theta', \eta')R(\omega). \tag{8.28}$$

The problem then is to determine the parameters θ', η', and ω. An interesting point is that the algebras required to determine parameters are identical to those given in Eqs. (8.16) and (8.19) for the case of the $Sp(2)$ group. This is precisely the content of the isomorphism between the two groups.

8.1.2 Wigner, Bargmann, and Iwasawa Decompositions of Sp(2)

The unimodular $Sp(2)$ matrix M can also be written as [51, 52]

$$M = R_1(\theta_1)B(-2\chi)R_2(\theta_2) \tag{8.29}$$

where

$$B(-2\chi) = \begin{pmatrix} \cosh \chi & -\sinh \chi \\ -\sinh \chi & \cosh \chi \end{pmatrix} \quad \text{and} \quad R_i(\theta_i) = \begin{pmatrix} \cos(\theta_i/2) & -\sin(\theta_i/2) \\ \sin(\theta_i/2) & \cos(\theta_i/2) \end{pmatrix} \tag{8.30}$$

which is known as the Bargmann decomposition [30]. The evolution of the Bargmann decomposition is shown in Fig. 8.2.

Here, the purpose is to decompose the original three-parameter matrix into three one-parameter matrices. We can rewrite the Bargmann decomposition of Eq. (8.29) as

$$M = R_1(\theta_1)B(-2\chi)R_2(\theta_2) = L(\delta)\,[R(\theta)B(-2\chi)R(\theta)]\,L^{-1}(\delta) \tag{8.31}$$

where

$$R_1 = LR, \qquad R_2 = RL^{-1}. \tag{8.32}$$

The rotation matrices are

$$R(\theta) = \begin{pmatrix} \cos\theta & -\sin\theta \\ \sin\theta & \cos\theta \end{pmatrix}, \qquad L(\delta) = \begin{pmatrix} \cos\delta & -\sin\delta \\ \sin\delta & \cos\delta \end{pmatrix} \tag{8.33}$$

with

$$\theta = \frac{\theta_1 + \theta_2}{2} \quad \text{and} \quad \delta = \frac{\theta_1 - \theta_2}{2}. \tag{8.34}$$

There are many strategies to keep the original four-momentum of a relativistic particle invariant.

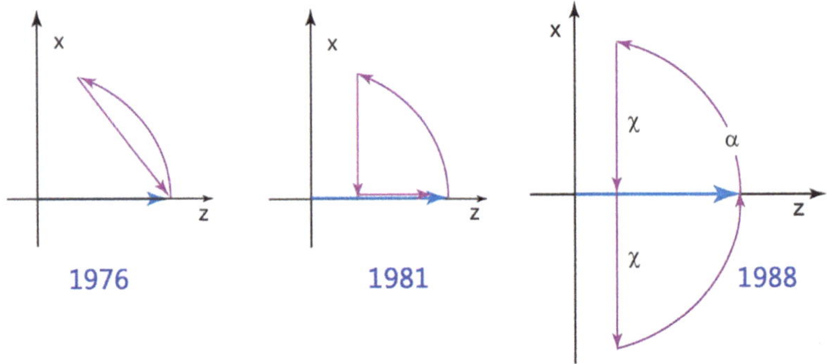

Fig. 8.2: Evolution of the Wigner loop. In 1976 [53], Kupersztych considered a rotation followed by a boost whose net result will leave the momentum invariant. In 1981 [54], Han and Kim considered the same problem with simpler forms for boost matrices. In 1988, Han and Kim [55] constructed the Lorentz kinematics corresponding to the Bargmann decomposition [30] consisting of one boost matrix sandwiched by two rotation matrices. In the present case, the two rotation matrices are identical [56].

For the purpose of decomposing $Sp(2)$ within the framework of little groups, we go over the momentum-preserving transformations of Eq. (3.24) from Sect. 3.2 of Chap 3 in conjunction with the Wigner transformation matrices W_i applicable to four-vector matrices. With this preparation, we consider two particular transformations that produce identical results, and therefore can be equated as

$$B(\eta)W_iB^{-1}(\eta) = R(\alpha)B(-2\chi)R(\alpha) \tag{8.35}$$

where W_i are provided in Table 3.2, with each i representing the case for massive, imaginary-mass, and massless particles. The right hand side of this equation is the core, RBR, of the Bargmann decomposition of Eq. (8.31), with both $\sin\alpha$ and $\sinh\chi$ chosen to be positive. The kinematics involved in here is depicted in Fig. 8.3. The above equation can explicitly be written as

$$
\begin{aligned}
&\begin{pmatrix} \cos(\theta/2) & -e^{\eta}\sin(\theta/2) \\ e^{-\eta}\sin(\theta/2) & \cos(\theta/2) \end{pmatrix} \\
&\begin{pmatrix} \cosh(\lambda/2) & e^{\eta}\sinh(\lambda/2) \\ e^{-\eta}\sinh(\lambda/2) & \cosh(\lambda/2) \end{pmatrix} \\
&\begin{pmatrix} 1 & -e^{\eta}\gamma \\ 0 & 1 \end{pmatrix} \\
&= \begin{pmatrix} (\cosh\chi)\cos\alpha & -(\cosh\chi)\sin\alpha - \sinh\chi \\ (\cosh\chi)\sin\alpha - \sinh\chi & (\cosh\chi)\cos\alpha \end{pmatrix}.
\end{aligned}
\tag{8.36}
$$

We have the following relations between the parameters:

(i) If the particle is massive, the off-diagonal elements should have opposite signs, and

$$
\cos(\theta/2) = (\cos\alpha)\cosh\chi \quad \text{and} \quad e^{2\eta} = \frac{(\sin\alpha)\cosh\chi + \sinh\chi}{(\sin\alpha)\cosh\chi - \sinh\chi},
\tag{8.37}
$$

with $(\sin\alpha)\cosh\chi > \sinh\chi$.

(ii) If the particle mass is imaginary, the off-diagonal elements have the same sign, and

$$
\cosh(\lambda/2) = (\cosh\chi)\cos\alpha \quad \text{and} \quad e^{2\eta} = \frac{(\cosh\chi)\sin\alpha + \sinh\chi}{-(\cosh\chi)\sin\alpha + \sinh\chi}.
\tag{8.38}
$$

(iii) If the particle is massless, the lower off-diagonal element should vanish, and

$$
\sinh\chi - (\sin\alpha)\cosh\chi = 0.
\tag{8.39}
$$

Thus, the diagonal element becomes $(\cos\alpha)\cosh\chi = 1$, and the non-vanishing off-diagonal element becomes $2\sinh\chi = \gamma e^{\eta}$. Finally, RBR matrix can be written as

$$
\begin{pmatrix} 1 & -2\sinh\chi \\ 0 & 1 \end{pmatrix}.
\tag{8.40}
$$

This special case is known as the Iwasawa decomposition [57].

The decompositions described in this section will prove to be advantageous in Chaps. 14 and 15, where we discuss the applications of the Poincaré and the Lorentz group in optical sciences.

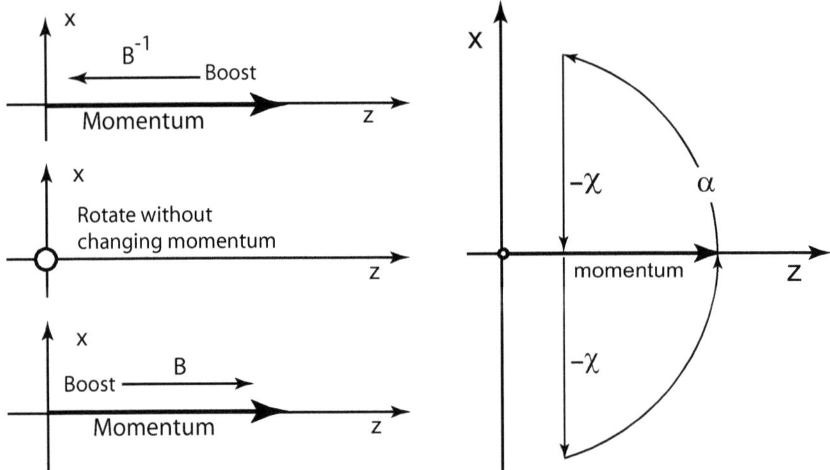

Fig. 8.3: Wigner decomposition (left) and Bargmann decomposition (right). Momentum-preserving transformations are illustrated in these diagrams. The Wigner transformation first brings the momentum of a massive particle to the rest frame, then the massive particle is rotated. The rotation process, in the rest frame, does not effect the momentum. The particle is then boosted back so that the particle is in the initial state. On the other hand, the Bargmann decomposition first rotates the momentum, then boosts it, and then rotates the momentum back to the original form [52].

8.1.3 Sp(2) Transformations in Euclidean Space and Ellipsoids

While the $Sp(2)$ transformations are area-conserving transformations, the proper $O(2, 1)$ transformation, whose determinant is unity, preserves the volume in the three-dimensional Euclidean space spanned by x, y, and t. In order to illustrate this point, let us start with a sphere of unit radius whose center is at the origin in the Lorentz frame whose coordinate variables are x^*, y^*, and t^*. The equation for this sphere takes the form

$$(x^*)^2 + (y^*)^2 + (t^*)^2 = 1 \,. \tag{8.41}$$

If we perform a Lorentz transformation along the x-direction:

$$x = x^* \cosh \eta + t^* \sinh \eta \,, \qquad t = x^* \sinh \eta + t^* \cosh \eta \,, \tag{8.42}$$

the sphere of Eq. (8.41) becomes an ellipsoid whose equation is

$$\frac{1}{2}\left\{ e^{-2\eta}(x+t)^2 - e^{2\eta}(x-t)^2 \right\} + y^2 = 1 \,. \tag{8.43}$$

If we perform the same Lorentz transformation along the θ-direction, the resulting ellipsoid becomes

$$\frac{1}{2}\left\{e^{-2\eta}(x\cos\theta + y\sin\theta + t)^2 + e^{2\eta}(x\cos\theta - y\sin\theta - t)^2\right\}$$

$$+ (x\sin\theta - y\cos\theta)^2 = 1 . \tag{8.44}$$

If we make another boost along the x-direction:

$$x' = x\cosh\xi + t\sinh\xi, \qquad t' = x\sinh\xi + t\cosh\xi , \tag{8.45}$$

the resulting equation will be that of another ellipsoid with the parameters θ' and η'. The algebra of determining these parameters is identical to that for the $Sp(2)$ ellipse given in Eq. (8.16).

While we are accustomed to associate Lorentz transformations with hyperbolas and hyperbolic surfaces, it is interesting to note that ellipsoids can also play an effective role in illustrating the $O(2, 1)$ group. This method of using ellipsoids is convenient in understanding and interpreting experimental data on Lorentz-deformed relativistic hadrons.

8.2 Finite-Dimensional Representations of O(2, 1)

We studied in Sect. 8.1 a simple geometrical picture of how transformations of $O(2, 1)$ and $SL(2, r)$ are different from the rotation group. This geometrical picture may be helpful in practical calculations. However, what we did in Sect. 8.1 is not necessarily the orthodox group theoretical approach of constructing irreducible representations. The standard approach is to construct first representations and representation spaces which are diagonal in the Casimir operators. The generators of $O(2, 1)$ in differential form are

$$K_1 = i\left(t\frac{\partial}{\partial x} + x\frac{\partial}{\partial t}\right) , \qquad K_2 = i\left(t\frac{\partial}{\partial y} + y\frac{\partial}{\partial t}\right) \tag{8.46}$$

and

$$J_3 = -i\left(x\frac{\partial}{\partial y} - y\frac{\partial}{\partial x}\right) . \tag{8.47}$$

The Casimir operator in this case is

$$C = J_3^2 - K_1^2 - K_2^2 , \tag{8.48}$$

which commutes with the three generators of the group. We are interested in constructing representations and representation spaces which are diagonal in this operator. In addition to this Casimir operator, we can choose the representation to be diagonal in J_3 which commutes with C. We can then draw some immediate conclusions from what we know from the three-dimensional rotation group.

We can obtain the above Casimir operator from that of $O(3)$ by replacing z by the imaginary time: (it). The representation space for the three-dimensional rotation group consists of tensors constructed from the coordinate variables x, y, z. For instance, the rotationally invariant combination in $O(3)$ is $(x^2 + y^2 + z^2)$. The eigenvalue of J^2 for this form is 0. For $\ell = 1$, there are three possible combinations. They are $(x \pm iy)$ and z. For $\ell = 2$ or higher, the representation space consists of the spherical harmonics: $Y_\ell^m(\theta, \phi)$. Let us see how these can be translated into the language of the $O(2, 1)$ group.

When $t > r = (x^2 + y^2)^{1/2}$, we use the parametrization:

$$\tan \phi = \frac{y}{x}, \qquad \cosh \eta = \frac{t}{\rho}, \qquad \sinh \eta = \frac{r}{\rho}, \tag{8.49}$$

where

$$\rho = (|t^2 - r^2|)^{1/2}.$$

In terms of the ϕ and η variables, the Casimir operator C takes the form

$$C = \frac{\partial}{\partial(\cosh \eta)} \left(\sinh^2 \eta \frac{\partial}{\partial(\cosh \eta)} \right) + \left(\frac{1}{\sinh \eta} \right)^2 \left(\frac{\partial}{\partial \phi} \right)^2. \tag{8.50}$$

When $t < r$, we have to use the parametrization:

$$\tan \phi = \frac{y}{x}, \qquad \sinh \eta = \frac{t}{\rho}, \qquad \cosh \eta = \frac{r}{\rho}. \tag{8.51}$$

The C operator becomes

$$C = \frac{\partial}{\partial(\sinh \eta)} \left(\cosh^2 \eta \frac{\partial}{\partial(\sinh \eta)} \right) - \left(\frac{1}{\cosh \eta} \right)^2 \left(\frac{\partial}{\partial \phi} \right)^2. \tag{8.52}$$

We can obtain the above forms by replacing z with (it) in the $(J)^2$ operator of $O(3)$:

$$(J)^2 = -\frac{\partial}{\partial(\cos \theta)} \left(\sin^2 \theta \frac{\partial}{\partial(\cos \theta)} \right) - \left(\frac{1}{\sin \theta} \right)^2 \left(\frac{\partial}{\partial \phi} \right)^2, \tag{8.53}$$

and the Casimir operator C of Eq. (8.50) by substituting $i\eta$ for θ in Eq. (8.53). C of Eq. (8.52) is obtained from the same substitution after a $90°$ rotation $\theta \rightarrow \theta + 90°$. For this reason, solving the eigenvalue equation:

$$C\psi = \lambda\psi \tag{8.54}$$

is just like constructing the spherical harmonics when the eigenvalue λ is $\ell(\ell + 1)$, where ℓ takes integer values. The resulting solutions of the above differential equation are

$$\psi_\ell^m(\eta, \phi) = (\sinh \eta)^m \left[\left(\frac{d}{d(\cosh \eta)} \right)^m P_\ell(\cosh \eta) \right] e^{im\phi}, \quad \text{for } |t| > r,$$

$$\psi_\ell^m(\eta, \phi) = (\cosh \eta)^m \left[\left(\frac{d}{d(\sinh \eta)} \right)^m P_\ell(\sinh \eta) \right] e^{im\phi}, \quad \text{for } |t| < r, \quad (8.55)$$

with

$$m = -\ell, -\ell + 1, \dots \ell - 1, \ell. \tag{8.56}$$

If we multiply these solutions by ρ which is an invariant quantity, they become finite homogeneous polynomials of the ℓth degree in x, y, and t.

The above solutions are written in the form convenient to represent rotations around the origin on the xy-plane. However, if we perform a Lorentz boost along the x-direction, the resulting polynomial will be a linear transformation. If $\ell = 0$, the representation space is one-dimensional, and the polynomial is 1 which remains invariant under transformations. If $\ell = 1$, the polynomials are $(x \pm iy)$ for $m = \pm 1$ respectively, and t for $m = 0$.

If $\ell = 2$, the polynomials will be $(x \pm iy)^2$, $t(x \pm iy)$, and $(x^2 + y^2 + 2t^2)$ for $m = \pm 2$, $m = \pm 1$ and $m = 0$ respectively. Each polynomial satisfies the eigenvalue equation with the eigenvalue of $\ell(\ell + 1) = 6$. It is very easy to rotate these functions around the origin on the xy-plane. If we boost the function $(x + iy)^2$ along the x-direction through the transformation resulting in the replacement of x and t by x' and t' respectively, where

$$x' = x \cosh \alpha - t \sinh \alpha, \quad y' = y, \quad t' = -x \sinh \alpha + t \cosh \alpha, \tag{8.57}$$

then

$$
\begin{aligned}
(x' + iy')^2 = {} & \left(\frac{1 + \cosh \alpha}{2} \right)^2 (x + iy)^2 \\
& + \left(\frac{\cosh \alpha - 1}{2} \right)^2 (x - iy)^2 \\
& - (\sinh \alpha)(\cosh \alpha + 1)(x + iy)t \\
& - (\sinh \alpha)(\cosh \alpha - 1)(x - iy)t \\
& + \frac{1}{2}(\sinh \alpha)^2 (x^2 + y^2 + 2t^2).
\end{aligned}
\tag{8.58}
$$

This expression is still an eigenstate of the Casimir operator C. However, it is a linear combination of different m states. The number of terms is not more than $(2\ell + 1)$. For this reason, what we did above constitutes a finite-dimensional representation.

The procedure for obtaining these finite dimensional representations is identical to that for the three-dimensional rotational group. Let us start with the usual differential forms of J_1, J_2, and J_3 given in Sect. 2.1 of Chap. 2. If we replace z by (it), then J_3 remains invariant, while J_1 and J_2 become:

$$J_1 \to iK_2 \quad \text{and} \quad J_2 \to iK_1 \, . \tag{8.59}$$

Thus the Casmir operator C of Eq. (8.48) is the same as that for the rotation group. The only difference is to replace the usual θ variable by $i\eta$. However, this replacement leads to the following non-trivial problems.

First, where can we use this representation in physics? Second, in the rotation group, the orthogonality relation is stated as

$$\int_1^1 d(\cos\theta) \int_0^{2\pi} (Y_{\ell'}^{m'}(\theta,\phi))^* Y_\ell^m(\theta,\phi)\delta_{\ell\ell'}\delta_{mm'} \, d\phi \, . \tag{8.60}$$

In the case of $SU(1,1)$, the corresponding integral is

$$\int_0^\infty \sinh\eta d\eta \int_0^{2\pi} (Y_{\ell'}^{m'}(\eta,\phi))^* Y_\ell^m(\eta,\phi) \, d\phi \tag{8.61}$$

where η ranges from 0 to infinity. The integral in this case does not converge for integer values of ℓ. For this reason, the generators K_1 and K_2 cannot be Hermitian, and the representation is not unitary. Is it possible to construct unitary representations as in the case of $E(2)$?

In spite of these questions, the study of the finite-dimensional representations is a preliminary step for further developments.

8.3 Complex Angular Momentum

We have seen in Sect. 8.2 that the study of $O(2,1)$ can start from the Legendre functions. The study of $O(3)$ is only a restricted case in which the appropriate boundary conditions are imposed on the solutions of the Legendre's differential equation. The boundary condition in this case is that it be analytic everywhere in the complex $\cos\theta$-plane. While $O(3)$ is concerned with the region in which $|\cos\theta| \leq 1$, we are in $O(2,1)$ interested in the region where $|\cos\theta|$ is greater than one. The Legendre function does not have to be analytic at $\cos\theta = \pm 1$. Indeed, the study of $O(2,1)$ is that of the Legendre function for $|\cos\theta| \geq 1$ [58, 59, 60].

The Legendre function is discussed extensively in standard textbooks [61, 62]. The fundamental definition of the Legendre function is given in an integral form:

$$P_\ell(z) = \frac{1}{2\pi i} \left(\frac{1}{2}\right)^\ell \int_C \frac{(t^2-1)^\ell}{(t-z)^{\ell+1}} dt, \tag{8.62}$$

where the counterclockwise contour C encloses $t = 1$ and $t = z$, but not $t = -1$ in the complex t-plane.

The above integral representation satisfies the Legendre differential equation:

$$\frac{d}{dt}\left((1 - z^2)\frac{d}{dz}P_\ell(z)\right) + \ell(\ell + 1)P_\ell(z) = 0 \tag{8.63}$$

and the symmetry property:

$$P_\ell(z) = P_{-\ell-1}(z). \tag{8.64}$$

For integer values of ℓ, the integral representation of Eq. (8.62) becomes a Legendre polynomial. Furthermore, the asymptotic behavior of the Legendre function for large values of z becomes

$$P_\ell(z) \rightarrow (z)^{Re(\ell)}. \tag{8.65}$$

In order that the representation be unitary, the solution of the differential equation given in Eq. (8.63) should be an infinite series in $\cosh\eta$ and $\sinh\eta$ to make the function vanish fast enough for increasing values of η. Thus ℓ cannot take integer values. Because of the symmetry of Eq. (8.64), the minimum value of $Re(\ell)$ expected from the Legendre function is -1/2. For this value, the orthogonality integral of Eq. (8.61) converges. Indeed, the functions ψ_ℓ^m derived from the Legendre function with $\ell = -1/2$ through the differentiation in Eq. (8.55) can form a representation space for infinite-dimensional unitary representations.

One important physical application of complex angular momentum is the Watson-Sommerfeld transformation of the non-relativistic scattering amplitude [27], and its application to the Regge-pole model for high-energy scattering [28]. Let us start with the scattering amplitude of the form:

$$f(k,\theta) = \frac{1}{2ik}\sum_{\ell=0}^{\infty}(2\ell + 1)(e^{2i\delta_\ell} - 1)P_\ell(\cos\theta), \tag{8.66}$$

where k and θ are the momentum and scattering angle respectively. We can write the above amplitude as

$$f(k,\theta) = \sum_{\ell}(2\ell + 1)\alpha(\ell, k)P_\ell(\cos\theta). \tag{8.67}$$

If the potential is of the Yukawa type, $V(r) \sim (1/r)e^{-\mu r}$, $\alpha(\ell, k)$ is analytic in the complex ℓ-plane in the region $Re(\ell) > -1/2$, except for the poles in the first quadrant [27]. $P_\ell(\cos\theta)$ is an entire function of ℓ. With this point in mind, we can write Eq. (8.67) as the contour integral:

$$f(k,\theta) = \frac{1}{2\pi i}\int_C \frac{e^{2\pi\ell}}{\sin\pi\ell}\alpha(\ell, k)P_\ell(\cos\alpha)d\ell. \tag{8.68}$$

The contour C can now be opened up and pushed to the vertical line along $\ell = 1/2+is$, where $-\infty < s < \infty$, provided that the integral vanishes along the infinite circular section in the region right of $Re(\ell) = 1/2$ [63]. During this process, the contour encloses the poles in the first quadrant. The result is

$$f(k,\theta) = \frac{1}{2\pi} \int_{-\infty}^{\infty} \frac{e^{i\pi(-1/2+is)}}{\sin\pi(-1/2+is)} \alpha(-1/2+is,k) P_{-1/2+is}(\cos\theta)ds$$

$$+ \sum_i \frac{e^{i\pi\alpha_i}}{\sin\pi\alpha_i} \alpha'(\alpha_i,k) P(\cos\theta), \tag{8.69}$$

where the second term in the above expression consists of contributions from the Regge poles. $\alpha'(\alpha,k)$ is the derivative of $\alpha(\alpha,k)$ with respect to α. The transformation from Eq. (8.68) to Eq. (8.69) is called the Watson-Sommerfeld transformation [64]. Let α_0 be the one with the largest real part. Then the scattering amplitude, in the limit of large $\cos\theta$, becomes

$$f(k,\theta) \rightarrow (\cos\theta)^{\mathrm{Re}[\alpha_0(k)]}. \tag{8.70}$$

The position of the Regge pole depends on k.

This asymptotic behavior inspired many physicists in the 1960's, as it was able to predict the high-energy scattering cross section in the cross channel in which $\cos\theta$ is proportional to the total energy. This approach is commonly known as the Regge-pole model. One of the building blocks of this model is the Chew-Frautchi plot [28] in which the hadron masses are plotted against their intrinsic angular momentum. As we noted in Chap. 5, the intrinsic angular momentum is one of the Casimir operators of the Poincaré group. The Chew-Frautchi plot eventually became the hadron mass spectrum in the harmonic oscillator model, and it is still not uncommon to call the harmonic oscillator spring constant the *slope of the Regge trajectory*. As as noted before, the study of the Legendre function is the study of $O(2,1)$ [58, 59, 60, 59, 65]. Indeed, the Regge-pole model has provided one of the stepping stones to the physicists' slow process of understanding the Poincaré group.

8.4 Unitary Representations of SU(1, 1)

Since $SU(1,1)$ is one of the simplest non-compact groups, study in this area is extensive [25]. Without a clear physical motivation, we are not going through all the papers published in the literature on this subject. We are not even going to scratch the surface of the field. The purpose of this section is to illustrate the scope of this subject by discussing the distribution of eigenvalues for infinite-dimensional unitary representations. Since the original work of Bargmann [30], there have been many reformulations of the same problem. Among the many excellent papers on this subject, we choose to follow closely the lucid presentation of Holman and Biedenharn [39] who emphasize the connection between $SU(2)$ and $SU(1,1)$.

Since $SU(1,1)$ is non-compact and contains no invariant subgroups, the dimension of its unitary representation is necessarily infinite. As before, the generators [25] of the group are K_1, K_2, and J_3. This time, these operators are Hermitian. The Casimir operator C takes the form of Eq. (8.48). As in the case of Sect. 8.3, we shall consider representations diagonal in J_3. The Hilbert space consists of normalized eigenstates of J_3:

$$J_3 \, |m\rangle = m \, |m\rangle \, , \tag{8.71}$$

where m has to be an integer or half integer. The $SU(1, 1)$ transformation on $|m\rangle$ will result in an infinite sum:

$$\sum_{n=-\infty}^{\infty} A_{nm} \, |n\rangle \, , \quad \text{with} \quad \sum_{k} (A_{nm})^* (A_{kn}) = \delta_{mn} \, . \tag{8.72}$$

unlike the case considered in Sect. 8.3 where the sum was finite. We are considering here a unitary infinite-dimensional representation.

For the rotation group, we know how eigenvalues of J_3 are distributed for a given value of the total angular momentum. The purpose of this section is to see how they are distributed in the case of $SU(1, 1)$. As in the case of the rotation group, let us consider the step-up and step-down operators:

$$K_\pm = K_1 \pm i K_2 \, . \tag{8.73}$$

then from the commutation relations $[J_3, K_\pm] = \pm J_3$,

$$J_3 K_- = (K_- - 1) J_3 \, , \qquad J_3 K_+ = (K_+ + 1) J_3 \, . \tag{8.74}$$

The application of $K_\pm$ on $|m\rangle$ results in $|m \pm 1\rangle$. Furthermore, from the definition of C in Eq. (8.48),

$$K_+ K_- = -C + J_3(J_3 - 1) \, , \qquad K_- K_+ = -C + J_3(J_3 + 1) \, , \tag{8.75}$$

so that

$$K_- K_+ - K_+ K_- = 2 J_3 \, . \tag{8.76}$$

Since K_1 and K_2 are Hermitian operators:

$$(K_\pm)^\dagger = K_\mp \, . \tag{8.77}$$

$K_+ K_-$ and $K_- K_+$ must be positive and Hermitian:

$$\langle m | \, K_\pm K_\mp \, |m\rangle = |K_\mp \, |m\rangle|^2 \geq 0 \, . \tag{8.78}$$

Hence the eigenvalues of

$$-C + J_3(J_3 - 1) \, , \qquad -C + J_3(J_3 + 1) \, , \tag{8.79}$$

must be positive or zero. By applying the above operators on $|m\rangle$, we derive the following inequalities:

$$-C + m(m - 1) \geq 0 \, , \qquad -C + m(m + 1) \geq 0 \, , \tag{8.80}$$

for sufficiently large $|m|$. The letter C is used also for the eigenvalue of the Casimir operator. For a given positive m, we can construct a chain of states:

$$|m\rangle, |m+1\rangle, |m+2\rangle, \ldots, \tag{8.81}$$

by applying repeatedly K_+ on $|m\rangle$. From Eq. (8.76), we obtain

$$|K_+|m\rangle|^2 = \langle m|(2J_3 + K_+K_-)|m\rangle$$

$$= 2m + |K_-|m\rangle|^2. \tag{8.82}$$

The chain does not terminate. If we apply the step-down operator K_- repeatedly to $|m\rangle$ to obtain

$$|m\rangle, |m-1\rangle, |m-2\rangle, \ldots, \tag{8.83}$$

one of the two alternatives occurs:

(a)$^+$ The chain never terminates.
(b)$^+$ The chain terminates.

Let us first consider the case (a)$^+$. The positivity conditions of Eq. (8.80) in terms of the eigenvalues m imply that

$$C < 0 \qquad \text{if m is an integer}, \tag{8.84}$$

$$C < -1/4 \quad \text{if m is a half} - \text{integer}. \tag{8.85}$$

These conditions can be restated as

$$C < -1/4 \qquad \text{for both integer and half} - \text{integer values of m}, \tag{8.86}$$

$$-1/4 < C < 0 \qquad \text{for integer values of m}. \tag{8.87}$$

If we write C as

$$C = \ell(\ell+1), \tag{8.88}$$

then the condition of Eq. (8.85) is translated into

$$\ell = -1/2 + is \tag{8.89}$$

where s is a real parameter. We have already seen one example for this case in Sect. 8.3.

In the case of (b)$^+$, let $|k\rangle$ be the last non-vanishing state in the descending chain. This means

$$K_-|k\rangle = 0, \tag{8.90}$$

and therefore

$$K_+K_-|k\rangle = [-C + J_3(J_3 - 1)] = 0,$$

or

$$C = k(k - 1).$$
(8.91)

Eq. (8.82) applied to $|k\rangle$ leads to $k = |K_+|k\rangle|^2/2$, which means that k must be positive or zero. The case $k = 0$ occurs only in the identity representation for which

$$K_+ |0\rangle = K_- |0\rangle = 0, \qquad \text{and} \qquad C = 0.$$
(8.92)

Let us next consider the case where the starting value of m is negative. Then the infinite chain of states

$$|m\rangle, |m - 1\rangle, \ldots$$
(8.93)

is obtained by the repeated application of K_- on $|m\rangle$. Then, as we apply the step-up operator K_+ repeatedly to $|m\rangle$ we again meet one of the two alternatives (a) and (b) above, which in this case we shall label $(a)^-$ and $(b)^-$:

$(a)^-$ the chain never terminates;
$(b)^-$ the chain terminates.

For case $(a)^-$ the positivity conditions on Eq. (8.80) again imply one of the conditions of Eqs. (8.86) and (8.87). In the case of $(b)^-$, let $|k\rangle$ be the last non-vanishing state in the ascending chain, i.e.,

$$K_+ |k\rangle = 0,$$
(8.94)

and

$$K_- K_+ |k\rangle = [-C + J_3(J_3 + 1)] = 0,$$

or

$$C = k(k + 1).$$
(8.95)

Now, Eq. (8.82) shows that k must be negative and non-zero. Again, the case $k = 0$ occurs only in the identity representation described by Eq. (8.92).

We have obtained a system of classification of the representations of $SU(1, 1)$ in terms of the cases $(a)^\pm$ and $(b)^\pm$. We shall call the representations obtained under cases $(a)^\pm$, for which all integral (or half-integral) states $|m\rangle$ exist, the continuous series of irreducible representations. We shall call those obtained under the cases $(b)^+$ and $(b)^-$, respectively, the positive and negative discrete series representations. In addition to these, of course, there exists the unique, one-dimensional identity representation.

Let us summarize what we did above for both positive and negative values of m. In the case of continuous series,

$$-1/4 < C < 0 \qquad \text{for integer values of m,}$$
(8.96)

$$C < -1/4 \quad \text{for both integer and half} - \text{integer values of m.}$$
(8.97)

For discrete series, there is a positive lower bound for positive values of m, while the upper bound on negative values of m is negative. The absolute value of m has to be equal to or greater than a positive number $|k|$ which may be integer or half integer. C in this case is $|k|(|k| - 1)$.

With this preparation, the reader can go to the the papers of Holman and Biedenharn [39, 66] and the paper of Basu and Wolf [48], where the subject was completely and thoroughly discussed. As for the mathematical methods, Bargmann in his original paper [30] uses the multiplier representation based on the conformal representation of $SL(2, c)$. This method has been explained thoroughly in Miller's book on Lie theory [44].

8.5 Exercises and Problems

Exercise 1. Special relativity deals with two coordinate frames which move with uniform velocity relative to each other. Therefore, for a particle in a circular orbit, we have to bring the particle back to the rest frame first, and then boost it by the same amount in a different direction. The net result is a pure boost which changes the initial velocity to the final velocity preceded by a rotation. The time rate of change of this rotation angle is called the Thomas precession. Calculate the Thomas precession from the expression given in Eq. (8.20).

In the calculation of Sect. 8.1, since the magnitude of velocity is the same, we can let

$$\eta = \xi = \frac{1}{2} \ln \frac{1 - \beta}{1 + \beta} , \tag{8.98}$$

where β is the magnitude of velocity. The final angle is 0, and we can let the initial angle be

$$\theta = \pi - \delta\theta .$$

Thus

$$\sin \theta = \delta\theta , \tag{8.99}$$

for small $\delta\theta$. In this limit, Eq. (8.20) becomes

$$\delta\alpha = 2 \left(\sinh \frac{\eta}{2} \right)^2 \delta\theta . \tag{8.100}$$

This means that the time rate of the Thomas precession is proportional to the time rate of the rotation of the velocity:

$$\omega_T = (\cosh \eta - 1) \frac{d\theta}{dt} . \tag{8.101}$$

For further group theoretical discussion of the Thomas precession, see Bargmann et al. [67], Gilmore [68] and Salingaros [69, 70].

Exercise 2. If a nucleus consists of two nucleons, it is very easy to study its structure. However, if there are many nucleons in the nucleus, then the effective force on a single nucleon is not necessarily isotropic. In fact, the surface of the nucleus can be described by the equation:

$$(x^2 + y^2)e^b + z^2 e^{-2b} = R^2 .$$ (8.102)

This is an equation for a spheroid whose volume remains unchanged as the parameter b takes different values. This equation indeed describes a nucleus undergoing pancake/football-like deformations. Is this deformation describable in terms of $Sp(2)$?

The deformation of Eq. (8.102) is achieved by an elongation/deformation on the xz-plane followed by the same deformation on the yz-plane. The elongation/deformation is achieved by the matrix

$$\begin{pmatrix} e^{b/2} & 0 \\ 0 & e^{-b/2} \end{pmatrix} .$$ (8.103)

If we allow full $Sp(2)$ transformations on both planes, the nucleus can also perform rotations around the y-axis and x-axis respectively. In this case, the transformation becomes the subset of volume-preserving transformation of $SL(3, r)$. This is called the collective model of nuclei [71]. For applications of the symplectic group in nuclear deformation, see the papers by Moshinsky and Winternitz [11], Arickx et al. [72], Rosensteel and Rowe [73], Deenen and Quesne [74].

Exercise 3. Let us next discuss another example of elliptic deformation. The concept of phase space plays an important role in statistical mechanics, semiclassical problems, and foundations of quantum mechanics. In these areas of physics, the Wigner function defined as

$$W(x, p) = \left(\frac{1}{\pi}\right) \int \psi^*(x + x')\psi(x - x')e^{ipx'} dx'$$ (8.104)

serves many useful purposes [75]. How does the symplectic group serve useful purposes for studying the Wigner function?

The simplest way to illustrate the Wigner function is to evaluate the integral of Eq. (8.104) for the one-dimensional harmonic oscillator, whose Hamiltonian takes the form

$$H = \frac{1}{2}mp^2 + \frac{1}{2}m\omega^2 x^2 .$$ (8.105)

The solutions of the Schrödinger equation with this Hamiltonian are well known. If we use the generating function of the Hermite polynomials, it is easy to carry out the integration in Eq. (8.104). In terms of the simplified variables:

$$u = (m\omega)^{1/2}x , \quad v = (1/m\omega)^{1/2}p ,$$ (8.106)

and

$$r^2 = 2(u^2 + v^2), \tag{8.107}$$

the Wigner function for the nth excited-state harmonic oscillator takes the form

$$W_n(u, v) = \left(\frac{n!}{\pi}\right)[\exp(-r^2/2)]\sum_{k=0}^{n}\frac{(-1)^{n-k}}{r^{2k}}\left[(n-k)!(k!)^2\right]. \tag{8.108}$$

The above expression is rotationally invariant in the uv-plane, and the function takes the same value on a circle centered at the origin.

From Eq. (8.106), it is clear that the story on the xp-plane is a symplectic deformation of the uv-plane. If we combine this with the rotation in the xp-plane, the net effect is the general linear canonical transformation. The basic symmetry of this Wigner function in phase space is therefore that of the $Sp(2)$ group.

The study of the Wigner function is so extensive in the literature that it is by now an almost independent branch of physics. The papers on this subject include those of Wigner [75, 76], Moyal [77], Bartlett and Moyal [78], Nix [79], Klauder and Sudarshan [80], Balasz and G. G. Zipel [81], Davies and Davies [82], O'Connel and Rajagopal [83], Carruthers and Zachariasen [84], and the references quoted in these papers.

Problem 1. Show that $W_n(r)$ of Eq. (8.108) satisfies the two-dimensional harmonic oscillator Schrödinger equation:

$$-\frac{1}{2r}\left(\frac{d}{dr}\right)^2(rW_n(r)) + \frac{1}{2}r^2W_n(r) = (2n+1)W_n(r). \tag{8.109}$$

W_n can naturally be normalized as

$$2\pi\int_0^{\infty}W_n(r)W_m(r)r\,dr = \delta_{nm}.$$

Show then that the following relation is satisfied for two different Wigner functions:

$$\frac{1}{2\pi}\left|\int\psi^*(x)\phi(x)dx\right|^2 = \int W_{\phi}(x, p)W_{\psi}(x, p)dx\,dp. \tag{8.110}$$

Problem 2. In Sect. 3.4 of Chap. 3, we discussed the irreducible representations of $SU(2)$ and the spinors and their symmetric combinations, particularly those for total spin 1/2, 1, and 3/2. Discuss the transformation properties of those spinors with all possible combinations of dotted and undotted α and β.

Problem 3. Show that it is possible to construct the ψ_{ℓ}^m of Eq. (8.55) starting from $m = 0$ for an arbitrary complex value of ℓ.

Problem 4. Prove that the Legendre functions $P_{-1/2+is}(\cosh\eta)$ form an orthonormal set with respect to the integral of Eq. (8.61).

Problem 5. Show that the asymptotic behavior of the Legendre function for large values of z is given by Eq. (8.65). See Morse and Feschbach [62].

Problem 6. Show that the integral representation given in Eq. (8.62) satisfies the Legendre differential equation of Eq. (8.63) for complex ℓ.

Problem 7. In Sect. A.4 of Appendix A, we noted that the transformation of Eq. (A.1) corresponds to the conformal mapping:

$$w = \frac{az + b}{b^* z + c^*} \; .$$

Find the little group which leaves i invariant: $w = z = i$. See Lang [47].

Problem 8. Explain why the Chew-Frautchi plot [28] is basically the eigenvalue distribution of the three-dimensional harmonic oscillator.

References

1. G. Feinberg, Possibility of Faster-Than-Light Particles, Physical Review **159**(5), 1089–1105 (1967). DOI 10.1103/PhysRev.159.1089. URL https://link.aps.org/doi/10.1103/PhysRev.159.1089

2. C. Schwartz, Some improvements in the theory of faster-than-light particles, Physical Review D **25**(2), 356–364 (1982). DOI 10.1103/PhysRevD.25.356. URL https://link.aps.org/doi/10.1103/PhysRevD.25.356

3. C. Schwartz, A Consistent Theory of Tachyons with Interesting Physics for Neutrinos, Symmetry **14**(6), 1172 (2022). DOI 10.3390/sym14061172. URL https://www.mdpi.com/2073-8994/14/6/1172

4. H. Weyl, *The classical groups: their invariants and representations*, 2nd edn. Princeton landmarks in mathematics and physics Mathematics (Princeton University Press, Princeton, N.J. USA, 1997). ISBN 978-0-691-07923-3;978-0-691-05756-9. (Originally published 1947.)

5. V.I. Arnold, *Mathematical methods of classical mechanics*, 2nd edn. No. 60 in Graduate texts in mathematics (Springer, New York, NY, USA, 1997). ISBN 978-0-387-96890-2. (This is an English translation by K. Vogtmann and A. Weinstein of the Russian original edition: Maremaricheskie Merody Klassicheskoi Mekhaniki, Nauka, Moscow, 1974.)

6. R. Abraham, E. Marsden, J, *Foundations of mechanics*, 2nd edn. (AMS Chelsea Pub./American Mathematical Society, Providence, RI, USA, 2008). ISBN 978-0-8218-4438-0. (Originally published 1978; OCLC: ocn191847156.)

7. R. Littlejohn, The semiclassical evolution of wave packets, Physics Reports **138**(4-5), 193–291 (1986). DOI 10.1016/0370-1573(86)90103-1. URL http://linkinghub.elsevier.com/retrieve/pii/0370157386901031

8. C.W. Horton, L.E. Reichl, V.G. Szebehely (eds.), *Long-time prediction in dynamics*. No. v. 2 in Nonequilibrium problems in the physical sciences and biology (John Wiley and Sons, New York, NY, 1983). ISBN 9780471864479

9. P.J. Morrison, J.M. Greene, Noncanonical Hamiltonian Density Formulation of Hydrodynamics and Ideal Magnetohydrodynamics, Physical Review Letters **45**(10), 790–794 (1980). DOI 10.1103/PhysRevLett.45.790. URL https://link.aps.org/doi/10.1103/PhysRevLett.45.790

10. C. Quesne, M. Moshinsky, Canonical Transformations and Matrix Elements, Journal of Mathematical Physics **12**(8), 1780–1783 (1971). DOI 10.1063/1.1665806. URL http://aip.scitation.org/doi/10.1063/1.1665806

11. M. Moshinsky, P. Winternitz, Quadratic Hamiltonians in phase space and their eigenstates, Journal of Mathematical Physics **21**(7), 1667–1682 (1980). DOI 10.1063/1.524615. URL http://aip.scitation.org/doi/10.1063/1.524615

12. F.J. Dyson, Statistical Theory of the Energy Levels of Complex Systems. I, Journal of Mathematical Physics **3**(1), 140–156 (1962). DOI 10.1063/1.1703773. URL http://aip.scitation.org/doi/10.1063/1.1703773

13. L.D. Faddeev, Symplectic structure and quantization of the Einstein gravitation theory, in *Actes Du Congres International Des Mathematiciens*, vol. 3 (Gathier-Villars, Paris, France, 1971), vol. 3, 35–39. (Nice, France, 1-10 September, 1970.)

14. A. Ashtekar, L. Bombelli, R. Koul, Phase space formulation of general relativity without a 3+1 splitting, in *The Physics of Phase Space Nonlinear Dynamics and Chaos Geometric Quantization, and Wigner Function*, vol. 278, ed. by Y.S. Kim, W.W. Zachary (Springer-Verlag, Berlin, Heidelberg, Germany, 1987), 356–359. DOI 10.1007/3-540-17894-5_378. ISBN 9783540178941,9783540479017. URL http://link.springer.com/10.1007/3-540-17894-5_378

15. C. Arias, R. Bonezzi, P. Sundell, Bosonic higher spin gravity in any dimension with dynamical two-form, Journal of High Energy Physics **2019**(3), 1 (2019). DOI 10.1007/JHEP03(2019)001. URL https://link.springer.com/10.1007/JHEP03(2019)001

16. N. Uzun, Reduced phase space optics for general relativity: symplectic ray bundle transfer, Classical and Quantum Gravity **37**(4), 045,002 (2020). DOI 10.1088/1361-6382/ab60b5. URL https://iopscience.iop.org/article/10.1088/1361-6382/ab60b5

17. H. Bacry, M. Cadilhac, Metaplectic group and Fourier optics, Phys. Rev. A **23**(5), 2533–2536 (1981). DOI 10.1103/PhysRevA.23.2533. URL https://link.aps.org/doi/10.1103/PhysRevA.23.2533

18. R. Simon, N. Mukunda, Iwasawa decomposition in first-order optics: universal treatment of shape-invariant propagation for coherent and partially coherent beams, J. Opt. Soc. Am. A **15**(8), 2146–2155 (1998). DOI 10.1364/JOSAA.15.002146. URL https://opg.optica.org/josaa/abstract.cfm?URI=josaa-15-8-2146

19. S. Başkal, Y. Kim, M. Noz, *Mathematical Devices for Optical Sciences*. (IOP Publishing, Bristol, UK, 2019). ISBN 978-0-7503-1612-5. URL https://dx.doi.org/10.1088/2053-2563/aafe78. (OCLC: 1034620988.)

20. G. Adesso, F. Illuminati, Entanglement in continuous–variable systems: recent advances and current perspectives, Journal of Physics A: Mathematical and Theoretical **40**(28), 7821–7880 (2007). DOI 10.1088/1751-8113/40/28/S01. URL http://stacks.iop.org/1751-8121/40/i=28/a=S01?key=crossref.653edcf3591291dd403ab925eac18fee

21. C. Weedbrook, S. Pirandola, R. García-Patrón, N.J. Cerf, T.C. Ralph, J.H. Shapiro, S. Lloyd, Gaussian quantum information, Reviews of Modern Physics **84**(2), 621–669 (2012). DOI 10.1103/RevModPhys.84.621. URL https://link.aps.org/doi/10.1103/RevModPhys.84.621

22. G. Ferrini, J. Roslund, F. Arzani, C. Fabre, N. Treps, Direct approach to Gaussian measurement based quantum computation, Physical Review A **94**(6), 062,332 (2016). DOI 10.1103/PhysRevA.94.062332. URL https://link.aps.org/doi/10.1103/PhysRevA.94.062332

23. G. Adesso, S. Ragy, A.R. Lee, Continuous Variable Quantum Information: Gaussian States and Beyond, Open Systems & Information Dynamics **21**(01n02), 1440,001 (2014). DOI 10.1142/S1230161214400010. URL https://www.worldscientific.com/doi/abs/10.1142/S1230161214400010

24. Y.S. Kim, M.E. Noz, Symplectic formulation of relativistic quantum mechanics, Journal of Mathematical Physics **22**(10), 2289–2293 (1981). DOI 10.1063/1.524763. URL http://aip.scitation.org/doi/10.1063/1.524763

25. Y.S. Kim, M.E. Noz, Dirac's light-cone coordinate system, American Journal of Physics **50**(8), 721–724 (1982). DOI 10.1119/1.12737. URL http://aapt.scitation.org/doi/10.1119/1.12737

26. Y.S. Kim, M.E. Noz, *Phase space picture of quantum mechanics: group theoretical approach*. No. 40 in Lecture notes in physics series (World Scientific Publishing Co., Singapore; Hackensack, NJ, USA, 1991). ISBN 978-981-02-0360-3,978-981-02-0361-0. URL https://doi.org/10.1142/1197

27. T. Regge, Introduction to complex orbital momenta, Il Nuovo Cimento **14**(5), 951–976 (1959). DOI 10.1007/BF02728177. URL http://link.springer.com/10.1007/BF02728177

28. G.F. Chew, S -Matrix Theory of Strong Interactions without Elementary Particles, Reviews of Modern Physics **34**(3), 394–401 (1962). DOI 10.1103/RevModPhys.34.394. URL https://link.aps.org/doi/10.1103/RevModPhys.34.394

29. P.D.B. Collins, E.J. Squires, *Regge Poles in Particle Physics*, *Springer Tracts in Modern Physics*, vol. 45 (Springer, Berlin, Heidelberg, 1968). DOI 10.1007/BFb0045799. ISBN 9783540043393~9783540359371. URL http://link.springer.com/10.1007/BFb0045799

30. V. Bargmann, Irreducible Unitary Representations of the Lorentz Group, The Annals of Mathematics **48**(3), 568–640 (1947). DOI 10.2307/1969129. URL http://www.jstor.org/stable/1969129?origin=crossref

31. A.O. Barut, C. Fronsdal, On non-compact groups. II. Representations of the 2+1 Lorentz group, Proceedings of the Royal Society of London. Series A. Mathematical and Physical Sciences **287**(1411), 532–548 (1965). DOI 10.1098/rspa.1965.0195. URL https://royalsocietypublishing.org/doi/10.1098/rspa.1965.0195

32. L. Pukánszky, On the Kronecker Products of Irreducible Representations of the 2×2 Real Unimodular Group. I, Transactions of the American Mathematical Society **100**(1), 116–152 (1961). DOI 10.2307/1993356. URL https://www.jstor.org/stable/1993356?origin=crossref

33. L. Pukánszky, The Plancherel formula for the universal covering group of SL(R, 2), Mathematische Annalen **156**(2), 96–143 (1964). DOI 10.1007/BF01359927. URL http://link.springer.com/10.1007/BF01359927

34. I.M. Gel'fand, G.E. Shilov, *Generalized functions*, 2016th edn. (American Mathematical Society : AMS Chelsea Publishing, Providence, RI, USA, 2016). ISBN 9781470426583. (Originally publised in Russian, 2nd Edition 1959, published in English 1963.)

35. P.J. Sally Jr., Uniformly bounded representations of the universal covering group of SL(2,R), Bulletin of the American Mathematical Society **72**(2), 269–274 (1966). DOI 10.1090/S0002-9904-1966-11489-1. URL http://www.ams.org/journal-getitem?pii=S0002-9904-1966-11489-1

36. N. Mukunda, Unitary Representations of the Group O (2, 1) in an O (1, 1) Basis, Journal of Mathematical Physics **8**(11), 2210–2220 (1967). DOI 10.1063/1.1705143. URL http://aip.scitation.org/doi/10.1063/1.1705143

37. N. Mukunda, Unitary Representations of the Lorentz Groups: Reduction of the Supplementary Series under a Noncompact Subgroup, Journal of Mathematical Physics **9**(3), 417–431 (1968). DOI 10.1063/1.1664595. URL http://aip.scitation.org/doi/10.1063/1.1664595

38. N. Mukunda, Matrices of finite Lorentz transformations in a noncompact basis. III. Completeness relation for O (2, 1), Journal of Mathematical Physics **14**(12), 2005–2010 (1973). DOI 10.1063/1.1666282. URL http://aip.scitation.org/doi/10.1063/1.1666282

39. W.J. Holman, L.C. Biedenharn, Complex angular momenta and the groups SU(1, 1) and SU(2), Annals of Physics **39**(1), 1–42 (1966). DOI 10.1016/0003-4916(66)90135-7. URL https://linkinghub.elsevier.com/retrieve/pii/0003491666901357

40. D. Han, E.E. Hardekopf, Y.S. Kim, Thomas precession and squeezed states of light, Physical Review A **39**(3), 1269–1276 (1989). DOI 10.1103/PhysRevA.39.1269. URL https://link.aps.org/doi/10.1103/PhysRevA.39.1269

41. N.J. Vilenkin, *Special functions and the theory of group representations*. No. 22 in Translations of mathematical monographs (AMS - American Mathematical Soc, Providence, RI, USA, 1968). ISBN 978-0-8218-1572-4

42. J.G. Kuriyan, N. Mukunda, E.C.G. Sudarshan, Master Analytic Representation: Reduction of O (2, 1) in an O (1, 1) Basis, Journal of Mathematical Physics **9**(12), 2100–2108 (1968). DOI 10.1063/1.1664551. URL http://aip.scitation.org/doi/10.1063/1.1664551

43. J.D. Talman, *Special Functions: a Group Theoretical Approach Based on Lectures By Eugene P Wigner* (W.A. Benjamin, Inc. (Benjamin Cummings), San Francisco, CA, USA, 1968). URL https://books.google.se/books?id=m-IbtAEACAAJ. 260 pp., lccn=68-54038.

44. W.J. Miller, *Lie Theory and Special Functions* (Academic Press Elsevier Science, New York, NY, USA, 1968). ISBN 978-0-08-095551-3. URL https://books.google.se/books/about/Lie_Theory_and_Special_Functions.html?id=XJ9RiMLiOigC&redir_esc=y

45. K.B. Wolf, Canonical transforms. I. Complex linear transforms, Journal of Mathematical Physics **15**(8), 1295–1301 (1974). DOI 10.1063/1.1666811. URL http://aip.scitation.org/doi/10.1063/1.1666811

46. E.G. Kalnins, W. Miller, Lie theory and separation of variables. 4. The groups SO (2,1) and SO (3), Journal of Mathematical Physics **15**(8), 1263–1274 (1974). DOI 10.1063/1.1666805. URL http://aip.scitation.org/doi/10.1063/1.1666805

47. S. Lang, *SL2(R)*, *Graduate Texts in Mathematics*, vol. 105 (Springer New York, New York, NY, 1985). DOI 10.1007/978-1-4612-5142-2. ISBN 9781461295815978146125142 2. URL http://link.springer.com/10.1007/978-1-4612-5142-2

48. D. Basu, K.B. Wolf, The unitary irreducible representations of SL(2, R) in all subgroup reductions, Journal of Mathematical Physics **23**(2), 189–205 (1982). DOI 10.1063/1.525337. URL http://aip.scitation.org/doi/10.1063/1.525337

49. Y.S. Kim, M.E. Noz, Illustrative examples of the symplectic group, American Journal of Physics **51**(4), 368–375 (1983). DOI 10.1119/1.13252. URL http://aapt.scitation.org/doi/10.1119/1.13252

50. D. Han, Y.S. Kim, M.E. Noz, Linear canonical transformations of coherent and squeezed states in the Wigner phase space, Physical Review A **37**(3), 807–814 (1988). DOI 10.1103/PhysRevA.37.807. URL https://link.aps.org/doi/10.1103/PhysRevA.37.807

51. S. Başkal, Y.S. Kim, Rotations associated with Lorentz boosts, Journal of Physics A: Mathematical and General **38**(29), 6545–6556 (2005). DOI 10.1088/0305-4470/38/29/009. URL http://stacks.iop.org/0305-4470/38/i=29/a=009?key=crossref.e99c952d1a8a20610d56358b253d04b4

52. S. Başkal, Y.S. Kim, M.E. Noz, Wigner's Space–Time Symmetries Based on the Two-by-Two Matrices of the Damped Harmonic Oscillators and the Poincaré Sphere, Symmetry **6**(3), 473–515 (2014). DOI 10.3390/sym6030473. URL http://www.mdpi.com/2073-8994/6/3/473/

53. J. Kupersztych, Is there a link between gauge invariance, relativistic invariance and electron spin?, Il Nuovo Cimento B Series 11 **31**(1), 1–11 (1976). DOI 10.1007/BF02730313. URL http://link.springer.com/10.1007/BF02730313

54. D. Han, Y.S. Kim, Little group for photons and gauge transformations, American Journal of Physics **49**(4), 348–351 (1981). DOI 10.1119/1.12509. URL http://aapt.scitation.org/doi/10.1119/1.12509

55. D. Han, Y.S. Kim, Special relativity and interferometers, Physical Review A **37**(11), 4494–4496 (1988). DOI 10.1103/PhysRevA.37.4494. URL https://link.aps.org/doi/10.1103/PhysRevA.37.4494

56. S. Başkal, Y.S. Kim, M.E. Noz, Loop Representation of Wigner's Little Groups, Symmetry **9**(7), 97–118 (2017). DOI 10.3390/sym9070097. URL http://www.mdpi.com/2073-8994/9/7/97

57. K. Iwasawa, On Some Types of Topological Groups, The Annals of Mathematics **50**(3), 507–558 (1949). DOI 10.2307/1969548. URL http://www.jstor.org/stable/1969548?origin=crossref

58. L. Sertorio, M. Toller, Complex angular momentum and three-dimensional Lorentz group, Il Nuovo Cimento **33**(2), 413–433 (1964). DOI 10.1007/BF02750202. URL http://link.springer.com/10.1007/BF02750202

59. M. Toller, On the group-theoretical approach to complex angular momentum and signature, Il Nuovo Cimento A **54**(2), 295–361 (1968). DOI 10.1007/BF02743789. URL http://link.springer.com/10.1007/BF02743789

60. A. Sciarrino, M. Toller, Decomposition of the Unitary Irreducible Representations of the Group SL (2 C) Restricted to the Subgroup SU (1, 1), Journal of Mathematical Physics **8**(6), 1252–1265 (1967). DOI 10.1063/1.1705341. URL http://aip.scitation.org/doi/10.1063/1.1705341

61. E.T. Whittaker, G.N. Watson, *A course of modern analysis: an introduction to the general theory of infinite processes and of analytic functions, with an account of the principal transcendental functions*, fifth reprinted edn. (Cambridge University Press, Cambridge, UK, 2013). ISBN 9780511608759. (Originally published 1902 ((Whitaker only), 2nd Ed. 1915, 3rd Ed. 1920, 4th Ed. 1927, online 2013; OCLC: 857769437.)

62. P.M. Morse, H. Feshbach, *Methods of theoretical physics*, reprinted edn. International series in pure and applied physics (McGraw-Hill Book Company, Boston, MA, USA, 1999). ISBN 978-0-07-043316-8. (Originally published 1953.)

63. L. Brown, D.I. Fivel, B.W. Lee, R.F. Sawyer, Fredholm method in potential scattering and applications to complex angular momentum, Annals of Physics **23**(2), 187–220 (1963). DOI 10.1016/0003-4916(63)90192-1. URL https://linkinghub.elsevier.com/retrieve/pii/0003491663901921

64. A. Sommerfeld, E.G. Straus, *Partial differential equations in physics*. No. 6 in Lectures in theoretical physics (Academic Press, London New York, 1964). ISBN 97801265465699780126546583

65. G. Soliani, M. Toller, A semigroup contained in SL (2 R) and complex angular momentum, Il Nuovo Cimento A **15**(3), 430–448 (1973). DOI 10.1007/BF02734681. URL http://link.springer.com/10.1007/BF02734681

66. W.J. Holman, L.C. Biedenharn, A general study of the Wigner coefficients of SU(1, 1), Annals of Physics **47**(2), 205–231 (1968). DOI 10.1016/0003-4916(68)90287-X. URL https://linkinghub.elsevier.com/retrieve/pii/000349166890287X

67. V. Bargmann, L. Michel, V.L. Telegdi, Precession of the Polarization of Particles Moving in a Homogeneous Electromagnetic Field, Phys. Rev. Lett. **2**(10), 435–436 (1959). DOI 10.1103/PhysRevLett.2.435. URL https://link.aps.org/doi/10.1103/PhysRevLett.2.435. Publisher: American Physical Society

68. R. Gilmore, *Lie groups, Lie algebras, and some of their applications* (Dover Publications, Mineola, NY, USA, 2005). ISBN 978-0-486-44529-8. (Originally published: 1974, John Wiley and Sons, New York, NY, USA.)

69. N. Salingaros, Particle in an external electromagnetic field, Physical Review D **28**(10), 2473–2476 (1983). DOI 10.1103/PhysRevD.28.2473. URL https://link.aps.org/doi/10.1103/PhysRevD.28.2473

70. N. Salingaros, Relativistic motion of a charged particle, the Lorentz group, and the Thomas precession, Journal of Mathematical Physics **25**(3), 706–716 (1984). DOI 10.1063/1.526179. URL http://aip.scitation.org/doi/10.1063/1.526179

71. A. Bohr, B.R. Mottelson, *Nuclear Structure: (In 2 Volumes)Volume I: Single-Particle Motion Volume II: Nuclear Deformations* (World Scientific Publishing Company, Singapore; Hachensack, NJ, USA, 1998). DOI 10.1142/3530. ISBN 9789810231972~9789812386601. URL https://www.worldscientific.com/worldscibooks/10.1142/3530

72. F. Arickx, J. Broeckhove, E. Deumens, The Sp(2,R) model applied to ^{8}Be, Nuclear Physics A **318**(3), 269–286 (1979). DOI 10.1016/0375-9474(79)90648-1. URL https://www.sciencedirect.com/science/article/pii/0375947479906481

73. D. Rowe, G. Rosensteel, On the algebraic formulation of collective models. II. Collective and intrinsic submanifolds, Annals of Physics **126**(1), 198–233 (1980). DOI 10.1016/0003-4916(80)90380-2. URL https://linkinghub.elsevier.com/retrieve/pii/0003491680903802

74. J. Deenen, C. Quesne, Dynamical group of microscopic collective states. I. One-dimensional case, Journal of Mathematical Physics **23**(5), 878–889 (1982). DOI 10.1063/1.525440. URL http://aip.scitation.org/doi/10.1063/1.525440

75. E.P. Wigner, On the Quantum Correction For Thermodynamic Equilibrium, Physical Review **40**(5), 749–759 (1932). DOI 10.1103/PhysRev.40.749. URL https://link.aps.org/doi/10.1103/PhysRev.40.749

76. E.P. Wigner, Quantum mechanical distribution functions revisited, in *Perspectives in Quantum Theory: Essays in Honor of Alfred Landé*, ed. by W. Yourgrau, A. van der Merwe (MIT Press, Cambridge, MA USA, 1971), 25–36. ISBN 978-0262240147

77. J.E. Moyal, Quantum mechanics as a statistical theory, Mathematical Proceedings of the Cambridge Philosophical Society **45**(01), 99–124 (1949). DOI 10.1017/S0305004100000487. URL http://www.journals.cambridge.org/abstract_S0305004100000487

78. M.S. Bartlett, J.E. Moyal, The exact transition probabilities of quantum-mechanical oscillators calculated by the phase-space method, Mathematical Proceedings of the Cambridge Philosophical Society **45**(04), 545 (1949). DOI 10.1017/S030500410002524X. URL http://www.journals.cambridge.org/abstract_S030500410002524X

79. J.R. Nix, Further studies in the liquid-drop theory on nuclear fission, Nuclear Physics A **130**(2), 241–292 (1969). DOI 10.1016/0375-9474(69)90730-1. URL http://linkinghub.elsevier.com/retrieve/pii/0375947469907301

80. J.R. Klauder, E.C.G. Sudarshan, *Fundamentals of quantum optics* (Dover Publications, Mineola, N.Y, 2006). ISBN 978-0-486-45008-7. (Originally published: New York, NY, USA : W.A. Benjamin, 1968.)

81. N. Balazs, G. Zipfel, Quantum oscillations in the semiclassical fermion μ-space density, Annals of Physics **77**(1-2), 139–156 (1973). DOI 10.1016/0003-4916(73)90412-0. URL https://linkinghub.elsevier.com/retrieve/pii/0003491673904120

82. R.W. Davies, K.T.R. Davies, On the Wigner distribution function for an oscillator, Annals of Physics **89**(2), 261–273 (1975). DOI 10.1016/0003-4916(75)90182-7. URL http://linkinghub.elsevier.com/retrieve/pii/0003491675901827

83. R.F. O'Connell, A.K. Rajagopal, New Interpretation of the Scalar Product in Hilbert Space, Physical Review Letters **48**(8), 525–526 (1982). DOI 10.1103/PhysRevLett.48.525. URL https://link.aps.org/doi/10.1103/PhysRevLett.48.525

84. P. Carruthers, F. Zachariasen, Quantum collision theory with phase-space distributions, Reviews of Modern Physics **55**(1), 245–285 (1983). DOI 10.1103/RevModPhys.55.245. URL https://link.aps.org/doi/10.1103/RevModPhys.55.245

Chapter 9
The Lorentz Group

Abstract We elaborate on an alternative approach to construct the representations of the Poincaré group to those that have been discussed in the previous chapters, which is mainly based on the construction of the little groups in a covariant manner. For this purpose, the quantities that remain invariant under Lorentz transformations are to be found along with the Casimir operators. The Casimir operators of the Lorentz group, which are quite different from those of the Poincaré group, commute with the six generators of $SL(2, c)$. Thus, all possible finite dimensional representations of $SL(2, c)$ are presented. We discuss the metric and the normalization useful for unitary representation of the Dirac spinors. The solutions of the covariant harmonic oscillator can serve both as the representations of Poincaré group and the Lorentz group depending on the coordinate system in which they are written along with their respective Casimir operators. It is pointed out that, as in $O(3)$ and $SL(2, c)$, the step-up and step-down operators may be useful in constructing representations of the Lorentz group. The hyperbolic coordinate system serves well for the Lorentz group. Traditional methods existing in the literature are also discussed.

We have so far discussed subgroups of the Lorentz group which can describe the internal space-time symmetries of relativistic particles. Transformations of these subgroups leave the four-momentum of a relativistic particle invariant. The four-momentum which remains invariant under transformations of the homogeneous Lorentz group (or simply the Lorentz group) has to vanish. Since there are no particles with vanishing four-momentum, this group does not serve any useful purpose as a little group. However, this fact alone cannot eliminate the Lorentz group from our consideration.

The construction of representations of the Poincaré group consists of two steps. The first step is to construct representations of the little group for a given four-momentum. The second step is to generalize the representation of the little group

© The Author(s), under exclusive license to Springer Nature Switzerland AG 2024
S. Başkal et al., *Theory and Applications of the Poincaré Group*, Fundamental Theories of Physics 217, https://doi.org/10.1007/978-3-031-64376-7_9

for all possible values of the four-momentum belonging to the same orbit. Lorentz transformations play the essential role during the process of orbit (discussed in Chap. 2) completion.

There are two different approaches to this orbit completion process. The first approach is to construct the representation and/or representation space of the little group in a given Lorentz frame, followed by the Lorentz generalization of the representation by performing necessary transformations on the representation space. The second approach is to construct the representation of the little group in a covariant manner. We have been using the first method throughout Chaps. 2 to 8. The purpose of this Chapter is to discuss the second method.

In order to proceed in a covariant manner, we have to know what quantities should remain Lorentz-invariant, in addition to the Casimir operators. For instance, in the rotation group, the total angular momentum is the sum of the orbital and spin angular momenta. The Casimir operator for the rotation group is (total angular momentum)^2. However, this is not the only quantity which remains invariant under rotations. The (total orbital angular momentum)^2 and the (total spin)^2 each are invariant quantities.

In the case of the Poincaré group, the six generators of spinor transformations form a Lie algebra of the Lorentz group. There are two Casimir operators for this subgroup, and they commute not only with the six generators applicable to spinors, but also with the ten generators of the Poincaré group. The purpose of this Chapter is to exploit this property.

In Sect. 9.1, we spell out the proposed procedure. Section 9.2 deals with a method of constructing all finite-dimensional representations of $SL(2, c)$. In Sect. 9.3, we discuss the metric and the normalization useful for unitary representation of the Dirac spinors.

In Sect. 9.4, we go back to the harmonic oscillator and discuss in detail when the method of Lorentz group works and when it does not. It is pointed out in Sect. 9.5 that, as in $O(3)$ and $SL(2, c)$, the step-up and step-down operators may be useful in constructing representations of the Lorentz group. We summarize in Sect. 9.6 the methods of constructing representations of the Poincaré group.

9.1 Statement of the Problem

As we stated in Chap. 2, the Poincaré group is generated by four translation generators and six generators of Lorentz transformations. We may again write these as

$$P_\mu \quad \text{and} \quad M_{\mu\nu}. \tag{9.1}$$

The operator $M_{\mu\nu}$ which generates Lorentz transformations is the sum of that applicable to the space-time coordinates and that for internal space-time or spin coordinates:

$$M_{\mu\nu} = L^*_{\mu\nu} + L_{\mu\nu}. \tag{9.2}$$

As in Chap. 5, we shall use for the particle (hadron or electron) the $L^*_{\mu\nu}$-coordinate. $L_{\mu\nu}$ is for internal degrees of freedom.

We observed in Chap. 2 the Poincaré group is a semi-direct product of the translation group generated by P_μ and the group of the Lorentz transformations generated by $M_{\mu\nu}$. The translation subgroup is an Abelian invariant subgroup, and, therefore, the above-mentioned Lorentz group is a quotient group consisting of cosets of the original Poincaré group over the translation group. The purpose of this section is to discuss *another* Lorentz group which is a subgroup of the Poincaré group.

The six $L^*_{\mu\nu}$ operators commute with $L_{\mu\nu}$, but they do not commute with P_μ. The $L_{\mu\nu}$ operators commute with both P_μ and $L^*_{\mu\nu}$. For this reason, we can regard the Poincaré group as a *direct product* of the group generated by P_μ and $L^*_{\mu\nu}$ and the Lorentz group generated by $L_{\mu\nu}$. This is precisely why we are interested in representations of the Lorentz group. After constructing a proposed representation, we should check whether the direct product should be diagonal in the Casimir operators of the Poincaré group [1]. As for the new Poincaré group generated by P_μ and $L^*_{\mu\nu}$, its representation is the Klein-Gordon equation.

The Casimir operators of the Lorentz group are

$$C_1 = \frac{1}{2} L^{\mu\nu} L_{\mu\nu} \quad \text{and} \quad C_2 = \frac{1}{2} \epsilon_{\mu\nu\alpha\beta} L^{\mu\nu} L^{\alpha\beta} . \tag{9.3}$$

These operators commute with all ten generators of the Poincaré group, but are quite different from the Casimir operators of the Poincaré group. Let us see how useful this Lorentz group is in studying representations of the original Poincaré group and in understanding internal space-time symmetry of relativistic particles.

9.2 Finite-Dimensional Representations of the Lorentz Group

By now, we are quite familiar with the fact that the Lorentz group is generated by the J_i and K_i which generate rotations and boosts respectively, and also with the commutation relations these generators satisfy. The commutation relations for these operators remain invariant under the sign change of the boost operators.

As we noted in Chap. 2, the Casimir operators for the Lorentz group are

$$C_1 = \left(J^2\right) - \left(K^2\right) , \quad C_2 = \mathbf{J} \cdot \mathbf{K} . \tag{9.4}$$

It is often convenient to define new operators:

$$F_i = \frac{1}{2}(J_i - iK_i) , \quad G_i = \frac{1}{2}(J_i + iK_i) . \tag{9.5}$$

Then these new operators satisfy the following commutation relations:

$$[F_i, G_j] = 0,$$
$$[F_i, F_j] = i\varepsilon_{ijk}F_k, \quad [G_i, G_j] = i\varepsilon_{ijk}G_k. \tag{9.6}$$

It is clear from the above commutation relations that F_i and G_i generate their own respective rotation groups. The Lorentz group can be regarded as a direct product of these two rotation groups. The Casimir operators for these groups are F^2 and G^2 respectively. Their eigenvalues are $f(f+1)$ and $g(g+1)$. The representation of the Lorentz group is $(2f+1)(2g+1)$-dimensional. In terms of F^2 and G^2, the Casimir operators of Eq. (9.4) are

$$C_1 = \frac{1}{2}(F^2 + G^2) \quad \text{and} \quad C_2 = -\frac{i}{4}(F^2 - G^2). \tag{9.7}$$

This means that the representations with eigenvalues of F^2 and G^2 give fixed values of C_1 and C_2. We can therefore label the representation by two numbers (f, g). Representations constructed in this way are finite-dimensional and non-unitary.

In addition to the Casimir operators, we can choose F_3 and G_3 to be diagonal. Since

$$J_i = F_i + G_i, \tag{9.8}$$

J_3 can be diagonal for the Lorentz group. Because the Lorentz group in this case is regarded as a direct product of the two rotation groups, the eigenvalues of J^2 is $j(j+1)$, with

$$j = (f+g), (f+g-1), ..., |f-g|. \tag{9.9}$$

J^2, F^2, and G^2 can be simultaneously diagonalized, but J^2 does not commute with F_3 or G_3.

As in the case of the three-dimensional rotation group, we should start constructing representations by making tensor products of spinors. In the $SL(2,c)$ case, as we studied in Chap. 3, the boost operators take two different signs. Indeed, this is why we are able to separate the generators into F_i and G_i.

Let us first study a Dirac particle with non-zero mass. When $J_i = \frac{1}{2}\sigma_i$, $K_i = \frac{i}{2}\sigma_i$, $F_i = \frac{1}{2}\sigma_i$, and $G_i = 0$. On the other hand, when $K_i = -\frac{i}{2}\sigma_i$, $F_i = 0$, and $G_i = \frac{1}{2}\sigma_i$. Since the boost matrices for upper and lower components of the Dirac wave functions have opposite signs, the Dirac particle is represented by the direct sum of $(f = \frac{1}{2}, g = 0)$ and $(f = 0, g = \frac{1}{2})$. Indeed the Dirac equation is a finite-dimensional non-unitary representation of the Lorentz group. This interpretation of the Dirac equation is quite consistent with the analysis given in Chap. 3.

Let us next consider four-vectors. According to Sect. 3.5 of Chap. 3 a four-vector can be constructed from a direct product of a spinor with one sign of the boost operator and another spinor with the opposite sign. Indeed, this direct product constitutes the representation $(\frac{1}{2}, \frac{1}{2})$. This is true for both massive and massless fields. Massless fields contain gauge degrees of freedom.

For the general case of (f, g), we know how to construct representations of each of the two rotation groups for higher values of f and g. The eigenvalues of J_3 will run from $(f+g)$ to $-(f+g)$. It is possible to construct a general theory starting from these observations [2]. However, we shall continue the practice of studying each case in detail as an illustrative example.

As for massless particles, in a series of papers published in 1964, Weinberg [2, 3, 4] was interested in constructing irreducible representations which are totally independent of gauge degrees of freedom and contain only the minimal number of variables. Gauge-dependent four-potentials are not included in this construction scheme. In order that the massless fields be gauge-independent, the state vectors should satisfy the conditions:

$$(F_1 - iF_2)\,|m\rangle = 0\,, \quad (G_1 - iG_2)\,|m\rangle = 0\,, \tag{9.10}$$

which are equivalent to

$$N_1^\pm\,|m\rangle = N_2^\pm\,|m\rangle = 0 \tag{9.11}$$

where the N operators defined in Sect. 4.2 of Chap. 4 are the generators of gauge transformations. These operators are the translation-like generators of the $E(2)$-like little group. As we shall see in Chap. 10, the above conditions lead to the polarization of the spin-1/2 particle. Indeed,

$$F_3\,|m\rangle = f\,|m\rangle\,, \quad G_3\,|m\rangle = -g\,|m\rangle\,, \tag{9.12}$$

where m is the eigenvalue of J_3. Since $J_3 = F_3 + G_3$,

$$m = f - g. \tag{9.13}$$

For a right-handed particle with $m = j$, the field can be one of the following representations:

$$(j, 0)\,, \quad (j + 1/2, 1/2)\,, \quad (j + 1, 1)\,, \quad ,\dots. \tag{9.14}$$

The left-handed particles can be associated with one of the representations:

$$(0, j)\,, \quad (\tfrac{1}{2}, j + \tfrac{1}{2})\,, \quad (1, j + 1)\,, \quad ,\dots. \tag{9.15}$$

According to this scheme, the simplest representation for photons is the direct sum of $(1, 0)$ and $(0, 1)$. This representation corresponds to the solutions of Maxwell's equations for plane waves. For spin-1/2 particles, they are represented by $(\tfrac{1}{2}, 0)$ and/or by $(0, \tfrac{1}{2})$ [2].

These problems will be discussed in detail in Chap. 10. As for photons, we shall discuss in detail the four-potential representation also in Chap. 10. In the above-mentioned representation of Weinberg, the photon state vectors can be constructed as a product of two spinors. The positive and negative helicity states are

$$|+\rangle = \alpha_1\alpha_2\,, \quad |-\rangle = \dot\beta_1\dot\beta_2\,, \tag{9.16}$$

respectively. Under rotations these state vectors can be identified as

$$|+\rangle = -(x + iy)\,, \quad |-\rangle = (x - iy)\,. \tag{9.17}$$

We shall study Lorentz transformation properties of the above spinor combinations in Sect. 10.6.

9.3 Pseudo-Unitary Representations for Dirac Spinors

We have seen in Sect. 9.2 and also in Sect. 3.5 of Chap. 3 that Lorentz boosts change the normalization of the $SL(2, c)$ spinors. This is a manifestation of the fact that finite-dimensional representations of the non-compact groups are not unitary, and that the Lorentz group is a non-compact group.

However, the boost generators of $SL(2, c)$ are double-valued. This property assures us that if a given boost changes the normalization by a number $\exp(\eta/2)$, there is another boost in the same representation which will change the normalization by $\exp(-\eta/2)$. Indeed, the normalization of the dotted spinor is the inverse of that of the undotted spinor. Therefore, if we define the inner product of two spinors in such a way that one normalization factor will cancel the other, then the unitarity will be maintained.

Let us see whether this is possible for Dirac particles in the Weyl representation. In this representation, the undotted and dotted spinors constitute the upper and lower components of the four-component Dirac spinor. If the inner product is defined in such a way that the upper component is multiplied by the lower component, the unitarity will be guaranteed during the process of Lorentz boost. In the Weyl representation, γ_0 takes the form (see Eq. (2.82) of Chap. 2)

$$\gamma_0 = \begin{pmatrix} 0 & I \\ I & 0 \end{pmatrix}. \tag{9.18}$$

The boost operator is

$$B(\eta) = \begin{pmatrix} e^{\eta \cdot \sigma/2} & 0 \\ 0 & e^{-\eta \cdot \sigma/2)} \end{pmatrix}. \tag{9.19}$$

This boost matrix is not unitary, but is clearly Hermitian. γ_0 is therefore Lorentz-invariant:

$$[B(\eta)]^\dagger \gamma_0 B(\eta) = \gamma_0\,. \tag{9.20}$$

Indeed, the Dirac spinors are unitary if γ_0 is placed between the two spinors when we take their inner product.

In the Dirac representation, γ_0 takes the form (see Eq. (3.139) of Chap. 3)

$$\gamma_0 = \begin{pmatrix} I & 0 \\ 0 & -I \end{pmatrix}. \tag{9.21}$$

Instead of taking the Hermitian conjugate of the spinor U, we often take the quantity $\bar{U} = U^{\dagger}\gamma_0$. Then the product of two spinors $\bar{U}V$ is known to be a Lorentz invariant quantity [5].

As we discuss in Sect. A.4 of Appendix A, the inner product with the metric of the form of Eq. (9.21) is called a pseudo-unitary representation. As was pointed out in Sect. 3.6 of Chap. 3, the Weyl representation is unitarily equivalent to the Dirac representation. Therefore, in this section, we have explained the physics of the above pseudo-unitary representation using the Weyl representation.

The pseudo-unitary representation as described above is useful only if there exist non-vanishing inner products between dotted and undotted spinors. There are non-vanishing spinor products in this pseudo-unitary space for massive Dirac particles. However, in the case of massless particles, the spin of the dotted spinor has to be opposite to that of the undotted spinor. For this reason, there is no non-zero pseudo-unitary inner product. The $SL(2, c)$ representations remain non-unitary for massless particles.

A similar explanation of the pseudo-unitary representation can be given in the Foldy-Wouthuysen picture of the Dirac equation in which the energy matrix takes a diagonal form, as was discussed in the literature [6, 7, 8, 9].

9.4 Harmonic Oscillator in the Lorentz Coordinate System

We study in this section finite-dimensional representations of the Lorentz group using covariant harmonic oscillators. In the covariant harmonic oscillator regime in which two quarks are bound together within a hadron, as was noted before, there are two space-time coordinates. One is for the hadron, and the other is the internal space-time separation between the quarks. This internal space-time coordinate acts like the spin coordinate in the present case. Thus, we use x_μ as the internal coordinate and use the notation $L_{\mu\nu}$ as

$$L_{\mu\nu} = i\left(x_\mu \frac{\partial}{\partial x_\nu} - x_\nu \frac{\partial}{\partial x_\mu}\right) \tag{9.22}$$

and

$$J_i = -i\epsilon_{ijk}x^j \frac{\partial}{\partial x_k}, \quad K_i = i\left(t\frac{\partial}{\partial x_i} - x_i\frac{\partial}{\partial t}\right). \tag{9.23}$$

In Chap. 5, we were primarily interested in the normalizable solutions of the harmonic oscillator equation whose forms depend on the four-momentum of hadrons. They constitute the representation space for the $O(3)$-like little group for massive particles. In order to obtain them, we used the coordinate system in which the t' variable is separated from the three space-like variables which span the three-dimensional Euclidean space. There are many other coordinate systems in which the oscillator equation is separable, and the coordinate system determines which quantities are diagonal [10]. We are interested in the coordinate system in which the sub-Casimir operators C_1 and C_2 of Eq. (9.4), which we would construct using the above generators, are diagonal.

If we evaluate C_2 using the explicit expression for J_i and K_i given in Eq. (9.23), this operator vanishes for the present case of spinless quarks. In order to obtain solutions diagonal in C_1, let us consider the hyperbolic coordinate system with the coordinate variables $\{x^\tau\} = \{\rho, \alpha, \theta, \phi\}$. They are related to Cartesian variables by

$$\rho = \left(|\, t^2 - r^2\, | \right)^{\frac{1}{2}} \tag{9.24}$$

where

$$r = \left(x^2 + y^2 + z^2 \right)^{\frac{1}{2}},$$

$$x = r \sin\theta \cos\phi, \qquad y = r \sin\theta \sin\phi, \qquad z = r \cos\theta,$$

$$t = \pm\rho \cosh\alpha, \qquad r = |\,\rho \sinh\alpha\,|, \qquad \text{when } |\,t\,| > r,$$

$$t = \rho \sinh\alpha, \qquad r = \rho \cosh\alpha, \qquad \text{when } |\,t\,| < r. \tag{9.25}$$

Among the above four variables, ρ is Lorentz invariant. For convenience, we shall call this coordinate system the *Lorentz coordinate system*.

In the Lorentz coordinate system, the sub-Casimir operator C_1 takes the form [11]

$$C_1 = \left(\frac{1}{\sinh\alpha} \right)^2 \frac{\partial}{\partial\alpha} \left(\sinh^2\alpha \frac{\partial}{\partial\alpha} \right) - \left(\frac{1}{\sinh\alpha} \right)^2 (\mathbf{L})^2 \tag{9.26}$$

in the time-like region. In the space-like region,

$$C_1 = \left(\frac{1}{\cosh\alpha} \right)^2 \frac{\partial}{\partial\alpha} \left(\cosh^2\alpha \frac{\partial}{\partial\alpha} \right) + \left(\frac{1}{\cosh\alpha} \right)^2 (\mathbf{L})^2 \tag{9.27}$$

where

$$(\mathbf{L})^2 = -\left(\frac{1}{\sin\theta} \right)^2 \frac{\partial}{\partial\theta} \left(\sin^2\theta \frac{\partial}{\partial\theta} \right) - \left(\frac{1}{\sin\theta} \right)^2 \left(\frac{1}{\sin\phi} \right)^2. \tag{9.28}$$

The harmonic oscillator differential equation of Eq. (5.10) of Chap. 5 then becomes

$$\left(\frac{1}{\rho} \right)^3 \frac{\partial}{\partial\phi} \left(\rho^3 \frac{\partial\Psi(x^\tau)}{\partial\phi} \right) + \left[\left(\frac{1}{\rho} \right)^2 C_1 - \rho^2 \right] \Psi(x^\tau) = \epsilon\Psi(x^\tau). \tag{9.29}$$

We can now consider the form of the solution

$$\Psi(x^\tau) = R(\rho)B(\alpha, \theta, \phi). \tag{9.30}$$

In terms of the R and B functions, the differential equation of Eq. (9.26) is separated into

$$\frac{1}{2}\left[-\left(\frac{1}{\rho}\right)^3\frac{\partial}{\partial\rho}\left(\rho^3+\frac{\partial}{\partial\rho}\right)+\eta\left(\frac{1}{\rho}\right)^2+\rho^2\right]R(\rho)=\epsilon R(\rho),\tag{9.31}$$

and

$$C_1 B(\alpha,\theta,\phi)=\eta B(\alpha,\theta,\phi)\,.\tag{9.32}$$

In order that the radial equation have regular solutions,

$$\eta=n(n+2)\tag{9.33}$$

where n takes integer values. The radial function in this case becomes

$$R^\ell_{\mu n}(\rho)=(2(\mu!)/[\Gamma(\mu+n+2)]^3)^{1/2}\rho^n L^{n+1}_\mu(\rho^2)\exp(-\rho^2/2)\,,\tag{9.34}$$

with

$$\epsilon=\pm(2\mu+n),\quad \mu=0,1,2,\quad,\dots.\tag{9.35}$$

where $L^{n+1}_\mu(\rho^2)$ is again the associated Laguerre function. This form of the radial function is the same as the one for the $O(4)$ case given in Sect. 5.5 of Chap. 5. The eigenvalue ϵ is positive in the space-like region, while it is negative in the time-like region.

From the form of C_1 given in Eqs. (9.26) and (9.27), it is clear that we can separate the B function as

$$B^{\ell m}_n(\alpha,\theta,\phi)=A^\ell_n(\alpha)Y^m_\ell(\theta,\phi)\,.\tag{9.36}$$

As is indicated in Eqs. (9.26) and (9.27), the A function will take different forms for space-like and time-like regions. For the time-like region, we use the notation

$$A^\ell_n(\alpha)=T^\ell_n(\alpha)\tag{9.37}$$

and for the space-like region

$$A^\ell_n(\alpha)=S^\ell_n(\alpha)\,.\tag{9.38}$$

Then T and S satisfy the following differential equations:

$$\frac{\partial^2}{\partial\alpha^2}(\sinh^2\alpha T^\ell_n(\alpha))-[n(n+2)+\ell(\ell+1)]T^\ell_n(\alpha)=0\,,$$

$$\frac{\partial^2}{\partial\alpha^2}(\cosh^2\alpha S^\ell_n(\alpha))-[n(n+2)+\ell(\ell+1)]S^\ell_n(\alpha)=0\,.\tag{9.39}$$

If $\ell=0$, the solutions to the above equations take the form

$$T^0_n(\alpha)=\sinh[n(n+2)\alpha]/\sinh(\alpha)\,,$$
$$S^0_n(\alpha)=\cosh[n(n+2)\alpha]/\cosh(\alpha)\,.\tag{9.40}$$

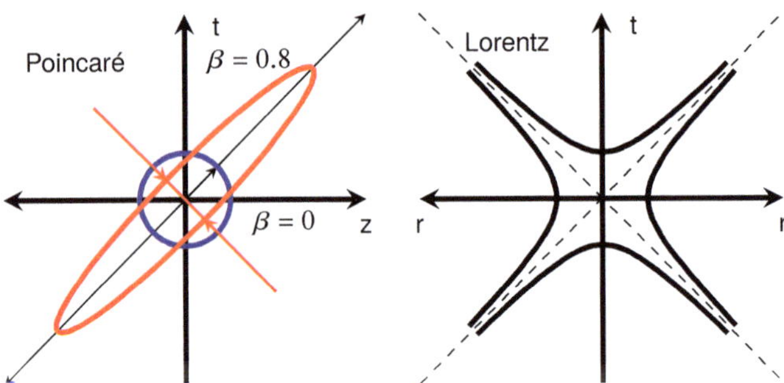

Fig. 9.1: Region of localization in the $O(4)$ and the $O(3, 1)$ coordinate systems. The normalizable $O(4)$ wave functions can be made diagonal in the Casimir operators of the Poincaré group after suitable linear combination. The wave functions in the $O(3, 1)$ coordinate system are not normalizable and cannot be made diagonal in the Casimir operators of the Poincaré group.

For non-vanishing values of ℓ ,

$$T_n^\ell(\alpha) = (\sinh \alpha)^\ell \left(\frac{1}{\sinh \alpha} \frac{d}{d\alpha} \right)^\ell T_n^0(\alpha) ,$$

$$S_n^\ell(\alpha) = (\cosh \alpha)^\ell \left(\frac{1}{\cosh \alpha} \frac{d}{d\alpha} \right)^\ell S_n^0(\alpha) . \tag{9.41}$$

We have discussed above the solutions of the oscillator differential equation which are diagonal in the sub-Casimir operators of the $O(3, 1)$ group. These solutions are quite different from the normalizable wave functions discussed in Chap. 5. As we can see in Fig. (9.1), the normalizable wave functions diagonal in the Casimir operators of the Poincaré group are localized within an elliptic region around the origin. However, the solutions of the oscillator equation discussed in this section are not normalizable and are localized hyperbolically around the light cones.

The question is then whether a solution with this hyperbolic distribution can be converted into a normalizable wave function with an elliptic distribution. If we replace t by (it), then the solutions given in this section which are diagonal in the sub-Casimir operators of Eq. (9.26) or (9.27) become the $O(4)$-based solutions given in Chap. 5. The $O(4)$-based solutions correspond to the circle in Fig. (9.1). The ellipse in the same figure corresponds to a Lorentz-deformed wave function which can be written as a linear expansion of orthonormal $O(4)$ wave functions. It in now clear that the connection between the solutions of the present section and the normalizable wave functions require more than a linear transformation.

Because of the Lorentz invariance of the Casimir operators, it is mathematically appealing to study solutions of the oscillator equation diagonal in the operators given in Eq. (9.23), instead of those of the original Poincaré group [1]. In fact, it is not difficult to find in the literature hadron models based on this representation [12, 13].

However, the crucial point is that it is very difficult to construct normalizable wave functions in these models.

It is also appealing to study unitary representations diagonal in C_1 and C_2. In fact, the analysis only for the part which does not depend on the ρ variable has been carried out by Gel'fand and Shilov [14]. In the case of harmonic oscillators, the study will also require that of the radial functions. The question is whether it is possible to construct normalizable solutions of the harmonic oscillator equation. This analysis would be very similar to the $O(2, 1)$ case where complex angular momentum plays a predominant role. In the present $O(3, 1)$ case, we still need a physical motivation for carrying out this analysis, even though it would produce beautiful mathematics.

9.5 Further Properties of the Lorentz Group

As we noted in Sects. 9.2 and 9.4, the study of the Lorentz group sometimes gives the state vectors which correspond to observable particles. It was Dirac, in 1945 [15], who observed that finite-dimensional representations of the Lorentz group are non-unitary and who suggested the study of unitary, though infinite-dimensional, representations of this group. For this purpose, Dirac suggested further the use of the harmonic oscillator wave functions normalizable in four-dimensional space-time. We discussed these wave functions in Chaps. 5 and 6.

Harish-Chandra [16] started the investigation of infinite-dimensional unitary representations of the Lorentz group. Harish-Chandra's method is based on the step-up and step-down operators which we use often in the rotation group and which serves useful purposes in studying unitary representations of $SU(1, 1)$. This method was developed further by Naimark [17], and has now been polished enough for a text book on quantum mechanics [18], as well as on mathematical physics [19].

Now, in terms of $J_\pm = J_1 \pm iJ_2$ and $K_\pm = K_1 \pm iK_2$, the commutation relations for the Lorentz group can be written as

$$[J_3, J_\pm] = \pm J_\pm , \quad [J_+, J_-] = 2J_3 ,$$

$$[J_3, K_3] = 0 , \quad [J_\pm, K_\pm] = 0 ,$$

$$[K_3, K_\pm] = \mp J_\pm , \quad [K_+, K_-] = -2J_3 ,$$

$$[J_\pm, K_3] = -K_\pm , \quad [J_\pm, J_\mp] = \pm 2K_3 ,$$

$$[J_3, K_\pm] = \pm K_\pm . \tag{9.42}$$

Because we are familiar with the rotation group, we construct the representation space in terms of the wave functions diagonal in J^2 and J_3. These wave functions have the well-known property:

$$\langle j', m'|J_3|j, m\rangle = m\delta_{mm'}\delta_{jj'} ,$$

$$\langle j', m'|J_\pm|j, m\rangle = [(j \pm m + 1)(j \mp m)]^{\frac{1}{2}}\delta_{jj'}\delta_{m',m\pm 1} . \tag{9.43}$$

Next, strictly in terms of the rotation group, the boost operators K_i form a three-dimensional vector, or a tensor operator of rank 1. The operators $Q_0 = \sqrt{2}K_3$ and $Q_\pm = \mp K_\pm$ constitute a spherical vector. We can thus apply the Wigner-Eckart theorem to write

$$\langle j', m'|K_3|j, m\rangle = \frac{1}{\sqrt{2}}N(j, j')C(j', m'|1, 0; j, m) \tag{9.44}$$

and

$$\langle j', m'|K_\pm|j, m\rangle = \mp N(j, j')C(j', m'|1, \pm 1; j, m) , \tag{9.45}$$

where $N(j, j')$ is independent of m and vanishes unless $j' = j$ or $j' = j \pm 1$
$C(j', m'|...)$ are the Clebsch-Gordan coefficients. These coefficients can be calculated or can be found in a standard table. The Wigner-Eckart theorem is discussed in textbooks on quantum mechanics, as well as standard reference books on the rotation group [20]. It is then possible to write the effect of the boost generators $|j, m\rangle$ as:

$$K_3 |j, m\rangle = [(j - m)(j + m)]^{\frac{1}{2}}A_j |j - 1, m\rangle$$
$$-mB_j |j, m\rangle - [(j + m + 1)(j - m + 1)]^{\frac{1}{2}}C_j |j + 1, m\rangle \tag{9.46}$$

$$K_\pm |j, m\rangle = \pm[(j \mp m)(j \mp m - 1)]^{\frac{1}{2}}A_j |j + 1, m \pm 1\rangle$$
$$-[(j \pm m + 1)(j \mp m)]^{\frac{1}{2}}B_j |j, m \pm 1\rangle$$
$$\pm[(j \pm m + 1)(j \pm m + 2)]^{\frac{1}{2}}C_j |j + 1, m \pm 1\rangle , \tag{9.47}$$

where A_j and B_j are coefficients which depend only on j. The remaining task is then to determine these coefficients.

After reducing the problem into that of determining the coefficients A_j, B_j, and C_j, we need a definite model to continue the study. For instance, these coefficients may be operators acting on the variables not depending on angles. In the case of the covariant oscillator model discussed in Sect. 5.4 of Chap. 5, the coefficients are differential operators of t and r. As we stated in Chap. 5, the complete study of this problem remains as a future research problem.

As for other models discussed in the literature, Naimark assumes that coefficients A_j, B_j and C_j are constants, and considers the case in which $C_j = A_j + 1$. The commutation relation $[K_\pm, K_3] = \pm J_\pm$ applied to $|j, m\rangle$ leads to:

$$\pm J_\pm |j, m\rangle = K_\pm(K_3 |j, m\rangle) - K_3(K_\pm |j, m\rangle) . \tag{9.48}$$

This means that the left-hand side can be calculated from Eq. (9.47), while the right-hand side, after repeated applications of the operations defined in Eqs. (9.46) and (9.47), becomes quadratic in A and B coefficients. After simplification, the equations for these coefficients become independent of m. They are:

$$[(j+1)B_j - (j-1)B_{j-1}]A_j = 0,$$

$$[(j+2)B_{j+1} - (j)B_j]A_{j+1} = 0,$$

$$(2j-1)(A_j)^2 - (2j+3)(A_{j+1})^2 - (B_j)^2 = 0. \tag{9.49}$$

The solution is

$$B_j = i(j_0 j_1)/[j(j+1)], \tag{9.50}$$

$$A_j = i[(j^2 - j_0^2)(j^2 - j_1^2)/(4j^2 - 1)]^{\frac{1}{2}}/j. \tag{9.51}$$

The above coefficients depend on the parameters j_0 and j_1 which specify the representation.

The concept of constructing models of elementary particles starting from representations of the Lorentz group still remains as an attractive proposition, and there are many papers in the literature on this subject [21, 22, 23, 24, 25].

9.6 Concluding Remarks

The purpose of this book was to study the representations of the Poincaré group for relativistic particles. The study consists of two steps. The first step is to work out the little groups for a given four-momentum. The second step is to complete the orbit by boosting and rotating the system. As we did in this Chapter, we can approach the problem by studying the representations of the Lorentz group for internal degrees of freedom. Table (9.1) summarizes the major conclusions.

For particles with non-zero mass, the little group is like $O(3)$. The little group applicable to electrons is $SU(2)$ which is locally homomorphic (two to one) to $O(3)$. The little group for massive vector mesons and also for the covariant harmonic oscillator is $O(3)$. For massive particles, it is convenient to study the little group in the Lorentz frame in which the particle is at rest.

There is no Lorentz frame in which the massless particle is at rest. For this reason, we align the coordinate system in such a way that the momentum is along the z-direction. The little group in this case is a three-parameter subgroup of the Lorentz group whose generators satisfy the Lie algebra of the two-dimensional Euclidean group. It was noted that a massless particle has three internal degrees of freedom. One of them is the rotation around the momentum associated with the helicity. The remaining two are the gauge degrees of freedom. The requirement of gauge invariance allows both positive and negative helicity states for massless particles with spin-1. However, for massless particles with spin-1/2, the gauge invariance leads to only one helicity (either positive or negative, not both) state.

We note in Chap. 10 that the little group for massless particles can be obtained from the large-momentum/small-mass limit of the massive case. We observe also

Table 9.1: Representation of the Poncaré group for relativistic particles. When we use the word *unitary*, it is important to specify where it applies. For instance, if we restrict ourselves to the little groups only, then all the particles have their own unitary representation. If we talk about the overall transformation properties, not many particles have unitary representations.

	Little groups	Orbit completion	Overall transformation
Massive particles			
Dirac	Unitary	Pseudo-unitary*	Pseudo-unitary*
Vector mesons	Unitary	Non-unitary	Non-unitary
Cov. Harmonic Oscillators	Unitary	Unitary	Unitary
Massless particles			
Spin-1/2	Unitary	Non-unitary	Non-unitary
Photons: A_μ	Non-unitary	Non-unitary	Non-unitary
		Unitary up to gauge transformation	
Photons: A_μ with helicity gauge	Unitary	Unitary	Unitary
Photons: electric and magnetic fields	Unitary	Non-unitary	Non-unitary

* The pseudo-unitarity of massive Dirac particles has been discussed in Sect. 9.3

that the four-vectors and Maxwell tensors can be constructed from direct products of two spinors for both massive and massless cases.

The little group for negative $(mass)^2$ does not appear to have direct application in physics, because there is no reason to expect that particles can move faster than light. However, $O(2,1)$ and $SU(1,1)$ have other physical applications. Perhaps the most important application is in the form of the symplectic group $Sp(2)$ which is becoming increasingly important in both new and traditional branches of physics. It is also an important fact that most of the practical calculations in special relativity are done in the $O(2,1)$ or $SU(1,1)$ regime, because we deal most often with motions of particles in a plane. $SU(1,1)$ is the simplest nontrivial non-compact group while $SU(2)$ is the simplest nontrivial compact group. Therefore $SU(1,1)$ occupies its unique place in the study of non-compact groups which are becoming more popular among physicists. We also take advantage of the fact that $SU(1,1)$ is unitarily equivalent to $SL(2,r)$ so that two-dimensional geometry can be used for computational purposes.

The Lorentz group plays three different roles in the study of the Poincaré group. The first role is to be the little group for particles with zero four-momentum. Since we do not expect to observe such particles, $O(3,1)$ or $SL(2,c)$ is not useful as a little group. Second, the Lorentz group is a factor group of the Poincaré group over the translation subgroup, and is the transformation group for the process of orbit completion. Third, as we discussed in this Chapter, this group is useful in studying the Lorentz-invariance properties of the internal degrees of freedom.

We have to be careful when we use the word *unitary*. As is indicated in Table (9.1), this word can apply to the representations of the little groups only, or to all Lorentz transformations including orbit completion. Some of the early original articles [26, 15, 27] appear to apply this word to the overall transformation, based on the probabilistic interpretation of quantum mechanics. However, the development of covariant quantum field theory does not require this overall unitarity. For example, the free-particle Dirac equation we use in quantum electrodynamics is not a unitary representation, but is a pseudo-unitary representation as is explained in Sect. 9.3.

In spite of this new development, we emphasize the traditional approach to the unitarity problem. Indeed, we study in detail the transformation properties of photon polarization vectors in Chap. 10. The representation is finite-dimensional and unitary in spite of the fact that the Lorentz group is non-compact. This is due to the fact that the $E(2)$-like little group contains an Abelian invariant subgroup which can be factored out, and the little group contains only one degree of freedom which is rotation around the momentum. This representation is often called Wigner's trivial representation [26, 28, 29, 30]. It will be observed in Chap. 10 that the electromagnetic four- potential in the helicity gauge is the correct description of the trivial representation.

Another unitary representation listed in Table (9.1) is the covariant harmonic oscillator model. This model was originally proposed by Dirac [15] for studying the covariant Fock space needed in quantum field theory. However, the model became more useful after Yukawa's proposal [31] to describe relativistic extended particles and its application to the quark model [32, 12]. It was shown in Chaps. 5 and 6 that this harmonic oscillator model has far reaching applications in physics and mathematics. It was shown in Chap. 5 that the solutions of the oscillator equation form a representation space for unitary irreducible representations of the Poincaré group. In Chap. 6, it was shown that this harmonic oscillator model meets Dirac's requirement for his *instant form* quantum mechanics [33]. Furthermore, the model has a very simple Lorentz deformation property in Dirac's light-cone coordinate system [33]. The next question about this harmonic oscillator model is whether it can explain observable features in the real world. We shall discuss these features in Chaps. 12 and 13.

References

1. L. Michel, Invariance in Quantum Mechanics and Group Extension, in *Group Theoretical Concepts and Methods in Elementary Particle Physics*, ed. by F. Gürsey (Gordon and Breach, New York, NY, USA, 1964), 135–200. ISBN 9780677101408. (Lectures of the Istanbul Summer School of Theoretical Physics, July 16-August 4, 1962, Istanbul, Turkey; OCLC: 948806935.)
2. S. Weinberg, Feynman Rules for Any Spin, Physical Review **133**(5B), B1318–B1332 (1964). DOI 10.1103/PhysRev.133.B1318. URL https://link.aps.org/doi/10.1103/PhysRev.133.B1318

3. S. Weinberg, Feynman Rules for Any Spin. II. Massless Particles, Physical Review **134**(4B), B882–B896 (1964). DOI 10.1103/PhysRev.134.B882. URL https://link.aps.org/doi/10.1103/PhysRev.134.B882

4. S. Weinberg, Photons and Gravitons in S -Matrix Theory: Derivation of Charge Conservation and Equality of Gravitational and Inertial Mass, Physical Review **135**(4B), B1049–B1056 (1964). DOI 10.1103/PhysRev.135.B1049. URL https://link.aps.org/doi/10.1103/PhysRev.135.B1049

5. J.D. Bjorken, S.D. Drell, *Relativistic quantum fields*. International series in pure and applied physics (McGraw-Hill, New York, NY, USA, 1965). ISBN 9780070054943. URL https://www.abebooks.com/9780486485881/Relativistic-Quantum-Fields-Dover-Books-0486485889/plp. (Reprinted in a Dover Edition 2014, ISBN 13: 9780486485881.)

6. R.F. Streater, A.S. Wightman, *PCT, spin and statistics, and all that*, 1st edn. Princeton landmarks in physics (Princeton University Press, Princeton, NJ, USA, 2000). ISBN 9780691070629. (Originally published W. A. Benjamin Publishing LTD, Boston, MA USA, 1964.)

7. N.N. Bogolubov, A.A. Logunov, I.T. Todorov, *Introduction to Axiomatic Quantum Field Theory*. No. 18 in Mathematical Physics Monograph Series (W.A. Benjamin, Inc., Boston, MA, 1975). URL https://www.osti.gov/biblio/4139916

8. A.O. Barut, J. McEwan, The four states of the Massless neutrino with pauli coupling by Spin-Gauge invariance, Letters in Mathematical Physics **11**(1), 67–72 (1986). DOI 10.1007/BF00417466. URL http://link.springer.com/10.1007/BF00417466

9. H.D. Doebner (ed.), *Differential geometric methods in mathematical physics. 1978: proceedings of the international conference held at the Techn. University of Clausthal, Germany, July 1978*. No. 139 in Lecture notes in physics (Springer, 1978). ISBN 9783540105787

10. E.G. Kalnins, W. Miller, Lie theory and the wave equation in space–time. I. The Lorentz group, Journal of Mathematical Physics **18**(1), 1–16 (1977). DOI 10.1063/1.523130. URL http://aip.scitation.org/doi/10.1063/1.523130

11. Y.S. Kim, M.E. Noz, S.H. Oh, Representations of the Poincaré group for relativistic extended hadrons, Journal of Mathematical Physics **20**(7), 1341–1344 (1979). DOI 10.1063/1.524237. URL http://aip.scitation.org/doi/10.1063/1.524237. See also, Physics Auxiliary Publication Service Document No. PAPS JMAPA-20-1336-12

12. R.P. Feynman, M. Kislinger, F. Ravndal, Current Matrix Elements from a Relativistic Quark Model, Physical Review D **3**(11), 2706–2732 (1971). DOI 10.1103/PhysRevD.3.2706. URL https://link.aps.org/doi/10.1103/PhysRevD.3.2706

13. H. Leutwyler, J. Stern, Harmonic confinement: A fully relativistic approximation to the meson spectrum, Physics Letters B **73**(1), 75–79 (1978). DOI 10.1016/0370-2693(78)90175-2. URL https://linkinghub.elsevier.com/retrieve/pii/0370269378901752

14. I.M. Gel'fand, G.E. Shilov, *Generalized functions*, 2016th edn. (American Mathematical Society : AMS Chelsea Publishing, Providence, RI, USA, 2016). ISBN 9781470426583. (Originally publised in Russian, 2nd Edition 1959, published in English 1963.)

15. P.A.M. Dirac, Unitary Representations of the Lorentz Group, Proceedings of the Royal Society A: Mathematical, Physical, and Engineering Sciences **183**(994), 284–295 (1945). DOI 10.1098/rspa.1945.0003. URL http://rspa.royalsocietypublishing.org/cgi/doi/10.1098/rspa.1945.0003

16. Harish-Chandra, Infinite irreducible representations of the Lorentz group, Proceedings of the Royal Society of London. Series A. Mathematical and Physical Sciences **189**(1018), 372–401 (1947). DOI 10.1098/rspa.1947.0047. URL https://royalsocietypublishing.org/doi/10.1098/rspa.1947.0047

17. M. Naimark, Linear Representation of the Lorentz Group, Usp.Mat. Nauk **9**, 19–93 (1954). ISBN 9781483169170. (Naimark M A 1957 Linear Representation of the Lorentz Group Am. Math. Soc. Transl. Ser. 2 6 379–458 Engl. transl.; Naimark M A Linear Representations of the Lorentz Group International Series of Monographs in Pure and Appliead Mathematics vol 63, First Edition 1964, reprinted 2014, series editor: Farahat, H. K., Oxford UK: Pergamon, Engl. transl.)

18. A. Böhm, M. Loewe, *Quantum mechanics: foundations and applications*, 3rd edn. Texts and monographs in physics (Springer-Verlag, New York, 1993). ISBN 9780387139852. (Originally published 1979; Second Edition 1986, Third Edition revised and enlarged prepared with M. Loewe, published in softcover 2001.)

19. W. Miller, *Symmetry groups and their applications*. No. 50 in Pure and applied mathematics; a series of monographs and textbooks (Academic Press, New York, NY, USA, 1972). ISBN 978-0-12-497460-9

20. A.R. Edmonds, *Angular momentum in quantum mechanics*. Princeton landmarks in physics (Princeton University Press, Princeton, 1996). ISBN 9780691025896. (Originally published 1957, Second Edition 1960, Revised printing 1968, Third printing with corrections 1974, Fourth printing 1996.)

21. A. Hanson, T. Regge, The relativistic spherical top, Annals of Physics **87**(2), 498–566 (1974). DOI 10.1016/0003-4916(74)90046-3. URL https://linkinghub.elsevier.com/retrieve/pii/0003491674900463

22. P. Roman, J. Haavisto, Relativistic quantum dynamical group for hadrons, Journal of Mathematical Physics **22**(2), 403–411 (1981). DOI 10.1063/1.524906. URL http://aip.scitation.org/doi/10.1063/1.524906

23. L.C. Biedenharn, H. van Dam, Galilean subdynamics and the dual resonance model, Physical Review D **9**(2), 471–486 (1974). DOI 10.1103/PhysRevD.9.471. URL https://link.aps.org/doi/10.1103/PhysRevD.9.471

24. R.R. Aldinger, A. Böhm, P. Kielanowski, M. Loewe, P. Magnollay, N. Mukunda, W. Drechsler, S.R. Komy, Relativistic rotator. I. Quantum observables and constrained Hamiltonian mechanics, Physical Review D **28**(12), 3020–3031 (1983). DOI 10.1103/PhysRevD.28.3020. URL https://link.aps.org/doi/10.1103/PhysRevD.28.3020

25. A. Böhm, M. Loewe, L.C. Biedenharn, van Dam. H., Relativistic rotator. II. The simplest representation spaces, Physical Review D **28**(12), 3032–3040 (1983). DOI 10.1103/PhysRevD.28.3032. URL https://link.aps.org/doi/10.1103/PhysRevD.28.3032

26. E.P. Wigner, On Unitary Representations of the Inhomogeneous Lorentz Group, The Annals of Mathematics **40**(1), 149–204 (1939). DOI 10.2307/1968551. URL http://www.jstor.org/stable/1968551?origin=crossref

27. V. Bargmann, E.P. Wigner, Group Theorectical Discussion of Relatistic Wave Equations, Proc. Nat. Acad. Sci. (USA) **34**(5), 211–223 (1948). DOI 10.1073/pnas.34.5.211. URL https://doi.org/10.1073/pnas.34.5.211

28. E.P. Wigner, Unitary representations of the inhomogeneous Lorentz group including reflections, in *Group Theoretical Concepts and Methods in Elementary Particle Physics*, ed. by F. Gürsey (Routledge, Abingdon, Oxfordshire, UK, 1964). ISBN 9780677101408. (Lectures of the Istanbul Summer School of Theoretical Physics, July 16-August 4, 1962, Istanbul, Turkey; OCLC: 948806935.)

29. G. Mackey, *Induced Representations of Groups and Quantum Mechanics* (W. A. Benjamin, Inc./Editore Boringhieri, Boston, MA, USA, 1968). (ASIN B001DOM62C.)

30. R.L. Lipsman, *Group representations: a survey of some current topics*. No. 388 in Lecture notes in mathematics (Springer-Verlag, Berlin, Germany; New York, NY, USA, 1974). ISBN 9783540384021

31. H. Yukawa, Structure and Mass Spectrum of Elementary Particles. I. General Considerations, Physical Review **91**(2), 415–416 (1953). DOI 10.1103/PhysRev.91.415.2. URL https://link.aps.org/doi/10.1103/PhysRev.91.415.2

32. K. Fujimura, T. Kobayashi, M. Namiki, Nucleon Electromagnetic Form Factors at High Momentum Transfers in an Extended Particle Model Based on the Quark Model, Progress of Theoretical Physics **43**(1), 73–79 (1970). DOI 10.1143/PTP.43.73. URL https://academic.oup.com/ptp/article-lookup/doi/10.1143/PTP.43.73

33. P.A.M. Dirac, Forms of Relativistic Dynamics, Reviews of Modern Physics **21**(3), 392–399 (1949). DOI 10.1103/RevModPhys.21.392. URL https://link.aps.org/doi/10.1103/RevModPhys.21.392

Chapter 10
Massless Particles

Abstract Here we study the internal space-time symmetries of massless particles, among which photons are particularly significant due to their involvement in the development of quantum mechanics and in quantum electrodynamics, which deals with the interaction of photons with charged particles. One of the difficulties associated with studying photons has been that the four-vector form for the photon wave function is not unique, depending on gauge degrees of freedom which are not measurable. The little group for massless particles is $E(2)$-like. A closer look at matrices of the $E(2)$-like group applicable to the photon four-vector indicates that they form a finite-dimensional non-unitary representation. We associate rotations of the $E(2)$-like group with the helicity and the translation-like transformations with gauge degrees of freedom. The four-momentum of relativistic particles is altered by Lorentz transformations, whose rotations are represented by unitary matrices, while boosts by non-unitary matrices. Therefore, we consider the possibility of constructing a unitary representation by combining the boost and gauge transformations.

We shall study in this Chapter the internal space-time symmetries of massless particles. As was noted in Chap. 2, the little group for massless particles is locally isomorphic to the group of Euclidean transformations on a two-dimensional plane which is often called $E(2)$. Therefore, the study of massless particles requires a careful investigation of the $E(2)$ group and its relation to the $E(2)$-like little group.

Among many massless particles, photons occupy the most prominent position. Indeed, the development of modern physics has been the process of understanding the properties of photons. The role of photons in the development of quantum theory is well known. The role of electromagnetism in the development of relativity is also well known. Quantum electrodynamics which constitutes the basis for most field theoretic models is a theory of photons interacting with charged point particles.

S. Başkal et al., *Theory and Applications of the Poincaré Group*, Fundamental Theories of Physics 217, https://doi.org/10.1007/978-3-031-64376-7_10

In spite of this overwhelming role in physics, we have been and are still behind in a full-fledged understanding of the internal space-time symmetry of photons. One of the difficulties associated with studying photons has been that the four-vector form for the photon wave function is not unique and is dependent on gauge degrees of freedom which are not measurable. It is of course possible to bypass four-potentials and use electric and magnetic fields which are independent of gauge parameters. However, this did not simplify our problem.

Unlike $O(3)$, $E(2)$ is a non-compact group, and its representations are infinite-dimensional. Certainly, this does not make the photon problem easier for us. In fact, the study of finite subsets of unitary representations of the $E(2)$ group in the past led to the restriction to the $O(2)$ subgroup of $E(2)$ [1]. This restriction of course freezes the translational degrees of freedom in the $E(2)$ group [2].

A closer look at the four-by-four matrices of the $E(2)$-like little group applicable to the photon four-vector indicates that they form a finite-dimensional non-unitary representation. Therefore, instead of restricting ourselves to unitary representations of the $E(2)$ group, we should study also its finite-dimensional non-unitary representations, and examine how these non-unitary representations will help us in understanding unitary representations of physical transformations. As is indicated in Table 10.1, it was not until 1982 that the desired non-unitary representations were discussed in the physics literature.

Table 10.1: Representations of the $E(2)$ group. In order that representations be useful in describing massless particles with definite helicity, they should be diagonal in the generators of rotations.

Diagonal in	Unitary infinite dimensional	Non-unitary finite dimensional
P_1 and P_2	Wigner [1]	Trivial
J_3	Inönü and Wigner [3]	Han et al. [4]

Many of the properties of the $E(2)$ group have already been discussed in Appendices A and B and also in Chap. 4. The present Chapter is devoted to the study of finite-dimensional non-unitary representations of the $E(2)$-like little group for massless particles and their relationship with the corresponding representations of the $E(2)$ group. We shall study first photons in detail. Neutrinos, having a mass at six orders of magnitudes smaller than the mass of an electron [5] and since $\frac{m_\nu}{E_\nu}$ is very small, given the necessary caution [6], were sometimes approximated as massless spin-1/2 particles at least for the sake of extending the formalism of $E(2)$-like groups. However, the current experimental evidence strongly confirms that neutrinos have mass so we shall not do that here.

In spite of what we said above, we are eventually interested in constructing unitary representations of physical transformations. Then what do we do with the non-unitary representations of the $E(2)$-like little group which we study in this Chapter? It is not difficult to associate rotations of the $E(2)$-like little group with the helicity. We shall see in this Chapter that the transnational degrees of freedom correspond to gauge transformations. Since the gauge transformation is not an observable transformation, its representation does not have to be unitary.

On the other hand, Lorentz transformations which change the four-momentum are physical transformations. Among those Lorentz transformations, there are rotations around and boosts along the three orthogonal directions. Rotations can be represented by unitary matrices. Boost matrices applicable to four-vectors are non-unitary. Therefore, when we boost four-potentials, we should consider the possibility of constructing a unitary representation by combining the boost and gauge transformations.

Section 10.1 contains the proof that gauge transformations are Lorentz-boosted rotations in the infinite-momentum/zero-mass limit. Section 10.2 is devoted to a detailed study of the little group for photons. It is shown that the translation-like degrees of freedom in the $E(2)$-like little group are gauge degrees of freedom.

Section 10.3 deals with Lorentz transformation properties of the four-potential. Because the choice of gauge is arbitrary, we usually choose a specific gauge for convenience. However, there is a unique gauge which makes the four-potential diagonal in the helicity operator. This gauge condition in one Lorentz frame is not always respected in other Lorentz frames. For this reason, we often say that the electromagnetic four-vector is not manifestly covariant. We study this problem in detail and show that it is possible to perform Lorentz transformations on the four-potential in a manner which preserves this gauge condition. It is shown in Sect. 10.4 that this gauge-preserving transformation is a unitary transformation with a one-dimensional trivial representation of the little group.

In spite of its importance in understanding physics, finite-dimensional representations of the $E(2)$ group have never been systematically discussed in the literature. Thus in Sect. 10.5 we derive the infinite and finite dimensional representations of E(2). In Sect. 10.6, the transformation property of Maxwell's equations is discussed as a further illustrative example of spinor representations.

In Sect. 10.7, we study the little group for massless spin-1/2 particles. Since the discovery of parity violation in weak interaction [7, 8] and the discovery of neutrino polarization [9], one of the most outstanding questions was what the physical or kinematical origin of neutrino polarization. We shall see in this Section that the polarization of a massless spin-1/2 particle is due to the requirement of the invariance under the $E(2)$-like little group. We give a very brief state of the art in neutrino physics. In the case of photons the $E(2)$-like little group corresponds to gauge transformations. We shall see that the gauge dependence of the photon four-potential can be traced to the spinor variance under the little group transformation.

In Chaps. 2 and 3 we studied the Dirac equation. Here in Sect. 10.8, we take the infinite-momentum/zero-mass limit of the Dirac equation.

For photons the internal space is purely a spin space, and is not derivable from the coordinate variables. Since the internal variable in the covariant harmonic oscillator discussed in Chap. 5 is directly derivable from the coordinate variables, we are led to consider whether such a study is possible. With this point in mind, we discuss in Sect. 10.9 the covariant harmonic oscillator wave functions and their $E(2)$-like symmetry for massless hadrons. While Sect. 10.10 lists exercises and problems, the primary purpose of this section is to discuss the items which are not contained in the main sections.

10.1 Infinite-Momentum/Zero-Mass Limit for Massive Particles

We studied in Sect. 4.2 of Chap. 4 the $E(2)$ group and determined its role in explaining the internal space-time symmetry of massless particles. In Chaps. 5 and 6, we studied in detail how the $O(3)$ group can explain the internal space-time symmetry of massive particles. We shall here see that the $E(2)$ group can be regarded as a limiting case of the $O(3)$ group.

Einstein's formula, $E = [(P)^2 + M^2]^{1/2}$, states that the energy-momentum relation for massive particles becomes that for massless particles in the limit of large momentum and/or zero-mass. For this reason, we should expect that the internal symmetry of massive particles becomes that of massless particles in the same limit. We should expect further that this limiting procedure will be very similar to the group contraction process discussed in Sect. 4.3 of Chap 4.

For simplicity, we shall use the rotation generators of $O(3)$ which can also be viewed as the angular momentum operators to illustrate this limiting procedure. If a massive particle is at rest, the symmetry group is generated by J_1, J_2, and J_3 given in Sect. 2.1 of Chap. 2. These operators do not change the four-momentum of the particle at rest. If this particle moves along the z-direction, J_3 remains invariant, and its eigenvalue is the helicity. However, a question arises as to what happens to J_1 and J_2, particularly in the infinite-momentum limit?

There are no Lorentz frames in which massless particles are at rest. The little group for a massless particle moving along the z-direction is generated by J_3, N_1, and N_2 given in Chaps. 2, 3, and 4. As we first noted in Eq. (2.45) of Chap. 2, the N_i generators are related to the boost and rotation operators by

$$N_1 = K_1 - J_2 , \quad N_2 = K_2 + J_1 . \tag{10.1}$$

The explicit matrix forms for K_1 and K_2 are also given in Sect. 2.1 of Chap. 2. The four-momentum of the massless particle remains invariant under transformations generated by the J_3, N_1, and N_2 operators. These generators of course satisfy the commutation relations for the generators of $E(2)$. J_3 is like the generator of rotations while N_1 and N_2 are like the generators of translations in the two-dimensional plane. As we discussed in Chap. 4, these translation-like operators generate gauge transformations.

Let us carry out an explicit calculation to justify the above expectation starting with a massive particle at rest whose $O(3)$-like little group is generated by J_1, J_2, and J_3. If we boost this massive particle along the z-direction, its momentum and energy will become P_z and $E = [P_z^2 + M^2]^{1/2}$ respectively. The boost matrix is

$$B(P) = \begin{pmatrix} E/M & P_z/M & 0 & 0 \\ P_z/M & E/M & 0 & 0 \\ 0 & 0 & 1 & 0 \\ 0 & 0 & 0 & 1 \end{pmatrix}. \tag{10.2}$$

P_z is the momentum along the z-direction. Under this boost operation, J_3 will remain invariant:

$$J_3' = BJ_3B^{-1} = J_3. \tag{10.3}$$

However, the boosted J_2 and J_1 become

$$J_2' = (E/M)J_2 - (P_z/M)K_1,$$

$$J_1' = (E/M)J_1 + (P_z/M)K_2. \tag{10.4}$$

Because the Lorentz boosts in Eqs. (10.3) and (10.4) are similarity transformations, the J' operators still satisfy the $O(3)$ commutation relations:

$$[J_i', J_j'] = i\varepsilon_{ijk}J_k'. \tag{10.5}$$

Since the quantities in Eq. (10.4) become very large as the momentum increases, we introduce new operators

$$G_1 = (M/E)J_1',$$

$$G_2 = -(M/E)J_2'. \tag{10.6}$$

In terms of these new operators, we can write the $O(3)$ commutation relations of Eq.(10.5) as

$$[J_3, G_1] = iG_2,$$

$$[J_3, G_2] = -iG_1,$$

$$[G_1, G_2] = -(M/E)^2J_3. \tag{10.7}$$

The quantity $(M/E)^2$ becomes vanishingly small if the mass becomes small or the momentum becomes very large. In this limit, G_1 and G_2 become N_1 and N_2 respectively, and the above commutation relations become those for the $E(2)$-like little group in the same manner as the $O(3)$ commutation relations become those for the $E(2)$ group through the group contraction process in Sect. 4.3 of Chap. 4. The quantity (M/E) acts like the radius of the sphere we used for the model of the contraction of $O(3)$ to $E(2)$. For completeness, let us write the commutation relations shared by the $E(2)$ group and the $E(2)$-like little group:

$$[J_3, N_1] = iN_2, \quad [J_3, N_2] = -iN_1, \quad [N_1, N_2] = 0. \tag{10.8}$$

We shall see in the following section that N_1 and N_2 are the generators of gauge transformations for the photon case. Indeed, rotations around the axes perpendicular to the momentum become gauge transformations in the infinite-momentum/zero-mass limit [10, 11].

We have so far worked out the large-momentum/zero-mass limit of the rotation using the four-by-four matrix applicable to four-vectors. Sect. 10.8 will deal with this limiting process for spinors.

10.2 E(2)-Like Little Group for Photons

Because the four-potential for the electromagnetic field is non-unique and dependent on gauge degrees of freedom, many attempts have been made in the past to construct an alternative form for the photon wave function. These attempts have been aimed at constructing a simpler form than the four-vector form. The most obvious alternative is the use of electric and magnetic fields which are directly measurable. However, we then have to deal with a quantity which behaves like a tensor under Lorentz transformations. Furthermore, we do not yet know any form other than the four-potential which is more convenient in describing the interaction of photons with charged particles.

It is therefore more productive to face the problem of gauge degrees of freedom than to avoid it. In approaching this problem, we note that the translation-like degrees of freedom were left unexplained in Wigner's original paper [1]. We note also that Wigner was only interested in constructing unitary representations even in his later papers [3, 12]. In this section, we shall study the photon problem by filling in these gaps.

Let us consider a single free photon moving along the z-direction. Then we can write the photon wave function as

$$A^\mu(x) = A^\mu e^{i\omega(z-t)} \tag{10.9}$$

where

$$A^\mu = (\, A_0\, ,\, A_3\, ,\, A_1\, ,\, A_2\,)\, . \tag{10.10}$$

As we saw in Eq. (4.21) of Chap. 4, the momentum four-vector is

$$p^\mu = (\, \omega\, ,\, \omega\, ,\, 0\, ,\, 0\,)\, . \tag{10.11}$$

Then, as we discussed in Chap. 2 and in Sect. 4.2 of Chap. 4, the little group applicable to the photon four-potential is generated by

$$J_3 = \begin{pmatrix} 0 & 0 & 0 & 0 \\ 0 & 0 & 0 & 0 \\ 0 & 0 & 0 & -i \\ 0 & 0 & i & 0 \end{pmatrix}, \quad N_1 = \begin{pmatrix} 0 & 0 & i & 0 \\ 0 & 0 & i & 0 \\ i & -i & 0 & 0 \\ 0 & 0 & 0 & 0 \end{pmatrix}, \quad \text{and} \quad N_2 = \begin{pmatrix} 0 & 0 & 0 & i \\ 0 & 0 & 0 & i \\ 0 & 0 & 0 & 0 \\ i & -i & 0 & 0 \end{pmatrix}. \tag{10.12}$$

These matrices satisfy the same commutation relations as given in Eq. (4.25):

$$[\,N_1\,,N_2\,] = 0\,, \quad [\,J_3\,,N_1\,] = iN_2\,, \quad [\,J_3\,,N_2\,] = -iN_1\,. \tag{10.13}$$

From the generators, we can construct the transformation matrices

$$D(u,v,\alpha) = D(u,0,0)D(0,v,0)D(0,0,\alpha)\,, \tag{10.14}$$

where

$$D(u,0,0) = D_1(u) = \exp[-iuN_1]\,,$$

$$D(0,v,0) = D_2(v) = \exp[-ivN_2]\,,$$

$$D(0,0,\alpha) = R(\alpha) = \exp[-i\alpha J_3]\,. \tag{10.15}$$

We can expand the above formulae in power series. The expansion for $D(0,0,\alpha)$ is a very familiar procedure for us, and the result is

$$R(\alpha) = \begin{pmatrix} 1 & 0 & 0 & 0 \\ 0 & 1 & 0 & 0 \\ 0 & 0 & \cos\alpha & -\sin\alpha \\ 0 & 0 & \sin\alpha & \cos\alpha \end{pmatrix}. \tag{10.16}$$

We expand D(u, 0, 0) and D(0, v, 0) as

$$D_1(u) = 1 - iuN_1 - (uN_1)^2/2\,,$$

$$D_2(v) = 1 - ivN_2 - (vN_2)^2/2\,, \tag{10.17}$$

with

$$(N_1)^3 = (N_2)^3 = N_1(N_2)^2 = N_2(N_1)^2 = 0\,. \tag{10.18}$$

Therefore, the four-by-four matrices for $D_1(u)$ and $D_2(v)$ are quadratic in u and v respectively,

$$D_1(u) = \begin{pmatrix} 1+u^2/2 & -u^2/2 & u & 0 \\ u^2/2 & 1-u^2/2 & u & 0 \\ u & -u & 1 & 0 \\ 0 & 0 & 0 & 1 \end{pmatrix}, \tag{10.19}$$

and

$$D_2(v) = \begin{pmatrix} 1+v^2/2 & -v^2/2 & 0 & v \\ v^2/2 & 1-v^2/2 & 0 & v \\ 0 & 0 & 1 & 0 \\ -v & v & 0 & 1 \end{pmatrix}. \tag{10.20}$$

The above D matrices commute with each other, and the product $D(u,v,0) = D_1(u)D_2(v)$ becomes the $T(u,v)$ matrix in Eq. (2.39) of Chap. 2.

The algebraic properties of the above D matrices have already been discussed in Sect. 2.2 of Chap. 2. It was noted there that these matrices leave the four-momentum given in Eq. (10.11) invariant. It was shown in Exercise 1 in Sect. 2.7 of Chap. 2 that each of the above matrices can be written as a product of one boost matrix and one rotation followed by another boost matrix. Therefore each of the D matrices is a transformation matrix belonging to the group of Lorentz transformations.

Recall that in the case of the $O(3)$-like little group, the four-by-four matrices of the little group can be reduced to a block diagonal form consisting of the three-by-three rotation matrix and one-by-one unit matrix. One major problem is that the D matrix is quadratic in the u and v variables. We thus have to get rid of these quadratic terms. In order to attack this problem, let us note that we are allowed to impose the Lorentz condition

$$\frac{\partial}{\partial x^\mu} A^\mu(x^\sigma) = p^\mu A_\mu(x^\sigma) = 0 , \qquad (10.21)$$

resulting in

$$A_0 = A_3 . \qquad (10.22)$$

Since the first and second components are identical, the N_1 and N_2 matrices of Eq. (10.12) can be replaced respectively by

$$N_1 = \begin{pmatrix} 0 & 0 & i & 0 \\ 0 & 0 & i & 0 \\ 0 & 0 & 0 & 0 \\ 0 & 0 & 0 & 0 \end{pmatrix} \quad \text{and} \quad N_2 = \begin{pmatrix} 0 & 0 & 0 & i \\ 0 & 0 & 0 & i \\ 0 & 0 & 0 & 0 \\ 0 & 0 & 0 & 0 \end{pmatrix} . \qquad (10.23)$$

At the same time the D matrices become [4]

$$D(u, v, 0) = \begin{pmatrix} 1 & 0 & u & v \\ 0 & 1 & u & v \\ 0 & 0 & 1 & 0 \\ 0 & 0 & 0 & 1 \end{pmatrix} . \qquad (10.24)$$

Consequently, all algebraic properties of the $E(2)$ group are directly applicable to the $E(2)$-like little group for photons. For instance, the concept of equivalence class in $E(2)$ is directly applicable to the u and v transformations in the little group [13]. This transformation has a geometrical meaning in the case of the $E(2)$ group. However, what does this mean in the case of the $E(2)$-like little group. We shall discuss this question in Sect. 10.3.

10.3 Transformation Properties of Photon Polarization Vectors

The use of scalar and vector potentials for electromagnetic fields plays an important role in all branches of physics. There are two important properties associated with these physical quantities. First, the electromagnetic fields derivable from these potentials are invariant under gauge transformations. Second, the scalar and vector potentials form a four-vector, and the electromagnetic field derivable from a

Lorentz-transformed four-vector potential is identical to that derivable from the Lorentz-transformed electromagnetic field tensor.

This means that we are free to choose a particular gauge condition to serve our convenience. The usual gauge condition for free electromagnetic waves is that the longitudinal and time-like components of the four-potential vanish in the observer's Lorentz frame. Do we use this condition only for convenience? The answer to this question is *No*. In order that the photon states be diagonal in the helicity operator, it is essential that the non-transverse components vanish. This condition remains invariant in Lorentz frames which move along the direction of the photon momentum. However, the time-like component does not remain zero when we perform a Lorentz transformation along the direction perpendicular to the photon momentum.

With this point in mind, let us write down the photon wave functions diagonal in the helicity operator J_3. The four-vector A^μ given in Eq. (10.9) should take the form

$$\epsilon^\mu_\pm = (0, 0, 1, \pm i), \tag{10.25}$$

where ϵ^μ_+ and ϵ^μ_- represent the photon states with positive and negative helicities respectively. In order that the system be diagonal in J_3, it is essential that the first and second components vanish. We shall call these four-component vectors the photon polarization vectors [13]. We can normalize the polarization vector of Eq. (10.25) by dividing every element by $\sqrt{2}$. However, this procedure is trivial and appears unnecessary.

We obtain the above polarization vector by imposing the transversality condition

$$\nabla \cdot \mathbf{A}(x) = 0, \quad \mathbf{p} \cdot \mathbf{A} = 0, \quad \text{or} \quad A_3 = 0, \tag{10.26}$$

in addition to the Lorentz condition of Eq. (10.21). These two conditions reduce the number of independent components in Eq. (10.9) from four to two. Indeed, if we combine the above condition with the Lorentz condition of Eq. (10.21),

$$A_0 = A_3 = 0. \tag{10.27}$$

Since the above condition is so essential for the photon four-vector to be an eigenstate of the helicty operator, we shall call the combined effect of the Lorentz and transversality conditions the *helicity gauge*. The Lorentz condition is Lorentz-invariant. However, the transversality condition of Eq. (10.26) is not Lorentz-invariant. Therefore we are faced with the question of whether helicity is a Lorentz-invariant concept in the case of photons.

With this point in mind, let us first examine the effect of the $D(u, v, 0)$ transformations given in Eq. (10.24) on the above polarization vectors. If we perform a straight-forward calculation:

$$\begin{pmatrix} \zeta_{\pm 0} \\ \zeta_{\pm 3} \\ \zeta_{\pm 1} \\ \zeta_{\pm 2} \end{pmatrix} = \begin{pmatrix} 1 & 0 & u & v \\ 0 & 1 & u & v \\ 0 & 0 & 1 & 0 \\ 0 & 0 & 0 & 1 \end{pmatrix} \begin{pmatrix} 0 \\ 0 \\ 1 \\ \pm i \end{pmatrix}, \tag{10.28}$$

the result is [14, 15],

$$\zeta_\pm^\mu = D(u, v, 0)\epsilon_\pm^\mu = \epsilon_\pm^\mu + \frac{u \pm iv}{\omega} p^\mu$$

$$= (u \pm iv, u \pm iv, 1, \pm i) \,, \tag{10.29}$$

where p^μ is the energy-momentum four-vector given in Eq. (10.11). Thus $D(u, v, 0)$ applied to the polarization vector results in the addition of a term which is proportional to the four-momentum. $D(u, v, 0)$ therefore performs a gauge transformation on ϵ^μ

Because $D(u, v, 0)$ is a Lorentz transformation,

$$\zeta^\mu \zeta_\mu = \epsilon^\mu \epsilon_\mu \,, \tag{10.30}$$

and because $D(u, v, 0)$ leaves the four-momentum invariant,

$$p^\mu \zeta_\mu = p^\mu \epsilon_\mu \,. \tag{10.31}$$

The first component of ζ^μ is equal to its second component, but they do not vanish. If we calculate the electric and magnetic fields from this four-potential [Problem 2 in Sect. 10.10], then they are identical to those obtained from ϵ^μ. ζ^μ is indeed a gauge transformation of ϵ^μ.

The four-component vector ζ^μ is no longer an eigenstate of J_3, but is diagonal in

$$J_3(u, v) = D(u, v, 0)J_3 D^{-1}(u, v, 0) \,. \tag{10.32}$$

The rotations by the same angle generated by J_3 and $J_3(u, v)$ belong to the same equivalence class. The gauge transformation is therefore a transformation within an equivalence class [13].

Let us next examine Lorentz transformations of the polarization vector ϵ^μ given in Eq. (10.25). If we boost this vector along the z-direction, it remains invariant. However, if we boost it along the x-direction,

$$\begin{pmatrix} A'_{\pm 0} \\ A'_{\pm 3} \\ A'_{\pm 1} \\ A'_{\pm 2} \end{pmatrix} = \begin{pmatrix} \cosh\eta & 0 & \sinh\eta & 0 \\ 0 & 1 & 0 & 0 \\ \sinh\eta & 0 & \cosh\eta & 0 \\ 0 & 0 & 0 & 1 \end{pmatrix} \begin{pmatrix} 0 \\ 0 \\ 1 \\ \pm i \end{pmatrix} \,, \tag{10.33}$$

or

$$A'^\mu_\pm = (\sinh\eta, 0, \cosh\eta, \pm i) \,. \tag{10.34}$$

Under this Lorentz boost, the momentum four-vector becomes

$$p'^\mu = (\omega \cosh\eta, \omega, \omega \sinh\eta, 0) \,, \tag{10.35}$$

The four-vectors p'^μ and $A'^\mu_\pm$ satisfy the Lorentz condition:

$$p'^\mu A'_{\pm\mu} = 0 \,. \tag{10.36}$$

However, the transversality condition is no longer satisfied:

$$\mathbf{p}' \cdot \mathbf{A}'_\pm = \omega \cosh \eta \sinh \eta \,, \tag{10.37}$$

which does not vanish. $\mathbf{A}'_\pm$ is not a helicity state.

On the other hand, if we boost the four-vector $\zeta^\mu_\pm$ given in Eq. (10.29) along the x-direction:

$$\begin{pmatrix} \zeta'_{\pm 0} \\ \zeta'_{\pm 3} \\ \zeta'_{\pm 1} \\ \zeta'_{\pm 2} \end{pmatrix} = \begin{pmatrix} \cosh \eta & 0 & \sinh \eta & 0 \\ 0 & 1 & 0 & 0 \\ \sinh \eta & 0 & \cosh \eta & 0 \\ 0 & 0 & 0 & 1 \end{pmatrix} \begin{pmatrix} \zeta_{\pm 0} \\ \zeta_{\pm 3} \\ \zeta_{\pm 1} \\ \zeta_{\pm 2} \end{pmatrix} \,, \tag{10.38}$$

the result is [15]:

$$\zeta'_{\pm 0} = \sinh \eta + (u \pm iv) \cosh \eta \,,$$

$$\zeta'_{\pm 3} = (u \pm iv) \,,$$

$$\zeta'_{\pm 1} = \cosh \eta + (u \pm iv) \sinh \eta \,,$$

$$\zeta'_{\pm 2} = \pm i \,. \tag{10.39}$$

The Lorentz condition $p'^\mu \zeta'_{\pm \mu} = 0$ is maintained. However,

$$\mathbf{p}' \cdot \zeta'_\pm = (\omega \cosh \eta)[\sinh \eta + (u \pm iv) \cosh \eta] \omega (u \pm iv) \,. \tag{10.40}$$

Since u and v are gauge parameters which do not affect the observable quantities, we are led to examine whether they can be chosen in such a way that $\zeta'_{\pm 0}$ of Eq. (10.39) and $\mathbf{p}' \cdot \zeta'_\pm$ of Eq. (10.40) vanish simultaneously. The answer to this question is definitely *Yes*. Indeed, both of them vanish if

$$u \pm iv = -\tanh \eta \,. \tag{10.41}$$

Thus

$$u = -\tanh \eta \qquad \text{and} \qquad v = 0 \,. \tag{10.42}$$

The Lorentz boost of the polarization vector $\epsilon^\mu_\pm$ along the x-direction does not lead to another polarization vector. However, it is still possible to construct the new polarization vector $\epsilon'^\mu_\pm$ by boosting $\epsilon^\mu_\pm$ after the $D(u, v, 0)$ transformation whose parameters are determined by the transversality condition of Eq. (10.41) or Eq. (10.42). Again if we use the form of Eq. (10.25), the multiplication of the gauge transformation matrix of Eq. (10.28) and the boost matrix of Eq. (10.38) leads to

$$\begin{pmatrix} 0 \\ \epsilon'_{\pm 3} \\ \epsilon'_{\pm 1} \\ \epsilon'_{\pm 2} \end{pmatrix} = \begin{pmatrix} \cosh \eta & 0 & 0 & 0 \\ 0 & 1 & -\tanh \eta & 0 \\ \sinh \eta & 0 & 1/\cosh \eta & 0 \\ 0 & 0 & 0 & 1 \end{pmatrix} \begin{pmatrix} 0 \\ 0 \\ 1 \\ \pm i \end{pmatrix} \,. \tag{10.43}$$

Thus

$$\epsilon_{\pm}^{\prime\mu} = (0, -\tanh\eta, 1/\cosh\eta, \pm i) . \tag{10.44}$$

This polarization vector satisfies the transversality condition

$$\mathbf{p}' \cdot \boldsymbol{\epsilon}_{\pm}' = 0 . \tag{10.45}$$

In addition, the transformation of Eq. (10.43) is a norm-preserving transformation:

$$|\epsilon_{\pm 3}'|^2 + |\epsilon_{\pm 1}'|^2 + |\epsilon_{\pm 2}'|^2 = |\epsilon_{\pm 3}|^2 + |\epsilon_{\pm 1}|^2 + |\epsilon_{\pm 2}|^2 . \tag{10.46}$$

We can carry out a similar calculation when the system is boosted along the y-direction. The gauge parameters in this case become

$$u = 0, \quad \text{and} \quad v = -\tanh\eta , \tag{10.47}$$

and the matrix which transforms the polarization vector becomes

$$\begin{pmatrix} \cosh\eta & 0 & 0 & 0 \\ 0 & 1 & 0 & -\tanh\eta \\ 0 & 0 & 1 & 1/\cosh\eta \\ \sinh\eta & 0 & 0 & 1 \end{pmatrix} . \tag{10.48}$$

We can also obtain this matrix by rotating the matrix of Eq. (10.43) by $90°$.

10.4 Unitary Transformation of Photon Polarization Vectors

We have seen in Sect. 10.3 that Lorentz boosts on photon polarization vectors are not in general helicity-preserving transformations. However, it is still possible to preserve the helicity gauge by performing a transformation of the $E(2)$-like little group before (or after) the boost. The net effect was a product of two non-unitary matrices. The resulting transformation matrix such as the one given in Eq. (10.43) is not unitary.

However, the result given in Eq. (10.46) is clearly that of a unitary transformation. Then there must be a way to transform the matrix of Eq. (10.43) to a unitary matrix. Can this matrix be decomposed further into a product of a unitary matrix and a non-unitary matrix which does not change the photon polarization vector? For example, the boost matrix along the z-direction is not unitary, but does not change the photon polarization vector of Eq. (10.25).

In order to tackle this problem, let us use the notation $B_x(\eta)$ for the four-by-four matrix we used in Eqs. (10.33) and (10.38). Under this boost operation, the four-momentum p became p'. However, in order to achieve the same purpose, we can consider a boost along the z-direction followed by a rotation around the y-axis as is shown in Fig. 10.1. In this case, the boost and rotation matrices are [See Exercise 2 in Sect. 2.7 of Chap. 2]

$$B_z(\eta) = \begin{pmatrix} \frac{1}{2}\dfrac{(1+\cosh^2\eta)}{\cosh\eta} & \dfrac{\sinh\eta}{2}\tanh\eta & 0 & 0 \\[2mm] \dfrac{\sinh\eta}{2}\tanh\eta & \frac{1}{2}\dfrac{(1+\cosh^2\eta)}{\cosh\eta} & 0 & 0 \\[2mm] 0 & 0 & 1 & 0 \\[1mm] 0 & 0 & 0 & 1 \end{pmatrix}, \qquad (10.49)$$

$$R_y(\eta) = \begin{pmatrix} 1 & 0 & 0 & 0 \\ 0 & 1/\cosh\eta & \tanh\eta & 0 \\ 0 & -\tanh\eta & 1/\cosh\eta & 0 \\ 0 & 0 & 0 & 1 \end{pmatrix}, \qquad (10.50)$$

with

$$(1/\cosh\eta)^2 + (\tanh\eta)^2 = 1. \qquad (10.51)$$

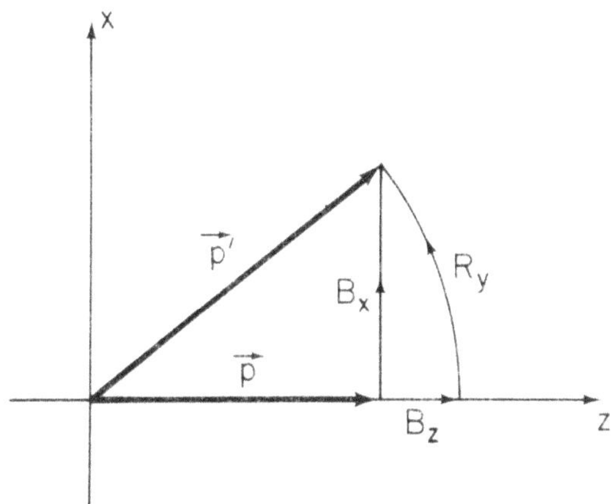

Fig. 10.1: Lorentz boost along the x-direction. The four-momentum can be boosted either directly by B_x or through the rotation R_y preceded by B_z along the z-direction. These operators produce two different four-vectors when applied to the polarization vector. However, they are connected by a gauge transformation.

The application of the transformation $[R_y(\eta)B_z(\eta)]$ on the four-momentum gives the same effect as that of the application of $B_x(\eta)$. Indeed, the matrix

$$D_x(\eta) = [B_x(\eta)]^{-1}R_y(\eta)B_z(\eta), \qquad (10.52)$$

leaves the four-momentum invariant, and is therefore an element of the $E(2)$-like little group for photons. The four-by-four matrix form for $D_x(\eta)$ is complicated, but it becomes very simple once the Lorentz condition of Eq. (10.21) is taken into account. Since $B_z(\eta)$ leaves $\epsilon_\pm^\mu$ invariant, we arrive at the conclusion that

$$\epsilon'^{\mu}_{\pm} = R_y(\eta)\epsilon^{\mu}_{\pm}. \tag{10.53}$$

Indeed, the Lorentz boost B_x on $\epsilon^{\mu}_{\pm}$ preceded by the gauge transformation $D_x(\eta)$ leads to the pure rotation $R_y(\eta)$. This rotation is a finite-dimensional unitary transformation.

The above result raises the following delicate mathematical question. The non-compact group of Lorentz transformations does not have Abelian invariant subgroups, and therefore there cannot be finite-dimensional unitary representations of this group. However, this theorem does not apply to trivial one-dimensional representations. The point is that, if we impose the helicity gauge, the little group has only one degree of freedom which in this case is the rotation around the momentum. The transformation with one-dimensional internal space can be described by a trivial representation [1, 12, 16, 17].

It is possible to approach the above problem through the construction of infinite-dimensional unitary representations. We note that the $E(2)$-like little group for photons is non-compact, and its unitary representation is necessarily infinite-dimensional [12]. However, its translation-like subgroup governing gauge degrees of freedom is an Abelian invariant subgroup. For this reason, we can write

$$D(u, v, \alpha) = D(0, 0, \alpha)D(u', v', 0), \tag{10.54}$$

where the gauge parameters u' and v' are determined from u, v, and α [Problem 4 in Sect. 10.10].

If the above operator is applied to the state which does not depend on the gauge parameters:

$$N_1 |\text{state}\rangle = N_2 |\text{state}\rangle = 0, \tag{10.55}$$

then $D(u', v', 0)$ can be replaced by I. The representation $D(u, v, \alpha)$ is simply that of the rotation around the momentum. This rotation is generated by the helicity operator. Its representation is that of $O(2)$ and is one-dimensional. Thus, again, we end up with the trivial representation.

Table 10.2: Representations of the $E(2)$-like little group for photons. If we freeze gauge degrees of freedom, both unitary infinite-dimensional and non-unitary finite-dimensional representations become the one-dimensional trivial representation.

Representations of the Little Group	Free Photons with definite helicity	Coupled Photons in QED
Unitary Infinite-dimensional	One-dimensional trivial representation	
Non-unitary Finite-dimensional	$\frac{\partial}{\partial x^{\mu}} A^{\mu} = \nabla \cdot A = 0$	Lorentz or Coulomb gauge

Let us summarize what we have done so far in this Chapter. The electromagnetic four-potential is commonly used in quantum electrodynamics (QED) as well as in classical physics. The usual gauge condition is either the Lorentz gauge of Eq. (10.21) or the Coulomb gauge of Eq. (10.26), but not both. In either case, photons are not in an eigenstate of the helicity operator.

If the photon is in a helicity eigenstate, it can be represented by a trivial representation of a unitary representation, or equivalently by the four-potential satisfying both the Lorentz and Coulomb gauge conditions. This situation can be summarized in Table 10.2.

10.5 Infinite and Finite Dimensional Representations of E(2)

As we noted in Appendices A and B, the two-dimensional Euclidean group, often called $E(2)$, consists of rotations and translations on a two-dimensional Euclidian plane. The coordinate transformation takes the form given in Eq. (4.34). This transformation was written in the four-by-four matrix form in Eq. (4.35) The algebraic properties of the above transformation matrix have been discussed in Appendices A and B.

As we saw in Sect. 4.2 of Chap. 4, when a very small xy-plane is considered, the matrix can be written in exponential form as in Eq. (4.36) where J_3 is the generator of rotations, and P_1 and P_2 are the generators of translations. These generators in the case of the above coordinate transformation are the rotation operator J_3, and the P_1 and P_2 given in matrix form in Eq. (4.37) and in differential form in Eqs. (4.38) and (4.39). These generators satisfy the following commutation relations:

$$[P_1, P_2] = 0, \qquad [J_3, P_1] = iP_2, \qquad [J_3, P_2] = -iP_1 . \tag{10.56}$$

The transformation described in Eqs. (4.34) and (4.35) is *active* in the sense that it transforms the object.

Let us next consider transformations of functions of x and y, and continue to use P_1 and P_2 as the generators of translations and J_3 as the generator of rotations. These generators take the form given in Eqs. (4.38) and Eq. (4.39) and seen below:

$$P_1 = -i\frac{\partial}{\partial x} , \qquad P_2 = -i\frac{\partial}{\partial y} ,$$

$$J_3 = -i\left(x\frac{\partial}{\partial y} - y\frac{\partial}{\partial x}\right) . \tag{10.57}$$

The above operators satisfy the commutation relations of Eq. (10.56). The transformation through these differential operators is *passive* in the sense that it is achieved through a coordinate transformation which is inverse to that for the active transformation given in Eq. (4.34).

As in the case of the rotation group, the standard method of studying this group is to find an operator which commutes with all three of the above generators. Using the P_i as defined in Eq. (10.57) it is easy to check that P^2, defined as

$$P^2 = P_1^2 + P_2^2 , \tag{10.58}$$

commutes with all three generators. Thus one way to construct representations of the $E(2)$ group is to solve the equation

$$[P_1^2 + P_2^2]\psi(x, y) = k^2 \psi(x, y) , \tag{10.59}$$

using the differential forms of P_1 and P_2 given in Eq. (10.57). This partial differential equation can be separated in the polar, Cartesian, parabolic, or elliptic coordinate system [18].

In his original paper [1], Wigner used the Cartesian coordinate system diagonal in P_1 and P_2. Inönü and Wigner [3] later studied the differential equation using the polar coordinate system. If we solve the above partial differential equation in the Cartesian coordinate system, the solution is

$$\psi(x, y) = \exp[-i(k_1 x + k_2 y)] , \tag{10.60}$$

where

$$k^2 = k_1^2 + k_2^2 . \tag{10.61}$$

However, in the polar coordinate system, the solution becomes

$$\psi(r, \phi) = J_m(kr)e^{\pm im\phi} , \tag{10.62}$$

where

$$r = (x^2 + y^2)^{1/2} ,$$

$$\phi = \tan^{-1}(y/x) , \tag{10.63}$$

and $J_m(kr)$ is the Bessel function of order m. The solution given in eq. (10.62) is known as the Helmholtz equation. These solutions are diagonal in P_1 and P_2, while the polar solution of (10.62) is diagonal in J_3. These results are summarized in Table 10.3.

It is easy to see that the Cartesian solution given in Eq. (10.60) forms the representation basis for the infinite-dimensional unitary representation of the $E(2)$ group. The application of $\exp(-iuP_1)$ or $\exp(-ivP_2)$ induces a solution different from the original one, and this process can be repeated to produce infinitely many different wave functions. The dimensionality problem for the polar solution is discussed in Exercise 1 in Sect. 10.10.

In addition to unitary representations, there are also finite-dimensional non-unitary representations. The coordinate transformation matrix of Eq. (4.35) is a four-by-four non-unitary matrix. If we compute P^2 using the generators given in

Table 10.3: Representation spaces for the $E(2)$ group. In order that representations be useful in describing massless particles with definite helicity, they should be diagonal in the generators of rotations.

Diagonal in	Unitary infinite dimensional	Non-unitary finite dimensional
P_1 and P_2	$\exp[-i(k_1 x + k_2 y)]$	1 (trivial)
J_3	$J_m(kr)e^{(\pm im\phi)}$	$r^m e^{(\pm im\phi)}$

Eq. (4.37), it vanishes. We are thus led to consider the fundamental differential equation of Eq. (10.59) with vanishing k^2 [19]:

$$\left[\left(\frac{\partial}{\partial x}\right)^2 + \left(\frac{\partial}{\partial y}\right)^2\right]\psi(x, y) = 0. \tag{10.64}$$

This is a two-dimensional Laplace equation, and its solutions are quite familiar to us. The analytic solution of this equation takes the form

$$\psi = r^m \exp(\pm im\phi)$$

$$= (x \pm iy)^m. \tag{10.65}$$

This is an eigenstate of J_3 or a rotation around the origin.

The effect of the rotation operator

$$R(\alpha) = \exp(-i\alpha J_3) \tag{10.66}$$

on ψ of Eq. (10.65) is that

$$R(\alpha)\psi = r^m e^{\pm im(\phi-\alpha)}. \tag{10.67}$$

If we translate the expression of Eq. (10.65) by applying the operator

$$T(u, v) = \exp[-i(uP_1 + vP_2)], \tag{10.68}$$

then the translated form becomes

$$T(u, v)\psi(x, y) = [(x - u) \pm i(y - v)]^m. \tag{10.69}$$

This is no longer an eigenstate of rotation around the origin but around the point at $x = u$ and $y = v$. The rotation operator in this case is

$$R_{uv}(\alpha) = T(u, v)e^{(-i\alpha J_3)}T^{-1}(u, v). \tag{10.70}$$

In Sect. A.3 of Appendix A, we discussed the explicit matrix representation for the coordinate transformations in the two-dimensional plane. All rotations by the same angle but not necessarily around the same point belong to the same equivalence class.

If $m = 1$, we are dealing with a linear form in x and y. Thus the transformation described above is simply a coordinate transformation. In order to establish the connection between the above analysis and the matrix transformation given in Eq. (10.65), let us consider the form:

$$V_1 = \begin{pmatrix} 1 \\ 1 \\ x + iy \\ x - iy \end{pmatrix} = \begin{pmatrix} 1 \\ 1 \\ re^{i\phi} \\ re^{-i\phi} \end{pmatrix}.$$

(10.71)

Then the passive transformation [Exercise 2 in Sect. B.6 of Appendix B] applicable to this vector is

$$E(u, v, \alpha) = \exp(-i\alpha J_3) \exp(-iuP_1 - ivP_2),$$

(10.72)

with

$$J_3 = \begin{pmatrix} 0 & 0 & 0 & 0 \\ 0 & 0 & 0 & 0 \\ 0 & 0 & -1 & 0 \\ 0 & 0 & 0 & 1 \end{pmatrix}, \quad P_1 = \begin{pmatrix} 0 & 0 & 0 & 0 \\ 0 & 0 & 0 & 0 \\ 0 & -i & 0 & 0 \\ 0 & -i & 0 & 0 \end{pmatrix}, \quad P_2 = \begin{pmatrix} 0 & 0 & 0 & 0 \\ 0 & 0 & 0 & 0 \\ 0 & 1 & 0 & 0 \\ 0 & -1 & 0 & 0 \end{pmatrix}.$$

(10.73)

The P_i matrices of Eq. (4.37) or Eq. (10.73) satisfy the commutation relations for the $E(2)$ group given in Eq. (4.26). In addition, these matrices satisfy

$$P_1^2 + P_2^2 = 0,$$

(10.74)

which is a reflection of the differential equation with $k^2 = 0$ in Eq. (10.64). In addition, P_1 and P_2 satisfy

$$P_1^2 = P_2^2 = P_1 P_2 = 0,$$

(10.75)

reflecting the fact that

$$\left(\frac{\partial}{\partial x}\right)^2 (x \pm iy) = \left(\frac{\partial}{\partial y}\right)^2 (x \pm iy) = \frac{\partial^2}{\partial x \partial y}(x \pm iy) = 0.$$

(10.76)

Because of Eq. (10.74) and Eq. (10.76), the power series expansion for $E(u, v, 0)$ terminates:

$$E(u, v, 0) = \exp(-iuP_1 - ivP_2)$$

$$= I - (iuP_1 + ivP_2).$$

(10.77)

The series expansion truncates because of Eq. (10.75).

10.6 Transformation Properties of Electric and Magnetic Fields

We stated in Sect. 3.5 of Chap. 3 that the spinor representations $\alpha_1\alpha_2$ and $\dot{\beta}_1\dot{\beta}_2$ are invariant under gauge transformations, their little group is simply $O(2)$ and unitary. We are now interested in their Lorentz transformation properties and in whether they are like those of electric and magnetic fields.

Let us first discuss the transformation properties of electromagnetic fields. If a plane wave propagates along the z-direction, the electric and magnetic fields are along the x- and y-axes respectively. In the instance of plane waves or free photons, the magnitude of the magnetic field (in our unit system) is the same as that of the electric field.

Under the Lorentz boost by velocity β, the longitudinal components of the electric and magnetic fields remain unchanged. The transverse components become

$$\mathbf{E}'_T = \frac{(\mathbf{E}_T - \beta \times \mathbf{B}_T)}{(1-\beta^2)^{\frac{1}{2}}} \quad \text{and} \quad \mathbf{B}'_T = \frac{(\mathbf{B}_T - \beta \times \mathbf{E}_T)}{(1-\beta^2)^{\frac{1}{2}}} \tag{10.78}$$

where $\mathbf{E}$ and $\mathbf{B}$ are the electric and magnetic fields respectively.

If the photon is boosted along the z-axis so that the momentum or energy is increased by

$$p'_0 = \frac{(p_0 + \beta p_0)}{(1-\beta^2)^{\frac{1}{2}}} = \left(\frac{1+\beta}{1-\beta}\right)^{\frac{1}{2}} p_0 , \tag{10.79}$$

then according to Eq. (10.78), the transverse electric field which stays in the x-direction is also increased by

$$E'_x = \left(\frac{1+\beta}{1-\beta}\right)^{\frac{1}{2}} E_x . \tag{10.80}$$

The magnetic field remains in the y-direction and is increased by the same amount.

Let us next boost along the x-direction, so that the momentum along the z-direction remains the same, but the new momentum has an x component:

$$p'_z = p_0 \quad \text{and} \quad p'_x = p_0 \sinh \eta , \tag{10.81}$$

then the magnitude of the new momentum or energy is

$$p'_0 = (\cosh \eta) p_0 . \tag{10.82}$$

The magnetic field stays in the y-direction, but becomes increased by

$$B'_y = (\cosh \eta) B_y . \tag{10.83}$$

The x component of the electric field remains unchanged: $E'_x = E_x$, but the new electric field has a z component:

$$E'_z = -(\sinh \eta)B_y = -(\sinh \eta)E_x , \tag{10.84}$$

so that it will be perpendicular to the momentum and the magnetic field. Its magnitude is also increased by the same amount as the magnetic field.

Let us identify the following spinor combinations as

$$\alpha_1\alpha_2 \simeq E_x + iE_y \quad \text{and} \quad \dot{\beta}_1\dot{\beta}_2 \simeq B_x - iB_y . \tag{10.85}$$

When the particles are at rest, adding two spin-1/2 spinors results in spin-zero and spin-one states. These two sets of spin-1/2 spinors behave differently when they are boosted. We can define the boost matrices along the z-axis by

$$B(\eta) = \exp\left(-i\eta K_3\right) = \begin{pmatrix} e^{\eta/2} & 0 \\ 0 & e^{-\eta/2} \end{pmatrix} ,$$

$$\dot{B}(\eta) = \exp\left(i\eta K_3\right) = \begin{pmatrix} e^{-\eta/2} & 0 \\ 0 & e^{\eta/2} \end{pmatrix} . \tag{10.86}$$

These boost matrices are applicable to the undotted and dotted spinors, respectively. We note that the two matrices defined in Eq. (10.86) commute with each other, and when applied to the spinors result in:

$$B(\eta)\alpha_1\alpha_2 = e^{\eta}\alpha_1\alpha_2 \quad \text{and} \quad B(\eta)\beta_1\beta_2 = e^{-\eta}\beta_1\beta_2 ,$$

$$\dot{B}(\eta)\dot{\alpha}_1\dot{\alpha}_2 = e^{-\eta}\dot{\alpha}_1\dot{\alpha}_2 \quad \text{and} \quad \dot{B}(\eta)\dot{\beta}_1\dot{\beta}_2 = e^{\eta}\dot{\beta}_1\dot{\beta}_2. \tag{10.87}$$

Then, when we boost along the z-direction, both the electric and magnetic fields become increased by e^{η}. This is consistent with the transformation law given in Eq. (10.80). For simplicity in what follows, we shall drop the designation one and two.

As we calculate in Sect. 10.10 (Exercise 3), when we boost along the x-direction, the spinor transformation matrix for each of the above α's is

$$\exp\left[-i(\eta/2)K_1\right] = \begin{pmatrix} \cosh\frac{\eta}{2} & \sinh\frac{\eta}{2} \\ \sinh\frac{\eta}{2} & \cosh\frac{\eta}{2} \end{pmatrix} . \tag{10.88}$$

and the matrix for $\dot{\beta}$ is the inverse of the above form. Then the new spinors become

$$\alpha' = \begin{pmatrix} \cosh\frac{\eta}{2} \\ \sinh\frac{\eta}{2} \end{pmatrix} = (\cosh \eta)^{\frac{1}{2}} \begin{pmatrix} \cos\frac{\theta}{2} \\ \sin\frac{\theta}{2} \end{pmatrix} , \tag{10.89}$$

$$\dot{\beta}' = \begin{pmatrix} -\sinh\frac{\eta}{2} \\ \cosh\frac{\eta}{2} \end{pmatrix} = (\cosh \eta)^{\frac{1}{2}} \begin{pmatrix} -\sin\frac{\theta}{2} \\ \cos\frac{\theta}{2} \end{pmatrix} , \tag{10.90}$$

where

$$\tan\theta = \sinh\eta \,. \tag{10.91}$$

New spinors α' and $\dot\beta'$ specify the positive and negative helicities with respect to the new momentum respectively. Each of the above spinors gained the normalization constant $(\cosh\eta)^{\frac{1}{2}}$. Thus both $\alpha_1\alpha_2$ and $\dot\beta_1\dot\beta_2$ became bigger by $(\cosh\eta)$. This is exactly the transformation property of the electric and magnetic fields.

With this preparation, we can now consider all the bilinear spinor combinations. Earlier, in Sect. 3.5 of Chap. 3 we have discussed combinations with one dotted and one undotted spinors. In order to deal with the electromagnetic field tensor we complete the list by including combinations with both undotted and both dotted spinors:

$$\alpha\alpha \,, \quad \frac{1}{\sqrt{2}}(\alpha\beta + \beta\alpha) \,, \quad \beta\beta \,, \quad \frac{1}{\sqrt{2}}(\alpha\beta - \beta\alpha) \tag{10.92}$$

and

$$\dot\alpha\dot\alpha \,, \quad \frac{1}{\sqrt{2}}(\dot\alpha\dot\beta + \dot\beta\dot\alpha) \,, \quad \dot\beta\dot\beta \,, \quad \frac{1}{\sqrt{2}}(\dot\alpha\dot\beta - \dot\beta\dot\alpha) \,. \tag{10.93}$$

For mixed combinations we have

$$\alpha\dot\alpha \,, \quad \frac{1}{\sqrt{2}}(\alpha\dot\beta + \beta\dot\alpha) \,, \quad \beta\dot\beta \,, \quad \frac{1}{\sqrt{2}}(\alpha\dot\beta - \beta\dot\alpha) \tag{10.94}$$

also

$$\dot\alpha\alpha \,, \quad \frac{1}{\sqrt{2}}(\dot\alpha\beta + \dot\beta\alpha) \,, \quad \dot\beta\beta \,, \quad \frac{1}{\sqrt{2}}(\dot\alpha\beta - \dot\beta\alpha) \,. \tag{10.95}$$

Here, in Eqs. (10.92) to (10.95), we observe that the symmetric combinations are for spin-1 and the antisymmetric combinations are for spin-0 particles. Also, both $\frac{1}{\sqrt{2}}(\alpha\beta-\beta\alpha)$ and $\frac{1}{\sqrt{2}}(\dot\alpha\dot\beta-\dot\beta\dot\alpha)$ are singlets and they are invariant both under rotations and boosts.

In order to construct the z component in the $O(3)$ space, let us first consider [20]:

$$f_z = \frac{1}{2}\left[(\alpha\beta + \beta\alpha) - (\dot\alpha\dot\beta + \dot\beta\dot\alpha)\right] \,, \quad g_z = \frac{1}{2i}\left[(\alpha\beta + \beta\alpha) + (\dot\alpha\dot\beta + \dot\beta\dot\alpha)\right] \,. \tag{10.96}$$

Here, f_z and g_z are respectively anti-symmetric and symmetric under the dot conjugation or the parity operation. These quantities are invariant under the boost along the z-direction. They are also invariant under rotations around the z-axis, but they are not invariant under boosts along or rotations around the x- or y-axis.

Next, in order to construct the x and y components, we start with $f_\pm$ and $g_\pm$ as:

$$f_+ = \frac{1}{\sqrt{2}}(\alpha\alpha - \dot\alpha\dot\alpha) \,, \quad f_- = \frac{1}{\sqrt{2}}(\beta\beta - \dot\beta\dot\beta) \,,$$

$$g_+ = \frac{1}{\sqrt{2}i}(\alpha\alpha + \dot\alpha\dot\alpha) \,, \quad g_- = \frac{1}{\sqrt{2}i}(\beta\beta + \dot\beta\dot\beta) \,. \tag{10.97}$$

Then:

$$f_x = \frac{1}{\sqrt{2}}\,(f_+ + f_-) = \frac{1}{2}\left[(\alpha\alpha + \beta\beta) - (\dot{\alpha}\dot{\alpha} + \dot{\beta}\dot{\beta})\right]\,,$$

$$f_y = \frac{1}{\sqrt{2}i}\,(f_+ - f_-) = \frac{1}{2i}\left[(\alpha\alpha - \beta\beta) - (\dot{\alpha}\dot{\alpha} - \dot{\beta}\dot{\beta})\right]\,, \qquad (10.98)$$

and

$$g_x = \frac{1}{\sqrt{2}}\,(g_+ + g_-) = \frac{1}{2i}\left[(\alpha\alpha + \beta\beta) + (\dot{\alpha}\dot{\alpha} + \dot{\beta}\dot{\beta})\right]\,,$$

$$g_y = \frac{1}{\sqrt{2}i}\,(g_+ - g_-) = -\frac{1}{2}\left[(\alpha\alpha - \beta\beta) + (\dot{\alpha}\dot{\alpha} - \dot{\beta}\dot{\beta})\right]\,. \qquad (10.99)$$

Here, f_x and f_y are anti-symmetric under dot conjugation, while g_x and g_y are symmetric.

In addition, the g_i components change signs under space inversion, while the f_i components remain invariant. They are thus like the electric and magnetic fields, respectively. Hence, they can be formed into a second rank tensor:

$$T = \begin{pmatrix} 0 & -f_z & -f_x & -f_y \\ f_z & 0 & -g_y & g_x \\ f_x & g_y & 0 & -g_z \\ f_y & -g_x & g_z & 0 \end{pmatrix}, \qquad (10.100)$$

whose Lorentz-transformation properties are well known.

For massless particles, such as photons with spin-1, when the system f_i and g_i is Lorentz-boosted, we keep only those values that become large by a factor of e^{η}, since values that become smaller tend to vanish when the particles speed approach to that of light [21]. Thus,

$$f_x \rightarrow \frac{1}{2}\left(\alpha\alpha - \dot{\beta}\dot{\beta}\right),\quad f_y \rightarrow \frac{1}{2i}\left(\alpha\alpha + \dot{\beta}\dot{\beta}\right)\,,$$

$$g_x \rightarrow \frac{1}{2i}\left(\alpha\alpha + \dot{\beta}\dot{\beta}\right)\,,\quad g_y \rightarrow -\frac{1}{2}\left(\alpha\alpha - \dot{\beta}\dot{\beta}\right) \qquad (10.101)$$

in the massless limit.

Then the tensor of Eq. (10.100) becomes [22]:

$$F = \begin{pmatrix} 0 & 0 & -E_x & -E_y \\ 0 & 0 & -B_y & B_x \\ E_x & B_y & 0 & 0 \\ E_y & -B_x & 0 & 0 \end{pmatrix}. \qquad (10.102)$$

Now, we have

$$E_x \simeq \frac{1}{2}\left(\alpha\alpha - \dot\beta\dot\beta\right), \quad E_y \simeq \frac{1}{2i}\left(\alpha\alpha + \dot\beta\dot\beta\right),$$

$$B_x \simeq \frac{1}{2i}\left(\alpha\alpha + \dot\beta\dot\beta\right), \quad B_y \simeq -\frac{1}{2}\left(\alpha\alpha - \dot\beta\dot\beta\right). \tag{10.103}$$

The electric and magnetic field components are perpendicular to each other. Furthermore,

$$B_x = E_y, \qquad B_y = -E_x. \tag{10.104}$$

In order to address symmetry of photons, let us go back to Eq. (10.97), where their massless limits reduce to

$$B_+ \simeq E_+ \simeq \alpha\alpha, \quad \text{and} \quad B_- \simeq E_- \simeq \dot\beta\dot\beta. \tag{10.105}$$

The B_+ and E_+ are for the photon spin along the z-direction, while B_- and E_- are for the opposite direction. Although Eq. (10.105) is only for the field propagating along the z-direction, expressions that prevail in any direction can be obtained by applying rotations [22].

It is important to note that an anti-symmetric second-rank tensor for massive particles, whose six components here are f_i and g_i, form a representation space for the $O(3)$-like little group, which is quite different from the representation space for massless particles. As we shall see in Sect. 10.7, the gauge transformations applicable to α and $\dot\beta$ are two-by-two matrices, and both α and $\dot\beta$ are invariant under gauge transformations, while $\dot\alpha$ and β are not. Therefore, for massless particles the only gauge-invariant products are $\alpha\alpha$ and $\dot\beta\dot\beta$, so only those products can appear in the components of the Maxwell tensor.

From the $SL(2, c)$ standpoint, there is a non-trivial difference between the four-potential and electromagnetic field. The four-potential is represented by a direct product of one undotted and one dotted spinor. The most crucial difference between the two different representations is that the transformations on the four-potential can be made unitary. Transformations on the electric and magnetic fields are non-unitary as in the case of the Dirac spinors. The amplitudes of the Dirac and Maxwell fields are proportional to $(p_0)^{\frac{1}{2}}$ and p_0 respectively.

Starting from Wigner's 1939 paper [1], in 1964 Weinberg [14] constructed gauge-invariant state vectors for massless particles. Then Weinberg's state vectors correspond to the bilinear spinors $\alpha\alpha$ and $\dot\beta\dot\beta$.

We have attempted to give here a more precise definition of Maxwell's tensor which can be derived from the zero-mass limit by using the Wigner excursion, given in Figure 10.2. Thus a more precise definition of the polarization of photons was able to be given.

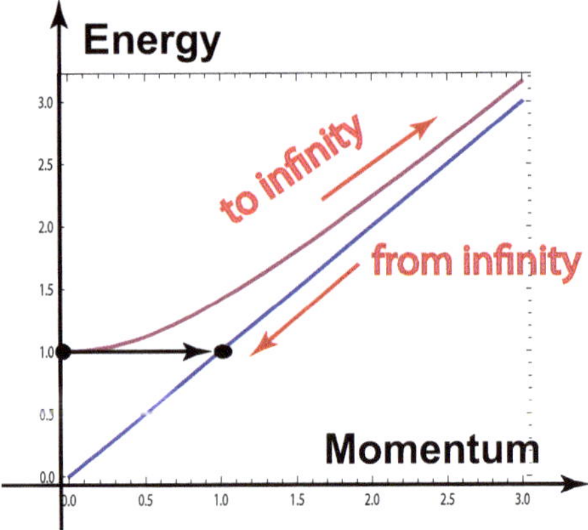

Fig. 10.2: Wigner excursion for massless particles. For a massive particle, the Wigner transformation matrix given in Table 3.2 can be boosted to the infinite momentum limit through the hyperbola to coincide with the light cone. At the infinite momentum limit it can come back for the massless particle through the light-cone line [20].

Wigner's original book of 1931 [23], discussed the rotation group, without Lorentz transformations. As we know, four states can be constructed from two spin-1/2 spinors, which leads to three spin-1 states and one spin-0 state. If we have three spinors, four spin-3/2 states and two spin-1/2 states can be constructed, resulting in six states. Although this partition process is much more complicated [24, 25] for three spinors, it is possible for all higher spin states.

In the relativistic Lorentz-covariant world, there are four states for each spin-1/2 particle. If we match two spinors, we get the sixteen states given in Eqs. (10.92) to (10.95). There should be 64 states for three spinors and for four spinors, 256 states. We illustrate in Figure 10.3 how spins for massive and massless particles can be bundled together.

Photons and gravitons, in this relativistic regime, are of great interest. From the preceding we can conclude that the observable components under gauge transformations, are invariant. Furthermore, these terms become largest for large values of η. We have also seen that the photon states consisting of $\alpha\alpha$ and $\dot\beta\dot\beta$ spinors, are parallel and anti-parallel to the momentum, respectively. Then it should be possible to conclude that the states must be $\alpha\alpha\alpha\alpha$ and $\dot\beta\dot\beta\dot\beta\dot\beta$, respectively for spin-2 gravitons.

Weinberg, when he constructed his states for massless particles [26], especially photons and gravitons [14], started with

$$N_1|\text{state}\rangle = 0 \quad \text{and} \quad N_2|\text{state}\rangle = 0, \tag{10.106}$$

where N_1 and N_2 are defined in Eq. 3.36 of Chap. 3, and we saw that these are the generators of gauge transformations. Therefore, the states defined by Weinberg are gauge-invariant states. We can then identify $\alpha\alpha$ and $\dot\beta\dot\beta$ as Weinberg's states for

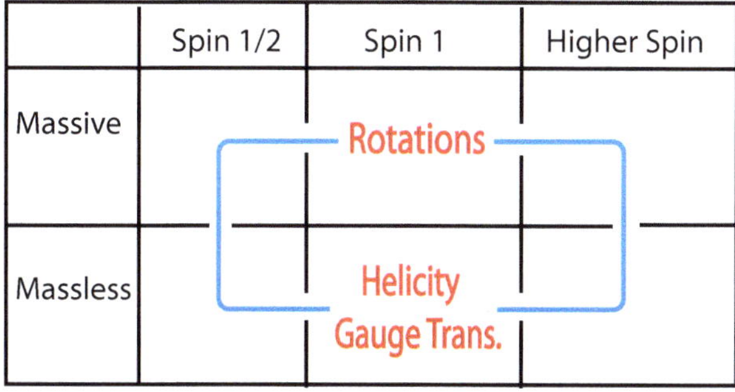

Fig. 10.3: Unified picture of massive and massless particles. The gauge transformation, being a Lorentz-boosted rotation matrix, is thus applicable to all massless particles. Higher-spin states in the Lorentz-covariant world, are capable of being constructed from the four states of the spin-1/2 particle [21, 20].

photons. Then $\alpha\alpha\alpha\alpha$ and $\dot\beta\dot\beta\dot\beta\dot\beta$ should be able to be identified as Weinberg's states for gravitons.

10.7 Massless Particles with Spin-1/2

As we noted in Eq. 2.45 of Chap. 2, for a massless particle moving along the z-direction, the little group is generated by

$$N_1 = K_1 - J_2, \quad N_2 = K_2 + J_1, \quad J_3, \tag{10.107}$$

where the above matrices in two-by-two form were introduced in Sect. 3.3 of Chap. 3. The N operators generate gauge transformations in the case of photons. What is then the significance of these operators for massless particles with spin-1/2?

The commutation relations for the generators of the Lorentz group remain invariant under the sign change in K_i. We can determine the sign of K_i unambiguously when they are applicable to the space-time coordinate variables and photon four-vectors. In the $SL(2, c)$ regime, we can choose $J_i = \frac{1}{2}\sigma_i$ as the generators of rotations. However, the boost generators can take two different signs: $K_i = \pm\frac{i}{2}\sigma_i$. We thus have to consider both $N_i^{(+)}$ and $N_i^{(-)}$:

$$N_1^{(+)} = \begin{pmatrix} 0 & i \\ 0 & 0 \end{pmatrix}, \qquad N_2^{(+)} = \begin{pmatrix} 0 & 1 \\ 0 & 0 \end{pmatrix}. \tag{10.108}$$

We can obtain $N_1^{(-)}$ and $N_2^{(-)}$ by taking the Hermitian conjugation of the above expressions respectively. The result is:

$$N_1^{(-)} = \begin{pmatrix} 0 & 0 \\ -i & 0 \end{pmatrix}, \qquad N_2^{(-)} = \begin{pmatrix} 0 & 0 \\ 1 & 0 \end{pmatrix}. \tag{10.109}$$

Then the transformation matrices become

$$D^{(+)}(u,v) = \exp(-i[uN_1^{(+)} + vN_2^{(+)}]) = \begin{pmatrix} 1 & u - iv \\ 0 & 1 \end{pmatrix},$$

$$D^{(-)}(u,v) = \exp(-i[uN_1^{(-)} + vN_2^{(-)}]) = \begin{pmatrix} 1 & 0 \\ -u - iv & 1 \end{pmatrix}. \tag{10.110}$$

Since these matrices correspond to gauge transformation matrices for photons, we shall hereafter call them the gauge transformation matrices in the $SL(2,c)$ regime [1, 27].

There are two sets of spinors in $SL(2,c)$. For spinors whose boosts are generated by $K_i = \frac{i}{2}\sigma_i$, we shall use the usual Pauli notation α and β for positive and negative helicity states respectively. This results in:

$$\alpha = \begin{pmatrix} 1 \\ 0 \end{pmatrix} \quad \text{and} \quad \beta = \begin{pmatrix} 0 \\ 1 \end{pmatrix}. \tag{10.111}$$

For those whose boosts are generated by $K_i = -\frac{i}{2}\sigma_i$, we shall use $\dot{\alpha}$ and $\dot{\beta}$. Then, the spinors are gauge-invariant in the sense that

$$D^{(+)}(u,v)\alpha = \alpha, \qquad D^{(-)}(u,v)\dot{\beta} = \dot{\beta}. \tag{10.112}$$

On the other hand, the $SL(2,c)$ spinors are gauge-dependent in the sense that

$$D^{(+)}(u,v)\beta = \beta + (u - iv)\alpha, \qquad D^{(-)}(u,v)\dot{\alpha} = \dot{\alpha} - (u + iv)\dot{\beta}. \tag{10.113}$$

Spinors of Eq. (10.112) are gauge-invariant [4, 28]. We see that

$$D(\alpha,\beta) = \begin{pmatrix} 1 & -\gamma \\ 0 & 1 \end{pmatrix} \quad \text{and} \quad \dot{D}(\alpha,\beta) = \begin{pmatrix} 1 & 0 \\ \gamma & 1 \end{pmatrix}, \tag{10.114}$$

in the massless limit where $D(\alpha,\beta)$ corresponds to Wigner's transformation matrix for massless particles given in Table 3.2 of Chap. 3.

Indeed, in the world of massless spin-1/2 particles and fields, gauge invariance prevails. As a result any massless spin-1/2 particles are polarized. This is shown in Figure 10.4. Here we see that the undotted spinor is left-handed, but the dotted spinor is right-handed [30, 4, 28, 31, 32].

Presently experiments do not observe any spin-1/2 massless particles in nature. For a very long time neutrinos were thought to be massless. They were and they still are quite perplexing particles having three flavors ν_α (namely, the electron ν_e, the muon ν_μ and the tau ν_τ neutrinos), with three masses $(m_i, \quad i = 1, 2, 3)$. To be more precise, each neutrino flavor eigenstate is the superposition of three neutrino mass eigenstates [33, 34]:

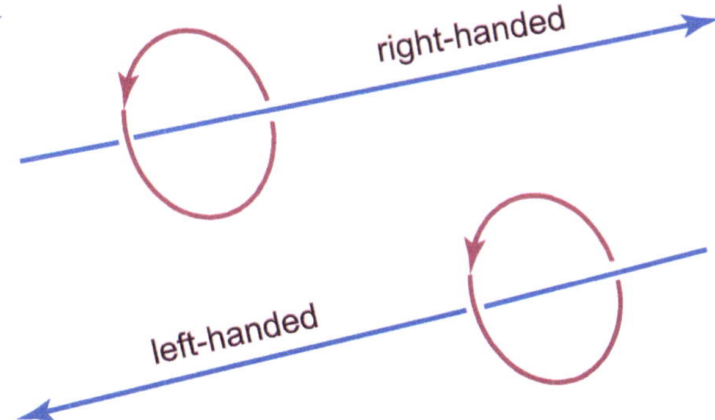

Fig. 10.4: Gauge invariance results in a massless spin-1/2 particle being polarized. This results in what we see for the left-handed and the right-handed particle [29].

$$|\nu_\alpha\rangle = \sum_i U_{\alpha i} |\nu_i\rangle, \tag{10.115}$$

where $U_{\alpha i}$ is the unitary 3×3 mass mixing matrix.

Currently there are many experiments carried out around the world to find answers to the questions pertinent to the nature of neutrinos [35, 36]. Since the exact mass value is not yet determined but only the differences of the mass squares, Δm_{ij}^2 are measured, the mass hierarchy problem i.e., whether $m_1 < m_2 < m_3$ or $m_3 < m_2 < m_1$ is the valid ordering has become the most pressing issue [37]. At this point it should also be pointed out that since three neutrino oscillations have been observed [38], we should not expect the smallest mass value could be zero. Some of the other questions queuing to be answered by the outcomes of the experiments are whether neutrinos are Dirac or Majorana particles [39, 40] and thus whether the neutrino is also its own antiparticle [41], and whether sterile neutrinos exist [42, 43].

All fermions, such as leptons and quarks, gain their mass through their interactions with the Higgs boson. However, neutrinos appear to defy that pattern. Hopefully, in the near future neutrino experiments will be able to contribute to new insights into theoretical descriptions of elementary particle physics. Consequently, how the neutrino has obtained mass will have a clear explanation.

Above, we have given an example as to how an E(2)-like group acts on a massless spin-1/2 particle. We think this is worthwhile from a representation theory of groups. In addition we have tried to elucidate a historical point of view regarding neutrinos.

Let us now consider where the above gauge-dependent spinors stand in the physics of particles. Are they really responsible for the gauge dependence of electromagnetic four-potentials when we construct a four-vector by taking a bilinear combination of spinors?

As we did in Eqs. (3.72) and (3.73) of Chap. 3, we can construct the following unit vectors in the Minkowski space by taking the direct products of two $SL(2, c)$ spinors:

$$-\alpha\dot{\alpha} = (0,0,1,i)\,, \qquad \beta\dot{\beta} = (0,0,1,-i)\,,$$

$$\alpha\dot{\beta} = (1,1,0,0)\,, \qquad \beta\dot{\alpha} = (1,-1,0,0)\,. \tag{10.116}$$

For $D(u,v)$ to be consistent with Eq. (10.116), we should choose

$$D(u,v) = D^{(+)}(u,v)D^{(-)}(u,v)\,, \tag{10.117}$$

where $D^{(+)}$ and $D^{(-)}$ are applicable to the first and second spinors of Eq. (10.116) respectively. Then

$$D(u,v)(-\alpha\dot{\alpha}) = -\alpha\dot{\alpha} + (u+iv)\alpha\dot{\beta}\,,$$

$$D(u,v)\beta\dot{\beta} = \beta\dot{\beta} + [u-iv)\alpha\dot{\beta}\,,$$

$$D(u,v)\alpha\dot{\beta} = \alpha\dot{\beta}\,. \tag{10.118}$$

The plane-wave photon four-potential does not depend on $\beta\dot{\alpha}$. The first two equations of the above expression correspond to the gauge transformations on the photon polarization vectors. The third equation corresponds to the effect of the D transformation on the four-momentum, confirming the fact that $D(u,v)$ is an element of the little group.

The remaining question is how the above analysis can be translated into the language of the Dirac equation. We shall study this problem in the next Section.

10.8 Infinite-Momentum/Zero-Mass Limit of the Dirac Equation

Let us start with the three generators of the $SU(2)$ subgroup of the $SL(2,c)$ group which were introduced in Chap. 3:

$$J_i = \frac{1}{2}\sigma_i\,. \tag{10.119}$$

The Lorentz boost along the z-direction is accomplished through the similarity transformation:

$$J'_i = B(P)J_i B^{-1}(P)\,, \tag{10.120}$$

where the applicable boost matrix is

$$B(P) = \begin{pmatrix} e^{\xi/2} & 0 \\ 0 & e^{-\xi/2} \end{pmatrix}\,, \tag{10.121}$$

with

$$e^{\xi/2} = \left(\frac{E+P_z}{E-P_z} \right)^{1/4}\,. \tag{10.122}$$

In the large-(momentum/mass) limit,

$$e^{\xi} \rightarrow \frac{2E}{M} \, . \tag{10.123}$$

Under the similarity transformation of Eq. (10.120), J_3 remains invariant. However, J_1 and J_2 become

$$J_1' = \begin{pmatrix} 0 & \frac{1}{2}e^{\xi} \\ \frac{1}{2}e^{-\xi} & 0 \end{pmatrix}, \quad J_2' = \begin{pmatrix} 0 & -\frac{i}{2}e^{\xi} \\ \frac{i}{2}e^{-\xi} & 0 \end{pmatrix} . \tag{10.124}$$

We can go through a limiting procedure similar to that given in Eqs. (10.6) and (10.7) to obtain

$$N_1 = -\frac{M}{E}J_2', \quad N_2 = \frac{E}{M}J_1' , \tag{10.125}$$

in the large-(momentum/mass) limit. The two-by-two matrix forms for N_1 and N_2 have been discussed in Chap. 3 and are given explicitly in Eq. (3.36) and Eqs. (10.108) and (10.109). They are upper triangular matrices with zero diagonal elements. The multiplication of the factor $e^{-\xi}$ is exactly like the procedure of getting G_1 and G_2 in Eq.(10.6) from the four-by-four representation of J_1' and J_2' respectively.

Unlike the case of coordinate transformations, we have to consider here both signs of the boost generators. The effect of changing the sign is the same as taking the Hermitian conjugate of the above N matrices. These Hermitian-conjugated matrices are lower triangular with vanishing diagonal elements.

Let us now discuss the large-momentum/zero-mass limit of the Dirac equation. In order to accommodate both signs of the boost operators, we can write down the generators of $SL(2, c)$ in the form:

$$J_1 = \begin{pmatrix} (\frac{1}{2})\sigma_1 & 0 \\ 0 & (\frac{1}{2})\sigma_1 \end{pmatrix}, \quad K_1 = \begin{pmatrix} (\frac{i}{2})\sigma_1 & 0 \\ 0 & (-\frac{i}{2})\sigma_1 \end{pmatrix} . \tag{10.126}$$

These are the generators, discussed in Sect. 3.6 of Chap. 3, applicable to the Dirac wave functions in the Weyl representation. Indeed, choosing the sign of the boost operator is equivalent to choosing the sign of γ_0. With this point in mind, we can construct the gauge transformation matrix:

$$D(u, v) = \begin{pmatrix} D^{(+)}(u, v) & 0 \\ 0 & D^{(-)}(u, v) \end{pmatrix} , \tag{10.127}$$

applicable to the Dirac spinors.

In order to understand the effect of the above D matrices, let us start with the eigenspinors J of Eq. (10.119) for a massive Dirac particle at rest:

$$U(0) = \begin{pmatrix} \alpha \\ \pm \dot{\alpha} \end{pmatrix}, \quad V(0) = \begin{pmatrix} \pm \beta \\ \dot{\beta} \end{pmatrix} . \tag{10.128}$$

The $+$ and $-$ signs in the above expression specify positive and negative energy states respectively. If we boost these spinors along the z-axis by applying the boost operator generated by K_3, then

$$U(\mathbf{P}) = \begin{pmatrix} [\exp(+\xi/2)]\alpha \\ \pm[\exp(-\xi/2)]\dot{\alpha} \end{pmatrix}, \quad V(\mathbf{P}) = \begin{pmatrix} \pm[\exp(-\xi/2)]\beta \\ [\exp(+\xi/2)]\dot{\beta} \end{pmatrix}. \qquad (10.129)$$

In the above expression, $\exp(\xi/2)$ becomes large and $\exp(-\xi/2)$ becomes small. From Eqs. (10.116) and (10.117), we can see that the large components are gauge-invariant while the small components are gauge-dependent. Therefore, in general, spin-1/2 particles with non-zero mass are not invariant under gauge transformations.

In the large-momentum/zero-mass limit, we can renormalize the above spinors, and write them as

$$U(\mathbf{P}) = \begin{pmatrix} \alpha \\ 0 \end{pmatrix}, \quad V(\mathbf{P}) = \begin{pmatrix} 0 \\ \dot{\beta} \end{pmatrix}. \qquad (10.130)$$

The spinors in Eq. (10.130) are invariant under the D transformation. The gauge-dependent spinors disappear in the large-momentum/zero-mass limit. This is why we did not talk about gauge transformations on massless spin-1/2 particles represented by the Dirac equation. As we saw in Sect. 10.6, this situation is similar to the case in which the electric and magnetic fields (not potentials), which are solutions of Maxwell's equations, are invariant under gauge transformations.

Let us summarize what we did above. We had to go through two different steps in taking the large-momentum/zero-mass limit of the Dirac equation and its solutions. The first step was to work out the limiting procedure for obtaining the two-by-two matrices of the $E(2)$-like subgroup of $SL(2, c)$ from $SU(2)$. The second step was to eliminate the small component in the Weyl representation of the Dirac equation.

10.9 Massless Composite Particles and the Harmonic Oscillator

Here we are concerned with the study of the internal space-time symmetry of composite particles in the harmonic oscillator regime. Indeed, we find this study is relegated to finding suitable separable coordinate systems. In Chap. 5, we were concerned with the representations of the $O(3)$-like little group for massive particles. The little group for massless particles is locally isomorphic to $E(2)$, and its generators are different from those for the $O(3)$-like little group for massive particles. It is therefore of interest to see whether there exists a separable coordinate system for studying internal space-time symmetries of massless composite particles.

In one of the thirty four-coordinate systems discussed by Kalnins and Miller [44], three of the space-time coordinate variables are conjugate to the generators of the $E(2)$-like little group for the massless composite particle. We can therefore construct representations of the $E(2)$-like little group by solving the differential equation using this Kalnins-Miller coordinate system.

Let us go back to the oscillator differential equation discussed in Chap. 5. If the composite particle is massless, this means that m_0 and λ given in Sect. 5.1 of Chap. 5 are such that p^2 is zero:

$$p^2 = m_0^2 + \lambda = 0. \qquad (10.131)$$

We assume without loss of generality that the momentum of the massless composite particle is in the z-direction. Then the little group is generated by J_3, N_1, and N_2, where

$$N_1 = K_1 - J_2$$

$$= i\left(x\frac{\partial}{\partial t} - t\frac{\partial}{\partial x}\right) + i\left(z\frac{\partial}{\partial x} - x\frac{\partial}{\partial z}\right),$$

$$N_2 = K_2 + J_1$$

$$= i\left(y\frac{\partial}{\partial t} - t\frac{\partial}{\partial y}\right) - i\left(z\frac{\partial}{\partial y} - y\frac{\partial}{\partial z}\right). \tag{10.132}$$

These generators satisfy the commutation relations for the generators of the $E(2)$ group given in Eq. (10.13); J_3 in this case is the helicity operator. In constructing representations, we should note that there are two maximal commuting sets of operators in the Lie algebra. They are [19, 45]

$$(a): \quad N^2, N_1 N_2 \quad \text{and} \quad (b): \quad N^2, J_3, \tag{10.133}$$

where $N^2 = N_1^2 + N_2^2$. Because we are interested in states with definite helicities, we have to construct wave functions which are diagonal in J_3 and N^2.

In order to construct solutions of the oscillator wave equation, we have to make a judicious choice of the coordinate system in which the differential equation is separable. We shall call this the Kalnins-Miller coordinate system.

If $x^\tau x_\tau$ is time-like with positive t, then the Kalnins-Miller coordinate variables $\{x^\mu\} = \{\rho, \xi, \alpha, \phi\}$, are related to $\{x^\tau\} = \{t, z, x, y\}$ by

$$t = (\rho/2)[e^\alpha + (\xi^2 + 1)e^{-\alpha}],$$

$$z = (\rho/2)[e^\alpha + (\xi^2 - 1)e^{-\alpha}],$$

$$x = \rho e^{-\alpha}\xi\cos\phi,$$

$$y = \rho e^{-\alpha}\xi\sin\phi. \tag{10.134}$$

These equations can also be written as

$$\rho = (t^2 - z^2 - r^2)^{1/2},$$

$$\xi = r/(t - z),$$

$$\alpha = -\ln\left\{(t - z)/(t^2 - z^2 - r^2)^{1/2}\right\},$$

$$\phi = \tan^{-1}(y/x), \tag{10.135}$$

where

$$r = (x^2 + y^2)^{1/2}. \tag{10.136}$$

In terms of the Kalnins-Miller variables, J_3, N_1, N_2, and N^2 take the form

$$J_3 = -i\frac{\partial}{\partial\phi}, \qquad N_1 = -i\frac{\partial}{\partial\xi_1}, \qquad N_2 = -i\frac{\partial}{\partial\xi_2},$$

$$N^2 = -\left(\frac{\partial}{\partial\xi}\right)^2 - \frac{1}{\xi}\left(\frac{\partial}{\partial\xi}\right) - \frac{1}{\xi^2}\left(\frac{\partial}{\partial\phi}\right)^2 = -\left\{\left(\frac{1}{\xi_1}\right)^2 + \left(\frac{1}{\xi_2}\right)^2\right\}, \quad (10.137)$$

where

$$\xi_1 = \xi\cos\phi, \qquad \xi_2 = \xi\sin\phi. \tag{10.138}$$

The harmonic oscillator differential equation can then be written as

$$\left(\frac{1}{\rho}\right)^3 \frac{\partial}{\partial\rho}\left[\rho^3\frac{\partial}{\partial\rho}\psi(x^\mu)\right] + \left(\frac{1}{\rho}\right)^2\left[\left(\frac{\partial}{\partial\alpha}\right)^2 - 2\left(\frac{\partial}{\partial\alpha}\right)\right]\psi(x^\mu)$$

$$- \left[\left(\frac{1}{\rho}\right)^2 e^{-2\alpha}N^2 + \rho^2\right]\psi(x^\mu) = 2\lambda\psi(x^\mu). \tag{10.139}$$

In order to separate this differential equation, let us write $\psi(x^\mu)$ in the form

$$\psi(x^\mu) = G(\rho,\alpha)F(\xi,\phi). \tag{10.140}$$

If $F(\xi,\phi)$ satisfies the eigenvalue equation

$$N^2 F(\xi,\phi) = b^2 F(\xi,\phi), \tag{10.141}$$

then, $G(\rho,\alpha)$ should satisfy the differential equation

$$\left(\frac{1}{\rho}\right)^3 \frac{\partial}{\partial\rho}\left(\rho^3\frac{\partial}{\partial\rho}G\right) + \left(\frac{1}{\rho}\right)^2\left[\left(\frac{\partial}{\partial\alpha}\right)^2 - 2\left(\frac{\partial}{\partial\alpha}\right) - e^{-2\alpha}b^2\right]G$$

$$-\rho^2 G = 2\lambda G. \tag{10.142}$$

The eigenvalue equation of Eq. (10.141) is a two-dimensional Helmholtz equation if b^2 does not vanish. It is a Laplace equation if $b^2 = 0$. As was discussed in Sect. 10.5, b^2 has to vanish in order that the system be physically interesting. The solution then becomes

$$F(\xi,\phi) = \xi^m \exp(\pm im\phi) \tag{10.143}$$

where m is an integer and is the magnitude of the angular momentum. Since $b^2 = 0$, the differential equation of Eq. (10.142) becomes

$$\left(\frac{1}{\rho}\right)^3 \frac{\partial}{\partial\rho}\left(\rho^3\frac{\partial}{\partial\rho}G\right) + \left(\frac{1}{\rho}\right)^2\left[\left(\frac{\partial}{\partial\alpha}\right)^2 - 2\left(\frac{\partial}{\partial\alpha}\right)\right]G - \rho^2 G = 2\lambda G. \tag{10.144}$$

The solution of the above differential equation will then take the form

$$G_{\mu n}(\rho, \alpha) = [\rho^n \exp(-\rho^2/2)]L_\mu^{(n+1)}(\rho^2)A_n^{\pm}(\alpha)\,, \qquad (10.145)$$

where

$$A_n^{(+)}(\alpha) = \exp[(n+2)\alpha] \quad \text{and} \quad A_n^{(-)}(\alpha) = \exp(-n\alpha)\,. \qquad (10.146)$$

$L_\mu^{(n+1)}(\rho^2)$ is the associated Laguerre function. The eigenvalue in Eq. (10.144) takes the values:

$$\lambda = -(n + 2\mu + 1)\,, \qquad (10.147)$$

where n and μ take integer values. In order that the composite particle be massless, the above eigenvalue and m_0^2 should satisfy the condition for massless particles:

$$p^2 = m_0^2 + \lambda = 0\,. \qquad (10.148)$$

We have considered so far only the case where x is time-like with positive values of t. If t is negative, we can reverse the sign of the Cartesian coordinate variables given in Eqs. (10.134) and (10.135). If x is a space-like vector, z and t of Eq. (10.134) have to be modified to

$$z = (\rho/2)[e^\alpha - (\xi^2 - 1)e^{-\alpha}]\,,$$

$$t = (\rho/2)[e^\alpha - (\xi^2 + 1)e^{-\alpha}]\,. \qquad (10.149)$$

Consequently, three of the equations in Eq. (10.135) are modified to

$$\rho = (r^2 + z^2 - t^2)^{1/2}\,,$$

$$\xi = r/(z - t)\,,$$

$$\alpha = -\ln[\rho/(z - t)]\,. \qquad (10.150)$$

The process of separating and solving the differential equation is the same as in the case of time-like region. We can use the form of Eq. (10.143) for $F(\xi, \phi)$, and Eq. (10.145) for $G(\rho, \alpha)$. However, the eigenvalue λ in this case takes the values:

$$\lambda = n + 2\mu + 1\,. \qquad (10.151)$$

It is important that the masslessness condition of Eq. (10.148) be satisfied for the above values of λ. Since λ's for the time-like and space-like regions have opposite signs, m_0^2 will also have different signs. This does not cause any conceptual difficulty, because the time-like region never mixes with the space-like region under Poincaré transformations.

In studying space-time symmetries of the solution of the wave equation obtained in Sect. 10.5, we note that the internal wave function $\psi(x)$ is a product of $G(\rho, \alpha)$ and $F(\xi, \phi)$, as is given in Eq. (10.140). This allows us to deal with F and G separately.

Let us first discuss the F function. It is not difficult to see that ϕ in Eq. (10.143) is the angle variable specifying the rotation around the z-axis with angular momentum $\pm m$. This is known as the helicity for the massless particle. Since physically observable states are expected to be helicity eigenstates, they are invariant under the rotation around the z-axis.

In terms of the ξ_1 and ξ_2 variables defined in Eq. (10.137), F can be written as

$$F(\xi, \phi) = F(\xi_1, \xi_2) = (\xi_1 \pm \xi_2)^m, \tag{10.152}$$

with

$$\xi_1 = x/(t - z), \qquad \xi_2 = y/(t - z), \tag{10.153}$$

for the time-like region. The operators N_1 and N_2 given in Eq. (10.132) now generate translations in the $\xi_1\xi_2$-plane. Since both of these *translation* operators commute with N^2, the differential equation of Eq. (10.139) is invariant under this transformation. We can replace ξ_1 and ξ_2 in $F(\xi_1\xi_2)$ of Eq. (10.152) by ξ_1' and ξ_2' respectively, where

$$\xi_1' = \xi_1 - v_1, \qquad \xi_2' = \xi_2 - v_2 \tag{10.154}$$

without changing the differential equation.

As was shown in Sects. 10.2 and 10.5, the above-mentioned N_1 and N_2 transformations are equivalent to gauge transformations. Then what is this gauge transformation in terms of the conventional space-time variables? The *translation* of Eq. (10.154) causes the following changes in the ξ and ϕ variables:

$$\xi \rightarrow \xi' = [(\xi_1 - v_1)^2 + (\xi_2 - v_2)^2]^{1/2},$$

$$\phi \rightarrow \phi' = \tan^{-1}\left[\frac{(\xi_2 - v_2)}{(\xi_1 - v_1)}\right]. \tag{10.155}$$

Another way to interpret the above transformation is to regard Eq. (10.152) as a rotation around the origin in the $\xi_1\xi_2$-plane. Then the transformation of Eq. (10.155) shifts the center of rotation from the origin to the coordinate point v_1v_2.

In order to see the effect of the N_1 and N_2 transformations in terms of the Cartesian space-time variables, let us write Eq. (10.134) for the time-like region as

$$r = \rho\xi e^{-\alpha}, \quad y/x = \tan\phi,$$

$$t + z = \rho[e^\alpha + \xi^2 e^{-\alpha}],$$

$$t - z = \rho e^{-\alpha}. \tag{10.156}$$

It is apparent that the $(t - z)$ variable remains invariant under the N transformation. The effects of Eq. (10.142) or (10.143) on other variables are

$$y'/x' = \tan\phi', \quad r'/r = \xi'/\xi,$$

$$\frac{(z'+t')}{(z+t)} = \frac{(e^{2\alpha} + \xi'^2)}{(e^{2\alpha} + \xi^2)}. \tag{10.157}$$

The variable $\rho = (t^2 - z^2 - r^2)^{1/2}$ is a N-invariant quantity.

Since $(t-z)$ is invariant under the N_1 and N_2 transformations, it is clear from Eq. (10.152) that the x- and y-coordinate variables are directly proportional to ξ_1 and ξ_2 respectively. Indeed, for $b^2 = 0$, the differential equation given in Eq. (10.141) can be written as

$$\left[\left(\frac{\partial}{\partial x}\right)^2 + \left(\frac{\partial}{\partial y}\right)^2\right] F(x,y) = 0, \tag{10.158}$$

with the solution

$$F(x,y) = (x \pm iy)^m. \tag{10.159}$$

The mathematics of this form is quite familiar to us, and does not require any further explanation. The point is that the N transformation parameters are now directly related to the x- and y-coordinate variables, and the spin of the massless composite particle is indeed due to the above orbit-like form. The N transformation in this case is a translation of the rotation axis from the origin to another point in the xy-plane.

As for the normalization of $F(\xi, \phi)$, the ϕ dependence is just like the case of hydrogen atom. The Hilbert space and the normalization of the wave function associated with this variable are well known. The ξ dependence is not normalizable, and there is no Hilbert space associated with this variable. As was noted before, this is due to the fact that the N transformation is not measurable.

Let us next discuss properties of the $G(\rho, \alpha)$ function given in Eq. (10.145). This function is a product of two separate functions. The ρ dependence is normalizable, and the wave function is concentrated within a hyperbolic region near the light cones. On the other hand, the α-dependence, which measures the $(t-z)$ variable for fixed ρ, is not normalizable. However, this does not introduce any additional difficulty to the overall wave function which is not normalizable due to the non-observability of N transformations.

The above discussion has so far been restricted to x in the forward light cone. By changing the sign of the Cartesian variables given in Eq. (10.134), we can give the same reasoning for the backward light cone. By replacing the z and t variables by those given in Eq. (10.149), we can give a similar treatment for the space-like region.

In this section, we have studied in detail the part of the solutions of the harmonic oscillator equation for massless particles containing the space-time symmetries of the $E(2)$-like little group. This part is independent of the form of potential.

Unlike the case of massive composite particles, the wave functions here are not normalizable. However, this should not alarm us. The transverse coordinates in this case are proportional to the parameters of the N transformation. According to the discussions given in Sects. 10.2, and 10.7, the N transformation can be identified as a gauge transformation which is not observable. In the case of massless composite particles, the fact that the N transformation is not observable is translated into the lack of a Hilbert space associated with the transverse coordinate variables.

We do not have answers to the fundamental question of whether massless composite particles exist in nature, or whether the existing massless particles such as photons and gravitons are ultimately composite. Yet, it is of interest to note that there are solutions of the harmonic oscillator differential equation which allow us to study the $E(2)$-like symmetry for massless composite particles in terms of the conventional space-time variables.

10.10 Exercises and Problems

Exercise 1. In Sect. 10.5, we noted that the two-dimensional Helmholtz equation has solutions with vanishing k^2 as well as with non-vanishing k^2. If k^2 does not vanish, the representation is unitary and infinite-dimensional. On the other hand, if $k^2 = 0$, the representation is finite-dimensional and non-unitary. Explain these.

The fact that the representation is finite-dimensional and non-unitary for $k^2 = 0$ has been seen in Sect. 10.5. If k^2 does not vanish, the solution of the Helmholtz equation is

$$\psi(x, y) = e^{im\phi} J_m(kr) \,. \tag{10.160}$$

Unlike the solutions of Laplace's equation which describe the finite-dimensional representations, the above solutions form a complete orthonormal basis for the two-dimensional space.

The rotation operation on this function is trivial. However, if we translate the origin of the coordinate system, so that [3]

$$x' = x - u, \qquad \text{and} \qquad y' = y - v \,. \tag{10.161}$$

Then

$$e^{im\phi} J_m(kr) = \sum_{m'=-\infty}^{\infty} A_{m'}^m e^{im'\phi} J_m(kr') . \tag{10.162}$$

The summation contains an infinite number of terms. Thus the form given in Eq. (10.160) is the basis for the infinite-dimensional unitary representation. We can use the Bessel function identity [46]:

$$e^{im\phi} J_m(kr) = \sum_{m=-\infty}^{\infty} e^{-i(m-m')\phi'} J_{m-m'}(kr') e^{im\phi'} J_m(kr') , \tag{10.163}$$

with

$$u + iv = r'' e^{i\phi''} , \tag{10.164}$$

to determine the coefficient $A_{m'}^m$ in Eq. (10.162).

Exercise 2. Show that the D matrices for massless spin-1/2 particles given in Eqs. (10.110) can also be obtained from the combined effect of boost and rotation described for photons in Sect. 10.2.

Let us start with the boost along the x-axis:

$$B_x^\pm(\eta) = \begin{pmatrix} \cosh(\eta/2) & \pm\sinh(\eta/2) \\ \pm\sinh(\eta/2) & \cosh(\eta/2) \end{pmatrix} . \tag{10.165}$$

The rotation around the y-axis takes the form

$$R_y(\eta) = [1/\cosh(\eta)]^{1/2} \begin{pmatrix} \cosh(\eta/2) & -\sinh(\eta/2) \\ \sinh(\eta/2) & \cosh(\eta/2) \end{pmatrix} . \tag{10.166}$$

Finally, the boost along the z-axis is

$$B_z^\pm(\eta) = \begin{pmatrix} [\cosh(\eta)]^{\pm 1/2} & 0 \\ 0 & [\cosh(\eta)]^{\pm 1/2} \end{pmatrix} . \tag{10.167}$$

We can now calculate

$$D_x^\pm(\eta) = [B_x^\pm(\eta)]^{-1} R_y(\eta) B_z^\pm(\eta) . \tag{10.168}$$

After matrix multiplications, we arrive at

$$D_x^{(+)} = \begin{pmatrix} 1 & -\tanh\eta \\ 0 & 1 \end{pmatrix} \quad \text{and} \quad D_x^{(-)} = \begin{pmatrix} 1 & 0 \\ \tanh\eta & 1 \end{pmatrix} , \tag{10.169}$$

which are gauge transformation matrices given in Eq. (10.110).

Exercise 3. Show that a Lorentz boost on the D-invariant spinor is a helicity preserving transformation, while the helicity is not preserved for a Lorentz boost of a spinor which is not D-invariant.

In order to see the effect of these transformations, let us apply the boost operator $B_x^{(+)}(\eta)$ of Eq. (10.165) on a right-handed massless particle moving along the z-direction. The resulting spinor is

$$B_x^{(+)}(\eta)\alpha = \begin{pmatrix} \cosh(\eta/2) \\ \sinh(\eta/2) \end{pmatrix}$$

$$= [\cosh(\eta/2)]^{1/2}\alpha' , \tag{10.170}$$

where α' is the *normalized* positive-helicity spinor along the momentum $\mathbf{p}'$. Indeed, when applied to α, $B_x^{(+)}(\eta)$ is a helicity-preserving transformation. However, this is not a unitary transformation.

On the other hand, if we apply the same boost operator on β, the result is

$$B_x^{(+)}(\eta)\beta = \begin{pmatrix} \sinh(\eta/2) \\ \cosh(\eta/2) \end{pmatrix} \tag{10.171}$$

This spinor is not orthogonal to that of Eq. (10.170), and therefore does not represent the negative helicity state. When applied to β, $B_x^{(+)}(\eta)$ is not a helicity-preserving transformation.

The question then is whether there is a transformation which will preserve the negative helicity state. The answer to this question is *yes*. One way is to use $B_x^{(-)}(\eta)$, so that

$$B_x^{(-)}(\eta)\beta = \begin{pmatrix} -\sinh(\eta/2) \\ \cosh(\eta/2) \end{pmatrix}$$

$$= \lfloor\cosh(\eta/2)\rfloor^{1/2}\beta' . \tag{10.172}$$

Another approach to this problem is to perform the D transformation before applying $B_x^{(+)}(\eta)$, in analogy to what we did in Sect. 10.7. Indeed,

$$B_x^{(+)}(\eta)D_x^{(+)}(\eta)\beta = [1/\cosh(\eta)]^{1/2}\beta' . \tag{10.173}$$

Since $D_x^{(+)}(\eta)$ leaves α unchanged, the effect of $[B_x^{(+)}(\eta)D_x^{(+)}(\eta)]$ is the helicity preserving transformation applicable to both α and β.

This is indeed the place where the D matrices play the decisive role. As in the case of spin-1, the boost preceded by the D transformation produces the desired spinor:

$$B_x^{(+)}(\eta)D_x^{(+)}(\eta)\alpha = [\cosh\eta]^{1/2}\alpha' . \tag{10.174}$$

The transformation $B_x^{(+)}(\eta)D_x^{(+)}(\eta)$ is therefore a helicity preserving transformation for both helicity states. It is not a unitary transformation because of the factors $[\cosh]^{\pm 1/2}$ in Eqs. (10.172) and (10.174).

This lack of unitarity should not alarm us. Unlike the case of massless particles with spin-1, the helicity-preserving Lorentz boost along the direction of the momentum is not a unitary transformation in the $SL(2,c)$ regime. We are in fact quite familiar with this in the Dirac equation, and we know how to take care of the problem.

Problem 1. Show that Eq. (10.160) becomes a transformation of a finite-dimensional non-unitary representation in the $k \to 0$ limit. Calculate the $A_{m'}^m$ coefficient in this limit using the form

$$J_m(kr) \to \frac{1}{m!}\left(\frac{kr}{2}\right)^m . \tag{10.175}$$

Problem 2. Calculate the electric and magnetic field from the four-vectors of Eq. (10.25) and Eq. (10.40), and show that these two different four-vectors give the same electric and magnetic fields. See Han and Kim [15].

Problem 3. Find solutions of the differential equation given in Eq. (10.142) when b^2 does not vanish. Do they form a basis for an infinite-dimensional unitary representation of the $E(2)$-like little group? See Kalnins and Miller [44].

Problem 4. Because the translation-like subgroup of the $E(2)$-like little group is an invariant subgroup, Eq. (10.54) is possible. Calculate the parameters u' and v' in terms of u, v, and α.

Problem 5. Consider the solutions of the differential equation of Eq. (10.59) which are diagonal in the the translation operators. They are not diagonal in the rotation operator, but can be expanded in terms of eigenfunctions of the rotation operator. Explain Eq. (10.54) in terms of this expansion.

Problem 6. Throughout this Chapter, we mostly concentrated our effort on the electromagnetic four-potentials. However, as we discussed in Sect. 10.6, since the electric and magnetic field are derivable from the potentials by differentiation, and since they are independent of gauge parameters, their transformation property is expected to be simpler than that of the four-potential. Is this true? Work out the Lorentz boost along the z- and x-directions, assuming that the photon propagates along the z-direction. See Weinberg [26].

Problem 7. It was noted in Chap. 3 that the $E(2)$-invariance condition alone leads to the solutions of the Dirac equation for massless particle. Show that the same invariance requirement will lead to gauge-invariant solutions of Maxwell's equations. See Weinberg [26].

Problem 8. What is the one-dimensional analog of the transition from Eq. (10.59) to Eq. (10.73) in the $k^2 \to 0$ limit? See Bowen and Coster [47].

Problem 9. The three-dimensional Euclidean space is the space in which we conduct our daily life. From the group theoretical point of view, $E(3)$ is a semi-direct product of the three-dimensional rotation and translation groups. Work out the irreducible representations of $E(3)$. See Rno [48].

Problem 10. We discussed in Chap. 5 the covariant harmonic oscillator wave functions for massive hadrons. In Sect. 10.9 we studied the harmonic oscillator wave functions for massless hadrons. Does the Lorentz-boosted oscillator wave function for a massive hadron behave like that of a massless hadron in the high-energy limit?

Problem 11. We learned that $D_1(u)$ and $D_2(v)$ of Eqs. (10.19) and (10.20) perform gauge transformations on the photon four-potential. What happens to these matrices when the Lorentz group is contracted to the Galilean group discussed in Chap. 4?

References

1. E.P. Wigner, On Unitary Representations of the Inhomogeneous Lorentz Group, The Annals of Mathematics **40**(1), 149–204 (1939). DOI 10.2307/1968551. URL http://www.jstor.org/stable/1968551?origin=crossref
2. S. Weinberg, Feynman Rules for Any Spin, Physical Review **133**(5B), B1318–B1332 (1964). DOI 10.1103/PhysRev.133.B1318. URL https://link.aps.org/doi/10.1103/PhysRev.133.B1318
3. E. Inönü, E.P. Wigner, On the Contraction of Groups and Their Representations, Proceedings of the National Academy of Sciences **39**(6), 510–524 (1953). DOI 10.1073/pnas.39.6.510. URL http://www.pnas.org/cgi/doi/10.1073/pnas.39.6.510

4. D. Han, Y.S. Kim, D. Son, E(2)–like little group for massless particles and neutrino polarization as a consequence of gauge invariance, Physical Review D **26**(12), 3717–3725 (1982). DOI 10.1103/PhysRevD.26.3717. URL https://link.aps.org/doi/10.1103/PhysRevD.26.3717

5. K. Olive, Review of Particle Physics, Chinese Physics C **38**(9), 090,001 (2014). DOI 10.1088/1674-1137/38/9/090001. URL https://iopscience.iop.org/article/10.1088/1674-1137/38/9/090001

6. F. Boehm, P. Vogel, *Physics of massive neutrinos*, 2nd edn. (Cambridge University Press, Cambridge UK; New York, NY, USA, 1992). ISBN 9780521418249,9780521428491. (Originally published 1987.)

7. T.D. Lee, C.N. Yang, Parity Nonconservation and a Two-Component Theory of the Neutrino, Physical Review **105**(5), 1671–1675 (1957). DOI 10.1103/PhysRev.105.1671. URL https://link.aps.org/doi/10.1103/PhysRev.105.1671

8. C.S. Wu, E. Ambler, R.W. Hayward, D.D. Hoppes, R.P. Hudson, Experimental Test of Parity Conservation in Beta Decay, Physical Review **105**(4), 1413–1415 (1957). DOI 10.1103/PhysRev.105.1413. URL https://link.aps.org/doi/10.1103/PhysRev.105.1413

9. M. Goldhaber, L. Grodzins, A.W. Sunyar, Helicity of Neutrinos, Physical Review **109**(3), 1015–1017 (1958). DOI 10.1103/PhysRev.109.1015. URL https://link.aps.org/doi/10.1103/PhysRev.109.1015

10. D. Han, Y.S. Kim, D. Son, Gauge transformations as Lorentz-Boosted rotations, Physics Letters B **131**(4-6), 327–329 (1983). DOI 10.1016/0370-2693(83)90509-9. URL http://linkinghub.elsevier.com/retrieve/pii/0370269383905099

11. D. Han, Y.S. Kim, M.E. Noz, D. Son, Internal space-time symmetries of massive and massless particles, American Journal of Physics **52**(11), 1037–1043 (1984). DOI 10.1119/1.13784. URL http://aapt.scitation.org/doi/10.1119/1.13784

12. E.P. Wigner, Invariant Quantum Mechanical Equations of Motion, in *Seminar on Theoretical Physics, Book1*, ed. by A. Salam (International Atomic Energy Commision, Vienna, Austria, 1963), 59–82. (16 July to 25 August, 1962; Treste Seminar; PUB–61; ISSN 0074-1884.)

13. A. Wightman, L'invariance dans la mecanique quantique relativiste, in *Relations de dispersion et particules elementaires*, ed. by C. DeWitt, R. Omnes (John Wiley & Sons Inc, New York, NY, USA, 1961), 150–226. URL https://inspirehep.net/literature/1345200. (Les Houches Lect. Notes, Volume 10, 1969; ASIN: B003IGFWDI.)

14. S. Weinberg, Photons and Gravitons in S -Matrix Theory: Derivation of Charge Conservation and Equality of Gravitational and Inertial Mass, Physical Review **135**(4B), B1049–B1056 (1964). DOI 10.1103/PhysRev.135.B1049. URL https://link.aps.org/doi/10.1103/PhysRev.135.B1049

15. D. Han, Y.S. Kim, Little group for photons and gauge transformations, American Journal of Physics **49**(4), 348–351 (1981). DOI 10.1119/1.12509. URL http://aapt.scitation.org/doi/10.1119/1.12509

16. G. Mackey, *Induced Representations of Groups and Quantum Mechanics* (W. A. Benjamin, Inc./Editore Boringhieri, Boston, MA, USA, 1968). (ASIN B001DOM62C.)

17. R.L. Lipsman, *Group representations: a survey of some current topics*. No. 388 in Lecture notes in mathematics (Springer-Verlag, Berlin, Germany; New York, NY, USA, 1974). ISBN 978-3-540-38402-1

18. P. Winternitz, I. Fris, Invariant Expansions of Relativistic Amplitudes and Subgroups of Proper Lorentz Group, Soviet Journal of Nuclear Physics-USSR **1**(5), 636–643 (1965). URL https://www.osti.gov/biblio/4671988

19. D. Han, Y.S. Kim, D. Son, Photon spin as a rotation in gauge space, Physical Review D **25**(2), 461–463 (1982). DOI 10.1103/PhysRevD.25.461. URL https://link.aps.org/doi/10.1103/PhysRevD.25.461

20. S. Başkal, Y.S. Kim, M.E. Noz, *Physics of the Lorentz Group (Second Edition): Beyond high-energy physics and optics* (IOP Publishing, Bristol, UK, 2021). DOI 10.1088/978-0-7503-3607-9. ISBN 978-0-7503-3607-9. URL https://iopscience.iop.org/book/978-0-7503-3607-9

21. S. Başkal, Y.S. Kim, M.E. Noz, Loop Representation of Wigner's Little Groups, Symmetry **9**(7), 97–118 (2017). DOI 10.3390/sym9070097. URL http://www.mdpi.com/2073-8994/9/7/97

22. S. Başkal, Y.S. Kim, Little groups and Maxwell-type tensors for massive and massless particles, Europhysics Letters (EPL) **40**(4), 375–380 (1997). DOI 10.1209/epl/i1997-00474-0. URL http://stacks.iop.org/0295-5075/40/i=4/a=375?key=crossref.1ab63191c5835dce7bd073be58e6a310

23. E.P. Wigner, *Group Theory: And its Application to the Quantum Mechanics of Atomic Spectra* (Academic Press, New York, NY, USA, 1959). ISBN 978-0127505503. (Originally published as: Gruppentheorie und ihre Anwendung auf die Quantenmechanik der Atomspektren, Springer Verlag, Braunscheig, Germany 1931.)

24. R.P. Feynman, M. Kislinger, F. Ravndal, Current Matrix Elements from a Relativistic Quark Model, Physical Review D **3**(11), 2706–2732 (1971). DOI 10.1103/PhysRevD.3.2706. URL https://link.aps.org/doi/10.1103/PhysRevD.3.2706

25. P.E. Hussar, Y.S. Kim, M.E. Noz, Three–particle symmetry classifications according to the method of Dirac, American Journal of Physics **48**(12), 1038–1042 (1980). DOI 10.1119/1.12301. URL http://aapt.scitation.org/doi/10.1119/1.12301

26. S. Weinberg, Feynman Rules for Any Spin. II. Massless Particles, Physical Review **134**(4B), B882–B896 (1964). DOI 10.1103/PhysRev.134.B882. URL https://link.aps.org/doi/10.1103/PhysRev.134.B882

27. D. Han, M.E. Noz, Y.S. Kim, D. Son, Space-time symmetries of confined quarks, Physical Review D **25**(6), 1740–1743 (1982). DOI 10.1103/PhysRevD.25.1740. URL https://link.aps.org/doi/10.1103/PhysRevD.25.1740

28. D. Han, Y.S. Kim, D. Son, Photons, neutrinos, and gauge transformations, American Journal of Physics **54**(9), 818–821 (1986). DOI 10.1119/1.14454. URL http://aapt.scitation.org/doi/10.1119/1.14454

29. Y.S. Kim, M.E. Noz, *New Perspectives on Einstein's E = mc²*. (World Scientific Publishing Co., Singapore; Hackensack, NJ, USA, 2018). ISBN 978-981-323-770-4. (OCLC: 1022620189.)

30. D. Han, Y.S. Kim, D. Son, Eulerian parametrization of Wigner's little groups and gauge transformations in terms of rotations in two–component spinors, Journal of Mathematical Physics **27**(9), 2228–2235 (1986). DOI 10.1063/1.526994. URL http://aip.scitation.org/doi/10.1063/1.526994

31. R.N. Mohapatra, A.Y. Smirnov, Neutrino Mass and New Physics, Annual Review of Nuclear and Particle Science **56**(1), 569–628 (2006). DOI 10.1146/annurev.nucl.56.080805.140534. URL http://www.annualreviews.org/doi/10.1146/annurev.nucl.56.080805.140534

32. Y.S. Kim, G.Q. Maguire Jr., M.E. Noz, Do Small-Mass Neutrinos Participate in Gauge Transformations?, Advances in High Energy Physics **2016**, 1–7 (2016). DOI 10.1155/2016/1847620. URL http://www.hindawi.com/journals/ahep/2016/1847620/

33. V. Barger, D. Marfatia, K.L. Whisnant, *The physics of neutrinos* (Princeton University Press, Princeton, NJ, USA, 2012). ISBN 9780691128535

34. F. Deppisch, *A modern introduction to neutrino physics*. IOP Concise Physics (Morgan & Claypool Publishers, San Rafael, CA USA, 2019). ISBN 9781643276793

35. S. Brice, M. Marshak, Co-chairs, The XXIX International Conference on Neutrino Physics and Astrophysics, in *ICNP XXIX 2020, avaiable online only* (Fermilab, Chicago, IL USA, 2020). URL http://nu2020.fnal.gov. (held as online only conference by Fermilab, June 22 - July 2.)

36. Neutrino 2022 SEOUL. URL https://neutrino2022.org/Neutrino2022SEOUL

37. I. Esteban, M. Gonzalez-Garcia, M. Maltoni, T. Schwetz, A. Zhou, The fate of hints: updated global analysis of three-flavor neutrino oscillations, Journal of High Energy Physics **2020**(9), 178 (2020). DOI 10.1007/JHEP09(2020)178. URL https://link.springer.com/10.1007/JHEP09(2020)178

38. I. Esteban, M.C. Gonzalez-Garcia, A. Hernandez-Cabezudo, M. Maltoni, T. Schwetz, Global analysis of three-flavour neutrino oscillations: synergies and tensions in the determination of θ_{23}, Δ_{CP}, and the mass ordering, Journal of High Energy Physics **2019**(1), 106 (2019). DOI 10.1007/JHEP01(2019)106. URL https://link.springer.com/10.1007/JHEP01(2019)106

39. ATLAS Collaboration, Search for heavy Majorana or Dirac neutrinos and right-handed W gauge bosons in final states with charged leptons and jets in pp collisions at $\sqrt{s}$=13 TeV with the ATLAS detector, The European Physical Journal C **83**(12), 1164 (2023). DOI 10.1140/epjc/s10052-023-12021-9. URL https://doi.org/10.1140/epjc/s10052-023-12021-9
40. E. Akhmedov, A. Trautner. Can quantum statistics help distinguish Dirac from Majorana neutrinos? (2024). URL http://arxiv.org/abs/2402.05172. ArXiv:2402.05172 [hep-ex, physics:hep-ph, physics:nucl-th]
41. A.B. Balantekin, B. Kayser, On the Properties of Neutrinos, Annual Review of Nuclear and Particle Science **68**(1), 313–338 (2018). DOI 10.1146/annurev-nucl-101916-123044. URL https://www.annualreviews.org/doi/10.1146/annurev-nucl-101916-123044
42. J.M. Berryman, L.A. Delgadillo, P. Huber, Future searches for light sterile neutrinos at nuclear reactors, Physical Review D **105**(3), 035,002 − 1 − 035,002 − 15 (2022). DOI 10.1103/PhysRevD.105.035002. URL https://link.aps.org/doi/10.1103/PhysRevD.105.035002
43. STEREO Collaboration, STEREO neutrino spectrum of ^{235}u fission rejects sterile neutrino hypothesis, Nature **613**, 257–261 (2023). DOI 10.1038/s41586-022-05568-2. URL https://arxiv.org/abs/2210.07664
44. E.G. Kalnins, W. Miller, Lie theory and the wave equation in space–time. I. The Lorentz group, Journal of Mathematical Physics **18**(1), 1–16 (1977). DOI 10.1063/1.523130. URL http://aip.scitation.org/doi/10.1063/1.523130
45. D. Han, Y.S. Kim, D. Son, Massless composite particles and space-time description of gauge transformations, Physical Review D **27**(10), 2348–2353 (1983). DOI 10.1103/PhysRevD.27.2348. URL https://link.aps.org/doi/10.1103/PhysRevD.27.2348
46. W.J. Miller, *Lie Theory and Special Functions* (Academic Press Elsevier Science, New York, NY, USA, 1968). ISBN 978-0-08-095551-3. URL https://books.google.se/books/about/Lie_Theory_and_Special_Functions.html?id=XJ9RiMLiOigC&redir_esc=y
47. M. Bowen, J. Coster, Infinite square well: A common mistake, American Journal of Physics **49**(1), 80–81 (1981). DOI 10.1119/1.12660. URL http://aapt.scitation.org/doi/10.1119/1.12660
48. J.S. Rno, Harmonic analysis on the Euclidean group in three-space, Journal of Mathematical Physics **26**(4), 675–677 (1985). DOI 10.1063/1.526606. URL http://aip.scitation.org/doi/10.1063/1.526606

Chapter 11
Special Relativity from Heisenberg's Uncertainty Relation

Abstract Heisenberg's uncertainty relation can be written in terms of the step-up and step-down operators in the harmonic oscillator representation. It is noted that the single-variable Heisenberg commutation relation contains the symmetry of the $Sp(2)$ group which is isomorphic to the Lorentz group applicable to two space-like dimensions and one time-like dimension known as the $O(2,1)$ group. This group has three independent generators. The one-dimensional step-up and step-down operators can be combined into one two-by-two Hermitian matrix which contains three independent operators. If we use a two-variable Heisenberg commutation relation, the two pairs of independent step-up, step-down operators can be combined into a four-by-four block-diagonal Hermitian matrix with six independent parameters. It is then possible to add one off-diagonal two-by-two matrix and its Hermitian conjugate to complete the four-by-four Hermitian matrix. This off-diagonal matrix has four independent generators. There are thus ten independent generators. It is then shown that these ten generators can be linearly combined with the ten generators for Dirac's two oscillator system leading to the $(3+2)$ de Sitter or $O(3,2)$ Lorentz group, which is isomorphic to $Sp(4)$. The $O(3,2)$ Lorentz group can then be contracted to the Poincaré group with four translation generators corresponding to the four-momentum in the Lorentz-covariant world. This Lorentz-covariant four-momentum is known as Einstein's $E = mc^2$.

In this Chapter we shall show that special relativity in the form of Einstein's $E = mc^2$ is derivable from Heisenberg's uncertainty relations which form the basis of quantum mechanics. This is one step consistent with Dirac's life long ambition to combine these two theories of modern physics. Heisenberg's uncertainty relation can be written in terms of the step-up and step-down operators. This means we can write the Heisenberg's uncertainty relation in terms of the harmonic oscillator representation. Additionally, in interpreting the Heisenberg uncertainty principle, a

S. Başkal et al., *Theory and Applications of the Poincaré Group*, Fundamental Theories of Physics 217, https://doi.org/10.1007/978-3-031-64376-7_11

fundamental role is played by the Gaussian wave function. Furthermore, the single-variable Heisenberg commutation relation can be represented, in the Wigner phase-space picture of quantum mechanics, as the minimum area in phase space by the uncertainty principle. Area-preserving linear transformations including rotations and squeezes are allowed in this two-dimensional phase space.

As we have seen in Chap. 8, this group of transformations forms the two-dimensional symplectic group $Sp(2)$. We know this group is locally isomorphic to the Lorentz group $O(2, 1)$ which applies to two space-like dimensions and one time-like dimension. Hence we see that the basic elements of the Lorentz group are already contained in the single-variable uncertainty relation.

We noted in Chap. 7, that the system of two oscillators leads to the four-dimensional symplectic group or $Sp(4)$. This group, as Dirac noted first [1], is isomorphic to the Lorentz group $O(3, 2)$, also known as the $(3 + 2)$ de Sitter group. This Lorentz group is applicable to three space and two time variables. Although we are familiar with the Lorentz group which has one time coordinate, the problem now is how to deal with the second time variable. The proper Lorentz group, as we recall, is generated by three rotation generators and three boost generators. This $O(3, 2)$ group has not only three additional boost generators but it also has one generator which produces rotations between the two time variables. In Chap. 7 we used the contraction or squeeze transformation introduced in Chap. 4 to convert these four extra variables to the translation variables of the Poincaré group. Here we shall derive a similar result from Heisenberg's uncertainty relation which form the basis for quantum mechanics.

Essentially Dirac's papers, published in 1949 and 1963 [2, 1], form the basis for this Chapter. We show in Fig. 11.1 that, as mentioned in Dirac's 1949 paper [2], the space-time symmetry of quantum mechanics can be derived from his two-oscillator system discussed in 1963 [1]. This is accomplished by using the group contraction procedure of Inönü and Wigner [3].

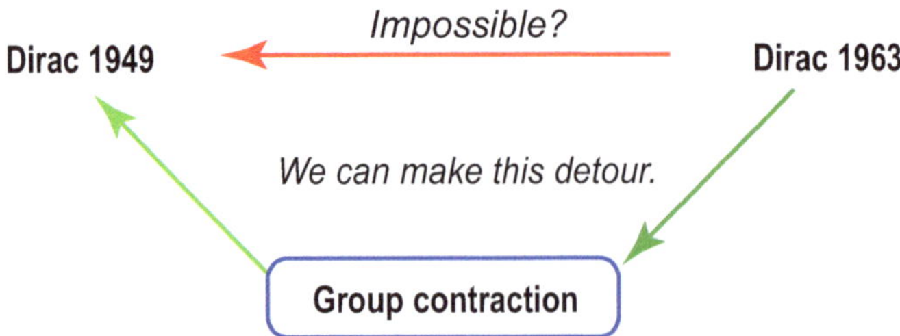

Fig. 11.1: Dirac's 1949 paper [2] tells us that constructing a representation of the Poincaré group is essential to the task of constructing quantum mechanics. In Dirac's 1963 paper [1] he constructed the Lie algebra of the $O(3, 2)$ Lorentz group. Dirac did this from the algebra of two harmonic oscillators. However, as we see in this Chapter, it is also a direct consequence of Heisenberg's uncertainty commutation relations. Using Inönü and Wigner's [3] group-contraction procedure, it is possible to derive from the $O(3, 2)$ Lorentz group, the Lie algebra of the Poincaré group [4, 5].

In Sect. 11.1, we define Heisenberg's uncertainty relation using a one oscillator system. We show that these definitions lead to the Galilean system which consists of space-time symmetry consisting of three translations and three rotations. We could call this symmetry the inhomogeneous rotation group or the Galilean group.

It is also possible that the uncertainty relation, can be stated, in Einstein's Lorentzian system in terms of the Lie algebra of the Poincaré group. The Lie algebra of the Poincaré group consists of three rotation generators and three boost generators, with the addition of four space-time translation generators. This Poincaré group is also known as the inhomogeneous Lorentz group [1]. The present form of quantum field theory forms a representation of this Poincaré group. This group was discussed in detail in Chap. 2.

Section 11.2, shows that using the Wigner function to represent the Gaussian function for the one harmonic oscillator state, is the best way to study the symmetry of the Heisenberg commutation relation. This function, in the Wigner phase space, contains the symmetry for the $Sp(2)$ group which is isomorphic to the $O(2, 1)$ Lorentz group. This $O(2, 1)$ group, as detailed many times before, applies to two space-like dimensions and one time-like dimension and has three generators. We can view this operation as equivalent to constructing, using quadratic forms of the step-up and step-down operators, a two-by-two block-diagonal Hermitian matrix.

In Sect. 11.3, we consider first two independent harmonic oscillators. With these independent harmonic oscillators, a four-by-four block diagonal matrix, can be constructed. Each block consists of a two-by-two matrix representing each operator defined in Sect. 11.2. Because the harmonic oscillators are uncoupled, the resulting four-by-four block-diagonal Hermitian matrix has six independent generators.

However, if the harmonic oscillators are coupled, it is necessary both to keep the block-diagonal Hermitian matrix, in addition to adding one off-diagonal block matrix and its Hermitian conjugate, which have four independent quadratic forms. Hence, there are now ten independent quadratic forms made up of the creation and annihilation operators in the resulting four-by-four matrix.

Furthermore, we can linearly relate these ten independent generators into those ten generators that Dirac constructed into the $O(3, 2)$ Lorentz group or $(3 + 2)$ de Sitter group [1]. This Lorentz group, as we have seen before, is applicable to three space-like dimensions and two time-like dimensions.

In Sect. 11.4, the boosts belonging to one of the time-like dimensions of $O(3, 2)$ plus the rotation between the two time-like dimensions, are contracted into four translations leading to the Lorentz-covariant four-momentum. Thus $O(3, 2)$ is converted into the homogeneous Lorentz group $O(3, 1)$ in addition to four translation operators thus forming the Poincaré group. The Lorentz-covariant four-momentum thus produced is commonly known as Einstein's $E = mc^2$.

Section 11.5 gives an overview of Dirac's quest to combine special relativity with quantum mechanics. We start from Dirac's 1949 and 1963 papers [2, 1].

11.1 Heisenberg's Commutation Relations

In this section we shall start with Heisenberg's commutation relations

$$[x_i, P_j] = i\,\delta_{ij}\,, \tag{11.1}$$

where

$$P_i = -i\frac{\partial}{\partial x^i}\,. \tag{11.2}$$

Here, in accordance with Eq. (2.9), x^i are represented by (z, x, y), with $i = 3, 1, 2$. The following three operators can be constructed with these x_i and P_i:

$$J_i = \varepsilon_{ijk}x^j P_k\,. \tag{11.3}$$

The closed set of commutation relations:

$$[J_i, J_j] = i\varepsilon_{ijk}J_k\,, \tag{11.4}$$

are satisfied by the three operators in Eq. (11.3). These commutation relations, of course, form the Lie algebra of the rotation group. It is interesting that this is a direct result of Heisenberg's commutation relations.

Each J_i corresponds, in quantum mechanics, to the angular momentum around the particular i-direction. This same Lie algebra can be used to construct two-by-two matrices, namely, the Pauli spin matrices. These matrices lead to the observable spin angular momentum which is not seen in classical mechanics.

Each P_i in Eq. (11.2) can be seen to generate a translation in the i-direction. The P_i have the following commutation relations with the rotation generators:

$$[J_i, P_j] = i\varepsilon_{ijk}P_k\,. \tag{11.5}$$

The commutation relations in Eq. (11.5) together with those of Eq.(11.3), form a closed set for both P_i and J_i. This closed set of commutation relations form the Lie algebra for the Galilei group. This Galilei group is, of course, the basic symmetry group for Schrödinger's non-relativistic quantum mechanics.

In the Schrödinger picture, we interpret the generator P_i as the particle's momentum along the i-direction. We can also obtain the time translation operator by

$$P_0 = i\frac{\partial}{\partial t}\,, \tag{11.6}$$

which also represents the energy operator.

If we now go to the Lorentzian world, with $x_\mu = (t, -z, -x, -y)$, we have to add the boost generators which include the time variable. The boost along each of the i-direction is generated by:

$$K_i = i\left(t\frac{\partial}{\partial x^i} - x_i\frac{\partial}{\partial t}\right)\,. \tag{11.7}$$

These generators satisfy the commutation relations:

$$\left[K_i, K_j\right] = -i\varepsilon_{ijk}J_k \,. \tag{11.8}$$

Therefore, these three boost generators alone do not form a closed set of commutation relations or a Lie algebra. However, together with the rotation generator J_i, the boost generators satisfy:

$$\left[J_i, K_j\right] = i\,\varepsilon_{ijk}K_k \,. \tag{11.9}$$

Additionally, the P_i operators satisfy the relations:

$$\left[K_i, P_j\right] = i\delta_{ij}P_0\,, \qquad [K_i, P_0] = iP_i\,, \quad \text{and} \quad [J_i, P_0] = 0\,. \tag{11.10}$$

Hence, Eqs. (11.4, 11.5, 11.8, 11.9, 11.10) form a closed set of the ten generators which is the Lie algebra of the Poincaré group.

As we saw above, from Heisenberg's commutation relations we can derive the rotation and translation generators. The time translation operator, on the other hand, comes from the Schrödinger equation. All of these rotation and translation operators are Hermitian and correspond to dynamical variables. The three boost generators of Eq. (11.7), however, cannot be derived from the Heisenberg relations. Furthermore, they seem not to correspond to observable quantities [2].

It is the purpose of this chapter to show that the Lie algebra of the Poincaré group can be derived from Heisenberg's commutation relations. We shall, with the use of the Wigner function in phase space, examine first the symmetry of Heisenberg's commutation relation. We find that the symmetry of the Lorentz group, $O(2, 1)$, applicable to two space-like dimensions and one time-like dimension is contained in the single-variable Heisenberg relation.

11.2 A Single-Mode Harmonic Oscillator and its Symmetries

In this section we aim to demonstrate that examining the symmetry of the Heisenberg commutation relation by considering the Wigner function in order to investigate the symmetries of the Gaussian function for the one harmonic oscillator state is a very effective method. The symmetry of this function is isomorphic to the $O(2, 1)$ Lorentz group and contains the $Sp(2)$ group in the Wigner phase space. In Chap. 4, the $O(2, 1)$ group was contracted to the two-dimensional Euclidean group. Here, we see that, a similar procedure can be implemented to contract the $Sp(2)$ group to $E(2)$.

11.2.1 Single-Mode Harmonic Oscillator and Sp(2)

Let us look at Heisenberg uncertainty relation for a single Cartesian variable. It has the form:

$$[x, P] = i ,\tag{11.11}$$

with

$$P = -i\frac{\partial}{\partial x} .\tag{11.12}$$

It is frequently convenient however, to use the step-up and step-down operators as:

$$a^\dagger = \frac{1}{\sqrt{2}}(x - iP) \quad \text{and} \quad a = \frac{1}{\sqrt{2}}(x + iP) ,\tag{11.13}$$

with

$$\left[a, a^\dagger\right] = 1 .\tag{11.14}$$

The harmonic oscillator representation is the basic language for the Fock [6] space for particle numbers. The Fock space, based on $a^\dagger$ and a, is known as the harmonic oscillator representation of the uncertainty relation and is the basic language for quantum optics.

If we consider next the quadratic forms: $aa, a^\dagger a^\dagger, aa^\dagger$, and $a^\dagger a$, then the linear combination together with the uncertainly relation, yields

$$aa^\dagger - a^\dagger a = 1 .\tag{11.15}$$

Here, there are three independent quadratic forms, which leads to the following two-by-two matrix:

$$\begin{pmatrix} \left(aa^\dagger + a^\dagger a\right)/2 & aa \\ a^\dagger a^\dagger & \left(aa^\dagger + a^\dagger a\right)/2 \end{pmatrix} .\tag{11.16}$$

Therefore we write the following three independent operators:

$$J_2 = \frac{1}{2}\left(aa^\dagger + a^\dagger a\right), \quad K_1 = \frac{1}{2}\left(a^\dagger a^\dagger + aa\right), \quad K_3 = \frac{i}{2}\left(a^\dagger a^\dagger - aa\right) .\tag{11.17}$$

The following set of closed commutation relations

$$[J_2, K_1] = -iK_3, \quad [J_2, K_3] = iK_1, \quad [K_1, K_3] = iJ_2\tag{11.18}$$

are produced. We commonly call this set of commutation relations the Lie algebra of the $Sp(2)$ group. This group, as we know from Chap. 8, is locally isomorphic to the Lorentz group $O(2, 1)$.

To study the symmetry property of these operators, it is best to use the Wigner function for the ground-state harmonic oscillator. This harmonic oscillator takes the form [7, 8, 9, 10]:

$$W(x, p) = \frac{1}{\pi}\exp\left[-\left(x^2 + p^2\right)\right] .\tag{11.19}$$

The distribution is concentrated in the circular region around the origin. The circle can, therefore, be defined as:

$$x^2 + p^2 = 1 . \tag{11.20}$$

The minimum uncertainty, derived from the area of this circle, can be used in the phase space of x and p. Even under rotations and squeezing, this minimum uncertainty is preserved in the phase space. We write these transformations in the phase space as:

$$\begin{pmatrix} \cos\theta & -\sin\theta \\ \sin\theta & \cos\theta \end{pmatrix} \begin{pmatrix} x \\ p \end{pmatrix} \quad \text{and} \quad \begin{pmatrix} e^{\eta} & 0 \\ 0 & e^{-\eta} \end{pmatrix} \begin{pmatrix} x \\ p \end{pmatrix} , \tag{11.21}$$

respectively. The rotation and squeeze are generated by

$$J_2 = -\frac{i}{2}\left(x\frac{\partial}{\partial p} - p\frac{\partial}{\partial x} \right) , \quad K_1 = -\frac{i}{2}\left(x\frac{\partial}{\partial x} - p\frac{\partial}{\partial p} \right) . \tag{11.22}$$

Taking the commutation relation of these two operators, yields a third operator:

$$[J_2, K_1] = -iK_3 , \tag{11.23}$$

where

$$K_3 = \frac{i}{2}\left(x\frac{\partial}{\partial p} + p\frac{\partial}{\partial x} \right) . \tag{11.24}$$

Thus these three generators form a closed set of commutation relations which as we said above form the Lie algebra of the $Sp(2)$ group. We saw also in Chap. 8 that this group is locally isomorphic to the Lorentz group, $O(2,1)$. Likewise, the $Sp(2)$ group, as we saw in Chap. 8, for the $O(2,1)$ group, consists of two boost generators and one rotation generator. We illustrate in Fig. 11.2 this isomorphic correspondence.

If we consider the Minkowski space of $x^{\mu} = (t, z, x, y)$ we can write four-by-four matrices satisfying the Lie algebra of Eq. (11.18) as:

$$J_2 = \begin{pmatrix} 0 & 0 & 0 & 0 \\ 0 & 0 & -i & 0 \\ 0 & i & 0 & 0 \\ 0 & 0 & 0 & 0 \end{pmatrix} , \quad K_3 = \begin{pmatrix} 0 & i & 0 & 0 \\ i & 0 & 0 & 0 \\ 0 & 0 & 0 & 0 \\ 0 & 0 & 0 & 0 \end{pmatrix} , \quad \text{and} \quad K_1 = \begin{pmatrix} 0 & 0 & i & 0 \\ 0 & 0 & 0 & 0 \\ i & 0 & 0 & 0 \\ 0 & 0 & 0 & 0 \end{pmatrix} . \tag{11.25}$$

We note that the fourth row and column of each of these matrices are null. Thus, they can generate Lorentz transformations applicable only to the three-dimensional space of (t, z, x), while the y variable remains invariant. Consequently it is not possible to describe what happens in the full four-dimensional Minkowski space with this single-oscillator system.

It is nevertheless interesting that this single oscillator system can produce, as shown in Table 11.1, three different representations which have the same Lie algebra as the $(2 + 1)$-dimensional Lorentz group.

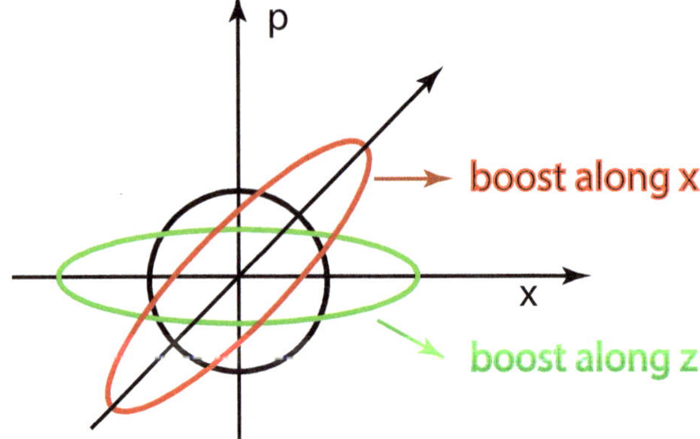

Fig. 11.2: $Sp(2)$ transformations produce rotations and squeezes in the phase space. Corresponding to the Lorentz boost along the z-direction is the squeeze along the x-direction. The boost along the x-direction corresponds to the squeeze along the $45°$ angle with respect to the x-coordinate of the phase space. Meanwhile the rotation by $45°$ corresponds to the rotation by $90°$ around the axis perpendicular to the plane of the boosts [4, 5]. As shown, the squeeze operations represent Lorentz boosts in two different directions in addition to a rotation around the axis perpendicular to the plane defined by two space-like directions. This constitutes the symmetry of the $O(2, 1)$ group [5].

Table 11.1: Gaussian function generators, shown in terms of the single harmonic oscillator, the two-dimensional phase space, and the four-dimensional Minkowski space (t, z, x, y) [4, 5].

Generators	Oscillator	Phase space	Lorentz
J_2	$\frac{1}{2}\left(aa^\dagger + a^\dagger a\right)$	$\frac{1}{2}\sigma_2$	$\begin{pmatrix} 0 & 0 & 0 & 0 \\ 0 & 0 & -i & 0 \\ 0 & i & 0 & 0 \\ 0 & 0 & 0 & 0 \end{pmatrix}$
K_1	$\frac{1}{2}\left(a^\dagger a^\dagger + aa\right)$	$\frac{i}{2}\sigma_1$	$\begin{pmatrix} 0 & 0 & i & 0 \\ 0 & 0 & 0 & 0 \\ i & 0 & 0 & 0 \\ 0 & 0 & 0 & 0 \end{pmatrix}$
K_3	$\frac{i}{2}\left(a^\dagger a^\dagger - aa\right)$	$\frac{i}{2}\sigma_3$	$\begin{pmatrix} 0 & i & 0 & 0 \\ i & 0 & 0 & 0 \\ 0 & 0 & 0 & 0 \\ 0 & 0 & 0 & 0 \end{pmatrix}$

11.2.2 Contraction of the Sp(2) group to E(2)

In Chap. 4, the $O(2, 1)$ group was contracted to the two-dimensional Euclidean group. Let us carry out this procedure again. We shall start from $Sp(2)$ for the single oscillator in phase space. The connection between the transformations of $O(2, 1)$ and $Sp(2)$ was illustrated in Fig. 11.2. The process that was used in Chap. 4 to contract $O(2, 1)$ to the $E(2)$ Euclidean group will again be used as the two groups are isomorphic. We want to study this procedure again in preparation for Sect. 11.4. For this purpose, we use the squeeze matrix

$$B(\eta) = \begin{pmatrix} e^{-\eta/2} & 0 & 0 & 0 \\ 0 & e^{\eta/2} & 0 & 0 \\ 0 & 0 & e^{\eta/2} & 0 \\ 0 & 0 & 0 & e^{\eta/2} \end{pmatrix}. \tag{11.26}$$

This squeeze matrix commutes with J_2. Using the same procedure as in Sect. 4.1 of Chap. 4, we shall first squeeze K_3 and apply the large η limit:

$$B(\eta)\, K_3\, B^{-1}(\eta) = \begin{pmatrix} 0 & ie^{-\eta} & 0 & 0 \\ ie^{\eta} & 0 & 0 & 0 \\ 0 & 0 & 0 & 0 \\ 0 & 0 & 0 & 0 \end{pmatrix} \rightarrow K_3' = \begin{pmatrix} 0 & 0 & 0 & 0 \\ ie^{\eta} & 0 & 0 & 0 \\ 0 & 0 & 0 & 0 \\ 0 & 0 & 0 & 0 \end{pmatrix}. \tag{11.27}$$

In the large η limit, the top row element becomes zero. We squeeze back after this procedure with $B^{-1}\, K_3'\, B$. Repeating the same procedure for K_1, we finally obtain

$$K_3 \rightarrow P_3 = \begin{pmatrix} 0 & 0 & 0 & 0 \\ i & 0 & 0 & 0 \\ 0 & 0 & 0 & 0 \\ 0 & 0 & 0 & 0 \end{pmatrix} \quad \text{and} \quad K_1 \rightarrow P_1 = \begin{pmatrix} 0 & 0 & 0 & 0 \\ 0 & 0 & 0 & 0 \\ i & 0 & 0 & 0 \\ 0 & 0 & 0 & 0 \end{pmatrix}. \tag{11.28}$$

The following commutation relations are satisfied by these new operators:

$$[P_3, P_1] = 0, \quad [J_2, P_1] = -iP_3, \quad \text{and} \quad [J_2, P_3] = iP_1. \tag{11.29}$$

This closed set of commentators is the Lie algebra for the $E(2)$ group or the two-dimensional Euclidean group.

The matrices P_3 and P_1 lead to the transformation matrix

$$\exp\left(-i\left[aP_3 + bP_1\right]\right) = \begin{pmatrix} 1 & 0 & 0 & 0 \\ a & 1 & 0 & 0 \\ b & 0 & 1 & 0 \\ 0 & 0 & 0 & 1 \end{pmatrix}. \tag{11.30}$$

When this matrix is applied to $(t,\ z,\ x,\ y)$ we obtain:

$$\begin{pmatrix} 1 & 0 & 0 & 0 \\ a & 1 & 0 & 0 \\ b & 0 & 1 & 0 \\ 0 & 0 & 0 & 1 \end{pmatrix} \begin{pmatrix} t \\ z \\ x \\ y \end{pmatrix} = \begin{pmatrix} t \\ z + a \\ x + b \\ y \end{pmatrix} . \tag{11.31}$$

As illustrated in Fig. 11.3, this is a translation matrix if we take the two-dimensional Euclidean subspace, with coordinates z and x. By eliminating the top row, we were able to contract $Sp(2)$ to $E(2)$. This was very similar to procedure used to contract $O(2, 1)$ to $E(2)$ detailed in Chap. 4. It is interesting that the phase space of a single oscillator can produce three different representations which share with $O(2, 1)$ the same Lie algebra. This was shown in Table 11.1.

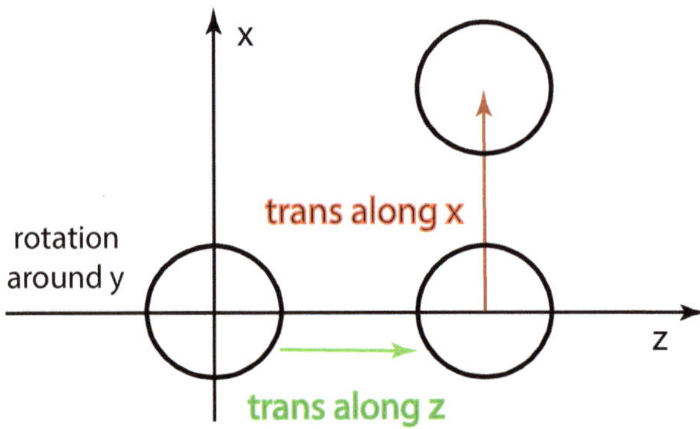

Fig. 11.3: Contraction of $Sp(2)$ to $E(2)$. The rotation around the y-direction remains unchanged during this contraction procedure. The boost along the z- and x-directions become translations along the z- and x-directions respectively [5].

11.3 Two Harmonic Oscillators and Their Symmetries

In 1963 [1], Dirac demonstrated that to generate Lorentz transformations applicable to the full Minkowski space, two Heisenberg commutation relations were needed. Dirac also showed that it is possible to use two harmonic oscillators to write these two uncertainty relations. These two harmonic oscillators can be expressed as:

$$\left[a_i, a^{\dagger}_j \right] = \delta_{ij} , \tag{11.32}$$

with

$$a_i = \frac{1}{\sqrt{2}} \left(x_i + i p_i \right) \quad \text{and} \quad a^{\dagger}_i = \frac{1}{\sqrt{2}} \left(x_i - i p_i \right) . \tag{11.33}$$

Therefore we can write:

$$x_i = \frac{1}{\sqrt{2}}\left(a_i + a_i^\dagger\right) \quad \text{and} \quad p_i = \frac{i}{\sqrt{2}}\left(a_i^\dagger - a_i\right), \tag{11.34}$$

where i and j could be 1 or 2.

In Eq. (11.16), we introduced a two-by-two matrix for the one harmonic oscillator case. Here, if the harmonic oscillators are not coupled, we can consider the following four-by-four block-diagonal matrix:

$$\begin{pmatrix} \left(a_1 a_1^\dagger + a_1^\dagger a_1\right)/2 & a_1 a_1 & 0 & 0 \\ a_1^\dagger a_1^\dagger & \left(a_1 a_1^\dagger + a_1^\dagger a_1\right)/2 & 0 & 0 \\ 0 & 0 & \left(a_2 a_2^\dagger + a_2^\dagger a_2\right)/2 & a_2 a_2 \\ 0 & 0 & a_2^\dagger a_2^\dagger & \left(a_2 a_2^\dagger + a_2^\dagger a_2\right)/2 \end{pmatrix}. \tag{11.35}$$

It is apparent that this matrix has six generators.

If now, we wish to couple these two harmonic oscillators, we need to fill in the off-diagonal blocks with a two-by-two matrix. We can accomplish this by using the most general form for this matrix and its Hermitian conjugate. These matrices take the form:

$$\begin{pmatrix} a_1^\dagger a_2 & a_1 a_2 \\ a_1^\dagger a_2^\dagger & a_1 a_2^\dagger \end{pmatrix} \quad \text{and} \quad \begin{pmatrix} a_2^\dagger a_1 & a_2 a_1 \\ a_2^\dagger a_1^\dagger & a_2 a_1^\dagger \end{pmatrix}. \tag{11.36}$$

There are four independent generators in the matrices in Eq. (11.36). Therefore, we can write the the following four-by-four matrix with ten $(6 + 4)$ generators:

$$\begin{pmatrix} \left(a_1 a_1^\dagger + a_1^\dagger a_1\right)/2 & a_1 a_1 & a_1^\dagger a_2 & a_1 a_2 \\ a_1^\dagger a_1^\dagger & \left(a_1 a_1^\dagger + a_1^\dagger a_1\right)/2 & a_1^\dagger a_2^\dagger & a_1 a_2^\dagger \\ a_2^\dagger a_1 & a_2 a_1 & \left(a_2 a_2^\dagger + a_2^\dagger a_2\right)/2 & a_2 a_2 \\ a_2^\dagger a_1^\dagger & a_2 a_1^\dagger & a_2^\dagger a_2^\dagger & \left(a_2 a_2^\dagger + a_2^\dagger a_2\right)/2 \end{pmatrix}. \tag{11.37}$$

With these ten generators, four rotation-like generators can now be constructed:

$$J_1 = \frac{1}{2}\left(a_1^\dagger a_2 + a_2^\dagger a_1\right), \quad J_2 = \frac{1}{2i}\left(a_1^\dagger a_2 - a_2^\dagger a_1\right),$$

$$J_3 = \frac{1}{2}\left(a_1^\dagger a_1 - a_2^\dagger a_2\right), \quad S_3 = \frac{1}{2}\left(a_1^\dagger a_1 + a_2 a_2^\dagger\right). \tag{11.38}$$

Six squeeze-like generators can also be constructed. The first three are:

$$K_1 = -\frac{1}{4}\left(a_1^\dagger a_1^\dagger + a_1 a_1 - a_2^\dagger a_2^\dagger - a_2 a_2\right),$$

$$K_2 = +\frac{i}{4}\left(a_1^\dagger a_1^\dagger - a_1 a_1 + a_2^\dagger a_2^\dagger - a_2 a_2\right),$$

$$K_3 = +\frac{1}{2}\left(a_1^\dagger a_2^\dagger + a_1 a_2\right). \tag{11.39}$$

The additional three are:

$$Q_1 - -\frac{i}{4}\left(a_1^\dagger a_1^\dagger - a_1 a_1 - a_2^\dagger a_2^\dagger + a_2 a_2\right),$$

$$Q_2 = -\frac{1}{4}\left(a_1^\dagger a_1^\dagger + a_1 a_1 + a_2^\dagger a_2^\dagger + a_2 a_2\right),$$

$$Q_3 = +\frac{i}{2}\left(a_1^\dagger a_2^\dagger - a_1 a_2\right). \tag{11.40}$$

There are now ten operators constructed from Eqs. (11.38, 11.39, 11.40). Dirac in 1963 [1] showed they satisfied the following set of commutation relations:

$$[J_i, J_j] = i\varepsilon_{ijk}J_k, \qquad [J_i, K_j] = i\varepsilon_{ijk}K_k,$$

$$[J_i, Q_j] = i\varepsilon_{ijk}Q_k, \qquad [K_i, K_j] = [Q_i, Q_j] = -i\varepsilon_{ijk}J_k,$$

$$[K_i, Q_j] = -i\delta_{ij}S_3, \qquad [J_i, S_3] = 0,$$

$$[K_i, S_3] = -iQ_i, \qquad [Q_i, S_3] = iK_i. \tag{11.41}$$

The Lie algebra, represented by the commutation relations in Eq. (11.41), as Dirac noted, is the same as the Lie algebra for the Lorentz group $O(3, 2)$ which is isomorphic to the $(3 + 2)$ de Sitter group. This $O(3, 2)$ Lorentz group has ten generators and is applicable to the three space and two time dimensions. A very important role is played by this group in space-time symmetries.

The ten generators for the $O(3, 2)$ group are represented by five-by-five matrices which are applicable to (s, t, z, x, y). Here the two time-like variables are s and t. The explicit form of these five-by-five matrices are given in Table 11.2.

With respect to the $(J_1, J_2, J_3, K_1, K_2, K_3)$, these generate rotations and boosts in the (t, z, x, y) space. It can be seen from Table 11.2 that they have zero elements in the first row and column, indicating that the s-coordinate remains invariant under these generators. They are essentially, the familiar four-by-four matrices of the Lorentz group $O(3, 1)$ in Minkowski space.

The three generators, Q_i, produce boosts along the s variable, while S_3, generates rotations between the two time variables s and t. It can be seen from Table 11.2 that these four matrices have elements only in the first row and column. It is these four generators that we shall, in Sect. 11.4, convert into the translation generators P_0 and P_i of the Poincaré group.

Table 11.2: Ten generators of the $O(3,2)$ or $(3 + 2)$ de Sitter group in the five-dimensional space (s, t, z, x, y). Rotations in the three-dimensional (z, x, y) space are generated by the J matrices. Lorentz boosts involving the time variable t are generated by the K matrices. The Q matrices generate the boosts involving the time variable s, while the S generator produces rotations in the (s, t)-plane.

$$J_1 = \begin{pmatrix} 0 & 0 & 0 & 0 & 0 \\ 0 & 0 & 0 & 0 & 0 \\ 0 & 0 & 0 & 0 & i \\ 0 & 0 & 0 & 0 & 0 \\ 0 & 0 & -i & 0 & 0 \end{pmatrix} \quad J_2 = \begin{pmatrix} 0 & 0 & 0 & 0 & 0 \\ 0 & 0 & 0 & 0 & 0 \\ 0 & 0 & 0 & -i & 0 \\ 0 & 0 & i & 0 & 0 \\ 0 & 0 & 0 & 0 & 0 \end{pmatrix} \quad J_3 = \begin{pmatrix} 0 & 0 & 0 & 0 & 0 \\ 0 & 0 & 0 & 0 & 0 \\ 0 & 0 & 0 & 0 & 0 \\ 0 & 0 & 0 & 0 & -i \\ 0 & 0 & 0 & i & 0 \end{pmatrix}$$

$$K_1 = \begin{pmatrix} 0 & 0 & 0 & 0 & 0 \\ 0 & 0 & 0 & i & 0 \\ 0 & 0 & 0 & 0 & 0 \\ 0 & i & 0 & 0 & 0 \\ 0 & 0 & 0 & 0 & 0 \end{pmatrix} \quad K_2 = \begin{pmatrix} 0 & 0 & 0 & 0 & 0 \\ 0 & 0 & 0 & 0 & i \\ 0 & 0 & 0 & 0 & 0 \\ 0 & 0 & 0 & 0 & 0 \\ 0 & i & 0 & 0 & 0 \end{pmatrix} \quad K_3 = \begin{pmatrix} 0 & 0 & 0 & 0 & 0 \\ 0 & 0 & i & 0 & 0 \\ 0 & i & 0 & 0 & 0 \\ 0 & 0 & 0 & 0 & 0 \\ 0 & 0 & 0 & 0 & 0 \end{pmatrix}$$

$$Q_1 = \begin{pmatrix} 0 & 0 & 0 & i & 0 \\ 0 & 0 & 0 & 0 & 0 \\ 0 & 0 & 0 & 0 & 0 \\ i & 0 & 0 & 0 & 0 \\ 0 & 0 & 0 & 0 & 0 \end{pmatrix} \quad Q_2 = \begin{pmatrix} 0 & 0 & 0 & 0 & i \\ 0 & 0 & 0 & 0 & 0 \\ 0 & 0 & 0 & 0 & 0 \\ 0 & 0 & 0 & 0 & 0 \\ i & 0 & 0 & 0 & 0 \end{pmatrix} \quad Q_3 = \begin{pmatrix} 0 & 0 & i & 0 & 0 \\ 0 & 0 & 0 & 0 & 0 \\ i & 0 & 0 & 0 & 0 \\ 0 & 0 & 0 & 0 & 0 \\ 0 & 0 & 0 & 0 & 0 \end{pmatrix}$$

$$S_3 = \begin{pmatrix} 0 & -i & 0 & 0 & 0 \\ i & 0 & 0 & 0 & 0 \\ 0 & 0 & 0 & 0 & 0 \\ 0 & 0 & 0 & 0 & 0 \\ 0 & 0 & 0 & 0 & 0 \end{pmatrix}$$

In this very same paper [1], Dirac noted that the set of commutation relations in Eq. (11.41) serves as the Lie algebra for $Sp(4)$ which is the four-dimensional symplectic group. As we said in Sect. 7.5 of Chap. 7, this $Sp(4)$ group plays an important role in polarization optics, squeezed states of light, and when special relativity is incorporated into quantum mechanics. As we did in Sect. 7.6 of Chap. 7 we can now use the four-dimensional phase space with the coordinate variables defined as [11, 12]

$$(x_1, p_1, x_2, p_2) \,. \tag{11.42}$$

These coordinates describe a dynamical system which consists of two pairs of canonical variables x_1, p_1 and x_2, p_2.

The transformation matrix M which applies to the four-component vector in Eq. (11.42) is canonical if [13, 14]

$$MJ\tilde{M} = J.\tag{11.43}$$

Here $\tilde{M}$ is the transpose of the M matrix. The matrix J is defined as

$$J = \begin{pmatrix} 0 & 1 & 0 & 0 \\ -1 & 0 & 0 & 0 \\ 0 & 0 & 0 & 1 \\ 0 & 0 & -1 & 0 \end{pmatrix}.\tag{11.44}$$

The matrix in Eq. (11.44) can be written in block-diagonal form as

$$J = i \begin{pmatrix} I & 0 \\ 0 & I \end{pmatrix} \sigma_2.\tag{11.45}$$

Here I is, of course, the unit two-by-two matrix.

Using this form of the J matrix, the phase space area for the x_1 and p_1 variables remains invariant, and this is also the same for x_2 and p_2 phase space.

Now it is possible to write the generators of the $Sp(4)$ group as [15]

$$J_1 = -\frac{1}{2}\begin{pmatrix} 0 & I \\ I & 0 \end{pmatrix}\sigma_2, \quad J_2 = \frac{i}{2}\begin{pmatrix} 0 & -I \\ I & 0 \end{pmatrix}I,$$

$$J_3 = \frac{1}{2}\begin{pmatrix} -I & 0 \\ 0 & I \end{pmatrix}\sigma_2, \quad S_3 = \frac{1}{2}\begin{pmatrix} I & 0 \\ 0 & I \end{pmatrix}\sigma_2,\tag{11.46}$$

and

$$K_1 = \frac{i}{2}\begin{pmatrix} I & 0 \\ 0 & -I \end{pmatrix}\sigma_1, \quad K_2 = \frac{i}{2}\begin{pmatrix} I & 0 \\ 0 & I \end{pmatrix}\sigma_3, \quad K_3 = -\frac{i}{2}\begin{pmatrix} 0 & I \\ I & 0 \end{pmatrix}\sigma_1,$$

$$Q_1 = -\frac{i}{2}\begin{pmatrix} I & 0 \\ 0 & -I \end{pmatrix}\sigma_3, \quad Q_2 = \frac{i}{2}\begin{pmatrix} I & 0 \\ 0 & I \end{pmatrix}\sigma_1, \quad Q_3 = \frac{i}{2}\begin{pmatrix} 0 & I \\ I & 0 \end{pmatrix}\sigma_3.\tag{11.47}$$

Six of the matrices in Eqs. (11.46) and (11.47) are in block-diagonal form. They are S_3, J_3, K_1, K_2, Q_1, and Q_2. Using the language of two harmonic oscillators, it is evident that the first and second harmonic oscillators are not mixed by these generators. The reason there are six of them is because each harmonic oscillator has three generators forming an $Sp(2)$ symmetry. In Table 11.3 we tabulate these generators and those for the harmonic oscillator representation.

As can be seen from Table 11.3, the J_2 matrix which is off-diagonal, couples the first and second oscillators but does not change the overall volume of the four-dimensional phase space. However, to construct a closed set of commutation relations, three additional generators, J_1, K_3, and Q_3, are needed. The commutation relations given in Eq.(11.41) are clearly consequences of Heisenberg's uncertainty relation.

Table 11.3: For the two harmonic oscillator system we list the transformation generators.

Generators	Two Oscillators	Phase space
J_1	$\frac{1}{2}\left(a_1^\dagger a_2 + a_2^\dagger a_1\right)$	$-\frac{1}{2}\begin{pmatrix} 0 & I \\ I & 0 \end{pmatrix}\sigma_2$
J_2	$\frac{1}{2i}\left(a_1^\dagger a_2 - a_2^\dagger a_1\right)$	$\frac{i}{2}\begin{pmatrix} 0 & -I \\ I & 0 \end{pmatrix} I$
J_3	$\frac{1}{2}\left(a_1^\dagger a_1 - a_2^\dagger a_2\right)$	$\frac{1}{2}\begin{pmatrix} -I & 0 \\ 0 & I \end{pmatrix}\sigma_2$
S_3	$\frac{1}{2}\left(a_1^\dagger a_1 + a_2 a_2^\dagger\right)$	$\frac{1}{2}\begin{pmatrix} I & 0 \\ 0 & I \end{pmatrix}\sigma_2$
K_1	$-\frac{1}{4}\left(a_1^\dagger a_1^\dagger + a_1 a_1 - a_2^\dagger a_2^\dagger - a_2 a_2\right)$	$\frac{i}{2}\begin{pmatrix} I & 0 \\ 0 & -I \end{pmatrix}\sigma_1$
K_2	$+\frac{i}{4}\left(a_1^\dagger a_1^\dagger - a_1 a_1 + a_2^\dagger a_2^\dagger - a_2 a_2\right)$	$\frac{i}{2}\begin{pmatrix} I & 0 \\ 0 & I \end{pmatrix}\sigma_3$
K_3	$\frac{1}{2}\left(a_1^\dagger a_2^\dagger + a_1 a_2\right)$	$-\frac{i}{2}\begin{pmatrix} 0 & I \\ I & 0 \end{pmatrix}\sigma_1$
Q_1	$-\frac{i}{4}\left(a_1^\dagger a_1^\dagger - a_1 a_1 - a_2^\dagger a_2^\dagger + a_2 a_2\right)$	$-\frac{i}{2}\begin{pmatrix} I & 0 \\ 0 & -I \end{pmatrix}\sigma_3$
Q_2	$-\frac{1}{4}\left(a_1^\dagger a_1^\dagger + a_1 a_1 + a_2^\dagger a_2^\dagger + a_2 a_2\right)$	$\frac{i}{2}\begin{pmatrix} I & 0 \\ 0 & I \end{pmatrix}\sigma_1$
Q_3	$\frac{i}{2}\left(a_1^\dagger a_2^\dagger - a_1 a_2\right)$	$\frac{i}{2}\begin{pmatrix} 0 & I \\ I & 0 \end{pmatrix}\sigma_3$

11.4 Contraction of O(3, 2) to the Poincaré Group

As we saw in Chap. 7, Dirac's two coupled oscillators lead to the Lorentz group of $O(3,2)$ [1]. In Fig. 11.4, it is illustrated that it is possible to contract the $O(3,2)$ group into the Poincaré group. This is accomplished by contracting one of those two time variables of the $O(3,2)$ group plus the extra rotation variable into the four translation generators. The remaining variables form the Lorentz group $O(3,1)$, and together with the four translation generators, form the Poincaré goup. The four translations generators correspond to the energy-momentum four-vector and thus to Einstein's energy-momentum relation $E = mc^2$.

We have seen two Heisenberg commutation relations can lead to both the $O(3, 2)$ group and the $Sp(4)$ group. We also showed that these groups are isomorphic to each other. Now we are interested in using group contraction to produce the Poincaré group from the $O(3, 2)$ group. From Table 11.2 it is clear there are only elements in the first row and first column for the matrices for the three boost generators, Q_i and the rotation generator S_3. We shall therefore convert Q_i and S_3 into the translation generators P_i and P_0, respectively which are applicable to (t, z, x, y).

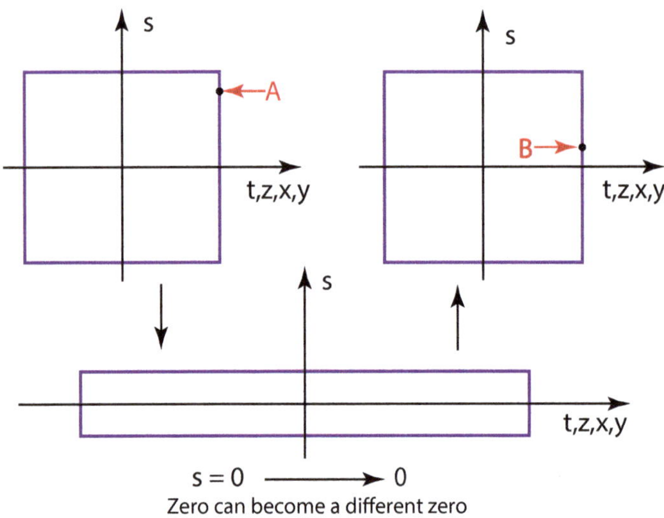

Fig. 11.4: Re-interpretation of the Inönü–Wigner contraction procedure [3] as a squeeze transformation. During the squeeze process, the square becomes a narrow rectangle. The point A can be moved to the horizontal axis after the rectangle becomes narrow enough. The rectangle is returned to the original shape by the inverse squeeze. The point A (denoted by B), however, remains on the horizontal axis. [4, 5].

Inönü and Wigner [3] originally used the group contraction procedure to transform the four-dimensional Lorentz group into the three-dimensional Galilei group. On the other hand, we shall use the squeeze technique introduced in Sect. 4.1 of Chap. 4 to accomplish the contraction of the $O(3, 2)$ to the Poincaré group. We shall contract the generators Q_i that boost the time-like s variable, together with the generator S_3 that rotates the two time-like variables.

For this purpose, we introduce the squeeze matrix

$$
B(\eta) = \begin{pmatrix} e^{-\eta/2} & 0 & 0 & 0 & 0 \\ 0 & e^{\eta/2} & 0 & 0 & 0 \\ 0 & 0 & e^{\eta/2} & 0 & 0 \\ 0 & 0 & 0 & e^{\eta/2} & 0 \\ 0 & 0 & 0 & 0 & e^{\eta/2} \end{pmatrix}, \tag{11.48}
$$

and the inverse matrix:

$$
B^{-1}(\eta) = \begin{pmatrix} e^{\eta/2} & 0 & 0 & 0 & 0 \\ 0 & e^{-\eta/2} & 0 & 0 & 0 \\ 0 & 0 & e^{-\eta/2} & 0 & 0 \\ 0 & 0 & 0 & e^{-\eta/2} & 0 \\ 0 & 0 & 0 & 0 & e^{-\eta/2} \end{pmatrix}. \tag{11.49}
$$

Although this matrix commutes with J_i and K_i, we shall use it to eliminate the element from the first row in Q_i and S_3. For instance, take Q_1,

$$
B(\eta)Q_1 B^{-1}(\eta) = \begin{pmatrix} e^{-\eta/2} & 0 & 0 & 0 & 0 \\ 0 & e^{\eta/2} & 0 & 0 & 0 \\ 0 & 0 & e^{\eta/2} & 0 & 0 \\ 0 & 0 & 0 & e^{\eta/2} & 0 \\ 0 & 0 & 0 & 0 & e^{\eta/2} \end{pmatrix} \begin{pmatrix} 0 & 0 & 0 & i & 0 \\ 0 & 0 & 0 & 0 & 0 \\ 0 & 0 & 0 & 0 & 0 \\ i & 0 & 0 & 0 & 0 \\ 0 & 0 & 0 & 0 & 0 \end{pmatrix}
$$

$$
\begin{pmatrix} e^{\eta/2} & 0 & 0 & 0 & 0 \\ 0 & e^{-\eta/2} & 0 & 0 & 0 \\ 0 & 0 & e^{-\eta/2} & 0 & 0 \\ 0 & 0 & 0 & e^{-\eta/2} & 0 \\ 0 & 0 & 0 & 0 & e^{-\eta/2} \end{pmatrix}, \tag{11.50}
$$

which results in

$$
Q_1' = \begin{pmatrix} 0 & 0 & 0 & ie^{-\eta} & 0 \\ 0 & 0 & 0 & 0 & 0 \\ 0 & 0 & 0 & 0 & 0 \\ ie^{\eta} & 0 & 0 & 0 & 0 \\ 0 & 0 & 0 & 0 & 0 \end{pmatrix}. \tag{11.51}
$$

In the limit of large η, the element in the first row becomes zero. Taking the inverse transformation as we did in Sect. 4.1 in Chap. 4 we obtain:

$$
B^{-1}(\eta)Q_1'B(\eta) =
\begin{pmatrix}
e^{\eta/2} & 0 & 0 & 0 & 0 \\
0 & e^{-\eta/2} & 0 & 0 & 0 \\
0 & 0 & e^{-\eta/2} & 0 & 0 \\
0 & 0 & 0 & e^{-\eta/2} & 0 \\
0 & 0 & 0 & 0 & e^{-\eta/2}
\end{pmatrix}
\begin{pmatrix}
0 & 0 & 0 & 0 & 0 \\
0 & 0 & 0 & 0 & 0 \\
0 & 0 & 0 & 0 & 0 \\
ie^{\eta} & 0 & 0 & 0 & 0 \\
0 & 0 & 0 & 0 & 0
\end{pmatrix}
$$

$$
\begin{pmatrix}
e^{-\eta/2} & 0 & 0 & 0 & 0 \\
0 & e^{\eta/2} & 0 & 0 & 0 \\
0 & 0 & e^{\eta/2} & 0 & 0 \\
0 & 0 & 0 & e^{\eta/2} & 0 \\
0 & 0 & 0 & 0 & e^{\eta/2}
\end{pmatrix}
\tag{11.52}
$$

yielding

$$
Q_1'' =
\begin{pmatrix}
0 & 0 & 0 & 0 & 0 \\
0 & 0 & 0 & 0 & 0 \\
0 & 0 & 0 & 0 & 0 \\
i & 0 & 0 & 0 & 0 \\
0 & 0 & 0 & 0 & 0
\end{pmatrix}.
\tag{11.53}
$$

Hence we see that the contraction process leads to

$$
Q_1 \to P_1
\tag{11.54}
$$

where the explicit five-by-five matrix for P_1 is given in Eq. (11.56). Hence we have in the limit of large η

$$
\begin{aligned}
Q_1 &\to P_1, & Q_2 &\to P_2 \\
Q_3 &\to P_3, & S_3 &\to P_0
\end{aligned}
\tag{11.55}
$$

As we did in Eq. (7.89) of Chap. 7, we write these four matrices in the following manner:

$$
S_3 = \begin{pmatrix} 0 & -i & 0 & 0 & 0 \\ i & 0 & 0 & 0 & 0 \\ 0 & 0 & 0 & 0 & 0 \\ 0 & 0 & 0 & 0 & 0 \\ 0 & 0 & 0 & 0 & 0 \end{pmatrix} \quad \rightarrow \quad P_0 = \begin{pmatrix} 0 & 0 & 0 & 0 & 0 \\ i & 0 & 0 & 0 & 0 \\ 0 & 0 & 0 & 0 & 0 \\ 0 & 0 & 0 & 0 & 0 \\ 0 & 0 & 0 & 0 & 0 \end{pmatrix},
$$

$$
Q_3 = \begin{pmatrix} 0 & 0 & i & 0 & 0 \\ 0 & 0 & 0 & 0 & 0 \\ i & 0 & 0 & 0 & 0 \\ 0 & 0 & 0 & 0 & 0 \\ 0 & 0 & 0 & 0 & 0 \end{pmatrix} \quad \rightarrow \quad P_3 = \begin{pmatrix} 0 & 0 & 0 & 0 & 0 \\ 0 & 0 & 0 & 0 & 0 \\ i & 0 & 0 & 0 & 0 \\ 0 & 0 & 0 & 0 & 0 \\ 0 & 0 & 0 & 0 & 0 \end{pmatrix},
$$

$$
Q_1 = \begin{pmatrix} 0 & 0 & 0 & i & 0 \\ 0 & 0 & 0 & 0 & 0 \\ 0 & 0 & 0 & 0 & 0 \\ i & 0 & 0 & 0 & 0 \\ 0 & 0 & 0 & 0 & 0 \end{pmatrix} \quad \rightarrow \quad P_1 = \begin{pmatrix} 0 & 0 & 0 & 0 & 0 \\ 0 & 0 & 0 & 0 & 0 \\ 0 & 0 & 0 & 0 & 0 \\ i & 0 & 0 & 0 & 0 \\ 0 & 0 & 0 & 0 & 0 \end{pmatrix},
$$

$$
Q_2 = \begin{pmatrix} 0 & 0 & 0 & 0 & i \\ 0 & 0 & 0 & 0 & 0 \\ 0 & 0 & 0 & 0 & 0 \\ 0 & 0 & 0 & 0 & 0 \\ i & 0 & 0 & 0 & 0 \end{pmatrix} \quad \rightarrow \quad P_2 = \begin{pmatrix} 0 & 0 & 0 & 0 & 0 \\ 0 & 0 & 0 & 0 & 0 \\ 0 & 0 & 0 & 0 & 0 \\ 0 & 0 & 0 & 0 & 0 \\ i & 0 & 0 & 0 & 0 \end{pmatrix}. \tag{11.56}
$$

The four contracted generators given in Eq. (11.55) lead to the five-by-five transformation matrix, which can be written as:

$$
\exp\left\{-i\left(aP_0 + bP_3 + cP_1 + dP_2\right)\right\}. \tag{11.57}
$$

Thus we see that Eq. (11.57) performs translations in the Minkowski space. We can then write this as:

$$
\begin{pmatrix} 1 & 0 & 0 & 0 & 0 \\ a & 1 & 0 & 0 & 0 \\ b & 0 & 1 & 0 & 0 \\ c & 0 & 0 & 1 & 0 \\ d & 0 & 0 & 0 & 1 \end{pmatrix} \begin{pmatrix} 1 \\ t \\ z \\ x \\ y \end{pmatrix} = \begin{pmatrix} 1 \\ t+a \\ z+b \\ x+c \\ y+d \end{pmatrix}. \tag{11.58}
$$

This contraction procedure is illustrated in Fig. 11.4. The square can be flattened into a rectangle or a horizontal line with $s = 0$. This means, as illustrated in Fig. 11.1, that the group $O(3,2)$, which we derived from Heisenberg's uncertainty relations, becomes the Poincaré group whose symmetries govern quantum mechanics and quantum field theory.

Now we shall translate this into differential operators with the translation operators:

$$
P_0 = i\frac{\partial}{\partial t}, \quad P_z = -i\frac{\partial}{\partial z}, \quad P_x = -i\frac{\partial}{\partial x}, \quad P_y = -i\frac{\partial}{\partial y}. \tag{11.59}
$$

In Table 11.3 we presented the generators J_i, K_i, Q_i, and S_3 derivable from the two harmonic system of Heisenberg's commutation relations together with the phase space equivalents. In Tables 7.4 and 7.5 of Chap. 7 the same twelve generators derived from Dirac's two harmonic oscillator system together with the initial uncontracted and in the instance of the Q_i and S_3 the contracted differential forms as well. Here, we have again converted these same four generators, now derived form Heisenberg's commutation relations, into the four translation generators. We again used the group contraction procedure, similar to that of Inönü and Wigner [3] interpreting it as a squeeze transformation as outlined in Sect. 4.1 of Chap. 4.

The contraction procedure produced four translation generators which correspond to the energy-momentum four-vector:

$$p_0^2 - p_3^2 - p_1^2 - p_2^2 = \text{constant}. \tag{11.60}$$

This energy-momentum relation is widely known as Einstein's $E = mc^2$ [15, 4, 5]. We note that Eqs. (7.90) through (7.93) of Chap. 7, derived from Dirac's two harmonic oscillator system are the same as Eqs. (11.57) through (11.60) produced this time from the Heisenberg commutation relations for the two harmonic oscillator system and both produce Einstein's energy-momentum relation.

In Table 7.6 we compared these three different systems: Galilean, Lorentz, and Dirac's two-oscillator system, which were derivable from two harmonic oscillators as well as from two Heisenberg commutation relations. These three systems have led us to the system of the Poincaré group.

11.5 Dirac's Quest to Unite Special Relativity and Quantum Mechanics

In Dirac's 1949 paper [2], he stated that the problem of finding a Lorentz-covariant quantum mechanics can be reduced to the problem of finding a representation of the Poincaré or inhomogeneous Lorentz group. Again, in Dirac's 1963 paper [1], he showed that it is possible to construct the Lie algebra of the group $O(3, 2)$. Earlier [15, 4, 5] we had shown that this $O(3, 2)$ group can be contracted to the Poincaré group by using the group contraction procedure introduced by Inönü and Wigner [3]. We discussed this in both Sect. 7.8 of Chap. 7 and in Sect. 11.4 in this Chapter.

Here, first we showed that the symmetry of a single harmonic oscillator consists of three generators. With two independent harmonic oscillators there are six generators. We therefore, needed four additional generators to couple the two harmonic oscillators resulting in ten generators. As was spelled out in Dirac's 1963 paper [1], these ten generators can be linearly combined to produce ten generators which form the $O(3, 2)$ group.

For the two harmonic oscillator system, there are four step-up and step-down operators. There are therefore sixteen quadratic forms [12]. However, Dirac's 1963 paper [1] only contains ten of them. Why are there only ten? In order to construct the Lie algebra for the $O(3, 2)$ group, Dirac only needed the chosen ten. In this paper, Dirac also stated that this Lie algebra applied as well to the $Sp(4)$ group. This group, isomorphic to $O(3, 2)$, preserves the minimum uncertainty for each harmonic oscillator.

In this Chapter, we started in Sect. 11.3 with the block-diagonal matrix given in Eq. (11.35). This was for two totally independent oscillators that had six independent generators. To couple the oscillators, in Eq. (11.36) a two-by-two matrix and its Hermitian conjugate are introduced. This produced four new independent generators for the off-diagonal blocks. This resulted in the four-by-four Hermitian matrix given in Eq. (11.37). The ten independent generators contained in this four-by-four Hermitian matrix can be combined linearly into the ten generators that Dirac chose. Thus, in this Chapter, we have shown how two harmonic oscillators can be coupled, and how this coupling introduces additional symmetries.

We note that P. A. M. Dirac made it his life-long ambition to make quantum mechanics consistent with special relativity. As is well known, special relativity and quantum mechanics developed in two separate pathways. As we saw in Chap. 6, in 1927 [16] Dirac was interested in understanding if these two scientific disciplines were compatible with each other. In his 1927 paper [16], Dirac noted that there existed a time-energy uncertainty relation without excitation. He referred to this concept as the *c-number time-energy uncertainty relation*. Dirac then noted that it is hard to construct the uncertainty relation in the Lorentz-covariant world because of the space-time asymmetry,

In 1945 [17], Dirac considered the four-dimensional harmonic oscillator wave functions applicable to four-dimensional space and time. At that time he was considering localized bound states. In Dirac's work the space and time variables are the separations between two constituents as, for example, the proton and the electron are separated in the hydrogen atom.

Later it was shown that because of the fact that a massive particle at rest has only three space-like dimensions [18], Dirac's concern about the c-number time-energy uncertainty relation was not necessary. As we saw in Chap. 2 the little group for the massive particle is isomorphic to $O(3)$ [19]. Therefore, we can illustrate this by using a circle in the zt-plane. We show this in Fig. 11.5. In that figure, z and t are the longitudinal and time separations respectively.

The light-cone coordinate system was introduced in Dirac's 1949 paper [2]. This coordinate system tells us that the Lorentz boost is a squeeze transformation. We illustrate this aspect also in Fig. 11.5. This then represents how the circle looks to a moving observer.

Then, we have to ask the question if this elliptic squeeze can be interpreted in the real world. In the early part of the last century, Niels Bohr and Albert Einstein would meet occasionally to discuss physics, even though they had different interests. Bohr was concerned about the electron orbit in the hydrogen atom. Einstein was concerned about how things look to moving observers. How the hydrogen atom would look to a moving observer was a metaphysical issue then, because the hydrogen atom could not move fast enough to exhibit this Einstein effect.

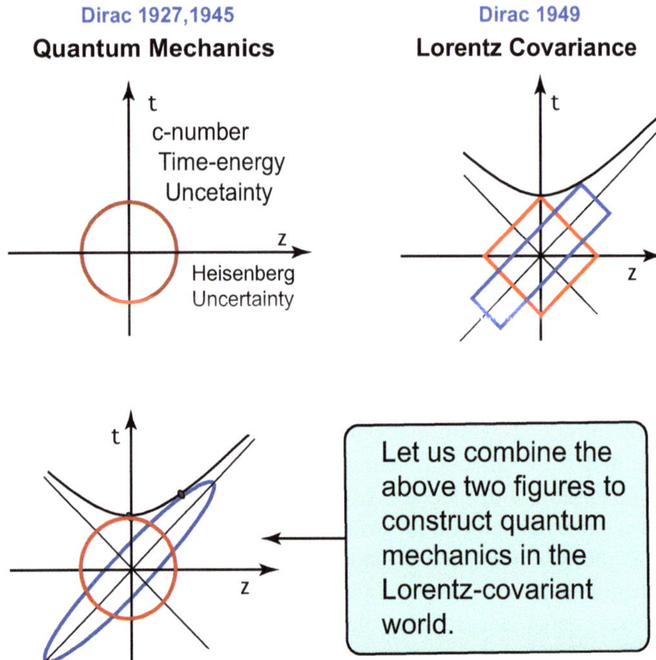

Fig. 11.5: Dirac's three papers. Dirac's 1927, 1945, and 1949 papers [16, 17, 2] are interpreted by the circle in the longitudinal and time-like coordinate while Dirac's light-cone coordinate system [2] can be interpreted as a squeeze transformation. These two figures, then can be synthesized to a squeezed circle or an ellipse. We shall explore in Chap. 13 how this squeeze appears in the real world.

Many years after these Bohr-Einstein discussions, a large number of high energy protons could be produced from particle accelerators. In 1955 [20] the discovery of Hofstadter and McAllister showed the proton charge has an internal distribution. In 1964 [21], Gell-Mann observed that the proton is a bound state of more fundamental particles that he called *quarks*. However, Feynman [22, 23], in 1969, observed that when the proton moves very fast, it appears like as a collection of a large number of free-moving light-like particles that he called *partons*. Feynman's parton picture was based entirely on what we can observe in laboratories [24].

The proton, therefore, unlike the hydrogen atom, can become accelerated to a speed very close to that of light. Thus the Bohr-Einstein issue became the Gell-Mann-Feynman issue. The question then is whether Gell-Mann's quark model and Feynman's parton picture are two different aspects of one Lorentz-covariant entity. This question was addressed by Kim and Noz in 1977 [25] and will be explored in Chap. 13.

References

1. P.A.M. Dirac, A Remarkable Representation of the 3 + 2 de Sitter Group, Journal of Mathematical Physics **4**(7), 901–909 (1963). DOI 10.1063/1.1704016. URL http://aip.scitation.org/doi/10.1063/1.1704016
2. P.A.M. Dirac, Forms of Relativistic Dynamics, Reviews of Modern Physics **21**(3), 392–399 (1949). DOI 10.1103/RevModPhys.21.392. URL https://link.aps.org/doi/10.1103/RevModPhys.21.392
3. E. Inönü, E.P. Wigner, On the Contraction of Groups and Their Representations, Proceedings of the National Academy of Sciences **39**(6), 510–524 (1953). DOI 10.1073/pnas.39.6.510. URL http://www.pnas.org/cgi/doi/10.1073/pnas.39.6.510
4. S. Başkal, Y.S. Kim, M.E. Noz, Einstein's E = mc^2 Derivable from Heisenberg's Uncertainty Relations, Quantum Reports **1**(2), 236–251 (2019). DOI 10.3390/quantum1020021. URL https://www.mdpi.com/2624-960X/1/2/21
5. S. Başkal, Y.S. Kim, M.E. Noz, *Physics of the Lorentz Group (Second Edition): Beyond high-energy physics and optics* (IOP Publishing, Bristol, UK, 2021). DOI 10.1088/978-0-7503-3607-9. ISBN 978-0-7503-3607-9. URL https://iopscience.iop.org/book/978-0-7503-3607-9
6. V. Fock, Quanten Elecktrodynamik, Physikalische Zeitschrift der Sovietunion **6**, 425–469 (1934)
7. D. Han, Y.S. Kim, M.E. Noz, Linear canonical transformations of coherent and squeezed states in the Wigner phase space, Physical Review A **37**(3), 807–814 (1988). DOI 10.1103/PhysRevA.37.807. URL https://link.aps.org/doi/10.1103/PhysRevA.37.807
8. Y.S. Kim, E.P. Wigner, Canonical transformation in quantum mechanics, American Journal of Physics **58**(5), 439–448 (1990). DOI 10.1119/1.16475. URL http://aapt.scitation.org/doi/10.1119/1.16475
9. Y.S. Kim, M.E. Noz, *Phase space picture of quantum mechanics: group theoretical approach*. No. 40 in Lecture notes in physics series (World Scientific Publishing Co., Singapore; Hackensack, NJ, USA, 1991). ISBN 978-981-02-0360-3,978-981-02-0361-0. URL https://doi.org/10.1142/1197
10. V.V. Dodonov, V.I. Man'ko (eds.), *Theory of nonclassical states of light* (Taylor & Francis, London ; New York, 2003). ISBN 978-0-415-28413-4. (OCLC: ocm49550380.)
11. D. Han, Y.S. Kim, M.E. Noz, L. Yeh, Symmetries of two–mode squeezed states, Journal of Mathematical Physics **34**(12), 5493–5508 (1993). DOI 10.1063/1.530318. URL http://aip.scitation.org/doi/10.1063/1.530318
12. D. Han, Y.S. Kim, M.E. Noz, O(3,3)–like symmetries of coupled harmonic oscillators, Journal of Mathematical Physics **36**(8), 3940–3954 (1995). DOI 10.1063/1.530940. URL http://aip.scitation.org/doi/10.1063/1.530940
13. R. Abraham, E. Marsden, J, *Foundations of mechanics*, 2nd edn. (AMS Chelsea Pub./American Mathematical Society, Providence, RI, USA, 2008). ISBN 978-0-8218-4438-0. (Originally published 1978; OCLC: ocn191847156.)
14. H. Goldstein, *Classical Mechanics*, 2nd edn. Addison-Wesley series in physics (Addison-Wesley Pub. Co, Reading, MA, USA, 1980). ISBN 978-0-201-02918-5. (Originally published 1952.)
15. S. Başkal, Y.S. Kim, M.E. Noz, Poincaré Symmetry from Heisenberg's Uncertainty Relations, Symmetry **11**(3), 409–1–9 (2019). DOI 10.3390/sym11030409. URL https://www.mdpi.com/2073-8994/11/3/409
16. P.A.M. Dirac, The Quantum Theory of the Emission and Absorption of Radiation, Proceedings of the Royal Society A: Mathematical, Physical, and Engineering Sciences **114**(767), 243–265 (1927). DOI 10.1098/rspa.1927.0039. URL http://rspa.royalsocietypublishing.org/cgi/doi/10.1098/rspa.1927.0039
17. P.A.M. Dirac, Unitary Representations of the Lorentz Group, Proceedings of the Royal Society A: Mathematical, Physical, and Engineering Sciences **183**(994), 284–295 (1945). DOI 10.1098/rspa.1945.0003. URL http://rspa.royalsocietypublishing.org/cgi/doi/10.1098/rspa.1945.0003

18. Y.S. Kim, M.E. Noz, S.H. Oh, Representations of the Poincaré group for relativistic extended hadrons, Journal of Mathematical Physics **20**(7), 1341–1344 (1979). DOI 10.1063/1.524237. URL http://aip.scitation.org/doi/10.1063/1.524237. See also, Physics Auxiliary Publication Service Document No. PAPS JMAPA-20-1336-12

19. E.P. Wigner, On Unitary Representations of the Inhomogeneous Lorentz Group, The Annals of Mathematics **40**(1), 149–204 (1939). DOI 10.2307/1968551. URL http://www.jstor.org/stable/1968551?origin=crossref

20. R. Hofstadter, R.W. McAllister, Electron Scattering from the Proton, Physical Review **98**(1), 217–218 (1955). DOI 10.1103/PhysRev.98.217. URL https://link.aps.org/doi/10.1103/PhysRev.98.217

21. M. Gell-Mann, A Schematic Model of Baryons and Mesons, Physics Letters **8**(3), 214–215 (1964). DOI 10.1016/S0031-9163(64)92001-3. URL http://linkinghub.elsevier.com/retrieve/pii/S0031916364920013

22. R.P. Feynman, The Behavior of Hadron Collisions at Extreme Energies, in *Proceedings of the 3rd International Conference on High Energy Collisions*, ed. by C. Yang, et al. (Gordon and Breach, New York, NY, USA, 1969), 237–249. (Stony Brook, New York, USA, 5-6-September.)

23. R.P. Feynman, Very High–Energy Collisions of Hadrons, Physical Review Letters **23**(24), 1415–1417 (1969). DOI 10.1103/PhysRevLett.23.1415. URL https://link.aps.org/doi/10.1103/PhysRevLett.23.1415

24. J.D. Bjorken, E.A. Paschos, Inelastic Electron–Proton and γ-Proton Scattering and the Structure of the Nucleon, Physical Review **185**(5), 1975–1982 (1969). DOI 10.1103/PhysRev.185.1975. URL https://link.aps.org/doi/10.1103/PhysRev.185.1975

25. Y.S. Kim, M.E. Noz, Covariant harmonic oscillators and the parton picture, Physical Review D **15**(1), 335–338 (1977). DOI 10.1103/PhysRevD.15.335. URL https://link.aps.org/doi/10.1103/PhysRevD.15.335

Chapter 12
Hadron Mass Spectra

Abstract Hadrons are universally considered quantum bound states of quarks. Employing the language of the direct product of $O(3)$, $SU(2)$, and $SU(3)$ we can obtain an understanding of this spectra. Even though $O(3)$ is not Lorentz-invariant, the little group for massive particle is locally isomorphic to $O(3)$. Indeed, $O(3)$ of the quark model is a manifestly relativistic concept. We shall start with non-relativistic models in the Lorentz frame in which the hadron is at rest. Among those models, the three-dimensional isotropic harmonic oscillator serves many useful purposes, as the harmonic oscillator formalism is mathematically simple, can be made relativistic, and the formalism readily accommodates three-particle kinematics and symmetry properties. The formalism contains the $O(3)$ symmetry. Here, we see whether there is evidence in the mass spectrum data for the existence of the harmonic oscillator excitations. Specifically, we study the (mass)^2 spectrum. The harmonic oscillator excitations should create an equal-spaced spectrum. To examine this linearity, at least three levels, including the ground state are needed. This is possible for non-strange baryons which are bound states of three quarks. Therefore, we go through the process of three-particle symmetry classification. We study mesons and conclude with a short summary of particles which are not hadrons.

Historically quantum mechanics was developed for the purpose of explaining discrete spectra. The first discrete spectrum was that of the hydrogen atom. The second discrete spectrum was that of nuclei which are quantum bound states of nucleons. Quantum mechanics has been very successful in explaining these two traditional spectra.

The third discrete spectrum is the mass spectrum of strongly interacting *elementary particles* commonly called *hadrons*. It is by now firmly established that hadrons are quantum bound states of quarks. One of the wonders of modern high-energy physics is that the use of non-relativistic quantum mechanics leads to a reasonable

S. Başkal et al., *Theory and Applications of the Poincaré Group*, Fundamental Theories of Physics 217, https://doi.org/10.1007/978-3-031-64376-7_12

understanding of hadron mass spectra. Then what is the theoretical basis for using non-relativistic bound-state picture for this highly relativistic situation?

The answer to the above question is very simple. When we study hadron mass spectra, we often employ the language of the direct product of $O(3)$, $SU(2)$, and $SU(3)$ [1, 2], where $SU(2)$ describes quark spins, and $SU(3)$ is used for other internal quantum numbers such as *flavor* and *color*. $O(3)$ means that non-relativistic quantum mechanics with rotational symmetry is used. Because $O(3)$ is not Lorentz-invariant, it is very easy to make a hasty conclusion that the quark model is inherently non-relativistic. This is not correct logic. As we studied in Chaps. 2 and 5, the little group for massive particle is locally isomorphic to $O(3)$. Indeed, $O(3)$ of the quark model is a manifestly relativistic concept.

With this point in mind, we shall start with non-relativistic models in the Lorentz frame in which the hadron is at rest. Among those models, the three-dimensional isotropic harmonic oscillator serves many useful purposes. As we discussed in Chap. 5, the harmonic oscillator formalism is mathematically simple and can be made relativistic. In addition, the formalism readily accommodates three-particle kinematics and symmetry properties. The formalism of course contains the above-mentioned $O(3)$ symmetry.

In this Chapter, we are interested in seeing whether there is evidence in the mass spectrum data for the existence of the harmonic oscillator excitations. Specifically, we should study the (mass)^2 spectrum derived in Chaps. 5 and 6. The harmonic oscillator excitations should create a equal-spaced spectrum. In order to examine this linearity, we need at least three levels, including the ground state. This is at present possible for non-strange baryons which are bound states of three quarks. For this reason, we have to go through the process of three-particle symmetry classification.

In Sect. 12.1, we briefly explain what the quark model is. In Sect. 12.2, we carry out explicit calculation for three-particle symmetry classification scheme following the procedure outlined by Dirac for the general case. The permutation operators which commute with the Hamiltonian and with one another are constructed and their eigenvalues are calculated. In Sect. 12.3, we construct wave functions which are diagonal in the permutation operators defined by Dirac. It is shown that this method leads to the three-quark wave functions [2]. Section 12.4 deals with the problem of combining two symmetrized wave functions.

In Sect. 12.5, the three-body spin wave functions are considered. It is pointed out that the three-quark spin wave functions are identical to those of the three-electron system. In Sect. 12.6, the unitary spin wave functions are considered for the three-quark system. It is shown that these wave functions specify the well-known $SU(3)$ multiplets. The $SU(3)$ states are then combined with the spin wave functions of Sect. 12.5 to generate the $SU(6)$ states.

In Sect. 12.7, we discuss three-body spatial wave functions using harmonic oscillators. The harmonic oscillator scheme produces reasonably accurate mass spectra and is mathematically simple for representing the symmetry properties. In Sect. 12.8, these harmonic oscillator wave functions are combined with the $SU(6)$ states to generate totally symmetric baryon wave functions.

In Sect. 12.9, we compare the mass spectrum for non-strange and strange baryons to those that are derived from the harmonic oscillator model. Section 12.10 presents a discussion of the current experimental classification for mesons and shows the relation between the unflavored mesons and the quark model. In Sect. 12.11 we make some concluding remarks. Section 12.12 contains exercises and problems which may be helpful in studying the $SU(3)$ group.

12.1 Quark Model

In the quark model, hadrons are quantum bound states of quarks and/or antiquarks. Baryons such as the proton and neutron are bound states of three quarks, and mesons such as the π and K mesons are bound states of a quark and antiquark. Like electrons, quarks are spin-1/2 fermions and have negligible size. Unlike electrons quarks carry two additional quantum numbers commonly called flavor and color. The traditional name for flavor is unitary spin.

In spite of the importance of understanding relativistic bound-state problems, the quark model was originally developed to explain selection rules in hadron processes. The idea started from the concept of isotopic spins in the system of π mesons and nucleons. Let us look at the nucleons. The proton and neutron have approximately the same mass, and further study of their properties had led us to believe that they belong to the same isotopic multiplet, and that the only difference between them is the electromagnetic property. For example, it is by now firmly believed that the neutron and the proton will have the same mass once the electromagnetic interaction is turned off. This isotopic spin is often called isospin for simplicity. What then is isospin?

The symmetry of electron spin is governed by the $SU(2)$ group. The spin can be up or down. Likewise, we can consider a Hilbert space of nucleonic states, and use I and I_3 to specify the total isospin and its third component. I for the nucleon system is 1/2. If the nucleon is proton, its I_3 is 1/2. The neutron's I_3 is -1/2. There are three mesons with approximately the same mass separated only by electromagnetic interaction. The total isospin for this meson system is 1, and the eigenvalues of I_3 for π^+, π^0, and π^- are 1, 0 and -1 respectively. The nucleonic multiplet and meson multiplets are represented by a spinor and a vector in isospin space respectively. From these vectors and spinors, we can construct scalar quantities which are invariant under rotations in isospin space. Indeed, the observed strong-interaction symmetries are consistent with the rotational symmetry in isospin space. This aspect of strong interaction is widely discussed in textbooks [3].

It was observed that there are, in addition to the nucleons, six more particles which may be put into the same multiplet, and that not all hadron transitions are due to strong interactions. It was absolutely necessary to add another dimension to the internal symmetry space, called the hypercharge, to explain the selection rules for these eight particles which are similar to the nucleon. If we add another dimension to the nucleonic multiplet, the symmetry group has to be enlarged from $SU(2)$ to

Table 12.1: Quantum numbers for quarks and antiquarks. BN and Q are baryon number and charge. I is the total isospin quantum number, and I_3 is its third component. S, C, B, and T denote strangeness, charm, bottomness, and topness as the flavor quantum number of quarks, respectively. The u and d quarks form a doublet in isospin space, while the s, c, b, and t quarks are isospin singlets.

Quantum numbers		BN	Q	I	I_3	S	C	B	T
Quarks	u	1/3	2/3	1/2	1/2	0	0	0	0
	d	1/3	-1/3	1/2	-1/2	0	0	0	0
	s	1/3	-1/3	0	0	-1	0	0	0
	c	1/3	2/3	0	0	0	1	0	0
	b	1/3	-1/3	0	0	0	0	-1	0
	t	1/3	2/3	0	0	0	0	0	1
Antiquarks	$\bar{u}$	-1/3	-2/3	1/2	-1/2	0	0	0	0
	$\bar{d}$	-1/3	1/3	1/2	1/2	0	0	0	0
	$\bar{s}$	-1/3	1/3	0	0	1	0	0	0
	$\bar{c}$	-1/3	-2/3	0	0	0	-1	0	0
	$\bar{b}$	-1/3	1/3	0	0	0	0	1	0
	$\bar{t}$	-1/3	-2/3	0	0	0	0	0	-1

$SU(3)$. However, since there are eight particles in the multiplet, it is not possible to describe this in terms of the fundamental representation of the $SU(3)$ group whose dimension is three. This line of reasoning had led to the concept of quarks [4, 5].

Quarks have fractional charges and fractional baryon numbers which are as listed in Table 12.1. Three quarks can form an integer charge and an integer baryon number to become an observable hadron. Likewise, a quark/antiquark pair have zero charge and zero baryon number and therefore form a meson to also become an observable hadron. For example, the proton consists of two u quarks and one d quark. The way in which the quarks form observable baryon multiplets is discussed in Sect. 12.6. It is not necessary to present here a full-fledged discussion of the quark model, because the model has been been extensively discussed in the physics literature [6, 7, 8, 9].

We are interested only in those aspects of the quark model which are needed in studying the bound-state property of hadrons. The key question is whether the quarks can be regarded as constituent particles within quantum bound states. For instance, the proton and electrons are clearly constituent particles in the hydrogen atom. The consequence of this bound-state picture is that the hydrogen atom has a localized probability of electron around the proton with the radius determined by the electron mass and the strength of interaction between the proton and electron. The localization condition imposed on the hydrogen wave function is responsible for the discreteness of the energy spectrum. The electron can sometimes be separated from the proton to become a free particle.

If the hadron is a bound state of quarks, it should have a non-zero radius determined by the interaction between the quarks, and should show evidence for the existence of a discrete mass spectra due to the localization condition. Since quarks have fractional charges, it is not advisable to consider unbound or free quarks. Therefore, unlike the case of the hydrogen atom, quarks in hadrons cannot be separated.

We do not yet know the exact form of force between the quarks. However, from both phenomenological and field theoretic approaches, the present indications are that the force is very weak at short distances and becomes very strong at large distances. The potential governing the mass is expected to be linear in distance between the quarks. It is therefore reasonable to study the harmonic oscillator potential for the (mass)^2 spectrum [10]. This means that we should see only equal-spaced (mass)^2 spectra in high-energy laboratories. This is not what is observed.

What we see in the real world is the perturbed mass spectra. Quarks have spins and unitary spins. In addition to the approximate harmonic oscillator force, their spins and unitary spins can remove the degeneracy of the harmonic oscillator system. Precisely for this reason, we have to know how to construct unperturbed symmetric wave functions and then perform perturbations on them.

Since mesons are two-body states of two different particles, there are no problems connected with identical particles, and the two-body problem should not cause any mathematical complications. However, baryons are bound states of three quarks. It is by now firmly established experimentally that the baryon wave functions be totally symmetric under the exchange of quarks. This exchange degeneracy is analogous to the case of many electrons. As Dirac pointed out in Sects. 55 and 56 of his classic book on quantum mechanics [11], the Hamiltonian should be invariant under the exchange of quarks, and physical states should therefore be eigenstates of the permutation operators whose eigenvalues correspond to constants of motion.

When we attempt to construct a totally symmetric wave function from spin-1/2 particles, we are led to the question of the Pauli exclusion principle. This is in fact a very serious question, and there have been many attempts to rectify the situation. At present, the prevailing view is that there is an additional quark quantum number called *color* [12]. There are three colors forming a unitary multiplet. The observable three-quark system always manifests itself in a color singlet or a totally antisymmetric state of this quantum number. For this reason, the hadron wave functions are totally symmetric in all other quantum numbers.

The hypothesis of the color space leads to very rich experimental and theoretical consequences [13, 14]. However, since it does not have a direct relation to what we plan to establish in this Chapter, we shall not discuss this subject further.

Once the quarks are identified, it is not difficult to add their quantum numbers to calculate the resulting quantum number for the hadron. The non-trivial aspect of constructing wave functions is to make symmetric combinations to construct irreducible representations. Indeed, this is also a widely discussed subject [1, 15]. The remarkable fact is that this symmetry problem was considered in depth by Dirac in his book on quantum mechanics [11] many years before the invention of the quark model in 1964. The method used by Feynman et al. [2] is along the line suggested by Dirac. In this Chapter, we shall give a full discussion of Dirac's symmetry classification method for the three-particle system.

12.2 Three-Particle Symmetry Classifications by the Method of Dirac

In his Sects. 55 and 56 entitled *Permutations as Dynamical Variables* and *Permutations as Constants of Motion*, respectively, Dirac [11] clearly spelled out his original ideas about the dynamical roles permutations play in quantum mechanics. The purpose of the present section is to work out a concrete illustrative example which might be helpful in understanding Dirac's original treatment. We shall carry out explicit calculations for the three-particle system.

Let us consider three similar objects labeled as 1,2,3 respectively. As we studied in Sect. A.7 of Appendix A, we can perform six different permutations on these three obejcts. First, there are three odd permutations of the form

$$(12), \quad (23), \quad (31), \tag{12.1}$$

where each number is replaced by the succeeding number in the bracket, while the first one goes to the last position. In addition, there are two even permutations of the form

$$(123), \quad (132). \tag{12.2}$$

The above five permutations together with the identity form the six permutations which can be performed on the three objects. The identity is an even permutation.

As we noted in Sect. A.7 of Appendix A, there are three operators which commute with all of the above permutations:

$$X_1 = I,$$

$$X_2 = [\,(12) + (23) + (31)\,]/3,$$

$$X_3 = [\,(123) + (132)\,]/2, \tag{12.3}$$

where I is the identity operator. X_1 can also be written as

$$X_1 = [\,I + (12) + (23) + (31) + (123) + (132)\,]/6. \tag{12.4}$$

X_1, X_2, and X_3 are the Casimir operators of the permutation group of three objects.

If the Hamiltonian is invariant under permutations, the above three X_i's can be simultaneously diagonalized. The next question is how to find eigenvalues for these operators. Here again, we follow the steps outlined by Dirac in his Eq. (14) of Sect. 56. By explicit calculation, we derive

$$X_1^2 = X_1 , \qquad X_1 X_2 = X_2 , \quad X_1 X_3 = X_3 ,$$

$$X_2^2 = (X_1 + 2X_3)/3 , \quad X_2 X_3 = X_2 ,$$

$$X_3^2 = (X_1 + X_3)/2 . \tag{12.5}$$

Following Dirac's Eq. (15) of Sect. 56 [11] for the general case, we consider the following arbitrary function of the X operators:

$$B = X_1 + X_2 + X_3 . \tag{12.6}$$

Then

$$B^2 = (11/6)X_1 + 4X_2 + (19/6)X_3 , \tag{12.7}$$

$$B^3 = (10/4)X_1 + 13X_2 + (37/4)X_3 . \tag{12.8}$$

Because X_1 is the identity operator, its eigenvalue is always 1. By eliminating X_2 and X_3 from Eqs. (12.6) to (12.8), we arrive at

$$B^3 = (9/2)B^2 + 5B - (3/2) = 0, \tag{12.9}$$

which is Dirac's Eq. (16) of Sect. 56 for the *arbitrary* B given in Eq. (12.6). The above cubic equation has three roots:

$$B_1 = 3 , \qquad B_2 = 1 , \qquad B_3 = 1/2 . \tag{12.10}$$

We can use each of these three numbers to calculate the left-hand side of Eqs. (12.6) to (12.8), which then become three simultaneous linear equations. The solutions to these linear equations will indeed be the eigenvalues for the X operators. They are given in Table 12.2.

The choice of the B function in Eq. (12.6) was arbitrary. However, the eigenvalues of the operators $X_1, ..., P$ are independent of the form of B. For instance, even if we choose $B = X_1 + 2X_2 + 3X_3$, the eigenvalue distribution would be the same as the one given in Table 12.2 [Problem 5 in Sect. 12.12].

If the Hamiltonian is invariant under exchange of particles, one of the five non-trivial permutations can also be simultaneously diagonalized, because it commutes with X_1, X_2, and X_3. Let us choose this particular permutation to be

$$P = (23) . \tag{12.11}$$

Other permutations which do not commute with the above P cannot be simultaneously diagonalized. Since $P^2 = I$, the eigenvalue of this operator is either +1 or -1. We are interested here in how these eigenvalues are distributed. For this purpose, we introduce the operator P' defined as

$$P' = 3X_2 - P = (12) + (31) , \tag{12.12}$$

Table 12.2: Eigenvalues of X_1, X_2, X_3 and P. There are three different sets of eigenvalues resulting in three different symmetry classifications.

B	X_1	X_2	X_3	P	Symbol
B_1	1	1	1	1	S
B_2	1	-1	1	-1	A
				1	α
B_3	1	0	-1/2		
				-1	β

or

$$P + P' = 3X_2 \, . \tag{12.13}$$

P and P' satisfy also the relation

$$P'P = 2X_3 \, . \tag{12.14}$$

For the symmetry classification corresponding to B_1, the eigenvalues already found, together with Eqs. (12.13) and (12.14), allow only P = 1. For the B_2 case, $P = -1$. However, for the B_3 case,

$$P + P' = 0, \qquad P'P = -1 \, . \tag{12.15}$$

P in this case can therefore have both values: +1 and -1. These results are also given in Table 12.2.

12.3 Construction of Symmetrized Wave Functions

We now construct three-particle wave functions having the symmetry properties summarized in Table 12.2. Let us consider three particles which can be in any of three quantum states x, y, and z, with one particle in state x, another in y, and another in z. We can then write the general state for such a system as

$$\psi = a \, |xyz\rangle + b \, |yxz\rangle + c \, |xzy\rangle + d \, |zyx\rangle + f \, |zxy\rangle + g \, |yzx\rangle \, , \tag{12.16}$$

where a, b,..., g are coefficients to be determined by the symmetry property of the wave function.

A state of classification S (totally symmetric) will obey

$$X_2\psi = X_3\psi = P\psi = \psi \, . \tag{12.17}$$

Explicit calculation shows that this requires $a = b = c = d = f = g$, resulting in the totally symmetric wave function:

$$|S\rangle = \left(\frac{1}{\sqrt{6}}\right) [\ |xyz\rangle + |yxz\rangle + |xzy\rangle + |zyx\rangle + |zxy\rangle + |yzx\rangle\ ]. \qquad (12.18)$$

If we want an A state, we must have $-X_2\psi = X_3\psi = -P\psi = \psi$, or $a = -b = -c = -d = f = g$, which results in the totally antisymmetric state:

$$|A\rangle = \left(\frac{1}{\sqrt{6}}\right) [\ |xyz\rangle - |yxz\rangle - |xzy\rangle - |zyx\rangle + |zxy\rangle + |yzx\rangle\ ]. \qquad (12.19)$$

If we want an α state, we need $X_2\psi = 0$, $X_3\psi = -(1/2)\psi$, and $P\psi = \psi$. These conditions will lead to

$$a + f + g = 0, \quad a = c, \quad b = g, \quad d = f. \qquad (12.20)$$

Hence, there are four equations and six unknowns. This means that there will be a two-dimensional subspace of the α state. We can pick two linearly independent α states as

$$|\alpha\rangle_1 = \left(\frac{1}{2\sqrt{3}}\right) [\ |xyz\rangle + |xzy\rangle + |yxz\rangle + |yzx\rangle - 2\,|zxy\rangle - 2\,|zyx\rangle\ ].$$

$$|\alpha\rangle_2 = \left(\frac{1}{2}\right) [\ |xyz\rangle - |yzx\rangle + |xzy\rangle - |yxz\rangle\ ]. \qquad (12.21)$$

The first α state is the one given by Feynman et al. [2], and the second is orthogonal to it.

Lastly, for a β state, the conditions from Table 12.2 are $X_2\psi = 0$, $X_3\psi = -(1/2)\psi$, $P\psi = -\psi$ or

$$a + f + g = 0, \quad a = -c, \quad b = -g, \quad d = -f. \qquad (12.22)$$

Here again, we have a two-dimensional subspace of possible states. We can pick the first β state to be that given by Feynman et al. [2], and the one orthogonal to it to be the second β state:

$$|\beta\rangle_1 = \left(\frac{1}{2}\right) [|xyz\rangle - |xzy\rangle + |yxz\rangle - |yzx\rangle],$$

$$|\beta\rangle_2 = \left(-\frac{1}{2\sqrt{3}}\right) [|xyz\rangle - |yxz\rangle - |xzy\rangle + |yzx\rangle + 2\,|zyx\rangle - 2\,|zxy\rangle](12.23)$$

We now have a complete set of six linearly independent states which are contained in the four symmetry classifications S, A, α, and β.

12.4 Symmetrized Products of Symmetrized Wave Functions

As was stated in Sect. 12.2, we have to combine spin, unitary spin, and spatial wave functions to construct the totally symmetric overall baryon wave function. For this purpose, we consider in this section products of two symmetrized three-particle states. We are interested in a product of wave functions with values in two separate spaces, for example, spin space and unitary spin space. Our wave function will be of the form

$$|ab\rangle = |a\rangle\,|b\rangle \,, \tag{12.24}$$

where a and b represent quantum numbers in two separate spaces. The state $|ab\rangle$ that can be made to conform to the results of Sects. 12.2 and 12.3 were derived from the properties of the operators X_1, X_2, X_3, and P, without any assumptions about the form the eigenstates would take. What we would like to do is to derive a complete set of states $|ab\rangle$ which conform to the symmetry classification obtained from the symmetrized $|a\rangle$ and $|b\rangle$.

A permutation operator acting on $|ab\rangle$ will permute the particles in $|a\rangle$ and $|b\rangle$ in an identical manner. Thus we can write

$$X_1 = I = I_a I_b = X_{1a} X_{1b} \,, \tag{12.25}$$

$$X_2 = (1/3)[\,(12) + (23) + (31)\,]$$

$$= (1/3)[\,(12)_a(12)_b + (23)_a(23)_b + (31)_a(31)_b\,]\,, \tag{12.26}$$

$$X_3 = (1/2)[\,(123) + (132)\,]$$

$$= (1/2)[\,(123)_a(123)_b + (132)_a(132)_b\,]\,, \tag{12.27}$$

$$P = P_a P_b = (23)_a(23)_b \,. \tag{12.28}$$

We are interested here in expressing symmetrized wave functions $|ab\rangle$ in terms of symmetrized $|a\rangle$ and $|b\rangle$. The simplest way to attack this problem is to write the right-hand sides of Eqs. (12.25) to [12.28] in terms of the operators which are diagonal in the symmetrized a and b spaces and/or other simple operators. X_1 of Eq. (12.25) and P of Eq. (12.28) are already in the desired form. The remaining problem is to work out X_2 and X_3. For this purpose, we carry out first the following simple calculations:

$$X_{2a}X_{2b} = (1/3)X_2 + (2/3)X_2(X_{3b})\,, \tag{12.29}$$

or equivalently

$$X_{2a}X_{2b} = (1/3)X_2 + (2/3)X_2(X_{3a})\,, \tag{12.30}$$

and

$$X_{3a}X_{3b} = (1/2)X_3 + X_3(X_{3b}) - (1/2)X_{3a}\,, \tag{12.31}$$

or

$$X_{3a}X_{3b} = (1/2)X_3 + X_3(X_{3a}) - (1/2)X_{3b}\,. \tag{12.32}$$

Let us consider a state which is a product of an S state in the a space with an S state in the b space:

$$\psi = |a\rangle_S\, |b\rangle_S \,, \tag{12.33}$$

and look at what the relations given in Eqs. (12.29) and (12.32) tell us about the wave function ψ of Eq. (12.33). Clearly,

$$X_{ia}\psi = X_{ib}\psi = \psi \,, \qquad i = 1,2,3\,, \tag{12.34}$$

so that

$$(X_{2b}X_{2b})\psi = (X_{3a}X_{3b})\psi = \psi \,, \tag{12.35}$$

giving

$$\psi = [\,(1/3)X_2 + (2/3)X_2(X_{3b})\,]\psi = X_2\psi \,, \tag{12.36}$$

$$\psi = [\,(1/2)X_3 + X_3(X_{3a}) - (1/2)X_{3b})\,]\psi \,. \tag{12.37}$$

Hence

$$X_3\psi = \psi \,. \tag{12.38}$$

Also, from P of Eq. (12.28),

$$P\psi = \psi \,. \tag{12.39}$$

Thus we have established that $\psi = |a\rangle_S\, |b\rangle_S$ is an eigenstate of X_1, X_2, X_3 and P, and that the wave function of Eq. (12.33) is an S state: $|ab\rangle_S$.

In a similar manner, we can use Eqs. (12.29) to (12.32) to show that the following states fall into the given symmetry classifications:

$$
\begin{aligned}
|a\rangle_S\,|b\rangle_S &= |ab\rangle_S \,, & |a\rangle_S\,|b\rangle_\alpha &= |ab\rangle_\alpha \,, \\
|a\rangle_S\,|b\rangle_\beta &= |ab\rangle_\beta \,, & |a\rangle_S\,|b\rangle_A &= |ab\rangle_A \,, \\
|a\rangle_A\,|b\rangle_S &= |ab\rangle_A \,, & |a\rangle_A\,|b\rangle_\alpha &= |ab\rangle_\beta \,, \\
|a\rangle_A\,|b\rangle_\beta &= |ab\rangle_\alpha \,, & |a\rangle_A\,|b\rangle_A &= |ab\rangle_S \,.
\end{aligned}
\tag{12.40}
$$

However, when ψ is taken as the product of an α or β state in the a space with an α or β state in the b space, the terms $X_2\psi$ and $X_3\psi$ appear in Eqs. (12.29) to (12.32) with zero coefficient. These equations thus reduce to the identities from which no information can be obtained. To handle these cases, we consider the operators which simply change an α to a β state, and vice versa. For this purpose, let us introduce the operator

$$R = (1/\sqrt{3})[\,(12) - (31)\,]\,. \tag{12.41}$$

Acting on an $|a\rangle$, $|b\rangle$, or $|ab\rangle$ state, R has the following properties:

$$[\,X_i, R\,] = 0\,, \qquad i = 1,2,3\,, \tag{12.42}$$

$$PR = -RP\,, \qquad R^2 = (2/3)[\,1 - X_3\,]\,. \tag{12.43}$$

From Eq. (12.43), we see that if a state $|a\rangle$ is an eigenstate of P_a, then $R_a |a\rangle$ will be an eigenstate of P_a with an eigenvalue opposite to that of $|a\rangle$. From Eq. (12.42), we see that $R_a |a\rangle$ will have the same eigenvalues under X_{1a}, X_{2a}, and X_{3a} as $|a\rangle$. The logic is the same for the b space.

Let us first consider the action of R upon an S state:

$$P(R| |\rangle_S) = -R\, |\rangle_S \, , \qquad X_i(R\, |\rangle_S) = R\, |\rangle_S \, , \qquad i = 1,2,3, \qquad (12.44)$$

where $|\rangle$ can be $|a\rangle$, $|b\rangle$, or $|ab\rangle$. Thus $R\, |\rangle_S$ will be an eigenstate of X_1, X_2, X_3, and P with eigenvalues 1, 1, 1, and -1 respectively. From the argument of Sect. 12.2, it is clear that no such state can possibly exist. We conclude, therefore, that $R\, |\rangle_S = 0$, and from a similar argument, $R\, |ab\rangle_A = 0$. This agrees with Eq. (12.43), which for S and A states reduces to $R^2 = 0$.

Next, we turn to the α and β states. R will again change the sign of the P eigenvalues, while leaving the X_1, X_2, and X_3 eigenvalues unchanged. This means that

$$P(R\, |\rangle_\alpha) = -R\, |\rangle_\alpha \, , \qquad X_1(R\, |\rangle_\alpha) = R\, |\rangle_\alpha \, ,$$

and

$$X_2(R\, |\rangle_\alpha) = 0 \, , \qquad\qquad X_3(R\, |ab\rangle_\alpha) = -(1/2)R\, |ab\rangle_\alpha \, . \qquad (12.45)$$

Thus $R\, |\rangle_\alpha$ will be a β state. Conversely, R operating on a β state will have an α state. Since $X_3 = -1/2$ for α and β states, we will have $R^2 = 1$, and

$$R^2\, |\rangle_\alpha\, | >=\, |\rangle_\alpha \quad \text{and} \quad R^2\, |\rangle_\beta = |\rangle_\beta \, . \qquad (12.46)$$

We can pick

$$|\rangle_\beta = R\, |\rangle_\alpha \qquad \text{and} \quad R\, |\rangle_\beta = |\rangle_\alpha \, . \qquad (12.47)$$

uniquely, given $|\rangle_\alpha$. In terms of the R operators acting on the a and b spaces, we can write X_2 of Eq. (12.26) as

$$X_2 = (1/3)P_a P_b + R_a R_b \, . \qquad (12.48)$$

In order to derive a similar formula for X_3, we introduce the operator

$$R' = (1/2)^{1/2}[\,(123) - (132)\,] \, . \qquad (12.49)$$

Here again

$$[\,R', X_i\,] = 0 \, , \quad i = 1,2,3, \quad \text{and} \quad R'P = -PR' \, . \qquad (12.50)$$

As for R defined in Eq.(12.41),

$$RR' = -R'R = P - X_2 \, . \qquad (12.51)$$

Thus for any α state $|\rangle_\alpha$,

$$R(R' \,|\rangle_\alpha) = (P - X_2) \,|\rangle_\alpha \, . \tag{12.52}$$

Since $R^2 = 1$ from Eq. (12.43),

$$RR' \,|\rangle_\alpha = R^2 \,|\rangle_\alpha \, , \qquad R' \,|\rangle_\alpha = R' \,|\rangle_\alpha \, . \tag{12.53}$$

From Eq. (12.51),

$$R'R \,|\rangle_\alpha = - \,|\rangle_\alpha \, . \tag{12.54}$$

From Eqs. (12.53) and (12.54), we derive, for $|\rangle_\alpha$ and $|\rangle_\beta$ obeying Eq.(12.47):

$$R' \,|\rangle_\alpha = \,|\rangle_\beta \qquad \text{and} \qquad R' \,|\rangle_\beta = - \,|\rangle_\alpha \, , \tag{12.55}$$

In terms of the R operators acting on the a and b spaces, we can write X_3 of Eq.(12.27) as

$$X_3 = (1/4)[\, 4X_{3a}X_{3b} + 3R_a R_b \,] \, . \tag{12.56}$$

Using the X_2 and X_3 operators given in Eqs. (12.48) and (12.56) respectively, together with X_1 and P of Eqs. (12.25) and (12.28) we can show that with $|\rangle_\beta = R \,|\rangle_\alpha$,

$$|ab\rangle_S = [|a\rangle_\alpha \,|b\rangle_\alpha + |a\rangle_\beta \,|b\rangle_\beta]/\sqrt{2} \, ,$$

$$|ab\rangle_\alpha = [-|a\rangle_\alpha \,|b\rangle_\alpha + |a\rangle_\beta \,|b\rangle_\beta]/\sqrt{2} \, ,$$

$$|ab\rangle_\beta = [\, |a\rangle_\alpha \,|b\rangle_\beta + |a\rangle_\alpha \,|b\rangle_\beta \,]/\sqrt{2} \, ,$$

$$|ab\rangle_\alpha = [\, - |a\rangle_\alpha \,|b\rangle_\beta + |a\rangle_\beta \,|b\rangle_\alpha \,]/\sqrt{2} \, . \tag{12.57}$$

The symmetrized wave functions given in Sect. 12.3 together with the combination formulas given in Eqs. (12.40) and (12.57) form the mathematical basis for constructing three-quark baryon wave functions for spin, unitary spin, and harmonic oscillator excitations, and for constructing the total wave function by making symmetrized combinations.

12.5 Spin Wave Functions for the Three-Quark System

We can now apply the results of Sect. 12.3 to obtain the spin wave functions for the symmetric quark model. We observe that, as in the case of electrons, the quark spin can be either up (+) or down (-). Therefore, in constructing spin wave functions for the three-quark system, the quantum numbers x, y, and z used in Sec 12.3 each take on the value of either + or -. As is specified in Schiff's book on quantum mechanics [16], the totally symmetric state $|\rangle_S$ represents a spin-3/2 state, while $|\rangle_{\alpha,\beta}$ are spin-1/2 states. Because at least two of the three quantum numbers are the same, the totally antisymmetric wave function does not exist.

We write here for completeness the spin wave functions. For total spin-3/2, the totally symmetric wave functions are

$$|3/2, 3/2\rangle = |+ + +\rangle \,,$$

$$|3/2, 1/2\rangle = [\, |+ + -\rangle + |+ - +\rangle + |- + +\rangle \,]/\sqrt{3} \,,$$

$$|3/2, -1/2\rangle = [\, |- - +\rangle + |- + -\rangle + |+ - -\rangle \,]/\sqrt{3} \,,$$

$$|3/2, -3/2\rangle = |- - -\rangle \,. \tag{12.58}$$

where the first number in the ket vectors on the left-hand side is the total spin and the second number is its third component. Because at least two quarks are in the same quantum state, there are only four totally symmetric states.

The α states with total spin-1/2 are

$$|1/2, 1/2\rangle_\alpha = [\, |+ + -\rangle + |+ - +\rangle - 2\,|- + +\rangle \,]/\sqrt{6} \,,$$

$$|1/2, -1/2\rangle_\alpha = [\, |- - +\rangle + |- + -\rangle - 2\,|+ - -\rangle \,]/\sqrt{6} \,. \tag{12.59}$$

The β states with total spin-1/2 are

$$|1/2, 1/2\rangle_\beta = [\, |+ + -\rangle - |+ - +\rangle \,]/\sqrt{2} \,,$$

$$|1/2, -1/2\rangle_\beta = [\, |- - +\rangle - |- + -\rangle \,]/\sqrt{2} \,. \tag{12.60}$$

α and β here are the α_1 and β_1 of Eqs. (12.21) and (12.23). We noted there that there are two α states. In the spin case in which there are only two different quantum states for each particle, the second α state either vanishes or becomes dependent on other states. For this reason, we consider only the first α state. The situation is the same for the β state.

The spin wave functions given in Eqs. (12.58) to (12.60) are usually covered in the standard quantum mechanics course and are discussed in Chap. 3 of this book. We wrote them down explicitly here in order to emphasize that they represent a particular case of the more general formulae given in Sect. 12.3, and that the unitary spin wave functions discussed in the next Section represent a more general case of the same symmetry classification method.

12.6 Three-Quark Unitary Spin and SU(6) Wave Functions

In addition to the spin, quarks carry unitary spin which is often called *flavor*. When the quark model was proposed by Gell-Mann in 1964 [17, 18, 4], it was believed that unitary spin could only take three different values, namely, up (u), down (d), and strange (s). It is now believed that there are three additional flavor quantum numbers which are called charm (c), bottom (b), and top (t). A summary of the currently known quarks and their quantum numbers is given in Table 12.1. As can be seen from the table, quarks carry fractional charges and baryon numbers. They are combined in such a way that the resulting hadrons have observable charges and baryon numbers.

We shall discuss here only the unitary spin wave functions of the three traditional quarks u, d, and s. It is straightforward to apply the method for these traditional quarks to states involving newer quarks. In constructing unitary spin wave functions for baryons consisting of three quarks, we can use the symmetrized forms given in Sect. 12.3.

The quark model utilizes the concept of isotopic spin, in which the proton and the neutron belong to the two-dimensional isospin multiplet. The mathematics for this isospin formalism is the same as that of $SU(2)$ which we discussed in Chap. 3. The fact that the strong interaction is invariant under rotations in this isospin space is well known and is discussed widely in textbooks in modern physics.

The conservation in strong interaction physics of isospin and strangeness has led us to believe that the three quarks should form a multiplet [4] under transformations of the $SU(3)$ group which is the group of unitary unimodular transformations in the three-dimensional complex space [Exercise 1 in Sect. 12.12]. Clearly, a rotation in unitary spin space will have no effect on the exchange symmetry of a given wave function. For instance, if the wave function is totally symmetric under exchange of particles, this symmetry is not affected by rotation in unitary spin space.

If we consider all possible combinations of unitary spins for three quarks, there are ten totally symmetric states, eight α states, and eight β states. There is one totally antisymmetric state. There are therefore all together $27\ (= 3^3)$ states. As is illustrated later in this section, these multiplets can be worked out explicitly. For example, the baryon Σ^0 belongs to the totally symmetric decuplet with the unitary spin wave function

$$|\Sigma^0\rangle = |uds\rangle_S$$

$$= [\ |uds\rangle + |dsu\rangle + |sud\rangle + |dus\rangle + |sdu\rangle + |usd\rangle\ ]/\sqrt{6}. \qquad (12.61)$$

The proton is a member of the α or β octet, and

$$|p\rangle = |uud\rangle_\alpha$$

$$= [\ |uud\rangle + |udu\rangle - 2\,|duu\rangle\ ]/\sqrt{6}, \qquad (12.62)$$

or

$$|p\rangle = |uud\rangle_\beta$$

$$= [\ |uud\rangle - |udu\rangle\ ]/\sqrt{2}. \qquad (12.63)$$

Likewise, we can construct all baryon unitary spin wave functions, and they are given in Table 12.3. The reason for the $\Sigma^0 - \Lambda^0$ degeneracy is that the (uds) states are the

Table 12.3: Baryon multiplets in the unitary spin space consisting only of (uds) quarks. There are ten totally symmetric and eight α or β states. When the spatial wave function is in the ground state and is totally symmetric, the decuplet combines with the spin-3/2 states, and the octets combine with the spin-1/2 states to form totally symmetric overall wave functions. The combination of spin and unitary spin states gives the $SU(6)$ multiplets. The $SU(6)$ wave functions, combined with totally symmetric ground-state spatial wave functions, form the familiar baryon octet and decuplet in a 56 $SU(6)$ multiplet. Q designates the charge and appears as a superscript in the particle name. S designates the number of strange particles in the baryon and is labelled with a minus sign. The numbers in parenthesis are the observed baryon masses measured in MeV.

				Octet: α or β	Decuplet: S
Quarks	Q	I_3	S	(spin = 1/2)	(spin = 3/2)
uuu	2	3/2	0		$\Delta^{++}(1232)$
uud	1	1/2	0	$p(938)$	$\Delta^{+}(1600)$
udd	0	-1/2	0	$n(939)$	$\Delta^{-}(1700)$
ddd	-1	-3/2	0		$\Delta^{-}(1920)$
uus	1	1	-1	$\Sigma^{+}(1189)$	$\Sigma^{+}(1385)$
uds	0	0	-1	$\Lambda^{0}(1115), \Sigma^{0}(1193)$	$\Sigma^{0}(1580)$
dds	-1	-1	-1	$\Sigma^{-}(1197)$	$\Sigma^{-}(1670)$
uss	0	1/2	-2	$\Xi^{0}(1314)$	$\Xi^{0}(1530)$
dss	-1	-1/2	-2	$\Xi^{-}(1320)$	$\Xi^{-}(1820)$
sss	0	0	-3		$\Omega^{0}(1675)$

only ones with three different quantum numbers so that the second α and β states do not vanish as they did in the discussion of Sect. 12.5 where two of the quantum numbers are the same.

The interesting question now is how these $SU(3)$ wave functions are combined with the $SU(2)$ spin wave functions of Sect. 12.5 to generate the $SU(6)$ multiplets. We can now use the combination formulae given in Sect. 12.4, using the notation $|10\rangle_S$ for the totally symmetric decuplet, and $|8\rangle_{\alpha,\beta}$ for the α and β octets, respectively. There are 56 totally symmetric $SU(6)$ wave functions, and they are

$$|3/2\rangle_S \, |10\rangle_S$$

and

$$[\, |1/2\rangle_\alpha \, |8\rangle_\alpha + |1/2\rangle_\beta \, |8\rangle_\beta \,]/\sqrt{2}. \tag{12.64}$$

The $SU(6)$ 56 state consists of the spin-3/2 decuplet (with 40 states) and the spin-1/2 octet (16 states). These $SU(6)$ wave functions combine with totally symmetric spatial wave functions. Since the ground-state spatial wave function is believed to be totally symmetric, the above multiplets indeed represent the most commonly observed baryon multiplet, and are usually called the $SU(6)$ multiplets.

There are 70 α states consisting of

$$32\,|3/2\rangle_S\,|8\rangle_\alpha\,,$$

$$20\,|1/2\rangle_\alpha\,|10\rangle_S\,,$$

$$16[\,-|1/2\rangle_\alpha\,|8\rangle_\alpha+|1/2\rangle_\beta\,|8\rangle_\beta\,]/\sqrt{2}\,,$$

$$2\,|1/2\rangle_\beta\,|1\rangle_A\,. \tag{12.65}$$

There are also 70 β states. They are

$$32\,|3/2\rangle_S\,|8\rangle_\beta\,,$$

$$20\,|1/2\rangle_\beta\,|10\rangle_S\,,$$

$$16[\,-|1/2\rangle_\alpha\,|8\rangle_\beta+|1/2\rangle_\beta\,|8\rangle_\alpha\,]/\sqrt{2}\,,$$

$$2\,|1/2\rangle_\alpha\,|1\rangle_A\,. \tag{12.66}$$

The above α and β type states are combined with the α and β type spatial wave functions to generate totally symmetric overall wave functions.

In addition, there are 20 totally antisymmetric states consisting of

$$4\,|3/2\rangle_S\,|1\rangle_A\,,$$

$$16[\,-|1/2\rangle_\alpha\,|8\rangle_\beta+|1/2\rangle_\beta\,|8\rangle_\alpha\,]/\sqrt{2}\,. \tag{12.67}$$

This multiplet can be combined with totally antisymmetric spatial wave functions. However, there has not been any experimental evidence to indicate the existence of this antisymmetric multiplet. We shall therefore omit this multiplet in the following discussions.

12.7 Three-Body Spatial Wave Functions

We are considering here the quantum relativistic bound state of three quarks in order to obtain the baryon mass spectra. As was noted in Chaps. 5 and 6, the harmonic oscillator potential can be made consistent with the principle of special relativity, and is mathematically simple and convenient for discussing symmetry properties. We shall therefore use the harmonic oscillator model for calculating the mass spectrum of baryons.

Let us consider that $\mathbf{x}_a$, $\mathbf{x}_b$, and $\mathbf{x}_c$ represent the coordinates of the three quarks. Then the *Hamiltonian* corresponding to the baryon (mass)^2 can be written as

$$H = -3(\nabla_a^2 + \nabla_b^2 + \nabla_c^2) + \frac{\Omega^2}{36}$$

$$\times [\ (\mathbf{x}_a - \mathbf{x}_b)^2 \ +(\mathbf{x}_a - \mathbf{x}_c)^2 + (\mathbf{x}_c - \mathbf{x}_b)^2\ ] + m_0^2 , \tag{12.68}$$

where Ω is the constant specifying the strength of the harmonic oscillator force, usually measured in (GeV)^2. It should be noted that the eigenvalue equation should be written for (mass)^2 in view of the representations of the Poincaré group discussed in Chaps. 5 and 6. The above form is totally symmetric under the exchange of quarks, and thus commutes with all six three-particle permutations discussed in Sect. 12.2.

For this three-quark system, it is more convenient to use the variables [19]:

$$\mathbf{R} = (\sqrt{\Omega}/3)(\mathbf{x}_a + \mathbf{x}_b + \mathbf{x}_c) ,$$

$$\mathbf{r} = (\sqrt{\Omega}/6)(-2\mathbf{x}_a + \mathbf{x}_b + \mathbf{x}_c) ,$$

$$\mathbf{s} = (\sqrt{\Omega}/2\sqrt{3})(\mathbf{x}_c - \mathbf{x}_b) . \tag{12.69}$$

The variable $\mathbf{R}$ is the overall center-of-mass coordinate for this three-quark system. The variables $\mathbf{r}$ and $\mathbf{s}$ are the internal quark coordinates and have α and β type symmetries respectively. The quantity $(r^2 + s^2)$ is totally symmetric under the exchange of quarks. In terms of the above coordinate variables, the Hamiltonian takes the form

$$H = \frac{\Omega}{2}(-\nabla_r^2 + r^2) + \frac{\Omega}{2}(-\nabla_s^2 + s^2) + m_0^2 . \tag{12.70}$$

We ignore here the trivial kinetic energy term associated with the coordinate $\mathbf{R}$. The coordinate variables r and s now have been completely separated, and the eigenvalue associated with Eq. (12.70) is

$$H = \Omega(N + 3) + m_0^2 , \tag{12.71}$$

where N is the *total* excitation number, and is the sum of the excitations in the r and s spaces:

$$N = n_r + n_s . \tag{12.72}$$

For each of the r- and s-coordinates, we can solve the eigenvalue problem according to the well-defined procedure for the three-dimensional isotropic harmonic oscillator, and the overall spatial solution is a product of the r and s solutions or linear combinations of them. In order to see this more clearly, let us work out the explicit forms for N = 0, 1, and 2.

When $N = 0$, the only possible choice for n_r and n_s is

$$n_r = n_s = 0 . \tag{12.73}$$

Therefore the total spatial wave function is

$$\exp[-(r^2 + s^2)/2].\tag{12.74}$$

We ignore here normalization constants. This form is totally symmetric under the exchange of quarks.

For $N = 1$, we have to consider the following two possibilities:

$$n_r = 1 \quad \text{and} \quad n_s = 0, \quad \text{or} \quad n_r = 0 \quad \text{and} \quad n_s = 1.\tag{12.75}$$

For the first case, the wave function is

$$|N = 1, L = 1\rangle_\alpha = r_i \exp[-(r^2 + s^2)/2], \quad i = 1, 2, 3.\tag{12.76}$$

This α-state solution corresponds to a total angular momentum $L = 1$ state. Likewise, the wave function for the second case is

$$|N = 1, L = 1\rangle_\beta = s_i \exp[-(r^2 + s^2)/2].\tag{12.77}$$

Both the r_i and s_i wave functions have three spatial components. This Cartesian representation can be transformed into spherical form and can be written in terms of the spherical harmonics.

Let us next consider the $n = 2$ case, with the following three possible degeneracies.

$$n_r = 2 \quad \text{and} \quad n_s = 0,$$

$$n_r = 1 \quad \text{and} \quad n_s = 1,$$

$$n_r = 0 \quad \text{and} \quad n_s = 2.\tag{12.78}$$

We have to combine these degenerate solutions to make the $L = 0$, 1, and 2 states.

For the $L = 0$ states, the wave functions are rotationally invariant and take the form

$$|N = 2, L = 0\rangle_S = (r^2 + s^2) \exp[-(r^2 + s^2)/2],$$

$$|N = 2, L = 0\rangle_\alpha = (-r^2 + s^2) \exp[-(r^2 + s^2)/2],$$

$$|N = 2, L = 0\rangle_\beta = (r_1 s_1 + r_2 s_2 + r_3 s_3) \exp[-(r^2 + s^2)/2].\tag{12.79}$$

For L = 1, the wave function becomes

$$|N = 2, L = 1\rangle_A = \varepsilon_{i,j,k} r_i s_k \exp[-(r^2 + s^2)/2].\tag{12.80}$$

Because total $L = 1$, there are three degenerate states.

For $N = L = 2$, we have

$$|N = 2, L = 2\rangle_A = [\, r_i s_j + r_i s_j - \delta_{ij}(r^2 + s^2)/3 \,] \exp[\, -(r^2 + s^2)/2 \,] \,,$$

$$|N = 2, L = 2\rangle_\alpha = [\, r_i s_j - r_i s_j - \delta_{ij}(r^2 + s^2)/3 \,] \exp[\, -(r^2 + s^2)/2 \,] \,,$$

$$|N = 2, L = 2\rangle_\beta = [\, r_i s_j + r_i s_j - \frac{2}{3}\delta_{ij}\mathbf{r}\cdot\mathbf{s} \,] \exp[\, -(r^2 + s^2)/2 \,] \,. \qquad (12.81)$$

The above expressions are symmetric in i and j, and thus there appears to be a six-fold degeneracy. However, owing to the δ_{ij} terms, only five of them are independent as in the case of $Y_2^m(\theta, \phi)$. The above wave functions can also be written in terms of the spherical harmonics [Problem 7 in Sect. 12.12].

12.8 Totally Symmetric Baryon Wave Functions

We combine in this section the $SU(6)$ wave functions of Sect. 12.6 and the spatial wave functions of Sect. 12.7 to generate the overall wave function. As we mentioned in Sect. 12.1, the quarks carry an additional quantum number called *color*. It is believed that there are three colors and that all observed baryons are in the color singlet state. Since the color wave function is totally antisymmetric, and since quarks are fermions, the rest of the wave function consisting of space, spin, and flavor has to be totally symmetric. We shall outline in this section how to construct totally symmetric baryon wave functions which do not include the color part.

We have to combine the harmonic oscillator wave functions discussed in Sect. 12.7 with the $SU(6)$ parts given in Sect. 12.6 using again the technique developed in Sect. 12.4. The useful formulae from Eq. (12.40) and Eq. (12.57) which lead to the totally symmetric state are

$$|ab\rangle_S = |a\rangle_S |b\rangle_S \,,$$

$$|ab\rangle_S = |a\rangle_A |b\rangle_A \,,$$

$$|ab\rangle_S = (|a\rangle_\alpha |b\rangle_\beta + |a\rangle_\beta |b\rangle_\alpha)/\sqrt{2}, \qquad (12.82)$$

where a and b in this case correspond to the $SU(6)$ and spatial parts respectively.

Let us start with $N = 0$. We have in this case only one spatial wave function given in Eq. (12.74) which is totally symmetric. Hence it can only be combined with the totally symmetric 56 multiplet of $SU(6)$ given in Eq. (12.64). Within this multiplet are the spin-3/2 decuplet containing the Δ, Σ, Ξ, resonances and the Ω^- particle, and the spin-1/2 octet containing the nucleons, Σ, Λ, and Ξ hyperons. By the $SU(6)$ scheme, we usually mean this totally symmetric 56 state. This scheme has been extensively discussed in the literature, and the explicit wave functions are given in the paper of Van Royen and Weisskopf [20].

Let us next look at the $N = 1$ states. Since r_i and s_i have the α and β type symmetries, respectively, they should be combined with the 70 α and 70 β states of Eqs. (12.65) and (12.66) to generate a totally symmetric overall wave function of the

form
$$|\text{overall}\rangle_S = (\, |70\rangle_\alpha \, r_i + |70\rangle_\beta \, s_i \,) \exp[-(r^2 + s^2)/2] \,. \tag{12.83}$$

All the baryons belonging to this multiplet are contained in the $SU(3)$ octets with spin-3/2 and spin-1/2, and the decuplet with spin-1/2.

For $N = 2$, we have to consider each value of L separately. For $L = 0$, we have S, α, and β type spatial wave functions. The S-type state is combined with the totally symmetric $SU(6)$ 56 multiplet which contains a spin-1/2 octet and spin-3/2 decuplet, just as in the $N = 0$ case. There is good experimental evidence for the existence of this $L = 0$ multiplet, and a more detailed explanation including the mass spectrum is given in Sect. 12.9.

For the $N = 2$, $L = 1$ state, we note that there is only a totally antisymmetric spatial wave function, which can be combined only with the antisymmetric $SU(6)$ wave functions given in Eq. (12.67). Baryons containing the totally antisymmetric unitary spin state $|1\rangle_A$ have not yet been observed. Thus we omit the states containing this $SU(3)$ singlet from the $SU(6)$ wave functions listed in Eqs. (12.65) and (12.66).

For $N = 2$, $L = 2$, we have again a totally symmetric spatial wave function which combines with the 56 $SU(6)$ totally symmetric wave function yielding again a spin-1/2 octet and a spin-3/2 decuplet. As is indicated in Eq. (12.81), there are also spatial wave functions with α and β type symmetries. They are combined with the $SU(6)$ 70 α and 70 β states containing spin-3/2 and spin-1/2 octets, and a spin-1/2 decuplet. As is discussed in Sect. 12.10, there is good experimental indication that this multiplet exists in nature.

In combining the spatial and $SU(6)$ wave functions, we also have to consider the addition of the orbital angular momentum in the spatial part and the spin angular momentum in the $SU(6)$ wave function. This angular momentum addition is a standard item to be covered in the standard quantum mechanics curriculum, and was discussed in Chap. 3.

12.9 Baryon Mass Spectra

If the hadron masses were determined from the Hamiltonian given in Eq. (12.68) and its counterpart for the mesons, they should depend only on the harmonic oscillator quantum numbers. This is not the case in the real world. There are indeed several perturbation terms to be added to the Hamiltonian which remove the $SU(6)$ and $SU(3)$ degeneracies. These symmetry breaking interactions are due to the mass difference between strange and nonstrange quarks, spin-spin interaction, interaction between two unitary spins, and the combined interaction of the spin and unitary spin. There are many excellent review articles on the $SU(3)$ and $SU(6)$ mass formulas. The review article of Levin and Frankfurt [21] contains the mass formula including the perturbation terms due to the above mentioned interactions.

Using the Levin-Frankfurt mass formula, we can predict the mass of the hadrons. For the $N = 0$ spatial state, mass spectra for nonstrange baryons are those of the

$SU(6)$ scheme and are in excellent agreement with the experimental observation. The mass formula indeed produces accurately the observed masses of the $N = 0, 1,$ and 2 nonstrange baryons given in Tables 12.3 and 12.4. It is customary to use the linear mass in calculating the $SU(6)$ and $SU(3)$ symmetry breaking for the baryon mass, while the harmonic oscillator spectrum is calculated for (mass)^2. Though this is an unfortunate convention, it should not cause any major confusion

As N becomes 3, the experimental situation is not yet as clear as for the $N = 0, 1,$ and 2. Table 12.4 summarizes the accuracy of the present experimental data in relation to the quark model multiplet scheme. In general, the baryon spectra are better understood than those of mesons, and nonstrange hadron spectra are better than those of strange hadrons. Among the four possible mass spectra, the spectrum of nonstrange baryons offers us a unique challenge. There are in this case three clean levels for us to test the linearity in the harmonic oscillator excitation.

Table 12.4: Mass spectrum of nonstrange baryons. The calculated masses are based on Eqs. (12.84) and (12.85). The experimental masses are from the *Particle Data Group 2023* [22]. The last column contains the identification code of the Particle Data Group. I is the total isospin and J^P is the total angular momentum and parity. For $N = 0$, 1, and 2, the quark model multiplet scheme is in excellent agreement with the experimental world. For $N = 3$, the model seems to work well, but more work is needed on both the theoretical and experimental fronts. Baryon masses are measured in MeV.

N	L	SU(6)	SU(3)	I	J^P	Calculated mass	Experimental mass	ID
0	0	56	8	1/2	1/2$^+$	940	938	p****
				1/2	1/2$^+$	940	940	n****
			10	3/2	3/2$^+$	1240	1232	Δ ****
1	1	70	8	1/2	1/2$^-$	1520	1535	N ****
				1/2	3/2$^-$	1520	1520	N ****
			8	1/2	1/2$^-$	1688	1650	N ****
				1/2	3/2$^-$	1688	1700	N***
				1/2	5/2$^-$	1688	1675	N ****
			10	1/2	1/2$^-$	1652	1650	N ****
				1/2	3/2$^-$	1652	1700	N ****
2	0	56	8	1/2	1/2$^+$	1480	1440	N****
			10	3/2	3/2$^+$	1780	1600	Δ ****
		70	8	1/2	1/2$^+$	1730	1710	N****
			8	3/2	3/2$^+$	1898	1920	Δ***
			10	1/2	1/2$^+$	1862	1880	N***
2	2	56	8	1/2	3/2$^+$	1660	1720	N ****
				1/2	5/2$^+$	1660	1680	N****
			10	3/2	1/2$^+$	1960	1910	Δ****
				3/2	3/2$^+$	1960	1920	Δ ***
				3/2	5/2$^+$	1960	1905	Δ ****
				3/2	7/2$^+$	1960	1950	Δ ****
2	2	70	8	1/2	3/2$^+$	1900	1900	N****
				1/2	5/2$^+$	1900	1860	N**

N	L	SU(6)	SU(3)	I	J^P	Calculated mass	Experimental mass	ID
				8	3/2	$1/2^-$	2078	1900	Δ***
				3/2	$3/2^-$	2078	1940	Δ**
				3/2	$5/2^+$	2078	2000	Δ**
				3/2	$7/2^-$	2078	2200	Δ***
				10	1/2	$3/2^+$	2030	2040	N*
				1/2	$5/2^-$	2030	2060	N***
3	1	70	8	1/2	$1/2^+$	2060	2100	N***
				1/2	$3/2^-$	2060		
			8	3/2	$1/2^-$	2228		
				3/2	3/2	2228		
				3/2	$5/2^+$	2228		
			10	1/2	$1/2^+$	2192	2300	N**
				1/2	$3/2^-$	2192	2120	N***
3	1	56	8	1/2	$1/2^-$	1810	1895	N****
				1/2	$3/2^+$	1810		
			10	3/2	$1/2^-$	2110	2150	Δ*
				3/2	$3/2^-$	2110		
				3/2	$5/2^-$	2110	1930	Δ***
3	2	70	8	1/2	$3/2^-$	2180		
				1/2	$5/2^+$	2180	2000	N**
			8	3/2	1/2	2348		
				3/2	3/2	2348		
				3/2	$5/2^-$	2348	2350	Δ*
				3/2	$7/2^+$	2348	2390	Δ*
			10	1/2	$3/2^-$	2312		
				1/2	5/2	2360		
3	3	70	8	1/2	$5/2^-$	2528	2570	N**
				1/2	$7/2^-$	2528		
			8	3/2	3/2	2528		
				3/2	$5/2^-$	2528		
				3/2	$7/2^+$	2528	2390	Δ*
				3/2	$9/2^-$	2528	2400	Δ**
			10	1/2	$5/2^-$	2492		
				1/2	$7/2^-$	2492		
3	3	56	8	1/2	$5/2^-$	2110	2060	N****
				1/2	$7/2^-$	2110	2190	N****
			10	3/2	3/2	2410		
				3/2	$5/2^-$	2410		
				3/2	$7/2^+$	2410		
				3/2	$9/2^-$	2410		

In addition, there are resonances which do not fit into this table. They are N(2220, $9/2^+$, ****), N(2250, $9/2^-$, ****), N(2600, $11/2^-$, ***), N(2700, $11/2^+$, **), Δ(1750, $1/2^+$, *), Δ(2300, $9/2^+$, **), Δ(2420, $11/2^+$, ****), Δ(2750, $13/2^-$, **), and Δ(2950, $15/2^+$, **). N and Δ denote the SU(3) octet and decuplet respectively. The first number in the bracket is the mass of the resonance. The next number is the total angular momentum with parity. Four stars mean that the accuracy in measurement is excellent. One star means the accuracy is poor. The baryons listed above presumably belong to $N = 4$ multiplet.

If we confine ourselves to a structure similar to that employed in Table 12.4, we can construct a table for the baryons where one or more quarks have been replaced

by a strange quark. We then arrive at similar results for the strange baryons. The
results for the strange baryons are listed in Table 12.5.

Table 12.5: Mass spectrum of strange baryons. The calculated masses are based
on Eqs. (12.84) and (12.85), even though it should be noted that these equations
were developed for the nonstrange baryons. The experimental masses are from the
Particle Data Group 2023 [22]. The last column contains the identification code of
the Particle Data Group. S is the number of strange particles in the baryon, I is the
total isospin, and J^P is the total angular momentum and parity. For $N = 0$, 1, and
2, the quark model multiplet scheme seems to be in fair/good agreement with the
experimental world. There are very few particles in $N = 3$ where only the first level
has been included. Baryon masses are measured in MeV.

N	L	SU(6)	SU(3)	S	I	J^P	Calculated mass	Experimental mass	ID
0	0	56	8	-1	1	$1/2^+$	940	1189	Σ^+****
				-1	1	$1/2^+$	940	1193	Σ^0****
			10	-1	1	$3/2^+$	1240	1385	Σ****
1	1	70	8	-2	1/2	$1/2^+$	1520	1315	Ξ^0****
				-1	0	$3/2^-$	1520	1520	Λ****
			8	-1	1	$1/2^+$	1688	1660	Σ***
				-1	0	$3/2^-$	1688	1690	Λ****
				-1	1	$5/2^-$	1688	1775	Σ****
			10	-1	1	$1/2^-$	1652	1620	Σ****
				-2	1/2	$3/2^-$	1652	1530	Ξ****
2	0	56	8	-2	1/2	$1/2^+$	1480	1322	Ξ^-****
			10	-3	0	$3/2^+$	1780	1672	Ω****
		70	8	-1	1	$1/2^-$	1730	1750	Σ***
			8	-1	1	$3/2^+$	1898	1890	Λ****
			10	-1	1	$1/2^+$	1862	1810	Λ***
2	2	56	8	-1	0	$3/2^-$	1660	1670	Σ****
				0	1	$5/2^-$	1660	1775	Σ****
			10	-1	1	$1/2^-$	1960	1900	Σ**
				-1	1/2	$3/2^-$	1960	1820	Ξ***
				-1	0	$5/2^+$	1960	1820	Λ****
				-1	1	$7/2^+$	1960	2030	Σ****
2	2	70	8	-1	1	$3/2^-$	1900	1910	Σ***
				-1	1	$5/2^+$	1900	1915	Σ****
			8	-1	0	$1/2^-$	2078	2000	Λ*
				-1	0	$3/2^-$	2078	2050	Λ*
				-1	1	$5/2^-$	2078	2080	Σ****
				-1	0	$7/2^-$	2078	2100	Λ****
			10	-1	0	$3/2^+$	2030	1890	Λ****
				-1	0	$5/2^+$	2030	1830	Λ****

N	L	SU(6)	SU(3)	S	I	J^P	Calculated mass	Experimental mass	ID
3	1	70	8	-1	1	$1/2^+$	2060	1900	Σ**
				-1	0	$3/2^-$	2060	1940	Σ*
			8	-1	1	$1/2^-$	2228	2110	Σ*
				-1	1	$3/2^+$	2228	2230	Σ*
				-1	0	$5/2^+$	2228	2110	Λ***
			10			1/2	2192		
						3/2	2192		
3	1	56	8			1/2	1810		
				-1	1	$3/2^+$	1810	1780	Σ*
			10			1/2	2110		
						3/2	2110		

There are a few more particles in each strangeness category (1, 2, or 3 strange particles), but the total angular momentum and parity has not yet been determined, so they are not included in this table.

By confining ourselves to nonstrange hadrons, there are only two unitary spin states for the quarks, and therefore the mathematics for the unitary spin becomes as simple as that for the spin. With this point in mind, Kim and Noz [23] simplified the mass formula with which we can study not only the $SU(3)$ and $SU(6)$ symmetry breaking but also the spatial excitation. Their mass formula is

$$M = M_0 + aN + b\left(S - \frac{1}{2}\right) + d\left(I - \frac{1}{2}\right)$$

$$+ f\left(\frac{T - 56}{14}\right) + gL(L + 1), \tag{12.84}$$

where N, S, I, T, and L are the total harmonic oscillator quantum number, spin, isospin, $SU(6)$ number (56 or 70), and the total orbital angular momentum, respectively. We are interested in the first two terms in the above expression which measure the linearity in the mass spectrum. However, the baryon masses we measure are perturbed values which include the remaining terms in the above mass formula. The following choice of the parameters gives a good agreement with the observed masses:

$$M_0 = 940 \text{ MeV}, \qquad a = 270 \text{ MeV}, \qquad b = 168 \text{ MeV},$$

$$d = 132 \text{ MeV}, \qquad f = 250 \text{ MeV}, \qquad g = (30 \pm 10) \text{ MeV}. \tag{12.85}$$

The crucial test is whether the mass spectrum exhibits the harmonic oscillator characteristics through the linearity in N in Eq. (12.84). The coefficient $a = 270$ MeV given above corresponds to the spring constant of $\Omega = 0.47$ $(\text{GeV})^2$.

We listed the calculated and experimental masses for the nonstrange baryons in Table 12.4. It is quite clear from Table 12.4 that the quark model multiplet scheme works well for $N = 0$, 1, and 2 levels. It is quite good for the $N = 3$ level, however, some more work is needed, particularly in the 2000 MeV region. It is too early to make a meaningful statement about the $N = 4$ states.

The study of baryon mass spectra is one of the long-lasting programs in modern physics. If we do not use the above simplification, the mass formula becomes much more complicated. In addition, there are many other sources of perturbation. There are also reasons to believe that the force between the quarks even before the symmetry breaking perturbation is not exactly of the harmonic oscillator type [24, 25, 26].

However, the situation is not as clear for the strange baryons which have unitary spin three. The Σ and Λ particles all have one strange quark and have isospin 1 and 0 respectively. The Ξ particle has two strange quarks and has isospin-1/2, whereas the Ω particle has three strange quarks and has isospin zero.

Additionally there are now baryons, namely for Λ, Σ, Ξ, and Ω, where the strange quark has been replaced by the charm quark. Additionally the Ξ particles has instances where both strange quarks have been replaced by charmed quarks. The set of baryons mentioned above also have the strange quark replaced by the bottom quark. More confirmation is currently needed to determine precisely the isospin and angular momentum states of these particles. Most of the baryons with charmed and bottom quarks are confirmed to exist at the three star level but the Λ_c^0, Σ_c, and Ξ_c^0 all with J^P equal to $1/2^+$ are confirmed at the four star level [22]. This quark replacement is also true for the mesons, as we shall see in the next section.

12.10 Mesons

Mesons differ from baryons in that they are bound states of two particles, a quark and an antiquark which we denote by q and $\bar{q}$, respectively. Since q and $\bar{q}$ are different particles, meson states are not subject to the symmetry considerations. This, however, should not cause any difficulty because we are dealing here with only two constituent particles.

Since the quark can take three different unitary spin quantum numbers, and the antiquark has the quantum numbers opposite to those of the quark, there are nine $\bar{q}q$ states, such as $\bar{u}u$, $\bar{u}d$, etc. Of these combinations, all define unique values for the resulting mesons. This means that we can make suitable linear combinations of these diagonal elements in order to identify the particles observed in nature. The first combination would be to construct $(\bar{u}u - \bar{d}d)/\sqrt{2}$ in order to complete the isospin triplet with the $d\bar{u}$ and $u\bar{d}$ states.

As for the total spin of the mesons, we are considering here the addition of two spins, which is a routine procedure. The resultant total spin is either 1 or 0, with the following spin wave functions:

$$|I\rangle_S = |++\rangle , \quad [\, |+-\rangle + |-+\rangle \,]/\sqrt{2}, \quad \text{or} \quad |--\rangle , \tag{12.86}$$

$$|0\rangle_A = [\, |+-\rangle - |-+\rangle \,]/\sqrt{2} . \tag{12.87}$$

We should note in Table 12.6 that for the u, d, and s quarks, the isospin can be one or zero. The mesons with isospin one or one half form isospin multiplets. For example the the π, K, ρ, and K^* mesons each form separate isospin multiplets. The mesons with isospin zero are singlets. For example, the η, ω, h, f, and ϕ mesons are singlets. When there are two mesons with isospin zero they are distinguished by having a prime added to the name, for example, η'. The spin J is added to the name as a subscript when it is more than zero. Light unflavored mesons where strange, charm, and bottom quarks are absence, are denoted by (S = C = B = 0). For isospin one, the quark combinations for (π, b, ρ, a) are $u\bar{d}$, $(u\bar{u} - d\bar{d})\sqrt{2}$, and $d\bar{u}$. For isospin zero the quark combinations for $(\eta, \eta', h, h', \omega, \phi, f, f')$ are $c_1(u\bar{u} + d\bar{d}) + c_2(s\bar{s})$. Note that mesons composed of an $s\bar{s}$ pair are not considered strange. For the light unflavored mesons the following notation is used: I^G where G parity $= (-1)^{(L+S+1)}$ and J^{PC} has the quantum numbers $P = (-1)^{(L+1)}$ and $C = (-1)^{(L+S)}$. Here, S, L, and J are the spin, orbital, and total angular momenta of the $q\bar{q}$ system. In this case, C stands for charge conjugation not for the charm quark. For the strange $(S = \pm 1)$, charmed $(C = \pm 1)$, bottom $(B = \pm 1)$, and the combinations of SC, SB and BC quarks, only I and J^P are used. However for resonances with quark content $c\bar{c}$ and $b\bar{b}$, I^G and J^{PC} are used. This nomenclature is described in the document *Naming Scheme for Hadrons* from the Particle Data Group [22].

The $SU(3)$ wave functions for the total angular momentum zero $(J = 0)$, such as η and f are different from those for the ω and ϕ mesons which have total angular momentum one $(J = 1)$. As we noted earlier, these are linear combinations of diagonal elements with vanishing quantum numbers. Therefore, the combination is made in accordance with what we observe in the experimental world.

We can construct the combined spin-unitary spin wave functions by taking products of spin and unitary spin wave functions. As in the case of baryons, we shall call these $SU(6)$ wave functions. The unitary spin wave functions in this case are also called $SU(3)$ wave functions.

These $SU(6)$ wave functions are then combined with the spatial wave function. The spatial component in this case is a two-body isotropic harmonic oscillator, and its form is well known. The spin singlet should not cause any problem in coupling the orbital and spin angular momenta. The spin triplet should be coupled to the orbital wave function according to the usual angular momentum addition rule.

The confirmed resonances for the light unflavored and strange mesons (where one of the u or d quarks has been replaced by a strange quark) are listed in Table 12.6. As can be seen, there are many more confirmed resonances for the light unflavored mesons than for the strange mesons. The same is true for one u or d quark combined with a charm or bottom quark as well as for pairs of two charm or bottom quarks. There are even fewer confirmed resonances for mesons which are combinations of bottom and strange, or charm and strange quarks. In addition there are confirmed resonances for mesons with two charm quarks or two bottom quarks. We should also keep in mind that there are some exotic meson states composed of three or four quarks. The evidence for such mesons is scarce at present, but is increasing.

Table 12.6: Light unflavored mesons together with the strange mesons are listed here. Because all quarks have baryon number 1/3 and antiquarks -1/3, mesons have baryon number 0. All the data in this table are from the *Particle Data Group 2023* [22]. The notation used here is explained above. The experimental mass is measured in MeV. Only confirmed resonances are listed here.

Light Unflavored Mesons $S = C = B = 0$								Strange Mesons $S = \pm 1, C = B = 0$		
I^G	J^{PC}	mass	ID	I^G	J^{PC}	mass	ID	I	J^P	mass	ID
1^-	0^-	140	$\pi^\pm$	0^+	0^{-+}	1475	η	1/2	0^-	494	$K^\pm$
1^-	0^{-+}	135	π^0	0^+	0^{++}	1500	f_0	1/2	0^-	498	K^0
0^+	0^{-+}	548	η	0^+	2^{++}	1525	f_2'	1/2	0^-	498	K^0_S
0^+	0^{++}	500	f_0	1^-	1^{-+}	1600	π_1	1/2	0^-	498	K^0_L
1^+	1^{--}	770	ρ	1^-	1^{++}	1640	a_1	1/2	0^+	700	K^*_0
0^-	1^{--}	782	ω	0^+	2^{-+}	1645	η_2	1/2	1^-	892	K^*
0^+	0^{-+}	958	η'	0^-	1^{--}	1650	ω	1/2	1^+	1270	K_1
0^+	0^{++}	980	f_0	0^-	3^{--}	1670	ω_3	1/2	1^+	1400	K_1
1^-	0^{++}	980	a_0	1^-	2^{-+}	1670	π_2	1/2	1^-	1410	K^*
0^-	1^{--}	1020	ϕ	0^-	1^{--}	1680	ϕ	1/2	0^+	1430	K^*_0
0^-	1^{+-}	1170	h_1	1^+	3^{--}	1690	ρ_3	1/2	2^+	1430	K^*_2
1^+	1^{+-}	1235	b_1	1^+	1^{--}	1700	ρ	1/2	0^-	1460	K
1^-	1^{++}	1260	a_1	1^-	2^{++}	1700	a_2	1/2	1^+	1650	K_1
0^+	2^{++}	1270	f_2	0^+	0^{++}	1710	f_0	1/2	1^-	1680	K^*
0^+	1^{++}	1285	f_1	1^-	0^{-+}	1800	π	1/2	2^-	1770	K_2
0^+	0^{-+}	1295	η	0^-	3^{--}	1850	ϕ_3	1/2	3^-	1780	K^*_3
1^-	0^{-+}	1300	π	0^+	2^{-+}	1870	η_2	1/2	2^-	1820	K_2
1^-	2^{++}	1320	a_2	1^-	2^{-+}	1880	π_2	1/2	2^+	1980	K^*_2
0^+	0^{++}	1370	f_0	0^+	2^{++}	1950	f_2	1/2	4^+	2045	K^*_4
1^-	1^{-+}	1400	π_1	1^-	4^{++}	1970	a_4				
0^+	0^{-+}	1405	η	0^+	2^{++}	2010	f_2				
0^-	1^{+-}	1415	h_1	0^+	0^{++}	2020	f_0				
0^+	1^{++}	1420	f_1	0^+	4^{++}	2050	f_4				
0^-	1^{--}	1420	ω	0^-	1^{--}	2170	ϕ				
1^-	0^{++}	1450	a_0	0^+	2^{++}	2300	f_2				
1^+	1^{--}	1450	ρ	0^+	2^{++}	2340	f_2				

12.11 Concluding Remarks

Although this chapter is concerned with hadrons, we thought that we should add some information about bosons and leptons. There are many currently active experiments involving these particles.

Included in the boson category are photons, gluons, and gravitons. The bosons also include the W, Z, and Higgs (H) bosons. At present, the mass of the right-handed W boson is under scrutiny [27]. It should also be noted that searches are underway for neutral Higgs bosons (H^0) [28] as well as charged Higgs bosons ($H^\pm, H^{\pm\pm}$) [29]. There are also experimental searches for new heavy bosons (W', Z') [30], leptoquarks and other heavy particles. Additionally there are searches on for axions (A^0) [31] and other very light bosons. The experimental evidence is inconclusive for all of these heavier or very light bosons.

The leptons consist of the electron, muon, and tau particles. There are also searches for heavy charged leptons. For each lepton, there is a corresponding neutrino. As we said in Sect. 10.7 of Chap 10, neutrinos are probably the most perplexing of all the known elementary particles [32, 33, 34], whose properties are constantly being explored [35, 36]. Especially, the neutrino mass hierarchy issue, namely the normal neutrino mass ordering with $m_1 < m_2 < m_3$ (or $\Delta_{31} > 0$), or the inverted neutrino mass ordering with $m_3 < m_1 < m_2$ (or $\Delta_{31} < 0$), seems to be most pressing [37]. There are many experiments trying to establish whether or not there exists in nature double beta decay (i.e., no neutrino involved in the decay), which would violate a number of conservation laws. Neutrinos are also involved in many other concerns such as the strong CP problem, matter–antimatter asymmetry, and the nature of dark matter and energy. These concerns bring into question the Standard Model view of (except for gravity) fundamental interactions and thereby producing physics beyond the Standard Model [38, 39]. There is also another topic which is revolving around the debate on whether neutrinos are Majorana or Dirac particles [40, 41]. Time and further experimental evidence will determine all these properties of neutrinos.

12.12 Exercises and Problems

Exercise 1. Next to the Poincaré group which contains $O(3)$ and $SU(2)$ as subgroups, $SU(3)$ has played the most important role in the development of modern elementary particle physics. What is $SU(3)$?

This is the group of unitary and unimodular transformations in the three-dimensional complex space. Its fundamental representation consists of three-by-three unitary matrices with unit determinant. This group is therefore generated by three-by-three Hermitian traceless matrices. There are eight linearly independent matrices which meet this specification. They are, in Gell-Mann's notation [17],

$$
\lambda_1 = \begin{pmatrix} 0 & 1 & 0 \\ 1 & 0 & 0 \\ 0 & 0 & 0 \end{pmatrix}, \qquad
\lambda_2 = \begin{pmatrix} 0 & -i & 0 \\ i & 0 & 0 \\ 0 & 0 & 0 \end{pmatrix}, \qquad
\lambda_3 = \begin{pmatrix} 1 & 0 & 0 \\ 0 & -1 & 0 \\ 0 & 0 & 0 \end{pmatrix},
$$

$$
\lambda_4 = \begin{pmatrix} 0 & 0 & 1 \\ 0 & 0 & 0 \\ 1 & 0 & 0 \end{pmatrix}, \qquad
\lambda_5 = \begin{pmatrix} 0 & 0 & -i \\ 0 & 0 & 0 \\ i & 0 & 0 \end{pmatrix}, \qquad
\lambda_6 = \begin{pmatrix} 0 & 0 & 0 \\ 0 & 0 & 1 \\ 0 & 1 & 0 \end{pmatrix},
$$

$$
\lambda_7 = \begin{pmatrix} 0 & 0 & 0 \\ 0 & 0 & -i \\ 0 & i & 0 \end{pmatrix}, \qquad
\lambda_8 = \frac{1}{\sqrt{3}} \begin{pmatrix} 1 & 0 & 0 \\ 0 & 1 & 0 \\ 0 & 0 & -2 \end{pmatrix}. \tag{12.88}
$$

These matrices are discussed in standard textbooks in elementary particle physics [3, 9] . The above eight matrices form the closed Lie algebra:

$$
[\lambda_i, \lambda_j] = 2i f_{ijk} \lambda_k \tag{12.89}
$$

where f_{ijk} is antisymmetric in ijk, and $f_{123} = 1$, $f_{147} = 1/2$, $f_{156} = 1/2$, $f_{246} = 1/2$, $f_{257} = 1/2$, $f_{345} = 1/2$, $f_{367} = -1/2$, $f_{458} = \sqrt{3}/2$, $f_{678} = \sqrt{3}/2$.

The most general form of the transformation matrix is

$$
U(\alpha) = \exp\left[-i\frac{1}{2} \sum_{i=1}^{8} (\alpha_i \lambda_i) \right], \tag{12.90}
$$

applicable to the $SU(3)$ spinors:

$$
x_1 = \begin{pmatrix} 1 \\ 0 \\ 0 \end{pmatrix}, \qquad
x_2 = \begin{pmatrix} 0 \\ 1 \\ 0 \end{pmatrix}, \qquad
x_3 = \begin{pmatrix} 0 \\ 0 \\ 1 \end{pmatrix}, \tag{12.91}
$$

and their linear combinations. As in the case of $SU(2)$, because the matrix $U(\alpha)$ is unitary, we can consider the Hermitian conjugates of the above column vectors:

$$
x_1^\dagger = (1, 0, 0), \qquad x_2^\dagger = (0, 1, 0), \qquad x_3^\dagger = (0, 0, 1). \tag{12.92}
$$

For the three-quark system, we take a direct product of three $SU(3)$ spinors $x_i x_j x_k$ and symmetrize the indices according to the procedures described in Sect. 12.6. For mesons consisting of a quark and antiquark, we take the product $x_i x_j^\dagger$.

Exercise 2. We defined in Eqs. (12.88) and (12.91) the $SU(3)$ spinors. What are the $SU(3)$ vectors? As in the case of $SU(2)$ or $SL(2, c)$, transformation matrices applicable to the above spinors form a spinor representation. We can next consider the three-by-three matrix X whose elements are $x_i x_j^\dagger$. The transformation of this matrix is carried out through

$$
X' = U(\alpha) X U^\dagger(\alpha). \tag{12.93}
$$

This is very similar to the procedure of recovering vectors from the spinors in $SU(2)$ and $SL(2, c)$ discussed in Chap. 3. The Hermitian matrix X is an $SU(3)$ vector, and the above transformation forms a vector representation. Furthermore, X can be written as a linear combination of the eight λ matrices given in Eq. (12.88) plus λ_0, where

$$\lambda_0 = \left(\frac{2}{3}\right)^{1/2}\begin{pmatrix} 1 & 0 & 0 \\ 0 & 1 & 0 \\ 0 & 0 & 1 \end{pmatrix}.\tag{12.94}$$

If we use the nine λ matrices as the basis vectors and define their inner product as

$$(\lambda_i, \lambda_j) = \frac{1}{2}Tr(\lambda_i^\dagger, \lambda_j) = \delta_{ij},\tag{12.95}$$

then, for

$$X = \sum_{i=0}^{8} \lambda_i \alpha_i,\tag{12.96}$$

the inner product is

$$(X', X) = \frac{1}{2}Tr[(X')^\dagger X] = \sum_{i=0}^{8} (\alpha_i')^\dagger \alpha_i.\tag{12.97}$$

Exercise 3. Discuss the little groups of $SU(3)$ using the vector representation.

The $U(2)$ subgroup generated by λ_1, λ_2, λ_3 leaves λ_8 invariant, when the transformation is performed according to Eq. (12.93). This is a little group. The subgroup generated by λ_2 and λ_8 leaves $(a\lambda_3 + b\lambda_8)$ invariant, where $a \neq \pm\sqrt{3}$. This subgroup is also a little group. Since λ_3 and λ_8 commute with each other, this little group is a direct product of $U(1)$ and $U(1)$ [42].

Problem 1. We noticed in Exercise 3 that there is a $SU(2)$ subgroup generated by λ_1, λ_2, λ_3. Show that there are two other $SU(2)$-like subgroups in $SU(3)$ and that they are unitarily equivalent.

Problem 2. In $SU(2)$, we can write the transformation matrix using the following form of $U = \exp[-i(\alpha/2)\mathbf{n}\cdot\boldsymbol{\sigma}]$. This can be expanded to

$$U = \left(\cos\frac{\alpha}{2}\right)I - i\left(\sin\frac{\alpha}{2}\right)\mathbf{n}\cdot\boldsymbol{\sigma}.\tag{12.98}$$

Is it possible to expand the expression in Eq. (12.90) in a finite polynomial in the λ matrices? See Kim and Markley [42].

Problem 3. Very often in physics, particularly in gauge theory [9], we deal with an eight-component field or wave function of the form

$$\psi(x) = \sum_{i=1}^{8} \lambda_i \psi_i(x).\tag{12.99}$$

Show that the above expression can be decomposed into a polar form

$$\psi(x) = e^{-i\theta(x)}\rho(x)e^{i\theta(x)}, \tag{12.100}$$

where

$$\theta(x) = \sum_{i=1}^{8} \lambda_i \theta_i(x), \quad \text{and} \quad \rho(x) = \sum_{i=1}^{8} \lambda_i \rho_i(x). \tag{12.101}$$

Show in particular that the generators of the little group mentioned in Exercise 3 can remain in $\rho(x)$, while others can be written in exponential form so that

$$\rho(x) = \lambda_1\rho_1(x) + \lambda_2\rho_2(x) + \lambda_3\rho_3(x) + \lambda_8\rho_8(x),$$
$$\theta(x) = \lambda_4\rho_4(x) + \lambda_5\rho_5(x) + \lambda_6\rho_6(x) + \lambda_7\rho_7(x), \tag{12.102}$$

for the $U(2)$ little group. Likewise, for the little group generated by λ_3 and λ_8,

$$\rho(x) = \lambda_8\rho_8(x) + \lambda_3\rho_3(x),$$
$$\theta(x) = \lambda_1\rho_1(x) + \lambda_2\rho_2(x)\lambda_4\rho_4(x) + \lambda_5\rho_5(x)$$
$$+\lambda_6\rho_6(x) + \lambda_7\rho_7(x). \tag{12.103}$$

See Kim and Markley [42].

Problem 4. In Eqs. (12.102) and (12.103), the λ matrices are separated into two distinctive sets. Show that this separation remains invariant under transformations of their respective little groups.

Problem 5. Repeat the calculations from Eq. (12.6) to Eq. (12.10) with $B = X_1 + 2X_2 + 3X_3$, instead of the form given in Eq. (12.6), to obtain the result given in Table 12.2.

Problem 6. Construct the Casimir operators for $SU(3)$ [43].

Problem 7. Express the wave functions given in Sect. 12.7 in terms of the spherical harmonics [44].

Problem 8. Estimate the numerical difference in calculating the mass versus the $(\text{mass})^2$ for baryons. See Kim and Noz [23].

Problem 9. Prove that the quantity in Eq. (12.96) transforms under Eq. (12.93) as an $SU(3)$ scalar (singlet) plus and an $SU(3)$ vector (octet), and that no $SU(3)$ transformation mixes them.

Problem 10. Since λ_3 and λ_8 are diagonal, they commute with each other. Let us call them the charge matrices. λ_3 and λ_8 are proportional to I_3 (third component of isospin) and Y (hypercharge) respectively.

(a) Compute the charges of the three spinor components of Eqs. (12.91) and (12.92) by

$$Qx_i = q_i x_i, \qquad i = 1, 2, 3, \tag{12.104}$$

where $Q = I_3, Y$.

(b) Compute the charges of the eight vector components of Eq. (12.88) by using the condition:

$$QQ\lambda_m - \lambda_m Q = q\lambda_m, \qquad (12.105)$$

where $m = 1, 2, ..., 8$.

(c) Use the result of part (b) to show that an $S(3)$-spinor (triplet) can be decomposed into an $SU_I(2)$-spinor (doublet) + an $SU_I(2)$ scalar (singlet). Show also that an $SU(3)$-vector (octet) decomposes into an $SU_I(2)$-vector (triplet) + an $SU_I(2)$-scalar (singlet), both with zero hypercharge and two $SU_I(2)$-spinors (doublet) with opposite hypercharges. Note that, since the $SU_I(2)$-vector and scalar generate the $U(2)$ of Exercise 2, the two $SU_I(2)$-spinors represent the $SU(3)/U(2)$ coset space.

(d) Verify that the direct product $x_i x_j^\dagger$ contains all the states found in (b) by explicit matrix multiplication and expansion in terms of Eq. (12.88) as well as by computing the I_3 and Y charges.

Problem 11.

(a) Do parts (a) and (b) of Problem 10 for the $SU(2)$-like subgroup found in Problem 1.

(b) Show that the $SU(3)$-spinor (triplet) can also serve as an $O(3)$-vector. Show also that an $SU(3)$-vector (octet) decomposes into an $O(3)$-vector (triplet) + an $O(3)$-second-rank tensor. Note that, while the $O(3)$-vector components found in the $SU(3)$ vector generate $O(3)$, the $O(3)$-second-rank tensor represents the $SU(3)/O(3)$ coset space.

Problem 12. Show that

(a)

$$x^a x_b^\dagger = \sum_{i=0}^{\infty} \alpha_i (\lambda_i)_b^a = \alpha_0 \delta_b^a + \sum_{i=0}^{\infty} \alpha_i (\lambda_i)_b^a, \qquad (12.106)$$

where a and b denote row and column indices of the three-by-three λ_i matrices.

(b) Note that λ_1, λ_2, and λ_3 operate only on the first two indices and that λ_8 is an identity matrix on them. Using this observation on splitting the set of indices a, b, and c into (1,2) and (3), carry out the following decomposition.

$$x^a \text{ and } x_b^\dagger,$$

$$x^{[a}x^{b]} = -\frac{1}{2}(x^a x^b - x^b x^a),$$

$$x^{[a}x^{b]} = -\frac{1}{2}(x^a x^b + x^b x^a),$$

$$x^a x_b^\dagger \text{ (with } \alpha_0 = 0). \qquad (12.107)$$

For an extensive use of this method for the symmetry breaking of $SU(N)$, see Hübsch *et al.* [45].

(c) Show that $\epsilon_{abc}x^b x^c$ has the same charge as $x_a^\dagger$.

References

1. O.W. Greenberg, M. Resnikoff, Symmetric Quark Model of Baryon Resonances, Physical Review **163**(5), 1844–1851 (1967). DOI 10.1103/PhysRev.163.1844. URL https://link.aps.org/doi/10.1103/PhysRev.163.1844
2. R.P. Feynman, M. Kislinger, F. Ravndal, Current Matrix Elements from a Relativistic Quark Model, Physical Review D **3**(11), 2706–2732 (1971). DOI 10.1103/PhysRevD.3.2706. URL https://link.aps.org/doi/10.1103/PhysRevD.3.2706
3. W.E. Frazer, *Elementary Particles* (Prentice-Hall, Englewood Cliffs, NJ, 1966). ISBN 13: 978-1114629868
4. M. Gell-Mann, A Schematic Model of Baryons and Mesons, Physics Letters **8**(3), 214–215 (1964). DOI 10.1016/S0031-9163(64)92001-3. URL http://linkinghub.elsevier.com/retrieve/pii/S0031916364920013
5. G. Zweig, An SU3 model for strong interaction symmetry and its breaking. CERN Technical Report TH412, CERN High Energy Collider, Geneva, Switerland (1964). Initial version, report number TH401, January, 1964.
6. J.J.J. Kokkedee, *The Quark Model.* Frontiers in Physics (W. A. Benjamin, Boston, MA, USA, 1969). URL https://cfs.cerm.ch/record/101899. (Based on a series of lectures given at CERN, in Autumn, 1967.)
7. D. Lichtenberg, *Unitary Symmetry and Elementary Particles* (Academic Press, New York, NY, USA, 1970)
8. O.W. Greenberg, Resource Letter Q-1: Quarks, American Journal of Physics **50**(12), 1074–1089 (1982). DOI 10.1119/1.12922. URL http://aapt.scitation.org/doi/10.1119/1.12922
9. K. Huang, *Quarks, leptons & gauge fields*, 2nd edn. (World Scientific, Singapore; Hachensack, NJ, USA, 1992). ISBN 978-981-02-0659-8978-981-02-0660-4. (Originally published 1982.)
10. C.L. Critchfield, Scalar potentials in the Dirac equation, Journal of Mathematical Physics **17**(2), 261–266 (1976). DOI 10.1063/1.522891. URL http://aip.scitation.org/doi/10.1063/1.522891
11. P.A.M. Dirac, *The principles of quantum mechanics*, 4th edn. No. 27 in International series of monographs on physics (Clarendon Press, Oxford University Press, Oxford, UK, 1967). ISBN 978-0-19-852011-5. (Originally published 1930, 2nd edn., 1935, 3rd edn., 1947, 4th edn., 1958, 4th edn., revised 1967.)
12. O.W. Greenberg, Spin and Unitary-Spin Independence in a Paraquark Model of Baryons and Mesons, Physical Review Letters **13**(20), 598–602 (1964). DOI 10.1103/PhysRevLett.13.598. URL https://link.aps.org/doi/10.1103/PhysRevLett.13.598
13. O.W. Greenberg, C.A. Nelson, Color models of hadrons, Physics Reports **32**(2), 69–121 (1977). DOI 10.1016/0370-1573(77)90035-7. URL http://linkinghub.elsevier.com/retrieve/pii/0370157377900357
14. W. Marciano, H. Pagels, Quantum chromodynamics, Physics Reports **36**(3), 137–276 (1978). DOI 10.1016/0370-1573(78)90208-9. URL https://linkinghub.elsevier.com/retrieve/pii/0370157378902089
15. J.A. Shapiro, A quarks-on-springs model of the baryons, Annals of Physics **47**(3), 439–467 (1968). DOI 10.1016/0003-4916(68)90209-1. URL https://linkinghub.elsevier.com/retrieve/pii/0003491668902091
16. L.I. Schiff, *Quantum Mechanics*, 3rd edn. International Series in Pure and Applied Physics (McGraw-Hill, New York, NY, USA, 1968). ISBN 978-0070856431. (Series editors: Condon, E. U. and Harrison, G. R. and Hutchisson, E. and Darrow, K. K.; Originally published 1949.)

17. M. Gell-Mann, *The Eightfold Way: A theory of strong interaction symmetry*, vol. TID-12608;CTSL-20 (California Institute of Technology, Synchrotron, Laboratory, Pasadena, CA, 1961). DOI 10.2172/4008239

18. M. Gell-Mann, Symmetries of Baryons and Mesons, Physical Review **125**(3), 1067–1084 (1962). DOI 10.1103/PhysRev.125.1067. URL https://link.aps.org/doi/10.1103/PhysRev.125.1067

19. M. Moshinsky, *Harmonic oscillator in modern physics: from atoms to quarks* (Gorden and Breach, New York, NY, 1969). ISBN 978-0-677-02450-9. (OCLC: 233550291.)

20. R. Van Royen, V.F. Weisskopf, Hardon decay processes and the quark model, Il Nuovo Cimento A **50**(3), 617–645 (1967). DOI 10.1007/BF02823542. URL http://link.springer.com/10.1007/BF02823542

21. E.M. Levin, L.L. Frankfurt, THE NONRELATIVISTIC QUARK MODEL, Soviet Physics Uspekhi **11**(1), 106–129 (1968). DOI 10.1070/PU1968v011n01ABEH003728. URL https://iopscience.iop.org/article/10.1070/PU1968v011n01ABEH003728. (translated from the Russian: Usp. Fiz. Nauk. 92, 243 (1968).)

22. R.L. Workman, Review of particle physics, Prog. Theo. Exp. Phys. **2022**, 083C01 (2022). DOI 10.1093/ptep/ptac097. (collaboration: Particle Data Group, updated 2023.)

23. Y.S. Kim, M.E. Noz, Radial excitations in the symmetric quark model. — II, Il Nuovo Cimento A **19**(4), 657–666 (1974). DOI 10.1007/BF02813412. URL http://link.springer.com/10.1007/BF02813412

24. N. Isgur, G. Karl, Positive-parity excited baryons in a quark model with hyperfine interactions, Physical Review D **19**(9), 2653–2677 (1979). DOI 10.1103/PhysRevD.19.2653. URL https://link.aps.org/doi/10.1103/PhysRevD.19.2653

25. D.P. Stanley, D. Robsen, Do Quarks Interact Pairwise and Satisfy the Color Hypothesis?, Physical Review Letters **45**(4), 235–238 (1980). DOI 10.1103/PhysRevLett.45.235. URL https://link.aps.org/doi/10.1103/PhysRevLett.45.235

26. K. Maltman, N. Isgur, Nuclear physics and the quark model: Six quarks with chromodynamics, Physical Review D **29**(5), 952–977 (1984). DOI 10.1103/PhysRevD.29.952. URL https://link.aps.org/doi/10.1103/PhysRevD.29.952

27. CMS collaboration, Search for a right-handed W boson and a heavy neutrino in proton-proton collisions at $\sqrt{s}$ = 13 TeV, Journal of High Energy Physics **2022**(4), 47 (2022). DOI 10.1007/JHEP04(2022)047. URL https://doi.org/10.1007/JHEP04(2022)047

28. CMS collaboration, Search for new neutral Higgs bosons through the H $\rightarrow$ ZA $\rightarrow$ $\ell^+\ell^-$ $b\bar{b}$ process in pp collisions at $\sqrt{s}$ = 13 TeV, Journal of High Energy Physics **2020**(3), 55 (2020). DOI 10.1007/JHEP03(2020)055. URL https://link.springer.com/10.1007/JHEP03(2020)055

29. CMS collaboration, Search for charged Higgs bosons decaying into a top and a bottom quark in the all-jet final state of pp collisions at $\sqrt{s}$ = 13 TeV, Journal of High Energy Physics **2020**(7), 126 (2020). DOI 10.1007/JHEP07(2020)126. URL https://link.springer.com/10.1007/JHEP07(2020)126

30. CMS collaboration, Search for new heavy resonances decaying to WW, WZ, ZZ, WH, or ZH boson pairs in the all-jets final state in proton-proton collisions at $\sqrt{s}$ = 13 TeV, Physics Letters B **844**, 137,813 (2023). DOI https://doi.org/10.1016/j.physletb.2023.137813. URL https://www.sciencedirect.com/science/article/pii/S0370269323001478

31. Du, N., and the ADMX Collaboration, Search for Invisible Axion Dark Matter with the Axion Dark Matter Experiment, Physical Review Letters **120**(15), 151,301 − 1 − 151,301 − 5 (2018). DOI 10.1103/PhysRevLett.120.151301. URL https://link.aps.org/doi/10.1103/PhysRevLett.120.151301

32. C. Giunti, C.W. Kim, *Fundamentals of neutrino physics and astrophysics* (Oxford University Press, Oxford UK; New York, NY, USA, 2007). ISBN 9780198508717. (OCLC: ocm76935610.)

33. V. Barger, D. Marfatia, K.L. Whisnant, *The physics of neutrinos* (Princeton University Press, Princeton, NJ, USA, 2012). ISBN 9780691128535

34. F. Deppisch, *A modern introduction to neutrino physics.* IOP Concise Physics (Morgan & Claypool Publishers, San Rafael, CA USA, 2019). ISBN 9781643276793

35. S. Brice, M. Marshak, Co-chairs, The XXIX International Conference on Neutrino Physics and Astrophysics, in *ICNP XXIX 2020, avaiable online only* (Fermilab, Chicago, IL USA, 2020). URL http://nu2020.fnal.gov. (held as online only conference by Fermilab, June 22 - July 2.)
36. Neutrino 2022 SEOUL. URL https://neutrino2022.org/Neutrino2022SEOUL
37. I. Esteban, M.C. Gonzalez-Garcia, A. Hernandez-Cabezudo, M. Maltoni, T. Schwetz, Global analysis of three-flavour neutrino oscillations: synergies and tensions in the determination of θ_{23}, Δ_{CP}, and the mass ordering, Journal of High Energy Physics **2019**(1), 106 (2019). DOI 10.1007/JHEP01(2019)106. URL https://link.springer.com/10.1007/JHEP01(2019)106
38. D. Alves et al., and LHC New Physics Working Group, Simplified models for LHC new physics searches, Journal of Physics G: Nuclear and Particle Physics **39**(10), 105,005 (2012). DOI 10.1088/0954-3899/39/10/105005. URL https://dx.doi.org/10.1088/0954-3899/39/10/105005
39. B. Allanach, Beyond the Standard Model, CERN Yellow Reports: School Proceedings **Vol 6**, 113–Pages (2019). DOI 10.23730/CYRSP-2019-006.113. URL https://e-publishing.cern.ch/index.php/CYRSP/article/view/937
40. ATLAS Collaboration, Search for heavy Majorana or Dirac neutrinos and right-handed W gauge bosons in final states with charged leptons and jets in pp collisions at $\sqrt{s}$=13 TeV with the ATLAS detector, The European Physical Journal C **83**(12), 1164 (2023). DOI 10.1140/epjc/s10052-023-12021-9. URL https://doi.org/10.1140/epjc/s10052-023-12021-9
41. E. Akhmedov, A. Trautner. Can quantum statistics help distinguish Dirac from Majorana neutrinos? (2024). URL http://arxiv.org/abs/2402.05172. ArXiv:2402.05172 [hep-ex, physics:hep-ph, physics:nucl-th]
42. Y.S. Kim, F.L. Markley, Spontaneously broken SU_3 and massive vector mesons, Il Nuovo Cimento A **63**(1), 60–74 (1969). DOI 10.1007/BF02963156. URL http://link.springer.com/10.1007/BF02963156
43. S. Okubo, Note on Unitary Symmetry in Strong Interactions, Progress of Theoretical Physics **27**(5), 949–966 (1962). DOI 10.1143/PTP.27.949. URL https://academic.oup.com/ptp/article-lookup/doi/10.1143/PTP.27.949
44. T. De, Y.S. Kim, M.E. Noz, Radial effects in the symmetric quark model, Il Nuovo Cimento A **13**(4), 1089–1101 (1973). DOI 10.1007/BF02804168. URL http://link.springer.com/10.1007/BF02804168
45. T. Hübsch, S. Meljanac, S. Pallua, Symmetry breaking of SU(n) gauge theories to maximal regular subgroups and fourth-rank tensors, Physical Review D **31**(2), 352–363 (1985). DOI 10.1103/PhysRevD.31.352. URL https://link.aps.org/doi/10.1103/PhysRevD.31.352

Chapter 13
Lorentz-Dirac Deformation in High-Energy Physics

Abstract Here we are interested in using Dirac's picture of Lorentz deformation to explain how hadrons appear to observers in Lorentz frames which are not at rest with respect to the hadron. We examine whether this picture is consistent with what we observe in high-energy laboratories. We are particularly interested in studying the nucleon form factors, Feynman's parton picture, and the hadron jet phenomenon. We concluded previously that hadrons are quantum bound states of quarks having a localized probability distribution. If the proton consists of quarks distributed within a finite space-time region, the virtual photon from high energy experiments will interact with quarks which carry fractional charges. The scattering amplitude will depend on the way in which quarks are distributed within the proton. We discuss the meaning of and calculate the proton form factor. In conjunction with high energy inelastic scattering, we discuss the process of scaling. Feynman observed that a fast-moving hadron can be regarded as a collection of many partons whose properties do not appear to be identical to those of quarks. We discuss how quarks and partons can appear different to an observer in a different Lorentz frame. Additionally, we discuss the jet phenomenon which occurs in high energy processes.

At the end of Chap. 11, we discussed the Bohr-Einstein question. Bohr was concerned about the electron orbit in the hydrogen atom and Einstein was concerned about how things look to moving observers. How the hydrogen atom would look to a moving observer was a metaphysical issue then, because the hydrogen atom could not and still cannot move fast enough to exhibit this Einstein effect.

Fifty years later, many high energy protons could be produced from particle accelerators. In 1955 [1] the discovery of Hofstadter and McAllister showed the proton charge has an internal distribution. In the early 1960's [2, 3, 4], Gell-Mann observed that the proton is a bound state of more fundamental particles that he called *quarks*. However, Feynman [5, 6, 7], in 1969, observed that when the proton moves

© The Author(s), under exclusive license to Springer Nature Switzerland AG 2024

S. Başkal et al., *Theory and Applications of the Poincaré Group*, Fundamental Theories of Physics 217, https://doi.org/10.1007/978-3-031-64376-7_13

very fast, it appears like as a collection of a large number of free-moving light-like particles that he called *partons*. Feynman's parton picture was based entirely on what we can observe in laboratories. Thus the Bohr-Einstein issue became the Gell-Mann-Feynman issue. This is illustrated in Fig. 13.1.

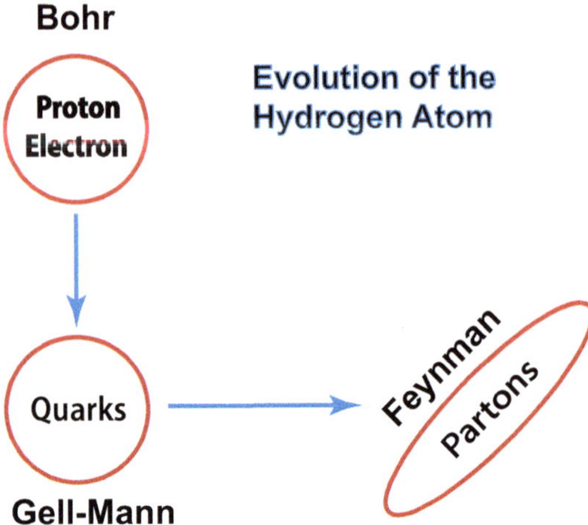

Fig. 13.1: Here we show how the hydrogen atom evolved. In quantum mechanics, a standing wave replaced the electron orbit. At that time the hydrogen atom could not and still cannot be accelerated close to the speed of light. The proton which was shown in 1955 [1] to have a distributed charge, was shown by Gell-Mann in the early 1960's to be a bound state of quarks and thus inherited the quantum mechanics of the hydrogen atom [2, 3, 4]. Furthermore, Feynman observed in 1969 [5, 6] that the proton appears like a collection of partons when moving with a velocity close to that of light [8].

In Chap. 12, we studied the problem of hadron mass spectra and concluded that it is safe to believe that hadrons are quantum bound states of quarks having localized probability distribution. As in all bound state cases, this localization condition is responsible for the existence of discrete mass spectra. Therefore the hadron mass spectra plays a very important role in demonstrating that hadrons are quantum bound states of quarks.

However, the above-mentioned picture of bound states is applicable only to observers in the Lorentz frame in which the hadron is at rest. How would hadrons appear to observers in other Lorentz frames? More specifically, can we use Dirac's picture of Lorentz deformation described in Chap. 6 to tackle this problem? The purpose of this Chapter is to examine whether this picture is consistent with what we observe in high-energy laboratories. We are particularly interested in studying the nucleon form factors, Feynman's parton picture, and the hadron jet phenomenon.

The size of the proton is 10^{-5} of that of the hydrogen atom. Therefore, it is not unnatural to assume that the proton behaves like a point charge in atomic physics. However, while carrying out experiments on electron scattering from proton targets, Hofstadter and McAllister in 1955 observed that the proton charge is spread out [1]. In this experiment, an electron emits a virtual photon, which then interacts with the proton. If the proton consists of quarks distributed within a finite space-time region, the virtual photon will interact with quarks which carry fractional charges. The scattering amplitude will depend on the way in which quarks are distributed within the proton.

The portion of the scattering amplitude which describes the interaction between the virtual photon and the proton is called the form factor. Although there have been many attempts to explain this phenomenon within the framework of quantum field theory, it is quite natural to expect that the wave function in the quark model will describe the charge distribution. In high-energy experiments, we are dealing with the situation in which the momentum transfer in the scattering process is large. We therefore have to know how to describe the quark-model wave functions for rapidly moving protons. The first application of the Lorentz-Dirac deformation described in Chap. 6 should be made on this form factor problem.

While this form factor is the quantity which can be extracted from the elastic scattering, it is important to realize that in high-energy processes, many particles are produced in the final state. They are called inelastic processes. While the elastic process is described by the total energy and momentum transfer in the center-of-mass coordinate system, there is, in addition, the energy transfer in inelastic scattering. Therefore, we would expect that the scattering cross section would depend on the energy, momentum transfer, and energy transfer. However, one prominent feature in inelastic scattering is that the cross section remains nearly constant for a fixed value of the momentum-transfer/energy-transfer ratio. This phenomenon is called *scaling*.

In order to explain the scaling behavior in inelastic scattering, Feynman in 1969 observed that a fast-moving hadron can be regarded as a collection of many *partons* whose properties do not appear to be identical to those of quarks. For example, the number of quarks inside a static proton is three, while the number of partons in a rapidly moving proton appears to be infinite. The question then is how the proton looking like a bound state of quarks to one observer can appear different to an observer in a different Lorentz frame? Can the Lorentz-deformation picture described in Chap. 6 explain this puzzle? We shall deal with this problem in the present Chapter.

Another peculiar behavior in high-energy processes is the jet phenomenon. In high-energy collisions, many hadrons are produced in the final state. Their momenta are not randomly distributed. They are bunched together to form a *jet*. The question then is whether the Lorentz deformation of hadron wave functions is responsible for hadrons coming out in the same direction. We shall also study this problem.

The Lorentz-Dirac deformation was extensively discussed in Chap. 6. In Sect. 13.1 we use the harmonic oscillator potential to illustrate the Lorentz deformation property of hadrons which are regarded as bound states of quarks. Section 13.2 explains the connection between the nucleon form factors and the development of the concept of extended hadrons having non-zero space-time extension which is shared by the quark model. In Sect. 13.3, we discuss how the Lorentz-Dirac deformation can explain the relativistic behavior of the form factors for large momentum transfer and calculate the proton form factor. We also show how the form factor calculation is consistent with Lorentz coherence, both in space-time and in momentum space.

Section 13.4 explains what Feynman's parton picture is. It is shown that, as the hadron moves very fast, the Lorentz deformation leads to widespread longitudinal spatial and momentum distributions for the constituent quarks. It is pointed out that these widespread distributions are responsible for those parton properties of the quarks in which they appear as free independent particles with a widespread momentum distribution.

In Sect. 13.5, we resolve the apparent paradox associated with simultaneous widespread distribution of the spatial and momentum wave functions. It is shown that this phenomenon, while consistent with Lorentz invariance of Planck's constant, is responsible for the infinite number of partons.

In Sect. 13.6, we discuss the present status of the efforts being made to determine quantitatively the distribution of partons inside the proton. In Sect. 13.7, it is shown that the Lorentz-Dirac deformation property can explain the formation of hadron jets. Section 13.8 contains exercises and problems connected with the derivation of some of the well-known formulae in high-energy physics.

13.1 Lorentz-Dirac Deformation of Hadron Wave Functions

The covariant harmonic oscillator formalism and its Lorentz-Dirac deformation property have been discussed in Chaps. 5 and 6 using the language of the Poincaré group. Since the purpose of this Chapter is to see whether the deformation property can be observed in high-energy laboratories, we translate the group-theoretical treatment of the Lorentz-Dirac deformation into a language suitable for studying experimental data.

Using the quark model as we discussed in Chap. 12, mesons are two-body bound states consisting of one quark and one anti-quark. Similarly, baryons are bound states consisting of three quarks. The quark model had many early successes which include the proton-neutron electromagnetic potential and magnetic moments ratio [9]. Feynman et al. in 1971 [10] demonstrated that three-dimensional harmonic oscillators were similar to that of the hadron mass spectra.

Considering the question then of how non-relativistic quantum mechanics relates to the mass spectrum for the hydrogen atom, we wish to understand how this non-relativistic calculation can be valid for the relativistic hadron mass spectrum. We know that in the hadron, it is not possible to ignore the time-separation variable. The answer given in Wigner's little group for massive particles is the three-dimensional rotation group. The rotation group was detailed previously. The role of the time-separation variable has also been previously discussed. The original question posed by the Bohr-Einstein issue was how the hydrogen atom would look to a moving observer. We now wish to see how much we can learn about this by using the three-dimensional harmonic oscillator. To accomplish this we use the Coulomb potential for the hydrogen atom, but we use the harmonic oscillator to model the binding force between quarks. Therefore, we make the point that those two different potentials are sharing the same quantum mechanics.

In order to do this, we will need a bound-state wave function, capable of being Lorentz-boosted. The choice of the harmonic oscillator wave function which we discussed in Chap. 5 is natural. Let us then start with the ground-state wave function. This wave function can be Lorentz boosted. This harmonic oscillator wave function can be separated in the Cartesian coordinate system. If the velocity, β, is along the z direction, we are able to drop the transverse components of the wave function. Hence, we consider only the longitudinal and time-like coordinates. Let us then write the wave function as (using the boost parameter η):

$$\psi_\eta(t, z) = \frac{1}{\sqrt{\pi}} \exp\left\{-\frac{1}{4}\left[\left(e^{2\eta}(t+z)^2 + e^{-2\eta}(t-z)^2\right)\right]\right\}. \tag{13.1}$$

This equation becomes

$$\psi_0(t, z) = \frac{1}{\sqrt{\pi}} \exp\left\{-\frac{1}{2}\left(t^2 + z^2\right)\right\}, \tag{13.2}$$

when $\beta = 0$.

To make this more concrete, we consider two quarks bound together inside a hadron. Let x_a and x_b be space-time coordinates for the first and second quarks respectively. Then it is more convenient to use the variables X and x defined as

$$X = (x_a + x_b)/2,$$

$$x = (x_a - x_b)/2\sqrt{2}. \tag{13.3}$$

The coordinate variable X specifies the space-time position of the hadron, and the x coordinate is for the space-time separation between the quarks. The spatial component of the four-vector X specifies where the hadron is, and its time component tells how old the hadron and the quarks become. The spatial components of the four-vector x specify the relative spatial separation between the quarks. Its time component is the time interval or separation between the quarks.

Let us start with a hadron at rest. As far as the spatial components are concerned, the relative quark motion is dictated by the attractive force between the quarks and by Heisenberg's uncertainty relation. If the attractive force is that of a harmonic oscillator, we can write down the wave function as

$$\psi(x, y, z) = H_k(\sqrt{\Omega}x)H_m(\sqrt{\Omega}y)H_n(\sqrt{\Omega}z)\exp[-\Omega(x^2 + y^2 + z^2)/2], \tag{13.4}$$

where Ω is the spring constant, and $H_n(z)$ is the Hermite polynomial of order n. We are not worrying about the normalization constant. This simple wave function is in fact the starting point for the quark model analysis of the observed mass spectra, as is shown in Chap. 12.

When the hadron moves, we have to boost the above wave function. Before carrying out the boost, we have to complete the wave function by taking into consideration the time-dependence. In the case of the harmonic oscillator formalism, we can accomplish this by multiplying the above non-relativistic wave function by $\exp(-\Omega t^2/2)$. As was discussed thoroughly in Chap. 6, this procedure is consistent with the existence of the time-energy uncertainty relation and with the non-existence of time-like excitation. If the hadron moves along the z direction, the transverse components are not affected, and can therefore be ignored. Furthermore, spatial excitations do not play any major role in discussion of the Lorentz-Dirac deformation property. For this reason, we shall drop all Hermite polynomials in Eq. (13.4). The relevant portion of the wave function is

$$\psi(t, z) = \exp[-\Omega(t^2 + z^2)/2] \, . \tag{13.5}$$

We emphasize again that t is the time-separation variable and is not the time variable appearing in the time-dependent Schrödinger equation. The above expression is localized within the region around $z = t = 0$, and vanishes rapidly as $(t^2 + z^2)$ becomes large. This circular region can be written as

$$(t^2 + z^2) < 1/\Omega \, . \tag{13.6}$$

If the hadron moves along the z direction with speed parameter β, t and z in Eq. (13.5) should be replaced by

$$z' = (z - \beta t)/(1 - \beta^2)^{1/2} \, ,$$

$$t' = (t - \beta z)/(1 - \beta^2)^{1/2} \, , \tag{13.7}$$

so that

$$\psi_\eta(t, z) = \psi(t', z') = \exp[-\Omega(t'^2 + z'^2)/2] \, . \tag{13.8}$$

This boost procedure becomes very convenient if we introduce the light-cone variables:

$$z_+ = (t + z)/\sqrt{2} \, ,$$

$$z_- = (t - z)/\sqrt{2} \, . \tag{13.9}$$

In terms of these variables, the Lorentz boost of Eq. (13.7) takes a simple form:

$$z'_+ = [(1 - \beta)/(1 + \beta)]^{1/2} z_+ \, ,$$

$$z'_- = [(1 + \beta)/(1 - \beta)]^{1/2} z_- \, . \tag{13.10}$$

The light-cone coordinate system consisting of the orthogonal variables z_+ and z_- is only a rotation of the zt system by $45°$. Thus

$$z^2 + t^2 = z_+^2 + z_-^2 \,. \tag{13.11}$$

In terms of the light-cone variable, the harmonic oscillator wave function for the moving hadron is

$$\psi_\eta(t, z) = \exp\left[-\Omega(z_+'^2 + z_-'^2)/2\right]$$

$$= \exp\left[-\left(\frac{\Omega}{2}\right)\left(\frac{1-\beta}{1+\beta}z_+^2 + \frac{1+\beta}{1-\beta}z_-^2\right)\right] \,. \tag{13.12}$$

The above expression becomes that of Eq. (13.5) when $\beta = 0$. When $\beta = 1$, the wave function becomes concentrated along one of the light-cone axes.

As for the momentum-energy wave function, let p_a and p_b be the four-momenta of the first and second quarks respectively. Then the four-momenta conjugate to X and x of Eq. (13.3) are

$$P = (p_a + p_b),$$

$$q = \sqrt{2}(p_a - p_b), \tag{13.13}$$

respectively. The momentum-energy wave function is identical to the space-time wave function if Ω is replaced by $1/\Omega$, and its form is given in Sect. 6.3 of Chap. 6.

While it is much easier to discuss the problem using a two-body bound state, hadrons of experimental interest, such as the nucleon, are bound-states of three quarks. In order to deal with the three quark nucleon, Feynman et al. [10] let x_a, x_b, and x_c be the space-time coordinates of the quarks, and introduced the more convenient variables:

$$X = \frac{1}{3}(x_a + x_b + x_c),$$

$$r = \frac{1}{6}(x_b + x_c - 2x_a),$$

$$s = \frac{1}{2\sqrt{3}}(x_b - x_c), \tag{13.14}$$

and their conjugate variables:

$$P = p_a + p_b + p_c,$$

$$q = p_b + p_c - 2p_a,$$

$$k = \sqrt{3}(p_b - p_c). \tag{13.15}$$

The above formulae were given in Eqs. (5.107) and (5.114) respectively of Chap. 5. The quadratic form of the r and s variables then becomes

$$18\left(r^2 + s^2\right). \tag{13.16}$$

The X variable, specifying the space-time coordinate of the proton, is not contained in the above equation. The covariant harmonic oscillator wave function for the three-particle bound system then takes the form:

$$\psi_\eta(r, s) = \exp\left[-\frac{\Omega}{2}\left(r_z'^2 + r_0'^2 + s_z'^2 + s_0'^2\right)\right] .$$ (13.17)

As in Eq. (13.12), the transverse components have been ignored.

The momentum-energy wave function is:

$$\phi_\eta(q, k) = \exp\left[-\frac{1}{2\Omega}\left(q_z'^2 + q_0'^2 + k_z'^2 + k_0'^2\right)\right] .$$ (13.18)

Again

$$18\left(q^2 + k^2\right) ,$$ (13.19)

was derived from Eq. (13.15). The variable P, which measures the energy and momentum of the proton, is not contained in this form. The Lorentz-Dirac deformation property of r and s and of their conjugate coordinate system are the same as x and q for the two-body system.

We should emphasize here that the above-mentioned Lorentz-Dirac deformation property is not restricted to the harmonic oscillator wave functions, and that it is not difficult to find the same deformation property in other theoretical and phenomenological models [11, 12, 13, 14, 15]. One important advantage of using the harmonic oscillator formalism is that it is mathematically simple. According Dirac [16], a theory possessing mathematical beauty is more likely to survive than those which do not.

13.2 Form Factors of Nucleons

The quark model is intimately tied to the idea that hadrons, such as nucleons, are bound states of more fundamental particles with non-zero spatial extension. The idea that nucleons have space-time extension has a much longer history than the quark model. As was noted in Chap. 5, Yukawa developed this idea as early as 1953 [17]. As Yukwawa pointed out in his 1953 paper, his work was also based on earlier ideas. Thus it is somewhat endless to trace the origin of the concept of extended model of elementary particles. Furthermore, this idea will persist whenever we find new particles, such as quarks. It was not difficult in the past to find articles on composite models of quarks. However, experimental discoveries of composite particles are rather rare.

The first experimental discovery of the non-zero size of the proton was made by Hofstadter and McAllister in 1955 [1], who used electron-proton scattering to measure the charge distribution inside the proton. Since then the study of form factors has been an important branch of physics. For instance, we use this method to distinguish newly discovered leptons from hadrons. Leptons are still point particles while hadrons are not.

For this reason, the study of form factors cannot be separated from the quark model. In Chap. 12, we studied the hadron mass spectra to believe that there is a spatial wave function whose localization condition provides discrete mass spectra. The question then is whether the same wave function can describe the charge distribution observed in form factor experiments. The purpose of this Section is to study whether we can explain the behavior of form factors in terms of the Lorentz-Dirac deformation properties of the nucleon.

Let us first see what effect the charge distribution has on the scattering amplitude, using non-relativistic scattering in the Born approximation. If we scatter electrons from a fixed charge distribution whose density is $e\rho(r)$, the scattering amplitude is

$$f(\theta) = -\frac{e^2 m}{2\pi} \int \frac{\rho(\mathbf{r}')}{R} e^{-i\mathbf{Q}\cdot\mathbf{x}} d^3x d^3x' , \tag{13.20}$$

where $r = |\mathbf{x}|$, $R = |\mathbf{r} - \mathbf{r}'|$, and $\mathbf{Q} = \mathbf{K}_f - \mathbf{K}_i$ which is the momentum transfer. This amplitude can be reduced to [Problem 3 in Sect. 13.8]

$$f(\theta) = [2me^2/Q^2]F(Q^2) . \tag{13.21}$$

$F(Q^2)$ is the Fourier transform of the density function

$$F(Q^2) = \int \rho(\mathbf{r})e^{-i\mathbf{Q}\cdot\mathbf{x}} d^3x . \tag{13.22}$$

The above quantity is called the form factor. It describes the charge distribution in terms of the momentum transfer. The charge density is normalized:

$$\int \rho(r)d^3x = 1 . \tag{13.23}$$

Then $F(0) = 1$ from Eq. (13.22). If the density function is a delta function $\delta(\mathbf{r})$ corresponding to a point charge, then $F(Q^2) = 1$ for all values of Q^2, and the scattering amplitude of Eq. (13.20) becomes the Rutherford formula for Coulomb scattering. The deviations from Rutherford scattering for increasing values of Q^2 give a measure of the charge distribution [1, 18]. This was precisely what Hofstadter found in the scattering of electrons from a proton target.

As the energy of incoming electrons becomes higher, we have to take into account the recoil effect of target protons, and formulate the problem relativistically. It is generally agreed that electrons and the electromagnetic interaction associated with them, can be described by quantum electrodynamics (QED), in which the method of perturbation theory using Feynman diagrams is often employed for practical calculations [19, 20]. In this perturbation approach the scattering amplitude is expanded in power series of the fine structure constant $\alpha = e^2/4\pi$. Therefore in lowest order in α, we can describe the scattering of an electron by a proton using the diagram given in Fig. 13.2. The corresponding matrix element is given in many textbooks on

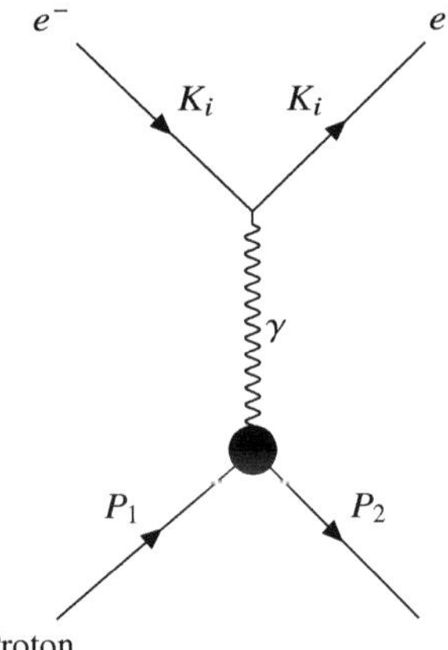

Fig. 13.2: Elastic electron-proton scattering. It is assumed that the electron behaves like a point charge. However, the proton has internal structure and the charge distribution is spread out.

elementary particle physics [21]. It is proportional to

$$\bar{U}(P_f)\Gamma_\mu(P_f, P_i)U(P_i)(1/Q^2)\bar{U}(K_f)\gamma^\mu U(K_i) , \tag{13.24}$$

where P_i, P_f, K_i, and K_f are the initial and final four-momenta of the proton and electron respectively. $U(P_i)$ is the Dirac spinor for the initial proton. Q^2 is (four-momentum transfer)^2 and is

$$Q^2 = (P_f - P_i)^2 = (K_i - K_f)^2 . \tag{13.25}$$

The $(1/Q^2)$ factor in Eq. (13.24) comes from the virtual photon being exchanged between the electron and the proton. It is customary to use the letter t for Q^2 in form factor studies, and this t should not be confused with the time separation variable used in Sect. 13.1. However, we shall not use t here. In the metric we use, this quantity, Q^2, is positive for physical values of the four-momenta for the particles involved in the scattering process.

The unknown function $\Gamma_\mu(P_f, P_i)$ in Eq. (13.24) represents the shaded circle in Fig. 13.2 and carries the effect of the nucleon structure. If the proton were a point charge, we would have $\Gamma_\mu = \gamma_\mu$. If the proton has an extended charge structure, it will be modified by a factor, so that $\Gamma_\mu = \gamma_\mu F(Q^2)$. This is not enough to explain the anomalous magnetic moments whose values are 2.79 and -1.91 in units of $e/2M$ for the proton and neutron respectively. We shall ignore the proton-neutron mass difference, and use M as the nucleon mass throughout this Chapter. If we include these observed anomalous magnetic moments, Γ_μ should be written as

$$\Gamma_\mu = \gamma_\mu F_1(Q^2) + i[\sigma_{\mu\nu}Q^\nu/2M]F_2(Q^2) . \tag{13.26}$$

At $Q^2 = 0$,

$$F_1^p(0) = 1, \qquad F_1^n(0) = 0,$$

$$F_2^p(0) = 2.79, \qquad F_2^n(0) = -1.91. \tag{13.27}$$

The above form factors are Lorentz-invariant functions in the Lorentz-invariant variable Q^2.

When we study the quark model, it is more convenient to use the following linear combinations:

$$G_M(Q^2) = F_1(Q^2) + F_2(Q^2),$$

$$G_E(Q^2) = F_1(Q^2) + (Q^2/4M^2)F_2(Q^2). \tag{13.28}$$

The scattering cross section takes a relatively simple form in terms of these form factors [Problem 2 in Sect. 13.8]. When $Q^2 = 0$,

$$G_{pM}(0) = \mu_p = 2.79, \qquad G_{pE}(0) = 1 \qquad \text{for proton},$$

$$G_{nM}(0) = \mu_n = -1.91, \qquad G_{nE}(0) = 0 \qquad \text{for neutron}. \tag{13.29}$$

These numbers are the magnetic moments of the proton and neutron respectively. The measured values of these moments were known for a long time [22]. There were many unsuccessful attempts to calculate these numbers, before the invention of the quark model. Indeed, one of the early successes of the quark model was the calculation of the magnetic ratio [Problem 4 in Sect. 13.8]:

$$\mu_n/\mu_p = -2/3. \tag{13.30}$$

However, the real challenge in modern physics is to calculate the form factors for increasing values of Q^2.

If the proton is a point charge, the form factor should be independent of Q^2. If the charge distribution is Gaussian, the non-relativistic calculation leads to an exponential decrease for increasing Q^2. The relativistic calculation gives a dipole behavior. We observe this dipole behavior in the real world.

At present, we can make the following experimental observations. Of the four form factors in the nucleon system given in Eq. (13.29), the neutron charge form factor is zero at $Q^2 = 0$, and remains small (not zero) for all values of Q^2. The three remaining form factors decrease like $1/Q^2$ as Q^2 increases beyond the value of (nucleon mass)^2. This behavior is usually called the dipole fit.

The question then is whether it is possible to calculate the above-mentioned dipole behavior using the wave function obtained from the static quark model, such as the harmonic oscillator wave function discussed in Chap. 12. If we are interested only in the overall behavior of the above-mentioned form factors, each of them can be derived for the single form factor $G(Q^2)$ by multiplication of a constant factor, where $G(Q^2)$ is normalized as

$$G(0) = 1 \,, \tag{13.31}$$

and is proportional to $1/Q^2$ for large values Q^2 as is illustrated in Fig. 13.3. In the case of the neutron charge form factor, we have to multiply the above $G(Q^2)$ by zero within the framework of the approximation we make.

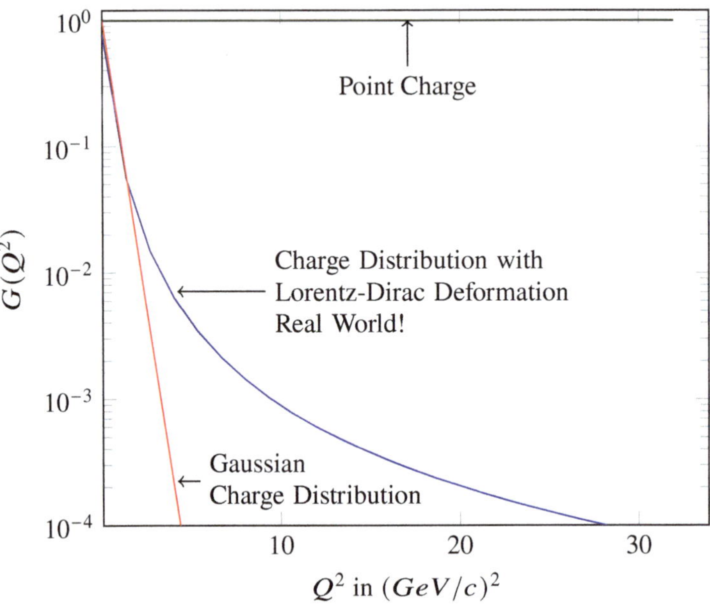

Fig. 13.3: Form factor behaviors for increasing values of Q^2. If the proton is a point charge, the form factor should be independent of Q^2. If the charge distribution is Gaussian, the non-relativistic calculation leads to an exponential decrease for increasing Q^2. The relativistic calculation gives a dipole behavior. We observe this dipole behavior in the real world. See Stanley and Robson [23] for the curves summarizing experimental data.

If we assume that the wave function for the nucleon in the quark model is Gaussian as is indicated in Chap. 12, the straight-forward Fourier transformation of the charge distribution function, according to Eq. (13.22), will lead to the form factor of the form

$$G(Q^2) = \exp(-Q^2/4\Omega) \,. \tag{13.32}$$

This expression is not consistent with the above-mentioned dipole fit. Indeed, it is a challenge to use the relativistic effect to convert the above expression into a dipole fit.

Before calculating the form factor behavior using relativistic wave functions, we should point out that there have been and there are still many attempts to do the same using other theoretical approaches. Perhaps, the single most important calculation which enhanced the development of the quark model is the work of Frazer and Fulco [24] based on dispersion relations. The concept of the ρ meson, which played the key role in the $SU(3)$ formulation of the meson multiplets, was initially invented to explain the form factor behavior for moderate values of Q^2.

13.3 Calculation of the Form Factors

In order to make a relativistic calculation of the form factor, let us go back to the definition of the form factor given in Eq. (13.22). The density function $\rho(r)$ depends only on the target atom, and is proportional to $|\psi(\mathbf{r})|^2$, where $\psi(\mathbf{r})$ is the wave function for electrons in the target atom. This expression is a special case of the more general case:

$$\rho(\mathbf{r}) = \psi_f^\dagger(\mathbf{r})\psi_i(\mathbf{r}) \tag{13.33}$$

where ψ_i and ψ_f are the initial and final wave function of the target atom. Indeed, the form factor of Eq. (13.22) can be written in the form

$$F(Q^2) = (\psi_f, e^{-i\mathbf{Q}\cdot\mathbf{x}}\psi_i). \tag{13.34}$$

Starting from this expression, we can make the required Lorentz generalization using the relativistic wave functions for hadrons. It is possible to obtain a Lorentz-invariant expression for the above quantity starting from a theoretical model based on the covariant Lagrangian formalism [Problem 5 in Sect. 13.8].

However, in order to see the details of the transition to relativistic physics, we should be able to replace each quantity in the expression of Eq. (13.34) by its relativistic counterpart. Let us go to the Lorentz frame in which the momenta of the incoming and outgoing nucleons have equal magnitude but opposite signs.

$$\mathbf{P}_f + \mathbf{P}_i = 0. \tag{13.35}$$

The Lorentz frame in which the above condition holds is usually called the *Breit frame*. We assume without loss of generality that the proton comes along the z direction before the collision and goes out along the negative z direction after the scattering process, as is illustrated in Fig. 13.4. In this frame, the four-vector $Q = (K_f - K_i) = (P_i - P_f)$ has no time-like component. Thus the exponential factor $\exp[-i\mathbf{Q}\cdot\mathbf{x}]$ in Eqs. (13.22) and (13.34) can be replaced by the Lorentz-invariant form $\exp[-iQ_\mu x^\mu]$. As for the wave functions for the protons, we can use the covariant harmonic oscillator wave functions discussed in Chaps. 5 and 6 assuming that the nucleons are in the ground state. Then the only difference between the non-relativistic and relativistic case is that the integral in the evaluation of Eq. (13.22) is four-dimensional, including that for the time-like direction. This integral in the time-separation variable does not interfere with the exponential factor which does not depend on the time-separation variable.

Let us now write down the integral:

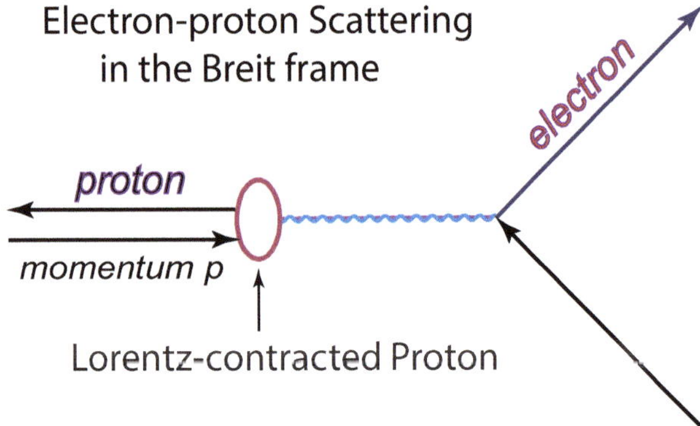

Fig. 13.4: Breit frame for electron-nucleon scattering. The outgoing and incoming nucleons have momenta equal in magnitude but opposite in sign [8].

$$g(Q^2) = \int \psi^{\dagger}_{-\eta}(x)\psi_{\eta}(x)e^{-iQ_{\mu}x^{\mu}} \, d^4x \tag{13.36}$$

where η is the boost parameter and β is the velocity parameter for the incoming proton, and the wave function $\psi_{\pm\eta}(x)$ takes the form:

$$\psi_{\pm\eta}(x) = \frac{\Omega}{\pi}(\exp[-\Omega(x^2 + y^2)/2])$$

$$\times \exp\left(-\frac{\Omega}{4}\left[\left(\frac{1 \mp \beta}{1 \pm \beta}\right)(t \pm z)^2 + \left(\frac{1 \pm \beta}{1 \mp \beta}\right)(t \mp z)^2\right]\right). \tag{13.37}$$

After the above decomposition of the wave functions, we can perform the integration in the x and y variables trivially. After dropping these trivial factors, we can write the product of the two wave functions as

$$\psi^{\dagger}_{-\eta}(t, z)\psi_{\eta}(t, z) = \frac{\Omega}{\pi}\exp(-\Omega[(1 + \beta^2)/(1 - \beta^2)](t^2 + z^2)). \tag{13.38}$$

Thus the t and z variables have been separated. Since the exponential factor in Eq. (13.34) does not depend on t, the t integral in Eq. (13.36) can also be trivially performed, and the integral of Eq. (13.36) can be written as

$$g(Q^2) = \left[\frac{\Omega}{\pi}((1 - \beta^2)/(1 + \beta^2))\right]^{1/2} \int e^{-2iP_z z} \exp\left(-\Omega\frac{1 + \beta^2}{1 - \beta^2}z^2\right) dz, \tag{13.39}$$

where P_z is the z component of the momentum of the incoming nucleon. The (momentum transfer)^2 variable Q^2 is $4P^2$. Indeed, the distribution of the hadron material along the longitudinal direction became contracted [25]. If we use the relation between β and Q^2:

$$\beta^2 = \frac{Q^2}{(Q^2 + 4M^2)}\,, \tag{13.40}$$

the evaluation of the above integral for $g(Q^2)$ in Eq. (13.39) leads to

$$g(Q^2) = \frac{2M^2}{(Q^2 + 2M^2)} \exp\left(\frac{-M^2Q^2}{[2\Omega(Q^2 + 2M^2)]}\right). \tag{13.41}$$

At $Q^2 = 0$, the above expression becomes $= 1$. It decreases as $1/Q^2$ for large values of Q^2.

We have so far carried out the calculation for an harmonic oscillator bound state of two quarks. The proton consists of two u and one d quarks, as we discussed in Chap. 12. There are one u and two d quarks in the neutron. If there are three quarks, there are two harmonic oscillator modes as was pointed out in Sect. 13.1. The generalization of the above calculation to the three-quark system is straightforward, and the result is that the form factor $G(Q^2)$ takes the form [Problem 6 in Sect. 13.8]:

$$G(Q^2) = \left(\frac{2M^2}{(Q^2 + 2M^2)}\right)^2 \exp\left(\frac{-2M^2Q^2}{[\Omega(Q^2 + 2M^2)]}\right), \tag{13.42}$$

which is 1 at $Q^2 = 0$, and decreases as $(1/Q^2)^2$ for large values of Q^2. Indeed, this function satisfies the requirement of the dipole behavior discussed in Sect. 13.2.

Let us re-examine the above calculation. If we replace β by zero in Eq. (13.39), $g(Q^2)$ will become the non-relativistic result of Eq. (13.32). The above-mentioned dipole behavior is due to the relativistic effect. Indeed, the longitudinal hadron distribution becomes contracted as β increases. As the momentum transfer increases, the width of the longitudinal distribution decreases. The wavelength of the photons also decreases, while the size remains constant in non-relativistic calculation. However, β is correlated to P_z in Eq. (13.39) or to the momentum transfer variable by Eq. (13.40). As the energy or the momentum transfer increases, both the longitudinal size and the wave-length of the virtual photon decrease. This correlation prevents a harsh high-frequency cut-off which happens in the non-relativistic case.

In order to gain a deeper understanding of the above-mentioned correlation, let us use the momentum-energy wave functions:

$$\phi_\eta(q) = \left(\frac{1}{2\pi}\right)^2 \int e^{-iq^\mu x_\mu} \psi_\eta(x)\, d^4x\,. \tag{13.43}$$

As before, we can ignore the transverse components. Then $g(Q^2)$ can be written as

$$\int \phi^\dagger_{-\eta}(q_0, q_z - P)\phi_\eta(q_0, q_z + P)\, dq_0 dq_z\,. \tag{13.44}$$

We have sketched the above overlap integral in Fig. 13.5. When $Q^2 = 0$ or $P = 0$, the two wave functions overlap completely in the $q_z q_0$-plane. As P increases, the wave functions become separated. However, they maintain a small overlapping region due

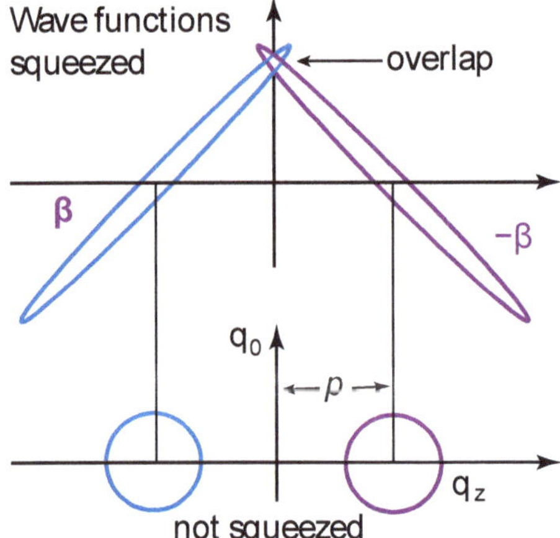

Fig. 13.5: Lorentz-Dirac deformation of the momentum-energy wave functions in the form factor calculation. As the momentum transfer increases, the two wave functions become separated. In the relativistic case, the wave functions maintain an overlapping region. Wave functions become completely separated in the non-relativistic calculation. This lack of overlapping region leads to an unacceptable behavior of the form factor.

to the Lorentz-Dirac deformation. In the non-relativistic case, where the deformation is not taken into account, there is no overlapping region. This is precisely why the relativistic calculation gives a slower decrease in Q^2 than in the non-relativistic case.

We have so far been interested only in the space-time behavior of the hadron wave function. We should not forget the fact that quarks are spin-1/2 particles. The effect of this spin manifests itself prominently in the baryon mass spectra. Since we are concerned here with the relativistic effects, we have to construct relativistic spin wave function for the quarks. This quark wave function should give the hadron spin wave function. In the case of nucleons, the quark spins should be combined in a manner to generate the vertex function of Eq. (13.26).

The most naive approach to this problem is to use free Dirac spinors for the quarks [10]. However, it was shown by Lipes [26] that the use of free-particle Dirac spinors leads to a wrong form factor behavior. Since quarks in a hadron are not free particles, the result of Lipes does not alarm us. The difficult problem is to find a suitable mechanism in which quark spins are coupled to orbital motion in a relativistic manner. This is a nontrivial research problem. It is interesting to note that there has been progress at least in the study of form factors [Problem 7 in Sect. 13.8].

13.3.1 Lorentz Coherence and the Proton Form Factor

In this Chapter we have so far been considering the elastic scattering of proton and electron with one photon exchange. The scattering cross section, if the proton is a point particle, can be calculated from a Feynman diagram with one-photon exchange. This calculation is straight-forward and is called Rutherford scattering. If the recoil of the proton is negligible, the cross section becomes the same as the classical Coulomb scattering

As indicated in Fig. 13.4, the cross section deviates from Rutherford scattering as the momentum transfer becomes substantial. As we said previously, this was first observed in 1955 by Hofstadter and McAllister [1]. Later it was observed that the decrease in the cross section can be written as:

$$\frac{1}{(\text{momentum transfer})^8}. \tag{13.45}$$

The reason for this deviation is that the proton is not a point particle. Additionally, the electric charge is distributed inside the proton with a finite radius. The proton form factor comes from a portion of the scattering amplitude and should decrease as

$$\frac{1}{(\text{momentum transfer})^4}. \tag{13.46}$$

As was mentioned earlier in this section, this decrease is known as the dipole cut-off in the literature. The dipole cut-off as well as possible deviations from it, has become a major branch of high-energy physics. In the past there were some far-reaching theoretical models which dealt with this problem [24].

Here we will approach the problem using the harmonic oscillator formalism developed in Chap. 5. The dipole cut-off will be shown to be a consequence of the coherence between the decrease in the wavelength of the incoming signal and the contraction of the proton wave function.

The formalism of Chap. 5 is based primarily on the papers of Dirac and Wigner. What is interesting is it is possible to derive the same harmonic oscillator functions from the authors attempting to understand the proton form factor, even though they were not aware of the papers written by Dirac and Wigner. We briefly review here what they did.

As said above Yukawa, as early as 1953, was interested in constructing harmonic oscillator wave functions which could be Lorentz-transformed [17]. Yukawa's main interest related to the Lorentz-invariant differential equation was in the mass spectrum produced. Unfortunately, his mass spectrum, at that time, did not seem to correspond to anything in the physical world.

However, when Hofstadter and McAllister [1] observed the non-zero charge radius of the proton, Markov, in 1956, used Yukawa's harmonic oscillator formalism to calculate the proton form factor [27]. At that time, however, there was no clear definition of the constituent particles contained in the harmonic oscillator wave functions.

The quark model's emergence in 1964 [4], prompted Ginzburg and Man'ko [28] to consider using relativistic harmonic oscillators to calculate the bound-states of quarks. Furthermore, without mentioning Yukawa's 1953 paper, Fujimura et al. used Yukawa's relativistic harmonic oscillator wave function to form a quark model. They then used this quark model to calculate the proton form factor, and obtained the dipole cut-off [29]. Licht and Pagnamenta derived the same result, in that same year, by using Lorentz-contracted harmonic oscillator wave functions. By using the Breit frame coordinate system they were able to bypass the time-separation variable appearing in the harmonic oscillator covariant formalism [25, 30].

In 1971, confirming the earlier suggestion made by Yukawa in 1953, Feynman et al. observed that the degeneracies of three-dimensional harmonic oscillators could be used to explain the hadron mass spectra [10]. Although they quoted the paper [29] of Fujimura et al., Yukawa's 1953 paper [17] was not mentioned.

Let us now go back to Chap. 5 and recall Eq. (5.63) which has the form:

$$\psi_\eta^{n,0}(t,z) = \left[\frac{1}{(2^n\pi)}\right]^{1/2} (1-\beta^2)^{(1+n)/2} \exp[-(t^2+z^2)/2]$$

$$\times \sum_{k=0}^{\infty} \left(\frac{1}{2}\right)^k \left(\frac{\beta}{2}\right)^k H_{n+k}(z)H_k(t) . \tag{13.47}$$

Here the $H_n(z)$ is the Hermite polynomial of order n.

Let us expand Eq. (13.47) as shown in Eq. (5.122) in terms of the functions of z and t variables [31, 32, 33]. The result is

$$\psi_\eta^{n,0}(t,z) = \left(\frac{1}{\cosh\eta}\right)^{(n+1)} \sum_k \left[\frac{(n+k)!}{n!k!}\right]^{1/2} (\tanh\eta)^k \chi_{n+k}(z)\chi_k(t) . \tag{13.48}$$

Here $\chi_n(z)$ is the nth excited harmonic state oscillator wave function and takes the form with which we are familiar:

$$\chi_n(z) = \left[\frac{1}{\sqrt{\pi}2^n n!}\right]^{1/2} H_n(z) \exp\left(\frac{-z^2}{2}\right) . \tag{13.49}$$

If $n=0$, we can simplify this formula to [31]

$$\psi_\eta^0(t,z) = \left(\frac{1}{\cosh\eta}\right) \sum_k (\tanh\eta)^k \chi_k(z)\chi_k(t) . \tag{13.50}$$

Let us now consider the overlap of the wave function $\psi_\eta^n(t,z)$ with $\beta=0$, as indicated in Fig. 13.5. We are interested in the integral

$$\int \left(\psi_\eta^{n'}(t,z)\right)^* \psi_0^n(t,z)dz\,dt , \tag{13.51}$$

where $\psi_0^n(t,z)$, given in Eq. (13.48), can be written as

$$\psi_0^n(t, z) = \chi_n(z)\chi_0(t) .$$
(13.52)

Then the evaluation of this integral is straight-forward from Eq. (13.48), and the result is [31]

$$\left(\psi_\eta^{n'}, \psi_0^n\right) = \left(\frac{1}{\cosh\eta}\right)^{(n+1)} \delta_{nn'} .$$
(13.53)

This means that the orthogonality relation is preserved between two wave functions in two different frames.

If $\beta = \beta'$, the inner product between two wave functions leads to the contraction given on the right-hand side of Eq. (13.53). In terms of the velocity parameter $\beta = v/c$, where v is the hadron velocity, we obtain using Eq. (5.121):

$$\frac{1}{\cosh\eta} = \sqrt{1 - \beta^2} .$$
(13.54)

This more familiar expression allows us to write Eq. (13.53) as:

$$\left(\psi_\eta^{n'}, \psi_0^n\right) = \left(\sqrt{1 - \beta^2}\right)^{(n+1)} \delta_{nn'} .$$
(13.55)

As we found in Chap. 5, for the ground-state wave function with $\beta = 0$, Eq. (13.54) is like the Lorentz contraction of a rigid rod. For the first excited state, it is like an additional rod. This was illustrated Fig. 5.2.

If the value of β in one of the wave functions is replaced by nonzero value β', $\cosh\eta$ in Eq. (13.54) should become $\cosh(\eta - \eta')$. Of particular interest is the case with $\eta' = -\eta$, as shown in Fig. 5.3. Then there is an overlap of two wave functions moving in the opposite directions, and the contraction factor is

$$\left(\frac{1}{\cosh(2\eta)}\right)^{(n+1)} ,$$
(13.56)

which becomes, using Eq. (5.55):

$$\left(\frac{1 - \beta^2}{1 + \beta^2}\right)^{(n+1)} .$$
(13.57)

Let us now consider once again the exchange of one photon in the process of scattering one electron and one proton. We are then free to choose the Lorentz frame where the incoming and outgoing protons move with the same speed in opposite directions. Then it is possible to assume the proton moves along the z-direction as pictured in Fig. 13.4. Then we can let P be the magnitude of the momentum. The initial and final energy-momentum four-vectors then become:

$$(E, P) \quad \text{and} \quad (E, -P) ,$$
(13.58)

respectively, where $E = \sqrt{1 + P^2}$. In this Breit frame the momentum transfer is:

$$(E, P) - (E, -P) = (0, 2P), \tag{13.59}$$

with zero energy component.

The proton form factor then becomes

$$F(P) = \int e^{2iPz} \left(\psi_\eta(t, z) \right)^* \psi_{-\eta}(t, z) \, dz \, dt. \tag{13.60}$$

Let us use the ground-state harmonic oscillator wave function. This integral then can be written as:

$$\frac{1}{\pi} \int e^{2iPz} \exp\left\{ -\cosh(2\eta) \left(z^2 + t^2 \right) \right\} dz \, dt. \tag{13.61}$$

The physics of $\cosh(2\eta)$ in this expression was explained in Eq. (13.56).

The exponential function in the Fourier integral of Eq. (13.61), does not depend on the t variable. Hence when the t integration is finished, Eq. (13.61) becomes

$$F(P) = \frac{1}{\sqrt{\pi \cosh(2\eta)}} \int e^{2iPz} \exp\left\{ -z^2 \cosh(2\eta) \right\} dz. \tag{13.62}$$

After completing this integral, the proton form factor becomes

$$F(P) = \frac{1}{\cosh(2\eta)} \exp\left\{ \frac{-P^2}{\cosh(2\eta)} \right\}. \tag{13.63}$$

Using the expression of $\cosh(2\eta)$ given in Eq. (13.57), the proton form factor we obtain is:

$$F(P) = \frac{1}{1 + 2P^2} \exp\left(\frac{-P^2}{1 + 2P^2} \right), \tag{13.64}$$

which decreases as $1/P^2$ for large values of P.

Let us illustrate in more detail the effect of the Lorentz contraction by performing the integral of Eq. (13.62) leaving out the contraction factor $\cosh(2\eta)$. Therefore, the wave function $\psi_\eta(t, z)$ of Eq. (13.60) should be replaced by the Gaussian form $\psi_0(t, z)$ of Eq. (13.51). Then the Fourier integral, with this non-squeezed wave function, becomes

$$G(P) = \int e^{2iPz} \left(\psi_0(t, z) \right)^* \psi_0(t, z) \, dz \, dt. \tag{13.65}$$

This integral results in

$$G(P) = \frac{1}{\sqrt{\pi}} \exp\left(-P^2 \right), \tag{13.66}$$

which leads to proton form factor having a Gaussian cut-off. The calculation of $G(P)$ is for the purpose of illustration only, as this Gaussian cut-off is not what happens in the real experimental world.

Let us return to the Fourier integrals of Eqs. (13.60) and (13.65). The only difference, after the integration over t, is the $\cosh(2\eta)$ factor in Eq. (13.60). This factor, which is in the normalization constant, does not affect the Fourier integral.

However, the $\cosh(2\eta)$ factor causes the Gaussian width, for large values of P, to shrink by $1/\sqrt{2}P$. The sinusoidal factor has a wavelength which is inversely proportional to the momentum $2P$. The result is that both the wavelength of the incoming signal and the Gaussian width shrink at the same rate of $1/P$ as P becomes large. If this coherence did not exist, the cut-off would be Gaussian, as noted in Eq. (13.66). The effect of this Lorentz coherence, Fig. 13.6 illustrates.

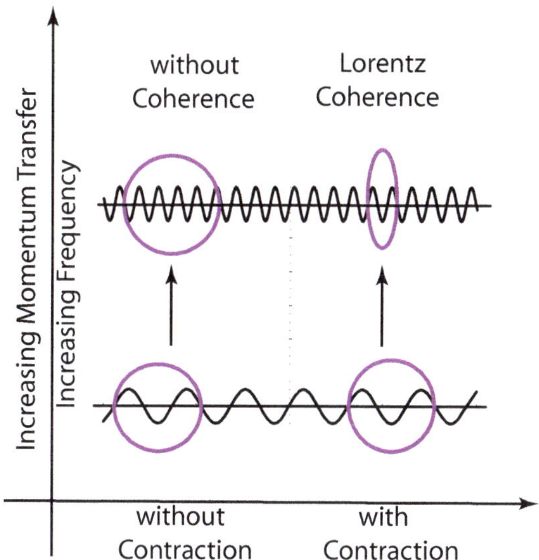

Fig. 13.6: The coherence between the wavelength and the proton size is illustrated. By recalling Fig. 13.4, that the proton sees the incoming photon is evident. The wavelength of this proton becomes smaller as the momentum transfer increases. If the proton size were to remain unchanged, this would result in a rapid cut-off for oscillation in the form factor. Thus the Fourier integral would lead to a Gaussian cut-off. However, there are no oscillation effects when the proton size decreases coherently with the wavelength. This decrease in the proton size leads to a polynomial decrease of the form factor [34, 8].

There is a gap still between the real world and $F(P)$ of Eq. (13.64). When we wish to compare this expression with the experimental data, it is necessary to realize that inside the proton there are two harmonic oscillator modes, which result from the three quarks inside the proton.

The Lorentz coherence process discussed in this section is gone through by one of these modes. The contraction process detailed in Eq. (13.56) is gone through by the other mode. The net effect is

$$F_3(P) = \left(\frac{1}{1 + 2P^2}\right)^2 \exp\left(\frac{-P^2}{1 + 2P^2}\right) . \tag{13.67}$$

This, of course, leads to the dipole cut-off of $\left(1/P^2\right)^2$ which is seen in the laboratory.

Deviations from the exact dipole cut-off have also been reported. Based on the four-dimensional rotation group when an imaginary time coordinate is included, studies of the proton form have also been attempted. Many papers based on lattice quantum chromodynamics (QCD) have been written. A brief set of references, given in [34], lists some of these efforts. This section was limited to the detailed study of the role of Lorentz coherence relative to keeping the momentum transfer variable in the proton form factor from the steep Gaussian cut-off. One of the primary issues in the current physics trends, deals with this coherence problem.

13.3.2 Coherence in Energy-Momentum Space

Let us see how, in energy-momentum space, Lorentz coherence manifests itself. We begin with the Lorentz-squeezed wave function. In energy-momentum space, this wave function can be written as

$$\phi_\eta\left(q_z, q_0\right) = \frac{1}{2\pi} \int e^{-i(q_z z - q_0 t)} \psi_\eta(t, z) dt\, dz . \tag{13.68}$$

Here, q_z and q_0 are the Fourier conjugate variables of z and t, respectively. The result of this integral is

$$\phi_\eta\left(q_z, q_0\right) = \frac{1}{\sqrt{\pi}} \exp\left\{-\frac{1}{4}\left[e^{-2\eta}\left(q_z + q_0\right)^2 + e^{2\eta}\left(q_z - q_0\right)^2\right]\right\} . \tag{13.69}$$

It is easy to see that Eq. (13.69) is a Fourier transformation of the Lorentz-squeezed wave function given in Eq. (13.1). Now the proton form factor of Eq. (13.60) can, in terms of this energy-momentum wave function, be written as

$$F(P) = \int \phi^*_{-\eta}\left(q_0, q_z - P\right) \phi_\eta\left(q_0, q_z + P\right) dq_0\, dq_z . \tag{13.70}$$

When the integral of Eq. (13.70) is evaluated, it results in the proton form factor $F(P)$ given in Eq. (13.64).

The effect of the Lorentz coherence, can be seen in the wave functions of Fig. 13.5 The integral is performed over the $q_z\, q_o$-plane. The two wave functions become separated as the momentum P increases. The wave functions would not overlap if the Lorentz squeeze were not present. This would result in the sharp Gaussian cut-off as in the case of $G(P)$ of Eq. (13.66).

It should be noted that the squeezed wave functions, as shown in Fig. 13.5, have an overlap which becomes smaller as P increases. This in turn leads to a slower polynomial cut-off [34, 8].

Indeed, a new era of physics was opened by the discovery of the non-zero size of the proton [1, 18]. Since the proton is no longer a point particle, by studying the dependence on the momentum transfer in proton-electron scattering amplitude where one photon is exchanged is one way to measure the proton internal structure. The deviation of the proton from a point particle is what we have called the proton form factor.

From the experimental point of view, the dipole cut-off has been firmly established. There are indeed experimental results indicating that there are deviations from this dipole behavior [35, 36]. However, a review of all the papers written to detail these corrections will not be given here. Those deviations, from the theoretical point of view, are corrections to the basic dipole behavior.

The study of the proton form factor remains a major subject in physics. It is noteworthy, however, to observe that the coherence between the decrease in the wavelength of the incoming signal and the Lorentz contraction of the longitudinal size of the proton is the major contributor to the dipole cut-off.

13.4 Scaling Phenomenon and the Parton Picture

The parton picture was developed originally by Feynman [5, 6] to explain the scaling behavior in deep inelastic electron-proton scattering. Let us consider the process $e + p \rightarrow e + \text{hadrons}$, as is described in Fig. 13.7. The matrix element for this process can be written as

$$M_\eta = (e^2/Q^2 U(K_f)\gamma_\mu U(K_i) \langle P'_{\text{all}}| J^\mu |P\rangle , \tag{13.71}$$

where $Q = (K_i - K_f)$ is the four-momentum transfer of the electron and J^μ is the current. We use P for the four-momentum of the initial nucleon. P' is the total four-momentum of the final-state hadrons. Here again, the electron-photon vertex is point-like. If we do not measure spins, the rate of transition is

$$\text{Rate} = \left(\frac{e^2}{4\pi}\right)^2 \int K_f \frac{m^2}{EE'} (1/Q^2)^2 \eta^{\mu\nu} W_{\mu\nu} \, d^4x , \tag{13.72}$$

where m is the electron mass, and

$$\eta^{\mu\nu} = \left(\frac{1}{2m}\right)^2 Tr[(\gamma \cdot K_f - m)\gamma^\mu (\gamma \cdot K_i - m)\gamma^\nu] . \tag{13.73}$$

The nontrivial part of physics is contained in

$$W_{\mu\nu} = (2\pi)^3 \sum_{\text{all}} \langle P| J_\mu |\text{all}\rangle \langle \text{all}| J_\nu |P\rangle \, \delta(P' - P - Q) . \tag{13.74}$$

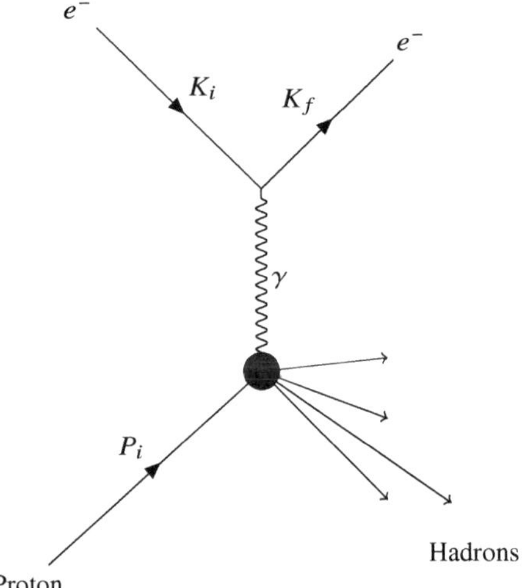

Fig. 13.7: Inelastic electron-proton scattering. The electron emits a virtual photon. After absorbing the virtual photon, the proton becomes many hadrons. Unlike the case of the elastic scattering described in Fig. 13.2, there is an energy transfer through the virtual photon, in addition to the momentum transfer.

The summation is over all hadrons in the final state. The above expression can be reduced to

$$W_{\mu\nu} = W_1(Q^2, \nu)(\eta_{\mu\nu} - Q_\mu Q_\nu / Q^2)$$

$$+ W_2(Q^2, \nu) \left(\frac{1}{M}\right)^2 (P_\mu - (P_\sigma Q^\sigma / Q^2) Q_\mu)$$

$$\times (P_\nu - (P_\sigma Q^\sigma / Q^2) Q_\nu), \tag{13.75}$$

where
$$\nu = Q_\mu P^\mu / M . \tag{13.76}$$

W_1 and W_2 are now the Lorentz-invariant functions of the two Lorentz-invariant variables.

Unlike the case of elastic scattering, the cross section depends also on the energy transfer. If the proton and electron spins are not measured, the cross section in the laboratory frame takes the form [Problem 8 in Sect. 13.8]:

$$\frac{d^2\alpha}{d\Omega dE'} = (\alpha^2 / 4E^2 \sin^4(\theta/2))(W_2(Q^2, \nu) \cos^2(\theta/2)$$

$$+ 2W_1(Q^2, \nu) \sin^2(\theta/2)), \tag{13.77}$$

where E and E' are the energies of the initial and final electrons, and Ω is the solid angle of the scattered electron. v is the energy transfer, and is thus

$$v = E - E'$$

$$Q^2 = (4EE') \sin \frac{\theta}{2} . \tag{13.78}$$

It should be emphasized here that W_1 and W_2 in Eq. (13.77) are basically different from the form factors G_1 and G_2 discussed in Sect. 13.2. The W functions depend not only on Q^2, but also on the energy transfer variable. The fundamental difference is that the W_1 and W_2 are proportional to the cross section, while the form factors are proportional to the scattering amplitude.

We call the above inelastic scattering *deep inelastic scattering* when both the momentum and energy transfer variables become very large. In this case, we can define the ratio:

$$\omega = \frac{-Q^2}{2Q_\mu P^\mu} . \tag{13.79}$$

It was observed [7] that both W_1 and vW_2 become functions of only one variable in the deep inelastic limit:

$$W_1 \to F_1(\omega) ,$$

$$vW_2 \to F_2(\omega) . \tag{13.80}$$

This is called the *scaling phenomenon*.

There are several different theoretical approaches to explain this scaling behavior in deep inelastic scattering. One of the first explanations was that of Feynman's parton picture, and the parton model still plays the major role in high-energy physics. In 1969 [5, 6], Feynman observed that a rapidly moving proton, which is now believed to be a bound state of three quarks, can be regarded as a collection of particles called partons which exhibit the following peculiar features.

(a) The picture is valid only for hadrons moving with velocity close to that of light.
(b) The interaction time between the quarks becomes dilated, and partons behave as free independent particles.
(c) The momentum distribution of partons becomes widespread as the hadron moves fast.
(d) The number of partons seems to be much larger than that of quarks, as is discussed in Fig, 13.8.

Because the hadron is believed to be a bound state of two or three quarks, each of the above phenomena appears as a paradox, particularly (b) and (c) together. This paradox will be resolved in Sect. 13.5. In the meantime, let us see how the parton picture will explain the scaling phenomenon.

We ignore for simplicity spins and assume that all the particles are scalar particles. Then $W_{\mu v}$ of Eq. (13.74) becomes

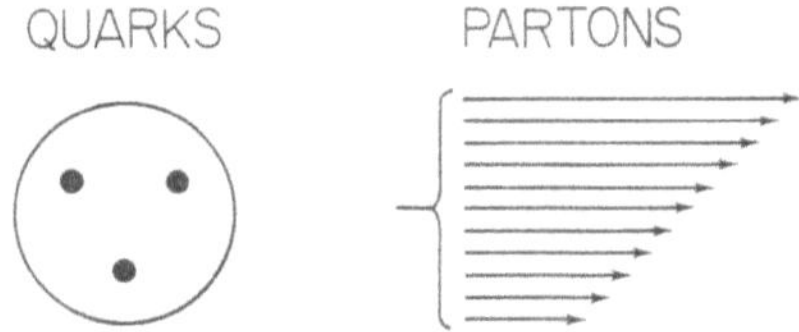

Fig. 13.8: Proton as it appears in the quark and parton models. Suppose that a proton is sitting quietly on the desk. According to the quark model, it appears like a bound state of three quarks to an observer who is sitting on the chair. However, to an observer who is on a jet plane which has a speed close to that of light, the proton would look like a collection of free particles with a wide momentum distribution. This is called Feynman's parton picture.

$$W = (2\pi)^2 \sum_{\text{all}} \langle P| J(0) |\text{all}\rangle \langle \text{all}| J(0) |P\rangle \, \delta(P' - P - Q) \,. \tag{13.81}$$

If we use the expression:

$$\delta(P' - P - Q) = \left(\frac{1}{2\pi}\right)^4 \int e^{-i(P'_\mu - P_\mu - Q_\mu)x^\mu} \, d^4x \,, \tag{13.82}$$

the W function can be transformed into

$$W = \frac{1}{2\pi} \int e^{iQ_\mu x^\mu} \langle P| [J(0), J(x)] |P\rangle \, d^4x \,. \tag{13.83}$$

Indeed, the *summation over all* in Eq. (13.81) drastically simplified the problem. All we have to do is to compute the commutator in the above expression. It is not difficult to generalize the above expression to the case where the spins are considered [Problem 8 in Sect. 13.8].

The calculation of the commutator requires the knowledge of dynamics, and a simple calculation is possible only for free point particles, and the nucleon is not a point particle. However, in the limit of large Q^2, the integrand makes a significant contribution to the integral only in the small-x^2 region near the light-cones. This is called the light-cone dominance. There is no reason to expect that the commutator will behave like a free-field commutator in this region. If we use the free-field commutators for the current in Eq. (13.83), we are still making a drastic assumption. Indeed, this is the place where the parton picture plays the decisive role. Let us go to the Lorentz frame in which the nucleon momentum is very large. Since partons are free particles, we can evaluate the integral of Eq. (13.83) using the free-field approximation. As is shown in Exercise 2 in Sect. 13.8, the evaluation of the integral leads to

$$\nu W(Q^2, \nu) = (1/2M^2)\delta(Q^2/2Q_\mu P^\mu - 1) \,. \tag{13.84}$$

If the nucleon is a collection of free partons, the relevant momentum variable in the above expression is the four-momentum of one of the partons. Therefore Eq. (13.84) should be replaced by

$$\nu W(Q^2, \nu) = (1/2M^2) \sum_{\text{all partons}} C_i^2 \delta(1 - Q^2/Q_\mu p_i^\mu), \tag{13.85}$$

where p_i and C_i^2 are the four-momentum and the (charge in unit of e)^2 of the ith parton respectively. If we parameterize the parton four-momentum by variable x_i :

$$p_i = x_i P, \tag{13.86}$$

then the formula in Eq. (13.84) becomes

$$\nu W(Q^2, \nu) = (1/2M^2) \sum_i C_i^2 x_i \delta(x_i - \omega) \tag{13.87}$$

where ω is given in Eq. (13.79). Indeed, νW is a function only of the ω variable.

In order to calculate the above quantity, we have to know the charge and momentum distributions of partons. Before starting this calculation, we are led to the question of whether partons are quarks. The question is then whether the quark model and the parton model are only two different manifestations of one physical entity which is a relativistic bound state of quarks. We shall study this problem in Sect. 13.6 using the covariant harmonic oscillator formalism.

13.5 Covariant Harmonic Oscillators and the Parton Picture

We examine in this Section how the Lorentz-Dirac deformation leads to the parton picture using for simplicity the covariant harmonic oscillator wave function for a two-body bound state [37]. By taking the Fourier transformation of Eq. (13.12), we get the momentum-energy wave function:

$$\phi(q) = \exp\left[-\left(\frac{1}{2\Omega}\right)\left(\frac{1+\beta}{1-\beta}q_-^2 + \frac{1-\beta}{1+\beta}q_+^2\right)\right], \tag{13.88}$$

with

$$q_- = (q_0 - q_z)/\sqrt{2},$$

$$q_+ = (q_0 + q_z)/\sqrt{2}. \tag{13.89}$$

Because we are using here the harmonic oscillator, the mathematical form of the above momentum-energy wave function is identical to that of the space-time wave function. The Lorentz deformation properties of these wave functions are given in Fig. 13.9. When $\beta = 0$, both wave functions behave like those for the static bound states of quarks. As β increases, the wave functions become continuously Lorentz-deformed until they become concentrated along their respective positive light-cone axes. When $\beta \to 1$, $z_- \to 0$ and $q_+ \to 0$. Therefore, in this limit,

$$z_+ = \sqrt{2}z = \sqrt{2}t \,,$$

$$q_- = \sqrt{2}q_z = \sqrt{2}q_0 \,. \tag{13.90}$$

If we use the approximation $1+\beta = 2$, and write the wave functions of Eqs. (13.12) and (13.88) in terms of the z and q_z variables only,

$$\psi_\eta(z) \to \exp[-\Omega(1 - \beta)z^2/2] \,,$$

$$\phi_\eta(q_z) \to \exp[-(1 - \beta)q_z^2/2\Omega] \,. \tag{13.91}$$

Both of the above wave functions become widespread.

Pictorially, we can obtain these distributions by taking the longitudinal projections of the elliptic deformations given in Fig. 13.9. As shown, the hadron matter becomes squeezed and therefore is concentrated in the elliptic region around the positive light-cone axis. This results in the interaction time of the quarks among themselves becoming dilated. The wide-spread wave function means there is an increasing distance between the ends of the harmonic oscillator well. This effect, universally observed in high-energy hadron experiments, was first noted by Feynman [5, 6]. On the other hand, the external signal, since it is moving in the direction opposite to the direction of the hadron, travels along the negative light-cone axis. If we consider that the energy of protons produced by the Large Hadron Collider at CERN is 13 TeV, then the ratio of the interaction time to the oscillator period becomes 1.25^{-9}. This small number means that the external signal cannot sense the interaction among the quarks inside the hadron. This is also illustrated in Fig. 13.9 which shows that the interaction time of the external signal with the bound state is much shorter than the period of oscillation of the quarks inside the hadron. This effect is often called Feynman's time dilation [5, 6, 39].

Let us consider first the longitudinal distribution. From Eq. (13.3), we can write the longitudinal coordinate for each quark as

$$z_a = Z + \sqrt{2}z \,,$$

$$z_b = Z - \sqrt{2}z \,, \tag{13.92}$$

where Z is the longitudinal coordinate of the hadron. The distribution in z given in Eqs. (13.90) and (13.91) and in Fig. 13.9 tells us that the position of each quark appears widespread to the observer in the laboratory frame, and that the quarks appear like free particles as the hadron velocity parameter β approaches 1. This effect, first noted by Feynman, is universally observed in high-energy hadron experiments.

Because the quarks appear almost free, we would normally expect that the momentum of each quark will appear sharply defined. However, let us consider the longitudinal momentum distribution derivable from Eq. (13.92).

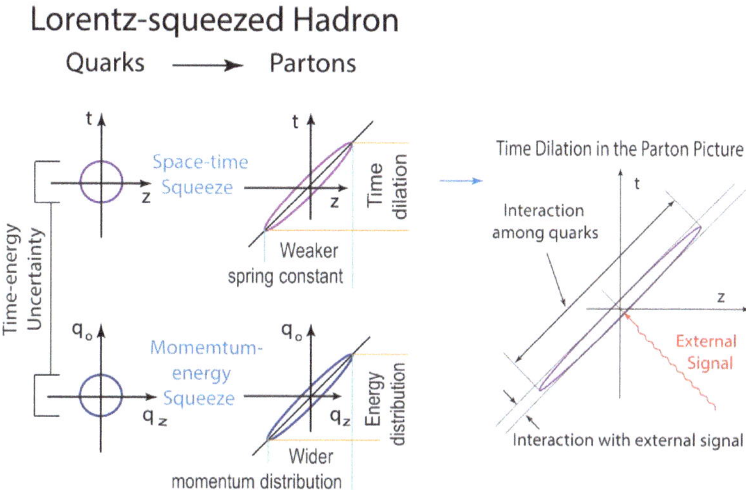

Fig. 13.9: Pictorial explanation of Feynman's parton picture based on the Lorentz-Dirac deformations in both the space-time and momentum-energy coordinate systems. In the harmonic oscillator model, the Lorentz deformation property of the momentum-energy wave function is identical to that of the space-time wave function. As the hadron moves very fast, both the space-time and the momentum-energy distribution become concentrated and elongated along their respective positive light-cone axes. This light-cone concentration is what leads to Feynman's parton picture [37, 38]. Here we also see that since the external signal which is moving in the direction opposite to that of the hadron travels along the negative light-cone axis. The quarks period of oscillation inside the hadron is thus much longer than the interaction time of the bound state with this signal. This effect is often called Feynman's time dilation [5, 6, 39].

We write the momentum of each quark as

$$p_{az} = P_z/2 + q_z/2\sqrt{2}\,,$$

$$p_{bz} = P_z/2 - q_z/2\sqrt{2}\,. \tag{13.93}$$

When the hadron moves very fast, we can use the approximation

$$(1 - \beta) = (P_0/M)^2/2\,, \tag{13.94}$$

in the expression given in Eq. (13.91) where M and P_0 are the hadron mass and energy respectively. Then q_z is spread over the region

$$-\sqrt{2\Omega}P_0/M \le q_z \le \sqrt{2\Omega}P_0/M\,, \tag{13.95}$$

which becomes widespread as P_0 becomes large. Consequently, the longitudinal momentum of the first quark given in Eq. (13.93) mostly lies within the interval between

$$p_z^{\max} = P_0(1/2 + \sqrt{\Omega}/2M),$$

and

$$p_z^{\min} = P_0(1/2 - \sqrt{\Omega}/2M), \tag{13.96}$$

where we assumed that $p_z = P_0$. If we assume that M is the proton mass, and use the value $\Omega - 1(\text{GeV})^2$, which is commonly used for the quark model analysis, then the quantity $\sqrt{\Omega}/2M$ is of the same order of magnitude as $1/2$. For this reason, the (almost) light-like four-momentum $p_{1\mu}$ can be written as:

$$p_{1\mu} = xP_\mu, \tag{13.97}$$

where the parameter x ranges approximately between zero and 1. This type of distribution, which was postulated first by Feynman, is commonly observed in high-energy hadron experiments.

We can make similar projections to the vertical axes in Fig. 13.9. When $\beta \to 1$, the relative quark motion is an oscillation along the positive light-cone axis. Fig. 13.9 clearly indicates that the oscillation time is dilated. Classically, the quarks make two *collisions* with each other within one period of oscillation. Therefore, the interquark collision time is one half of the oscillation time. If the hadron is at rest with $\beta = 0$, the collision time is $1/2\sqrt{\Omega}$. If the hadron velocity is close to that of light, and if we make the same approximation as the one used for the derivation of Eq. (13.91) the collision time becomes $1/[2\sqrt{\Omega(1-\beta)}]$ to an observer in the laboratory frame. If we use the expression of Eq. (13.94) the collision time ratio becomes

$$\frac{\text{Collision time for } \beta \to 1}{\text{Collision time for } \beta = 0} = P_0/2M. \tag{13.98}$$

As the hadron moves very fast, the above ratio becomes very large. This is also consistent with what we observe in the real world, and also with our previous observation that quarks appear as almost free particles to observers in the laboratory frame. This time dilation allows us to make an incoherent sum of cross sections due to each parton.

The widespread momentum distribution shown in Fig. 13.9 appears to contradict our initial expectation that, based upon the widespread spatial distribution, the quark be a free particle with a sharply defined momentum. This apparent contradiction presents to us the following two fundamental questions:

(a) If both the spatial and momentum distributions become widespread as the hadron moves, and if we insist on Heisenberg's uncertainty relation, is Planck's constant dependent on the hadron velocity?

(b) Is this apparent contradiction related to another apparent contradiction that the number of partons is infinite while there are only two or three quarks inside the hadron?

The answer to the first question is *No*, and that for the second question is *Yes*. The purpose of this section is to explain these answers.

Let us answer the first question which is related to the Lorentz invariance of Planck's constant. If we take the product of the width of the longitudinal momentum distribution and that of the spatial distribution, we end up with the relation

$$\langle z^2 \rangle \langle q_z^2 \rangle = (1/4)[(1+\beta)/(1-\beta)]^2 . \tag{13.99}$$

The right-hand side increases as the velocity parameter increases. This could lead us to an erroneous conclusion that Planck's constant becomes velocity dependent. This is not correct, because the product of two longitudinal coordinates is not Lorentz invariant.

In order to maintain the Lorentz-invariance of the uncertainty product, we have to work with a conjugate pair of variables whose product does not depend on the velocity parameter. We noted in Sect. 6.3 of Chap. 6 that the use of the light-cone variables will solve this problem. The content of the argument given there is contained in Fig. 13.9. The major axis of the space-time ellipse is conjugate to the minor axis of the momentum-energy ellipse, as is illustrated in Fig. 13.9. Therefore, the exact calculation gives

$$\langle z_+^2 \rangle \langle q_+^2 \rangle = 1/4 , \quad \langle z_-^2 \rangle \langle q_-^2 \rangle = 1/4 . \tag{13.100}$$

Planck's constant is indeed Lorentz invariant. It was noted in Sect. 6.5 of Chap. 6 the manner in which the above uncertainty products can be stated in terms of the conventional longitudinal and time-like variables.

Let us next resolve the puzzle of why the number of partons appears to be infinite while there are only a finite number of quarks inside the hadron. It is not difficult to see from Fig. 13.9 that both the x and q distributions become concentrated along the positive light-cone axis, as the hadron velocity approaches the speed of light. This means that the quarks also move with velocity very close to that of light. We have in fact shown that the quark four-momentum is proportional to that of the hadron in Eq. (13.97). Quarks in this case behave like massless particles.

We then know from statistical mechanics that the number of massless particles is not a conserved quantity. For instance, in black-body radiation, free light-like particles have a widespread momentum distribution. However, this does not contradict the known principles of quantum mechanics, because the massless photons can be divided into infinitely many massless particles with a continuous momentum distribution.

Likewise, in the parton picture, massless free quarks have a wide-spread momentum distribution. They can appear as a distribution of an infinite-number of free particles. These free massless particles are the partons. We shall see in Sect. 13.6 whether the widespread momentum distribution can be measured experimentally.

13.6 Calculation of the Parton Distribution Function for the Proton

We studied in Sects. 13.4 and 13.5 the kinematics through which the peculiarities of Feynman's parton picture can be derived from the Lorentz-Dirac deformation of hadron wave functions. A more interesting problem is to calculate the parton momentum distribution function in the proton and compare it with the distribution measured in high-energy experiments.

Before carrying out this program, let us see what we do in the Lamb-shift calculation. Because calculations in QED are much more complicated than those in non-relativistic quantum mechanics, we often forget the fact that the starting point in the Lamb-shift calculation is the Rydberg formula for unperturbed hydrogen energy levels. The Rydberg formula is obtained from the solution of the Schrödinger or Dirac equation for the hydrogen atom. The discreteness of the energy levels is a consequence of the boundary condition on the wave function requiring that the probability distribution be localized around the proton. QED then makes corrections to the Rydberg formula by taking into account the detailed interaction mechanism between photons and electrons.

Likewise, in calculating the parton distribution function, we have to know first how the partons are distributed. After that, we may make corrections due to the detailed interaction between the free partons and probing particles. Like QED in the Lamb shift calculation, there is at present a field theoretic method called QCD for calculating the corrections to the distribution function. Since it will require a separate book to explain QCD, and since perturbative QCD is discussed exhaustively in the literature [40, 41, 42], we shall not go into this subject here.

The purpose of this section is to provide a calculation of the parton distribution function using the covariant harmonic oscillator wave function. The nucleon in the rest frame is regarded as the ground state of the three-quark bound state. As before, we name the three quarks a, b, c, respectively, and use three independent four-momentum variables P, q, and k given in Eq. (13.15). P represents the four-momentum of the hadron, and is on the mass shell. q and k are the relative internal momenta. Here again, we ignore the transverse components. In terms of the q and k variables, the ground state wave function takes the form of Eq. (13.18). The distribution function is the square of the wave function.

In the parton model calculation, only one of the quarks interacts with the virtual photon. Since the wave function is totally symmetric under the exchange of quarks, we can pick quark c as the one interacting with the virtual photon. As can be seen from Eq. (13.15), the four-momentum of this quark does not depend on k. Thus we can integrate over the k variables. After this integration, the probability distribution function for the ground-state takes the form

$$\left(\frac{1}{\pi\Omega}\right)\exp\left\{-\left(\frac{1}{\Omega}\right)\left[\left(\frac{1-\beta}{1+\beta}\right)q_+^2+\left(\frac{1+\beta}{1-\beta}\right)q_-^2\right]\right\},\qquad(13.101)$$

where

$$q_{\pm} = q_0 \pm q_z \, . \tag{13.102}$$

In the limit $\beta \to 1$, the q_- integral can be performed as in the case of the two-quark system. If we introduce the x variable defined as

$$p_{c0} = xP_0 \, , \tag{13.103}$$

then the parton distribution function becomes

$$\rho(x) = \frac{3M}{\sqrt{2\pi\Omega}} \exp[-(M^2/2\Omega)(3x - 1)^2] \, . \tag{13.104}$$

The above distribution function is normalized as

$$\int_{-\infty}^{\infty} \rho(x)dx = 1 \, . \tag{13.105}$$

This normalization is of course that of the three-quark wave function in the hadron rest frame.

We are now interested in relating the above distribution function to the measurable quantity $\nu W(Q^2, \nu)$ given in Eq. (13.87). Since this function depends on only one variable, and since the scaling variable ω is equal to x, this quantity is usually written as $F_2(x)$ which is called the *structure function*. If we assume that partons interact with the virtual photons like an electron, then the definition given in Eq. (13.87) and the distribution given in Eq. (13.104) lead to the following form of the proton structure function:

$$F_2^{ep} = \langle C^2 \rangle \{x\rho(x)\} = \frac{1}{3}\left(\frac{3Mx}{\sqrt{2\pi\Omega}}\right)\exp\left(-\frac{M^2}{2\Omega}(3x - 1)^2\right) \, . \tag{13.106}$$

$\langle C^2 \rangle$ is the average of the (quark charge)^2 and is $1/3$ for the proton.

The numerical value of the proton mass M is approximately 940 MeV. Thus the only parameter in the above expression is the spring constant Ω. This number can be estimated from the mass spectrum discussed in Sect. 12.9 of Chap. 12. A reasonable estimate for this number is approximately $M^2/2$. This simplifies the expressions of Eq. (13.104) and Eq.(13.106) to

$$\rho(x) = \frac{3}{\sqrt{\pi}}e^{-(3x-1)^2} \, , \tag{13.107}$$

$$F_2^{ep}(x) = (x/\sqrt{\pi})e^{-(3x-1)^2} \, , \tag{13.108}$$

respectively. If we calculate the above form of $F_2^{ep}(x)$ and compare with the measured structure function, the agreement is only qualitative. The calculated structure function is much smaller than the measured curve [43].

Let is look more closely at the proton structure function. We know that in energy-momentum space, the quark distribution which involves one-photon exchange, is able to be measured from inelastic electron-proton scattering [7]. We have said this measured distribution is the proton structure function. Let us now see how close the Gaussian form of Eq. (13.8) is to the experimental world.

In the large-Ω limit, the proton wave function forms a narrow elliptic region where $q_z = q_0$, Thus we are left with only one variable upon which the wave function depends. This one-variable wave function takes the form

$$\phi_\eta(q_z) = \left(\frac{1}{\pi}\right)^{1/4} \exp\left\{-e^{-2\eta}(q_z)^2\right\}. \tag{13.109}$$

We have for the two quark model using Eq. (13.13):

$$p_{az} = \left(\frac{P_z}{2} + \frac{q_z}{2\sqrt{2}}\right) \quad \text{and}$$

$$p_{bz} = \left(\frac{P_z}{2} - \frac{q_z}{2\sqrt{2}}\right). \tag{13.110}$$

Let us now introduce the parameter

$$x = \frac{p_{az}}{P_z}, \tag{13.111}$$

which is the ratio of the quark to hadron momentum. Indeed, the parton distribution in high-energy laboratories is frequently measured using this variable.

It now is possible to write the Gaussian form of Eq. (13.109) using this x variable. The quark distribution then takes the form:

$$\rho(x) = \exp\left[-\Omega\left(x - \frac{1}{2}\right)^2\right], \tag{13.112}$$

where the constant Ω, as said before, is determined from the level separation in the hadron mass spectra [10]. The range of the variable x goes from a minimum value of zero to the maximum value 1 and peaks at $x = 1/2$.

However, to make contact with the proton, we need to use the three quark model defined in Eqs. (13.14) and (13.15). If we suppose the external signal interacts with quark c, this means that the momentum of the quark c depends only on the q variable. This allows us to write the q variable as:

$$q = P - 3p_c. \tag{13.113}$$

Now we define the x variable as

$$x = \frac{p_{cz}}{P_z}. \tag{13.114}$$

This means that the quark distribution takes the form

$$\rho(x) = \left(\frac{1}{\pi\Omega}\right)^{1/2} \exp\left[-\Omega\left(x - \frac{1}{3}\right)^2\right]. \tag{13.115}$$

Fig. 13.10 shows this Gaussian form which is compared with what is observed in experimental laboratories. Inelastic electron-proton scattering [7], as said above is used to measure the experimental data. The reason these two curves are somewhat different comes from the fact that the proton and thus the quarks do not interact as a point particle with the incoming photon.

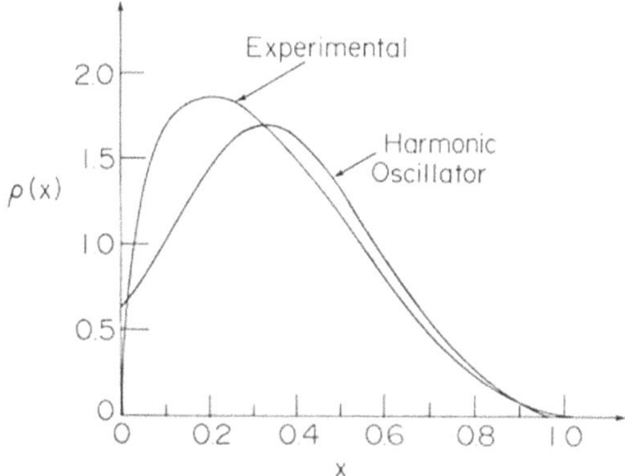

Fig. 13.10: Parton momentum distribution inside a rapidly moving proton in the harmonic oscillator model. The x variable is the ratio of the parton momentum to that of the proton. The experimental distribution is obtained from the measured structure function. For $x \geq 0.3$, the calculated curve is very close to the experimental distribution. For $x \leq 0.3$, the agreement does not appear to be impressive. However, it should be noted that the error bars are still very large in this region, and that strong-interaction dynamics is not yet completely understood for small values of x. This figure is reprinted, with permission, from Paul Hussar's paper [44], which has a detailed discussion.

To explain this difference in the curves, a simple model is to put all the non-point effects into one additional quark. Hence, the proton becomes a bound state of four quarks and it is possible to attribute to this fourth quark all the unexplained effects [43]. A better way is to use the valon model [45, 46]. This model allows all those non-point effects to be screened out. Hussar [44] used this valon model to derive the experimental curve in Fig. 13.10 which is to be compared with the Gaussian form.

Although this graph is not as accurate as we desire, it is nevertheless remarkable that the Gaussian form was calculated from the proton at rest. So is the constant Ω which came from the level spacing in the hadron mass spectra. It should be noted that these two features manifest themselves when the speed of the proton is very close to that of light.

Many other models also deal with providing corrections to the parton distribution. The starting point for the model here, however, is provided by the covariant harmonic oscillator function. The model provided by QCD, particularly perturbative QCD, is like that of QED. QCD does not provide the starting point for the form-factor model, while QED, although it produces the Lamb shift in the hydrogen energy spectrum, does not produce the Rydberg energy levels in the hydrogen atom.

13.7 Jet Phenomenon

We shall study another universally observed high-energy phenomenon. In high-energy collisions, many hadrons are produced in the final state. A peculiar phenomenon is that these hadrons are bunched together to form two or three jets. This is called the jet phenomenon. What makes these hadron momenta to line up?

Let us consider for simplicity a high-energy e^+e^- collision producing many hadrons in the final state. In the center-of-mass system where the total momentum is zero, we can consider a plane containing the initial e^+e^- beams. This plane divides the entire space into two subspaces which we call L and R respectively. The total momentum carried by hadrons in L should be equal in magnitude and opposite in sign to that carried by hadrons in R.

Since we are interested only in the qualitative behavior, we shall consider the case where only two hadrons are produced in R, and restrict ourselves to this region. Energy-momentum conservation laws do not put any restriction on the magnitude of the total momentum in either region. Therefore, there is no restriction on the angle between the hadron momenta from these conservation laws. We shall see, however, that the Lorentz-Dirac deformation makes this angle become smaller as the hadrons move faster, and is thus consistent with the formation of jets in hadron production.

The idea that the Lorentz deformation may be responsible for the origin of jets was made originally by Kitazoe and Hama [47] who used the Bethe-Salpeter wave functions, and followed up by Kim et al. [48] and by Kitazoe and Morii [49]. We use here the harmonic oscillator wave functions to study this problem. For simplicity, we assume that all hadrons are mesons which are regarded as bound states of a quark and antiquark, and use the ground-state harmonic oscillator wave function which contains all the essential features of the Lorentz-Dirac deformation. One important difference between the calculations of the present Section and those in earlier Sections is that we have to consider here a hadron moving not necessarily along the z direction. We therefore need a more general expression for the wave function than those for the hadron moving along the z direction. Indeed, the generalization to an arbitrary direction of the hadron wave function given in Eq. (13.8) or Eq. (13.12) is [Problem 9 in Sect. 13.8]

$$\psi(x, P) = \frac{1}{\pi \Omega} \exp\left\{ -\frac{\Omega}{2} \left[x_\mu x^\mu + 2 \left(\frac{P_\mu x^\mu}{M} \right)^2 \right] \right\}, \tag{13.116}$$

where x is, as defined before, the space-time separation between the quarks, and P is the four-momentum of the hadron.

Since our primary purpose is to extract the kinematical effect of the Lorentz-Dirac deformation, we work with the simplest dynamical model, in which a quark-antiquark pair is initially created for the hadrons in region R and then this pair is combined with a newly created pair to form two hadrons. The simplest possible dynamics for creation of the original pair is a point interaction. Then the dependence of the production amplitude on the angle between the two hadron momenta becomes:

$$F(\theta) = \int \psi(x, P)\psi(x, P') \exp[-i(P_\mu + P'_\mu)x^\mu]\, d^4x \qquad (13.117)$$

where P and P' are the four-momenta of the first and second hadrons respectively.

The above form is not unlike the expression for the electromagnetic form factor discussed in Sect. 13.3. Except for the sign of P' in the exponential factor, the integrand of the above expression is identical to that of Eq. (13.36). Here again the integral can be illustrated by the two overlapping momentum wave functions. The overlap is maximal when the two hadron momenta are lined up as illustrated in Fig. 13.11. After the integration, $F(\theta)$ becomes

$$F(\theta) = -\frac{M^2}{P_\mu P'^\mu} \exp\left(-\frac{M^2(P+P')^2}{4\Omega P_\mu P'^\mu}\right). \qquad (13.118)$$

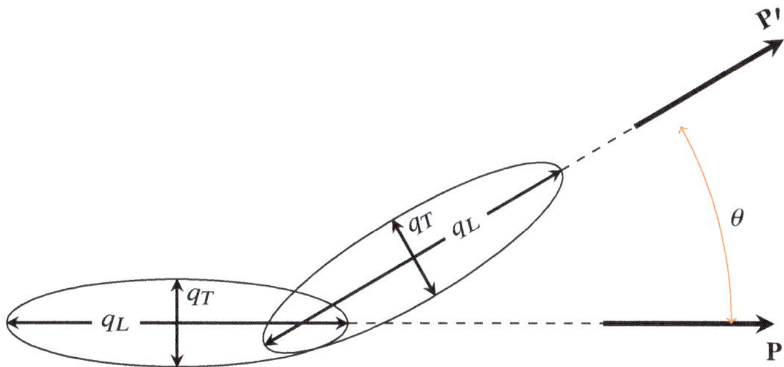

Fig. 13.11: Two overlapping Lorentz-deformed wave functions in momentum space. As the hadron momentum becomes large, the wave function becomes more eccentric and the overlap becomes a rapidly decreasing function of θ.

In order to extract the θ dependence, we consider the symmetric case where $|P| = |P'|$. Then

$$F(\theta) = \frac{1}{1 + 2b^2 \sin^2(\theta/2)} \exp\left[-\frac{M^2}{\Omega}\left(1 + \frac{1}{1 + 2b^2 \sin^2(\theta/2)}\right)\right], \qquad (13.119)$$

where $b = \frac{|\mathbf{P}|}{M}$. The first factor in the expression of Eq.(13.119) is like a Coulomb scattering amplitude and has a sharp peak at $\theta = 0$ as b becomes very large. On the other hand the exponential factor has its minimum value at $\theta = 0$, and increases to a finite asymptotic value as θ becomes large. This exponential factor therefore tempers the sharp Coulomb-like peak at $\theta = 0$.

This means that the scattering amplitude is maximum when the two momenta are parallel and decreases rapidly when the angle between them increases. This explains qualitatively why the hadrons in the final state have to form a jet.

13.8 Exercises and Problems

Exercise 1. We discussed in Sect. 13.3 how to effect a transition from a non-relativistic calculation of the form factor into a relativistic expression. In scattering theory also, such a transition exists. Show how the Born approximation amplitude for a Yukawa potential scattering can be generalized to the Feynman amplitude for the scattering with one-meson exchange.

The Born approximation amplitude for the Yukawa potential:

$$V(r) = (G^2/2m^2)\frac{e^{-\lambda r}}{r} \tag{13.120}$$

is

$$f(\theta) = -\frac{G^2}{Q^2 + \lambda^2}, \tag{13.121}$$

where $Q^2 = (\mathbf{K}_f - \mathbf{K}_i)^2$. If we calculate the lowest-order Feynman amplitude for the scattering of two spinless nucleons through exchange of one meson with mass λ, in the center-of-mass frame, the expression becomes identical to the above formula. In the center-of-mass system, the energy transfer for this scattering process is zero. Therefore Q^2 is also the (four-momentum transfer)^2. This is the way in which Yukawa was originally led to formulate the idea of a meson being exchanged to generate the force between nucleons.

Exercise 2. Show that the expression of Eq. (13.84) for W leads to scaling if partons are free and independent particles, by deriving Eq. (13.85) from Eq. (13.84). Assume for simplicity that the proton is a scalar particle.

For the scalar field $\phi(x)$, we use for the current $J(x)$ the normal-ordered product

$$J(x) =: \phi'(x)\phi(x) : . \tag{13.122}$$

Then

$$[J(x), J(0)] = i\Delta(x)\Delta_1(x)$$

$$+i\Delta(x)\{: \phi^\dagger(x)\phi(0) : + : \phi^\dagger(0)\phi(x) :\}, \tag{13.123}$$

with

$$\Delta(x) = -i(1/2\pi)^3 \int e^{-iq_\mu x^\mu} \epsilon(q_0)\delta(q^2 + M^2)\, d^4q \, ,$$

$$\Delta_1(x) = (1/2\pi)^3 \int x e^{-iq_\mu x^\mu} \delta(q^2 + M^2)\, d^4q \, . \tag{13.124}$$

For space-like Q, it can be shown that

$$\int e^{-iQ_\mu x^\mu} \Delta(x)\Delta_1(x)\, d^4x = 0 \, . \tag{13.125}$$

Thus

$$W(Q^2, v) = \left(\frac{1}{2\pi}\right)^2 \int e^{-i(Q_\mu - q_\mu)x^\mu} \epsilon(q_0)\delta(q^2 + M^2)\, d^4x d^4q$$

$$+ \langle P| \{: \phi^\dagger(x)\phi(0) : + : \phi^\dagger(0)\phi(x) :\} |P\rangle \, . \tag{13.126}$$

The evaluation of this integral is straight-forward, and the result is

$$W(Q^2, v) = \frac{1}{M}\epsilon(Q_0 \pm P_0)\delta[(Q \pm P)^2 + M^2]$$

$$= (-1/2MQ_\mu p^\mu)\delta(\omega \pm 1) \, . \tag{13.127}$$

Thus

$$vW(Q^2, v) = (1/2M^2)\delta(\omega - 1) \, . \tag{13.128}$$

Problem 1. The Lamb shift is the energy difference between the $2s$ and $2p$ state due to the delta function potential at the origin caused by the effect of QED. Since the behavior of the wave function at the origin is different for $2s$ and $2p$ states, the perturbing potential near the origin will give rise to the energy splitting between the two states. We now know that the proton charge is spread out. Assuming that the proton charge distribution is of the form $\exp(-r/r_0)$ with $r_0 = 2.3 \times 10^{-14}$ cm, calculate the contribution to the Lamb shift due to this charge distribution. Would this significantly change the measured value of the Lamb shift which is approximately 1058 megacycles per second?

Problem 2. Show that, in terms of the form factor $G_M(Q^2)$ and $G_E(Q^2)$ given in Eq. (13.28), the electron proton elastic cross section in the laboratory frame (in which the initial proton is at rest) takes the form:

$$\frac{d\sigma}{d\Omega} = \sigma_0 \left(\frac{G_E^2(Q^2) - G_M^2(Q^2)}{1 - Q^2/4M} - \frac{1}{2m^2}G_M^2(Q^2)\tan^2\left(\frac{\theta}{2}\right) \right) , \tag{13.129}$$

with

$$\sigma_0 = \frac{\sigma^2}{4E^2 \sin^2(\theta/2)} \left(\frac{\cos^2(\theta/2)}{1 + (2E/M)\sin^2(\theta/2)} \right), \qquad (13.130)$$

where E is the incident electron energy, and θ is the scattering angle.

Problem 3. Show that the Rutherford scattering amplitude in the non-relativistic Born approximation is $f(\theta) = 2m^2/Q^2$ if the electron is scattered by a point charge. Show next that, when the electron is scattered by a charge whose distribution is $e\rho(\mathbf{x})$, the scattering amplitude becomes

$$f(\theta) = [2m^2/Q^2] \int \rho(\mathbf{x}) e^{i\mathbf{Q}\cdot\mathbf{x}} d^3 x . \qquad (13.131)$$

See the discussion given in Sect. 13.2.

Problem 4. Assuming that the electromagnetic field couples directly and minimally to the quarks, calculate the neutron/proton magnetic moment ratio. See Bég, Lee, and Pais [9].

Problem 5. It is possible to derive formally the expression for the factor given in Eq. (13.34) from the Lagrangian formalism. Carry out this derivation using the minimal electromagnetic coupling to the quark charge. See Lipes [26].

Problem 6. Show that the form factor for the three-quark nucleon takes the form of Eq. (13.42). See Fujimura et al. [29] and Lipes [26].

Problem 7. Show that the use of free Dirac spinors for the quarks leads to a wrong asymptotic behavior for the nucleon form factor [26]. Show however that it is possible to construct spinor models in which the quarks do not, but the nucleon does, satisfy the free Dirac equation [50, 51, 52].

Problem 8. Carry out the calculation from Eq. (13.71) to Eq. (13.83), and repeat the calculation of Exercise 2 for spin-1/2 partons.

Problem 9. Show that the wave function given in Eq. (13.116) is the covariant form for the harmonic oscillator wave function of Eq. (13.8) for the hadron moving along the z direction.

Problem 10. Show that the weak interaction coupling ratio $-g_A/g_V$ is 5/3 in the quark model if there is no interaction between the quarks. Show however that the internal quark motion, such as that of the harmonic oscillator, brings down this number to an acceptable value of approximately 1.16 [53].

Problem 11. This Chapter is concerned with the relativistic bound-state problem. This has been and still is regarded as one of the most difficult problems in physics. For this reason, there has been an attempt to understand this bound-state problem using the S matrix which can be formulated covariantly. In this method, bound-states correspond to poles in the complex energy plane. Show that approximations in the S matrix approach do not guarantee the localization of bound-state wave functions. Show however that it is possible to construct a non-relativistic bound-state wave function from the S matrix. See Kim and Noz [54].

References

1. R. Hofstadter, R.W. McAllister, Electron Scattering from the Proton, Physical Review **98**(1), 217–218 (1955). DOI 10.1103/PhysRev.98.217. URL https://link.aps.org/doi/10.1103/PhysRev.98.217
2. M. Gell-Mann, *The Eightfold Way: A theory of strong interaction symmetry*, vol. TID-12608;CTSL-20 (California Institute of Technology, Synchrotron, Laboratory, Pasadena, CA, 1961). DOI 10.2172/4008239
3. M. Gell-Mann, Symmetries of Baryons and Mesons, Physical Review **125**(3), 1067–1084 (1962). DOI 10.1103/PhysRev.125.1067. URL https://link.aps.org/doi/10.1103/PhysRev.125.1067
4. M. Gell-Mann, A Schematic Model of Baryons and Mesons, Physics Letters **8**(3), 214–215 (1964). DOI 10.1016/S0031-9163(64)92001-3. URL http://linkinghub.elsevier.com/retrieve/pii/S0031916364920013
5. R.P. Feynman, Very High–Energy Collisions of Hadrons, Physical Review Letters **23**(24), 1415–1417 (1969). DOI 10.1103/PhysRevLett.23.1415. URL https://link.aps.org/doi/10.1103/PhysRevLett.23.1415
6. R.P. Feynman, The Behavior of Hadron Collisions at Extreme Energies, in *Proceedings of the 3rd International Conference on High Energy Collisions*, ed. by C. Yang, et al. (Gordon and Breach, New York, NY, USA, 1969), 237–249. (Stony Brook, New York, USA, 5-6-September.)
7. J.D. Bjorken, E.A. Paschos, Inelastic Electron–Proton and γ-Proton Scattering and the Structure of the Nucleon, Physical Review **185**(5), 1975–1982 (1969). DOI 10.1103/PhysRev.185.1975. URL https://link.aps.org/doi/10.1103/PhysRev.185.1975
8. S. Başkal, Y.S. Kim, M.E. Noz, *Physics of the Lorentz Group (Second Edition): Beyond high-energy physics and optics* (IOP Publishing, Bristol, UK, 2021). DOI 10.1088/978-0-7503-3607-9. ISBN 978-0-7503-3607-9. URL https://iopscience.iop.org/book/978-0-7503-3607-9
9. M.A.B. Bég, B.W. Lee, A. Pais, SU(6) and Electromagnetic Interactions, Physical Review Letters **13**(16), 514–517 (1964). DOI 10.1103/PhysRevLett.13.514. URL https://link.aps.org/doi/10.1103/PhysRevLett.13.514
10. R.P. Feynman, M. Kislinger, F. Ravndal, Current Matrix Elements from a Relativistic Quark Model, Physical Review D **3**(11), 2706–2732 (1971). DOI 10.1103/PhysRevD.3.2706. URL https://link.aps.org/doi/10.1103/PhysRevD.3.2706
11. V.N. Gribov, Glauber corrections and the interaction between high-energy hadrons and nuclei, Sov. Phys. JETP **29**(3), 483–108 (1969). (English translation, originally published Zh. Eksp. Teor. Fiz. 56, 892-901 (March, 1969).)
12. B. Ioffe, Space-time picture of photon and neutrino scattering and electroproduction cross section asymptotics, Physics Letters B **30**(2), 123–125 (1969). DOI 10.1016/0370-2693(69)90415-8. URL https://linkinghub.elsevier.com/retrieve/pii/0370269369904158
13. S.D. Drell, T.M. Yan, Partons and their applications at high energies, Annals of Physics **66**(2), 578–623 (1971). DOI 10.1016/0003-4916(71)90071-6. URL http://linkinghub.elsevier.com/retrieve/pii/0003491671900716
14. Y.S. Kim, R. Zaoui, Lorentz Contraction of the Bethe-Salpeter Green's Function, Physical Review D **4**(6), 1764–1769 (1971). DOI 10.1103/PhysRevD.4.1764. URL https://link.aps.org/doi/10.1103/PhysRevD.4.1764
15. G. Preparata, N.S. Craigie, A space-time description of quarks and hadrons, Nuclear Physics B **102**(3), 478–496 (1976). DOI 10.1016/0550-3213(76)90433-8. URL http://linkinghub.elsevier.com/retrieve/pii/0550321376904338
16. P.A.M. Dirac, Can equations of motion be used in high-energy physics?, Physics Today **23**(4), 29–31 (1970). DOI 10.1063/1.3022063. URL http://physicstoday.scitation.org/doi/10.1063/1.3022063
17. H. Yukawa, Structure and Mass Spectrum of Elementary Particles. I. General Considerations, Physical Review **91**(2), 415–416 (1953). DOI 10.1103/PhysRev.91.415.2. URL https://link.aps.org/doi/10.1103/PhysRev.91.415.2

18. R. Hofstadter, Electron Scattering and Nuclear Structure, Reviews of Modern Physics **28**(3), 214–254 (1956). DOI 10.1103/RevModPhys.28.214. URL https://link.aps.org/doi/10.1103/RevModPhys.28.214

19. S.S. Schweber, *An Introduction to Relativistic Quantum Field Theory* (Dover Books on Physics, Dover Publications, Inc, New York, NY, USA, 2005). ISBN 978-0-486-44228-0. (Originally published 1961, Harper & Row, Publishers, New York, NY, USA.)

20. C. Itzykson, J.B. Zuber, *Quantum field theory*, dover edn. Dover books on physics (Dover Publications, Mineola, NY, USA, 2005). ISBN 978-0-486-44568-7. (With new preface and list of errata. Originally published: McGraw-Hill, New York, NY, USA 1980, in series: International series in pure and applied physics.)

21. W.E. Frazer, *Elementary Particles* (Prentice-Hall, Englewood Cliffs, NJ, 1966). ISBN 13: 978-1114629868

22. J.M. Blatt, V.F. Weisskopf, *Theoretical nuclear physics* (Dover Publications, New York, 1991). ISBN 978-0-486-66827-7. (Original publication: John Wiley & Sons, New York, NY, USA 1952; Unabridged: - Springer-Verlag, New York, NY USA 1979.)

23. D.P. Stanley, D. Robson, Relativistic form factors for hadrons with quark-model wave functions, Physical Review D **26**(1), 223–232 (1982). DOI 10.1103/PhysRevD.26.223. URL https://link.aps.org/doi/10.1103/PhysRevD.26.223

24. W.R. Frazer, J.R. Fulco, Effect of a Pion–Pion Scattering Resonance on Nucleon Structure. II, Physical Review **117**(6), 1609–1614 (1960). DOI 10.1103/PhysRev.117.1609. URL https://link.aps.org/doi/10.1103/PhysRev.117.1609

25. A.L. Licht, A. Pagnamenta, Wave Functions and Form Factors for Relativistic Composite Particles. I, Physical Review D **2**(6), 1150–1156 (1970). DOI 10.1103/PhysRevD.2.1150. URL https://link.aps.org/doi/10.1103/PhysRevD.2.1150

26. R.G. Lipes, Electromagnetic Excitations of the Nucleon in a Relativistic Quark Model, Physical Review D **5**(11), 2849–2863 (1972). DOI 10.1103/PhysRevD.5.2849. URL https://link.aps.org/doi/10.1103/PhysRevD.5.2849

27. M. Markov, On dynamically deformable form factors in the theory of elementary particles, Il Nuovo Cimento **3**(S4), 760–772 (1956). DOI 10.1007/BF02746074. URL http://link.springer.com/10.1007/BF02746074

28. V.L. Ginzburg, V.I. Man'ko, Relativistic oscillator models of elementary particles, Nuclear Physics **74**(3), 577–588 (1965). DOI 10.1016/0029-5582(65)90203-8. URL http://linkinghub.elsevier.com/retrieve/pii/0029558265902038

29. K. Fujimura, T. Kobayashi, M. Namiki, Nucleon Electromagnetic Form Factors at High Momentum Transfers in an Extended Particle Model Based on the Quark Model, Progress of Theoretical Physics **43**(1), 73–79 (1970). DOI 10.1143/PTP.43.73. URL https://academic.oup.com/ptp/article-lookup/doi/10.1143/PTP.43.73

30. Y.S. Kim, M.E. Noz, Covariant Harmonic Oscillators and the Quark Model, Physical Review D **8**(10), 3521–3527 (1973). DOI 10.1103/PhysRevD.8.3521. URL https://link.aps.org/doi/10.1103/PhysRevD.8.3521

31. M.J. Ruiz, Orthogonality relation for covariant harmonic-oscillator wave functions, Physical Review D **10**(12), 4306–4307 (1974). DOI 10.1103/PhysRevD.10.4306. URL https://link.aps.org/doi/10.1103/PhysRevD.10.4306

32. Y.S. Kim, M.E. Noz, S.H. Oh, A simple method for illustrating the difference between the homogeneous and inhomogeneous Lorentz groups, American Journal of Physics **47**(10), 892–897 (1979). DOI 10.1119/1.11622. URL http://aapt.scitation.org/doi/10.1119/1.11622

33. F.C. Rotbart, Complete orthogonality relations for the covariant harmonic oscillator, Physical Review D **23**(12), 3078–3080 (1981). DOI 10.1103/PhysRevD.23.3078. URL https://link.aps.org/doi/10.1103/PhysRevD.23.3078

34. Y.S. Kim, M.E. Noz, Lorentz Harmonics, Squeeze Harmonics, and Their Physical Applications, Symmetry **3**(4), 16–36 (2011). DOI 10.3390/sym3010016. URL http://www.mdpi.com/2073-8994/3/1/16/

35. R. Alkofer, A. Höll, M. Kloker, A. Krassnigg, C.D. Roberts, On Nucleon Electromagnetic Form Factors, Few-Body Systems **37**(1-2), 1–31 (2005). DOI 10.1007/s00601-005-0110-6. URL http://link.springer.com/10.1007/s00601-005-0110-6

36. H.H. Matevosyan, A.W. Thomas, G.A. Miller, Study of lattice QCD form factors using the extended Gari–Krümpelmann model, Physical Review C **72**(6), 065,204 – 1 065,204 – 5 (2005). DOI 10.1103/PhysRevC.72.065204. URL https://link.aps.org/doi/10.1103/PhysRevC.72.065204

37. Y.S. Kim, M.E. Noz, Covariant harmonic oscillators and the parton picture, Physical Review D **15**(1), 335–338 (1977). DOI 10.1103/PhysRevD.15.335. URL https://link.aps.org/doi/10.1103/PhysRevD.15.335

38. Y.S. Kim, Observable gauge transformations in the parton picture, Physical Review Letters **63**(4), 348–351 (1989). DOI 10.1103/PhysRevLett.63.348. URL https://link.aps.org/doi/10.1103/PhysRevLett.63.348

39. Y.S. Kim, M.E. Noz, Coupled oscillators, entangled oscillators, and Lorentz–covariant harmonic oscillators, Journal of Optics B: Quantum and Semiclassical Optics **7**(12), S458–S467 (2005). DOI 10.1088/1464-4266/7/12/005. URL http://stacks.iop.org/1464-4266/7/i=12/a=005?key=crossref.8970fab58458b312ba68e38d4ed3c24a

40. G. Altarelli, G. Parisi, Asymptotic freedom in parton language, Nuclear Physics B **126**(2), 298–318 (1977). DOI 10.1016/0550-3213(77)90384-4. URL http://linkinghub.elsevier.com/retrieve/pii/0550321377903844

41. W. Marciano, H. Pagels, Quantum chromodynamics, Physics Reports **36**(3), 137–276 (1978). DOI 10.1016/0370-1573(78)90208-9. URL https://linkinghub.elsevier.com/retrieve/pii/0370157378902089

42. A.J. Buras, Asymptotic freedom in deep inelastic processes in the leading order and beyond, Reviews of Modern Physics **52**(1), 199–276 (1980). DOI 10.1103/RevModPhys.52.199. URL https://link.aps.org/doi/10.1103/RevModPhys.52.199

43. Y.S. Kim, M.E. Noz, Quarks, Partons, and Lorentz-Deformed Hadrons, Progress of Theoretical Physics **60**(3), 801–816 (1978). DOI 10.1143/PTP.60.801. URL https://academic.oup.com/ptp/article-lookup/doi/10.1143/PTP.60.801

44. P.E. Hussar, Valons and harmonic oscillators, Physical Review D **23**(11), 2781–2783 (1981). DOI 10.1103/PhysRevD.23.2781. URL https://link.aps.org/doi/10.1103/PhysRevD.23.2781

45. R.C. Hwa, Evidence for valence-quark clusters in nucleon structure functions, Physical Review D **22**(3), 759–764 (1980). DOI 10.1103/PhysRevD.22.759. URL https://link.aps.org/doi/10.1103/PhysRevD.22.759

46. R.C. Hwa, M.S. Zahir, Parton and valon distributions in the nucleon, Physical Review D **23**(11), 2539–2553 (1981). DOI 10.1103/PhysRevD.23.2539. URL https://link.aps.org/doi/10.1103/PhysRevD.23.2539

47. T. Kitazoe, S. Hama, Jet structure in e^+e^- meson production, Physical Review D **19**(7), 2006–2017 (1979). DOI 10.1103/PhysRevD.19.2006. URL https://link.aps.org/doi/10.1103/PhysRevD.19.2006

48. Y.S. Kim, M.E. Noz, S.H. Oh, Lorentz deformation and the jet phenomenon, Foundations of Physics **9**(11-12), 947–954 (1979). DOI 10.1007/BF00708703. URL http://link.springer.com/10.1007/BF00708703

49. T. Kitazoe, T. Morii, Jet-production mechanism in lepton-hadron reactions, Physical Review D **21**(3), 685–694 (1980). DOI 10.1103/PhysRevD.21.685. URL https://link.aps.org/doi/10.1103/PhysRevD.21.685

50. A. Henriques, B. Kellett, R. Moorhouse, General three-spinor wave functions and the relativistic quark model, Annals of Physics **93**(1-2), 125–151 (1975). DOI 10.1016/0003-4916(75)90209-2. URL https://linkinghub.elsevier.com/retrieve/pii/0003491675902092

51. S. Ishida, K. Takeuchi, S. Tsuruta, M. Watanabe, M. Oda, Electromagnetic interactions of hadrons in the relativistic harmonic-oscillator quark model, Physical Review D **20**(11), 2906–2922 (1979). DOI 10.1103/PhysRevD.20.2906. URL https://link.aps.org/doi/10.1103/PhysRevD.20.2906

52. M.L. Haberman, Weak form factors in an extended hadron model, Physical Review D **29**(7), 1412–1416 (1984). DOI 10.1103/PhysRevD.29.1412. URL https://link.aps.org/doi/10.1103/PhysRevD.29.1412

53. M.J. Ruiz, Covariant harmonic oscillators and the axial-vector coupling constant, Physical Review D **12**(9), 2922–2924 (1975). DOI 10.1103/PhysRevD.12.2922. URL https://link.aps.org/doi/10.1103/PhysRevD.12.2922
54. Y.S. Kim, M.E. Noz, Relativistic harmonic oscillators and hadronic structures in the quantum-mechanics curriculum, American Journal of Physics **46**(5), 484–488 (1978). DOI 10.1119/1.11240. URL http://aapt.scitation.org/doi/10.1119/1.11240

Chapter 14
Decoherence and the Poincaré Sphere

Abstract The two-by-two density matrix constitutes a representation of the Lorentz group. However, the Lorentz group preserves the determinant of the density matrix and therefore it cannot describe the evolution of the decoherence process. Stokes vectors obtained from the density matrix are known to be a four-vector in the Lorentzian regime. Consequently, within this regime the radius of the associated Poincaré sphere has a constant value. It is noted that the $O(3,2)$ group contains two Lorentz subgroups. The change in the determinant in one Lorentz subgroup can be compensated by the other. It is thus possible to describe the decoherence process as well as the radius variation of the Poincaré sphere, as a symmetry transformation in the $O(3,2)$ space. It is shown also that these two coupled Lorentz groups provide a concrete example of Feynman's rest of the universe. The two-by-two matrix representations of the energy-momentum four-vectors in the two Lorentz subgroups of the $O(3,2)$ are also related through the sum of their determinants. These matrix representations are mathematically analogous to those of the density matrices. While density matrices provide a suitable description of decoherence, the energy-momentum matrices deal with the change of mass. It is indicated that, the coherence parameter depends on the environment, thus is not a fundamental quantity, while mass is most fundamental.

Among many original fundamental contributions to mathematics and physics, Henri Poincaré in addition to the formulation of the mathematics of special relativity, also introduced a geometrical representation for different states of light by assigning points on a sphere, which is now called the Poincaré sphere [1]. In quantum mechanics and computing, there is an analogous sphere, the eponymous Bloch sphere, a geometrical visualization of a pure state space of a two-level quantum mechanical system, namely qubits in this context, which are the basic units of quantum information [2].

© The Author(s), under exclusive license to Springer Nature Switzerland AG 2024
S. Başkal et al., *Theory and Applications of the Poincaré Group*, Fundamental Theories
of Physics 217, https://doi.org/10.1007/978-3-031-64376-7_14

In the conventional representation, three of the four Stokes parameters are located on the surface of the Poincaré sphere. However, in the Lorentzian regime there are four Stokes parameters, which behave like four-vectors [1]. The fourth parameter, represents intensity and is conventionally set to unity. If the radius of the Poincaré sphere is allowed to vary, the insertion of the fourth parameter into the system gains an additional physical meaning.

The equivalence between Stokes parameters and the coherency matrix is well known [3]. The off-diagonal elements accommodate the decoherence parameter. The two-by-two Lorentz group leaves the determinant thus the decoherence parameter of this matrix is also left invariant. To give a complete account of the decoherence process it is worthwhile to study the effects of a larger Lorentz group.

In Sect. 14.1, we introduce the density matrix which is equivalent to the coherency matrix, that can foster the variation in the degree of decoherence. We see that this variation is time dependent and the system evolves from an initial pure state to a total decoherence.

In Sect. 14.2, we discuss the interplay between the coherency matrix and the four-component Stokes vector. We embed a *decoherency angle* into the coherency matrix to account for the minimum and maximum values of decoherence between two beams. We formulate the entropy problem in terms of the coherency matrix which also takes the form of a density matrix. In this regard we discuss Feynman's rest of the universe.

In Sect. 14.3, the Poincaré sphere and its relation with the Stokes vectors are given. We introduce two concentric spheres. The difference in their respective radii can continuously be changed from a maximum value to zero, which can then be regulated by an extra parameter borrowed from the Lorentz group $O(3, 2)$, also known as the $(3 + 2)$ de Sitter group.

In Sect. 14.4, we explore how mass variation, which is invariant in the Lorentzian regime, becomes attainable when the Lorentz group is enlarged to $O(3, 2)$. We discuss mathematical similarities with Sect. 14.2, along with contrasts about their physical significances.

14.1 Polarization Optics and Decoherence

In Chap. 15 we shall see that the Jones vector formalism provides physical examples of the two-by-two representation of the Lorentz group. In this context it is possible to identify the attenuation matrix, the phase shifter matrix, and the rotation matrix as the Lorentz group transformation matrices. This formulation is not solely confined within the contents of polarization optics, but can be applied to all two-beam systems with coherent or partially coherent phases, such as interferometers [4]. However, Jones vectors are not suitable for dealing with a loss of coherence between two beams. On the other hand this issue can be managed within the formalism of the density matrix, Stokes parameters, and Poincaré sphere [5].

We can start with a pair of complex numbers a and b and construct the density matrix :

$$\rho = \begin{pmatrix} aa^* & ab^*e^{-\lambda t} \\ a^*be^{-\lambda t} & bb^* \end{pmatrix} . \tag{14.1}$$

The decay in the off-diagonal components of this matrix is essential for the decoherence process. The determinant of this matrix is

$$aa^*bb^*(1 - e^{-2\lambda t}) . \tag{14.2}$$

When $t = 0$, the system is in a pure state, and the determinant is zero, however as t increases, the determinant in Eq. (14.1) increases from zero to aa^*bb^*, while the system gradually becomes decoherent. Therefore, the two-by-two coherency matrix which takes the form of a density matrix incorporates the concept of coherency as well as the degree of decoherence between two beams.

The elements of this matrix has four parameters which can then be redefined to form Stokes parameters. The traditional Poincaré sphere needs only three of those parameters like the Euler angles. It is therefore important to study the significance of the fourth Stokes parameter. Consequently, we can demonstrate that the radius of the Poincaré sphere can change.

The two-component Jones vectors transform like the $SL(2, c)$ spinors, and are recombined to form Stokes parameters, which behave like a four-vector under the Lorentz transformations. Accordingly the two-by-two coherency matrix should transform like the two-by-two form of the space-time four-vector discussed extensively in Chap. 3. Therefore, the Lorentz group should leave the determinant of the coherency matrix invariant. However, the determinant of the coherency matrix changes according to the degree of coherency. The decoherency process is governed by an extra-Lorentzian variable. We shall study its implications in Einstein's energy-momentum relation where the particle mass is a Lorentz-invariant quantity. Thus the symmetry group has to be extended to the $O(3, 2)$ group which was discussed in Chap. 7.

14.2 Coherency Matrix, Stokes Parameters, and Entropy

In this section, we give the relation between the coherency matrix and the four-component Stokes vector. A *decoherency angle* is introduced into the coherency matrix to account for the minimum and maximum values of decoherence between two beams. The coherency matrix which also takes the form of a density matrix is shown to be well suited in the formulation of the entropy problem. In this context we discuss Feynman's rest of the universe.

14.2.1 The Coherency Matrix and Stokes Parameters

We study first the question of coherency between two beams, denoted by Ψ_1 and Ψ_2[4] :

$$\begin{pmatrix} \Psi_1 \\ \Psi_2 \end{pmatrix} = \begin{pmatrix} a \exp\{i(kz - \omega t + \phi_1)\} \\ b \exp\{i(kz - \omega t + \phi_2)\} \end{pmatrix} . \tag{14.3}$$

For this purpose, we consider the coherency matrix, defined as [3, 6]

$$C = \begin{pmatrix} S_{11} & S_{12} \\ S_{21} & S_{22} \end{pmatrix} . \tag{14.4}$$

Starting with the two-component object in the form of Eq. (14.3), which behaves like a spinor in the framework of the two-by-two representation of the Lorentz group [7], we can associate the elements of the coherence matrix as :

$$S_{11} = < \Psi_1^* \Psi_1 > = a^2 , \qquad S_{12} = < \Psi_1^* \Psi_2 > = ab\, e^{-(\lambda t + i\phi)},$$

$$S_{21} = < \Psi_2^* \Psi_1 > = ab\, e^{-(\lambda t - i\phi)} , \qquad S_{22} = < \Psi_2^* \Psi_2 > = b^2 . \tag{14.5}$$

Here, $< \dots >$ denotes

$$< \Psi_i^* \Psi_j > = \frac{1}{T} \int_0^T \Psi_i^*(t + \tau)\Psi_j(\tau)\, d\tau \tag{14.6}$$

the time averages, with T being a sufficiently long time interval. The absolute values of Ψ_1 and Ψ_2 respectively are given by the diagonal elements. If the two beams are not completely coherent, the off-diagonal elements could be smaller than the product of Ψ_1 and Ψ_2. The degree of decoherence in the system is minimum when $t = 0$, and becomes maximum when $t \to \infty$.

In view of Eq. (14.5) the coherency matrix becomes

$$C = \begin{pmatrix} a^2 & ab\, e^{-(\lambda t + i\phi)} \\ ab\, e^{-(\lambda t - i\phi)} & b^2 \end{pmatrix} . \tag{14.7}$$

The symmetry properties of this matrix with the transformation matrices applicable to it, is of particular interest here.

The determinant and the trace of the above coherency matrix is calculated to be

$$\det(C) = (ab)^2 \left(1 - e^{-2\lambda t}\right) ,$$

$$\mathrm{tr}(C) = a^2 + b^2 . \tag{14.8}$$

We note that $e^{-2\lambda t}$ is always smaller than one, so we introduce an angle χ that can be defined as

$$\cos\chi = e^{-\lambda t} \qquad \text{or} \qquad \sin\chi = e^{-\lambda t} . \tag{14.9}$$

This angle will be called the *decoherency angle*. Depending on whether the $\sin\chi$ or the $\cos\chi$ function is chosen, the minimum and the maximum values for the decoherence are reversible at $\chi = 0$ or at $\chi = 90°$. In order to keep the symmetric correlation between the sine and cosine, we shall use them both, with the physical interpretation becoming clear when we discuss the entropy problem in the following section.

The coherency matrix of Eq. (14.7) can then be written as

$$C_c = \begin{pmatrix} a^2 & ab(\cos\chi)e^{-i\phi} \\ ab(\cos\chi)e^{i\phi} & b^2 \end{pmatrix} \tag{14.10}$$

or

$$C_s = \begin{pmatrix} a^2 & ab(\sin\chi)e^{-i\phi} \\ ab(\sin\chi)e^{i\phi} & b^2 \end{pmatrix} . \tag{14.11}$$

Specifically, when the beams correspond to the two orthogonal components of an electric field E_x and E_y, and so the two-component vector of Eq. (14.3) is referred as the Jones vector, the degree of polarization can be defined as [8]:

$$f = \sqrt{1 - \frac{4\,\det(C)}{(\mathrm{tr}(C))^2}} \tag{14.12}$$

with

$$f_c(\chi) = \sqrt{1 - \frac{4(ab)^2 \sin^2\chi}{(a^2 + b^2)^2}} \quad \text{or} \quad f_s(\chi) = \sqrt{1 - \frac{4(ab)^2 \cos^2\chi}{(a^2 + b^2)^2}} . \tag{14.13}$$

For the case f_c, when $\chi = 0$ the degree of polarization becomes one and when $\chi = 90°$, the degree attains the value

$$\frac{a^2 - b^2}{a^2 + b^2} . \tag{14.14}$$

On the other hand, for the case of f_s, the degree of polarization becomes one when $\chi = 90^o$, and takes the value of Eq. (14.14) when $\chi = 0$. We can always assume without loss of generality that a is greater than b. Should they be equal, the minimum degree of polarization is zero.

Let us consider the Lorentz transformation defined for Eq. (14.3), then the coherency matrix transforms as

$$C_i' = W\, C_i\, W^\dagger , \tag{14.15}$$

where W is the two-by-two representation of the $SL(2, c)$ Lorentz group as was discussed in Chap. 3.

From the correspondence between four-vectors and the two-by-two matrix representation we introduce quantities

$$S_0 = \frac{1}{2}(S_{11} + S_{22}), \qquad S_3 = \frac{1}{2}(S_{11} - S_{22}),$$

$$S_1 = \frac{1}{2}(S_{12} + S_{21}), \qquad S_2 = \frac{i}{2}(S_{12} - S_{21}). \tag{14.16}$$

whose components in view of Eq. (14.7) read as:

$$S_0 = \frac{1}{2}(a^2 + b^2), \qquad S_3 = \frac{1}{2}(a^2 - b^2),$$

$$S_1 = ab\,(\cos\phi)e^{-\lambda t}, \qquad S_2 = ab\,(\sin\phi)e^{-\lambda t} \tag{14.17}$$

and are called Stokes vectors or more conventionally Stokes parameters.

Mueller matrices are four-by-four transformation matrices, which represent some optical elements such as polarizers, waveplates, and scatterers that can be applied to Stokes vectors [9, 10, 6]. While Stokes vectors behave as the components of a four-vector, Mueller matrices perform Lorentz transformations on them [1].

14.2.2 Von Neumann Entropy and the Rest of the Universe

The quantum counterpart of the coherency matrix serves as the density matrix with $\rho = C/2$ [11, 12, 13, 14, 15]. The amplitudes can be redefined as

$$a = \sqrt{2q}\,\cos(\delta/2) \qquad \text{and} \qquad b = \sqrt{2q}\,\sin(\delta/2). \tag{14.18}$$

Thus the density matrix $\rho = C/2$ is rewritten

$$\rho_c(\chi) = 2q \begin{pmatrix} \cos^2(\delta/2) & \sin(\delta/2)\cos(\delta/2)e^{-i\phi}(\cos\chi) \\ \sin(\delta/2)\cos(\delta/2)e^{i\phi}(\cos\chi) & \sin^2(\delta/2) \end{pmatrix}. \tag{14.19}$$

This becomes

$$\rho_c(\chi) = \frac{1}{2} \begin{pmatrix} 1 & e^{-i\phi}\cos\chi \\ e^{i\phi}\cos\chi & 1 \end{pmatrix} \tag{14.20}$$

when the angle δ is conveniently chosen to be 90°. In addition the normalization condition $Tr(\rho) = 1$ is maintained. Following a similar procedure for ρ_s, these can now be diagonalized to take the form

$$\rho_c(\chi) = \frac{1}{2} \begin{pmatrix} 1 + \cos\chi & 0 \\ 0 & 1 - \cos\chi \end{pmatrix} \tag{14.21}$$

and

$$\rho_s(\chi) = \frac{1}{2} \begin{pmatrix} 1 + \sin\chi & 0 \\ 0 & 1 - \sin\chi \end{pmatrix}. \tag{14.22}$$

Von Newman defined the entropy by [16]

$$S = -Tr(\rho \ln \rho). \tag{14.23}$$

For the density matrix ρ_c, it becomes [17, 18, 19]:

$$S_c(\chi) = -\frac{1 + \cos \chi}{2} \ln \left(\frac{1 + \cos \chi}{2} \right) - \frac{1 - \cos \chi}{2} \ln \left(\frac{1 - \cos \chi}{2} \right) \tag{14.24}$$

which can be simplified to

$$S_c(\chi) = -[\cos^2(\chi/2)] \ln[\cos^2(\chi/2)] - [\sin^2(\chi/2)] \ln[\sin^2(\chi/2)]. \tag{14.25}$$

The entropy S_c of this space is a monotonically increasing function of χ. We also see that if the entropy is zero the system is completely coherent with $\chi = 0°$. When the system is totally incoherent with $\chi = 90°$ the entropy takes the maximum value of $\ln(2)$.

In a similar manner, the entropy of the second space can be expressed as

$$S_s(\chi) = -\frac{1 + \sin \chi}{2} \ln \left(\frac{1 + \sin \chi}{2} \right) - \frac{1 - \sin \chi}{2} \ln \left(\frac{1 - \sin \chi}{2} \right). \tag{14.26}$$

The entropy S_s of this second space, unlike the first one, is a decreasing function of χ. Feynman's rest of the universe [20, 21, 22] is discussed within the context of coupled harmonic oscillators. The first of those belongs to the part of the universe in which we make measurements, while the second is in the rest of the universe. The increase of entropy in the first space can be interpreted as the loss of energy in the second. There is a decoherence and a recoherence process that is taking place in these coupled universes. Yet, the sum of the entropies is not a conserved quantity of the total system. On the other hand we observe that the sum of the determinants of the density matrices is

$$\det \rho_c + \det \rho_s = 1 \tag{14.27}$$

which is independent of the decoherency angle χ and thus is a conserved quantity.

14.3 Poincaré Sphere

Originally the Poincaré sphere proposed by Henri Poincaré was a three dimensional object for representing the polarization state of light. Three of the Stokes vector S_1, S_2, S_3 correspond to Cartesian coordinates. The radius of the Poincaré sphere is given in terms of these three Stokes vector as:

$$r = \sqrt{S_1^2 + S_2^2 + S_3^2}. \tag{14.28}$$

While S_0 is known to represent light intensity, currently it is considered to be the time-like component of a four-vector. We shall use this property of the Stokes vector to devise a second concentric sphere, the radius of which is taken to be $s = S_0$.

Within the framework of the Lorentz group the relation between these two radii, namely $s^2 - r^2$, is an invariant. This suggests that the determinant of the density matrix is likewise constrained and cannot deal with the decoherence process by the decaying of the off-diagonal components. As a result, one would question if this process can be adequately reformalized by incorporating the symmetry features of a larger Lorentz group, particularly the $O(3, 2)$ group. In this section we shall discuss and answer these issues.

14.3.1 Two Concentric Poincaré Spheres

In view of Eq. (14.17) and Eq. (14.28) the radius of the sphere is rewritten as

$$r = \frac{1}{2}\sqrt{(a^2 - b^2)^2 + 4(ab)^2 e^{-2\lambda t}} \qquad (14.29)$$

which describes the conventional Poincaré sphere. However, the four-vector in Eq. (14.17), apart from the space-like components, has a time-like component which provides another radius

$$s = \frac{1}{2}\left(a^2 + b^2\right) \qquad (14.30)$$

defining an outer sphere concentric to the inner sphere. The quantity $s^2 - r^2$ is Lorentz-invariant and is equal to the value of the determinant in Eq. (14.7). The t parameter cannot be changed in the Lorentzian regime. However, this restriction can be released, when a larger Lorentz group is introduced. In that case, when $t = 0$, the inner radius is equal to the outer radius, and becomes

$$S_3 = \frac{1}{2}\left(a^2 - b^2\right) \qquad (14.31)$$

when t attains very large values.

Let us introduce a spherical coordinate system with

$$r_z = (a^2 - b^2)/2 = r(\cos\theta),$$

$$r_x = ab(\cos\phi)e^{-\lambda t} = r(\sin\theta)\cos\phi,$$

$$r_y = ab(\sin\phi)e^{-\lambda t} = r(\sin\theta)\sin\phi. \qquad (14.32)$$

Since, the Lorentz symmetry allows rotations in this three-dimensional scheme, with an appropriate rotation the four-vector can be brought to

$$(s, r, 0, 0). \qquad (14.33)$$

The rotations do not change the radii of the outer and inner spheres, thus s and r remain invariant. On the other hand, if the above four-vector is boosted with

$$\begin{pmatrix} \cosh\eta & -\sinh\eta & 0 & 0 \\ -\sinh\eta & \cosh\eta & 0 & 0 \\ 0 & 0 & 1 & 0 \\ 0 & 0 & 0 & 1 \end{pmatrix} \tag{14.34}$$

it becomes

$$\begin{pmatrix} s(\cosh\eta) - r(\sinh\eta) \\ r(\cosh\eta) - s(\sinh\eta) \\ 0 \\ 0 \end{pmatrix}. \tag{14.35}$$

This transformation changes the outer and inner radii, but keeps $(s^2 - r^2)$ invariant. One can choose the value of η such that

$$r(\cosh\eta) - s(\sinh\eta) = 0, \tag{14.36}$$

which leads to $\tanh\eta = r/s$. Then, using Eqs. (14.29) and (14.30), the four-vector of Eq. (14.35) becomes

$$\begin{pmatrix} \sqrt{s^2 - r^2} \\ 0 \\ 0 \\ 0 \end{pmatrix} = \begin{pmatrix} ab\sqrt{1 - e^{-2\lambda t}} \\ 0 \\ 0 \\ 0 \end{pmatrix}. \tag{14.37}$$

Lorentz transformations bring the Poincaré sphere to a one-number system. How will it be possible to change the value of $(s^2 - r^2)$ in the above expression by changing the time variable t, while it is clearly an invariant quantity in the $O(3, 1)$ regime?

Let us consider the rotation matrix between the two time-like coordinates (t_1, t_2, z, x, y) of $O(3, 2)$

$$\begin{pmatrix} \cos\chi & -\sin\chi & 0 & 0 & 0 \\ \sin\chi & \cos\chi & 0 & 0 & 0 \\ 0 & 0 & 1 & 0 & 0 \\ 0 & 0 & 0 & 1 & 0 \\ 0 & 0 & 0 & 0 & 1 \end{pmatrix} \tag{14.38}$$

and a five-vector of the form $(v, 0, 0, 0, 0)$. Then the rotation matrix transforms this vector to $(v\cos\chi, v\sin\chi, 0, 0, 0)$. The invariant quantity in this larger regime is v^2. The $O(3, 2)$ group has two Lorentzian subgroups, where their respective four-vectors are expressed as

$$\begin{pmatrix} v\cos\chi \\ 0 \\ 0 \\ 0 \end{pmatrix} \quad \text{and} \quad \begin{pmatrix} v\sin\chi \\ 0 \\ 0 \\ 0 \end{pmatrix}. \tag{14.39}$$

We compare the four-vector in Eq. (14.37) with, for instance, the second vector above and identify their non-zero components $(ab\sqrt{1 - e^{-2\lambda t}})$ and $(v\sin\chi)$. Then we have

$$\left(s^2 - r^2\right) = v^2 \sin^2\chi \tag{14.40}$$

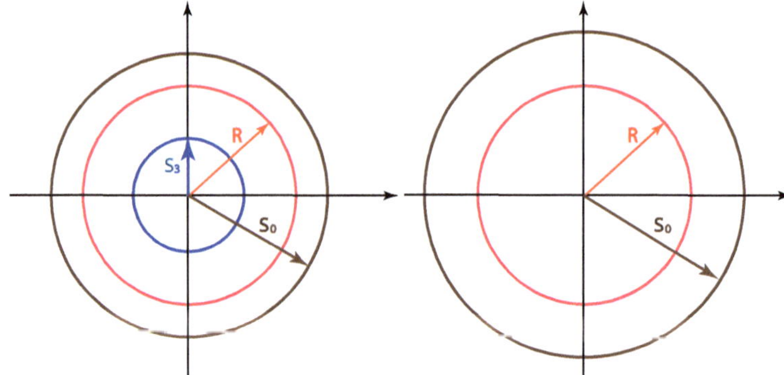

Fig. 14.1: Variable radius of the Poincaré sphere. From Eq. (14.30) and Eq. (14.42), the variable radius R takes its maximum value S_0, when $\chi = 0°$. It becomes minimum when the decoherency angle reaches $90°$. Its minimum value is S_3 as is illustrated in the left figure. The degree of polarization is maximum when $R = S_0$, and is minimum when $R = S_3$. According to Eq. (14.31), S_3 becomes 0 when $a = b$, thus the minimum value of R becomes zero, as is indicated in the right figure. Its maximum value is still S_0 [23].

which allows variable radii for the Poincaré spheres as illustrated in Fig. 14.1.

Equivalently, from Eq. (14.37) we also have

$$(ab)^2(1 - e^{-2\lambda t}) = v^2 \sin^2 \chi \tag{14.41}$$

rendering

$$\cos \chi = e^{-\lambda t} \tag{14.42}$$

by redefining $ab \doteq v$. It is also possible to make the identification of components with the first vector in Eq. (14.39), which would yield

$$\sin \chi = e^{-\lambda t} . \tag{14.43}$$

Besides what has already been said about the concept of Feynman's rest of the universe, the reasoning above provides a further explanation as to why both of the sine and cosine of χ in Eq. (14.9) is introduced in the very beginning of the discussion about decoherence. Thus, the t parameter that cannot be changed in the Lorentzian regime becomes a continuous variable quantity within the framework of a larger Lorentz group, namely $O(3, 2)$.

14.3.2 Dimensionally Reduced Sphere

In order to visualize how the decoherency angle is also responsible for the variation of the radius of Poincaré sphere we go to a lower dimension. For this purpose the Stokes vector is simplified by letting $a = b = 1$. Then we have

$$S_0 = 1, \qquad S_3 = 0, \qquad S_1 = (\cos \chi) \cos \phi, \qquad S_2 = (\cos \chi) \sin \phi. \qquad (14.44)$$

Now, since $S_3 = 0$, the Poincaré sphere is dimensionally reduced and becomes a circle with radius

$$r_2 = \sqrt{S_1^2 + S_2^2} = \cos \chi. \qquad (14.45)$$

The determinant $\sin^2 \chi$ of C in Eq. (14.8) would remain invariant in the Lorentzian regime, as is the invariance of the Stokes vector of Eq. (14.44)

$$S_0^2 - S_1^2 - S_2^2 = \sin^2 \chi. \qquad (14.46)$$

As discussed above it is the introduction of a larger Lorentz group that allows the radius r_2 to vary.

14.3.3 The Bloch Ball and Mixed States

To conclude this section it is worthwhile to mention how the Poincaré sphere and the properties we have developed here can be related to other areas of physics. In quantum mechanics and computing, an analogous sphere exists, known as the Bloch sphere, a geometrical visualization of a pure state space of a two-level quantum mechanical system. In the context of quantum computing states are replaced by qubits, which are the basic units of quantum information [2]. Quantum states of the Bloch sphere, to be precise of the Bloch ball, are not necessarily pure. As a matter of fact, pure states eventually interact with the environment leading to the emergence of mixed states. From a mathematical point of view, the unitarity condition of the states can be relaxed to generate mixed states. In general, the density matrix is $\rho = \frac{1}{2}(\mathbf{I} + \mathbf{r} \cdot \boldsymbol{\sigma})$, where $|\mathbf{r}|$ is the radius of the Bloch sphere. For pure states $Tr(\rho^2) = 1$, while $Tr(\rho^2) < 1$ for mixed states [24]. Since $Tr(\rho^2) = \frac{1}{2}(1 + \mathbf{r} \cdot \mathbf{r})$, then $|\mathbf{r}| < 1$. Therefore, mixed quantum states are located interior of the unit sphere, meaning that they are contained within the Bloch sphere. In order to realize a variable radius we described a procedure in Sect. 14.3.1 providing a possible strategy to produce mixed states from pure states.

To stretch the idea of mixed states a little further, the Bures metric [25, 26]

$$ds_B^2 = d\chi^2 + \sin^2 \chi (d\theta^2 + \sin^2 d\varphi^2) \qquad (14.47)$$

can be assigned to the ball, describing distance for two infinitesimally close density matrix operators ρ and $\rho + d\rho$. Now let us consider the spatial part of the Robertson-Walker metric [27]:

$$ds^2_{RW/s} = \frac{dr^2}{1 - k\,r^2} + r^2(d\theta^2 + \sin^2\theta\,d\varphi^2)\,. \tag{14.48}$$

Introducing another radial coordinate χ defined by $d\chi = dr/\sqrt{1 - k\,r^2}$ and taking $k = 1$ for positive curvature, this yields

$$ds^2_{RW/s} = d\chi^2 + \sin^2\chi(d\theta^2 + \sin^2 d\varphi^2)\,, \tag{14.49}$$

which is precisely the same as Eq. (14.47). A word of caution is in order here, regarding the current cosmological model that the universe is considered to have a flat geometry i.e., $k = 0$ on large scales, although the possibility of a closed universe with positive curvature is not completely ruled out.

Here we see another example of how seemingly unrelated concepts of physics such as the Stokes vectors of classical optics, the quantum bits of quantum information and a cosmological model describing a closed universe can be connected through the mathematics we have been discussing in this book.

14.4 Mass Variation within O(3, 2) Symmetry

We have seen in Chap. 7 that the group $O(3,2)$ is the Lorentz group applicable to a five-dimensional space which contains three space dimensions and two time dimensions. In this space the five-component energy-momentum vector is

$$(E_1, E_2, p, 0, 0) \tag{14.50}$$

when the momentum is chosen to be only in the z-direction. In general, momentum has three components p_z, p_x, and p_y. However, the $O(3)$ rotations in the three dimensional subspace will transform p_i so that the five-vector takes the above form.

In the two Lorentzian subgroups of the $O(3,2)$ group, the corresponding four-vectors are

$$(E_1, p, 0, 0) \quad \text{and} \quad (E_2, p, 0, 0)\,. \tag{14.51}$$

In this format these two energy variables take the form

$$E_1 = \sqrt{p^2 + m_1^2} \quad \text{and} \quad E_2 = \sqrt{p^2 + m_2^2}\,. \tag{14.52}$$

We can then define the two mass variables: [28, 5]

$$m_1 = m\,\cos\chi \quad \text{and} \quad m_2 = m\,\sin\chi\,. \tag{14.53}$$

This results in

$$E_1^2 + E_2^2 = m^2 + 2p^2\,. \tag{14.54}$$

This remains constant for a fixed value of p^2. This leads to the fact the there is, in the E_1 and E_2 two-dimensional space, a rotational symmetry. Indeed, we have thus defined the $O(3, 2)$ or $(3 + 2)$ de Sitter symmetry [5, 29]. Now we can write the two-by-two matrix representations as:

$$P_1 = \begin{pmatrix} E_1 + p & 0 \\ 0 & E_1 - p \end{pmatrix} \quad \text{and} \quad P_2 = \begin{pmatrix} E_2 + p & 0 \\ 0 & E_2 - p \end{pmatrix}, \tag{14.55}$$

with determinants equal to $m^2 \cos^2 \chi$ and $m^2 \sin^2 \chi$, respectively. If we choose $p = E \cos \chi$ for P_1 and $p = E \sin \chi$ for P_2, then

$$P_c = E \begin{pmatrix} 1 + \cos \chi & 0 \\ 0 & 1 - \cos \chi \end{pmatrix} \quad \text{and} \quad P_s = E \begin{pmatrix} 1 + \sin \chi & 0 \\ 0 & 1 - \sin \chi \end{pmatrix}. \tag{14.56}$$

Here, we note that, these are like the matrices ρ_c and ρ_s of Eq. (14.21) and Eq. (14.22), respectively.

Let us recall Eq. (3.18) of Chap. 3. Then for the P_1 matrix, when $\chi = 0$, $E_1 = p$, the particle becomes massless. On the other hand in the second Lorentzian space with P_2 we see that if $\chi = 0$, the matrix represents the four-momentum of a particle at rest with a mass value equated to E_2. This relation is illustrated in Fig. 14.2.

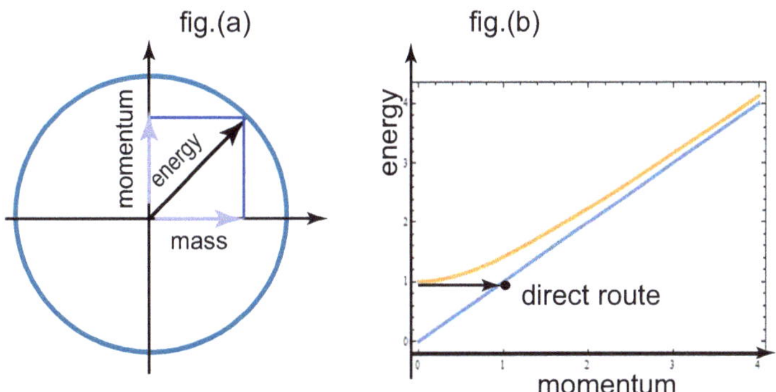

Fig. 14.2: The relation between momentum and mass for a fixed value of energy is shown in fig.(a). As the angle χ increases, the momentum also increases while mass decreases. The particle loses all of its mass at $\chi = 90°$ and thus $E = p$. fig.(b) shows how the $O(3, 2)$ group allows the transition from a massive energy-momentum plot to that of massless, not being admissible in the context of the Lorentz group. The $O(3, 2)$ group allows this transition in a continuous way [30].

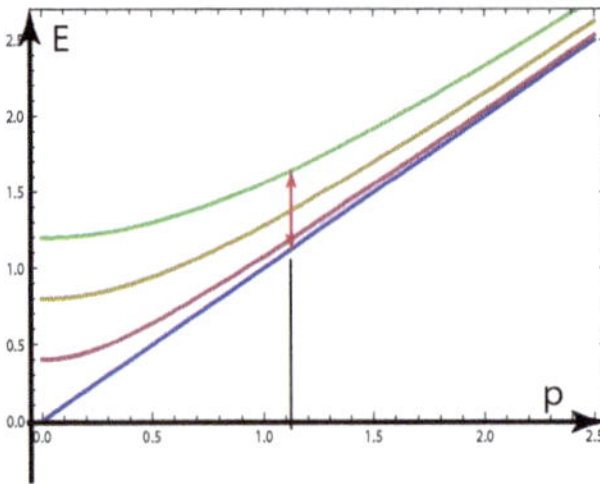

Fig. 14.3: This figure illustrates the energy-momentum hyperbolas for various values of mass and the transition between them while the magnitude of the momentum is kept constant. Moving from one hyperbola to another is not possible within the framework of the Lorentz group; however it becomes possible when the $O(3,2)$ or $(3+2)$ de Sitter symmetry is employed [30].

Table 14.1: Special relativity and polarization optics utilize identical mathematical underpinnings. The function of each matrix is well-defined both in optics and relativity. The two-by-two matrix representing the Stokes vector and the energy-momentum four-vector both have a determinant which remains invariant under Lorentz transformations. We note that the decoherence parameter, which is least fundamental in optics, corresponds to the mass, which is most fundamental in particle physics [31].

Polarization Optics	Transformation Matrix	Particle Symmetry
Phase shift ϕ	$\begin{pmatrix} e^{-i\phi/2} & 0 \\ 0 & e^{i\phi/2} \end{pmatrix}$	Rotation around z
Rotation around z	$\begin{pmatrix} \cos(\theta/2) & -\sin(\theta/2) \\ \sin(\theta/2) & \cos(\theta/2) \end{pmatrix}$	Rotation around y
Squeeze along x and y	$\begin{pmatrix} e^{\mu/2} & 0 \\ 0 & e^{-\mu/2} \end{pmatrix}$	Boost along z
$\sin^2 \chi$	Determinant	$(\text{mass})^2$

It is discussed in Chap. 10 that, within the framework of little groups, a massive particle at rest can become a massless particle having the same energy. The transition from the massive particle to the massless particle was shown to be a singular transformation, since mass remains invariant under Lorentz transformations. Although the Lorentz group cannot give an account of a variable mass, the $O(3,2)$ group has an extra variable χ which we can exploit to change the mass value of this four-vector.

Furthermore, the symmetry of $O(3,2)$ allows us to jump from one mass hyperbola to another. This additional feature is illustrated in Fig. 14.3.

Eq. (14.21), Eq. (14.22) and Eq. (14.56) are based on identical mathematical formulations. The decoherency angle χ in Eq. (14.21) and Eq. (14.22) depends on the environment and so in that sense, is not a fundamental quantity. On the other hand, in the case of Eq. (14.56) this angle is responsible for the value of the mass, which is most fundamental. We encapsulate this aspect of the decoherency angle in Table 14.1. In Chap. 15 we define the explicit form for the transformation matrices in the second column of the Table 14.1.

References

1. D. Han, Y.S. Kim, M.E. Noz, Stokes parameters as a Minkowskian four-vector, Physical Review E **56**(5), 6065–6076 (1997). DOI 10.1103/PhysRevE.56.6065. URL https://link.aps.org/doi/10.1103/PhysRevE.56.6065
2. M.A. Nielsen, I.L. Chuang, *Quantum computation and quantum information*, 10th edn. (Cambridge University Press, Cambridge, UK; New York, NY, USA, 2010). ISBN 9781107002173. (Originally published 2000.)
3. M. Born, E. Wolf, *Principles of optics: electromagnetic theory of propagation, interference and diffraction of light*, 7th edn. (Cambridge University Press, Cambridge. UK; New York, NY, USA, 1999). ISBN 978-0-521-64222-4978-0-521-63921-7. (Originally published 1980.)
4. D. Han, Y.S. Kim, M.E. Noz, Interferometers and decoherence matrices, Physical Review E **61**(5), 5907–5913 (2000). DOI 10.1103/PhysRevE.61.5907. URL https://link.aps.org/doi/10.1103/PhysRevE.61.5907
5. S. Başkal, Y.S. Kim, de Sitter group as a symmetry for optical decoherence, Journal of Physics A: Mathematical and General **39**(24), 7775–7788 (2006). DOI 10.1088/0305-4470/39/24/014. URL http://stacks.iop.org/0305-4470/39/i=24/a=014?key=crossref.2381d47174712f9115293516c91a3b1e
6. C. Brosseau, *Fundamentals of polarized light: a statistical optics approach* (John Wiley and Sons, New York, NY, USA, 1998). ISBN 978-0-471-14302-4
7. D. Han, Y.S. Kim, M.E. Noz, Jones–matrix formalism as a representation of the Lorentz group, Journal of the Optical Society of America A **14**(9), 2290–2298 (1997). DOI 10.1364/JOSAA.14.002290. URL https://www.osapublishing.org/abstract.cfm?URI=josaa-14-9-2290
8. B.E.A. Saleh, M.C. Teich, *Fundamentals of photonics*, 2nd edn. Wiley series in pure and applied optics (Wiley Interscience; A John Wiley & Sons, Inc., Publication, New NY, USA, Hoboken, NJ, USA, 2007). ISBN 978-0-471-35832-9. (Originaly published 1991.)
9. H. Mueller, Memorandum on the polarization optics of the photo elastic shutter. OSRD project OEMsr-576 2, Office of Scientific Research and Development of the United States (1943)
10. R.M.A.G. Azzam, N.M. Bashara, *Ellipsometry and polarized light*, 4th edn. North-Holland personal library (Elsevier, Amsterdam, NL, 1999). ISBN 978-0-444-87016-2. (Originally published 1977; OCLC: 247501433.)
11. D.L. Falkoff, J.E. MacDonald, On the Stokes Parameters for Polarized Radiation, Journal of the Optical Society of America **41**(11), 861–862 (1951). DOI 10.1364/JOSA.41.000861. URL https://opg.optica.org/abstract.cfm?URI=josa-41-11-861
12. U. Fano, A Stokes-Parameter Technique for the Treatment of Polarization in Quantum Mechanics, Physical Review **93**(1), 121–123 (1954). DOI 10.1103/PhysRev.93.121. URL https://link.aps.org/doi/10.1103/PhysRev.93.121
13. E. Wolf, Optics in terms of observable quantities, Il Nuovo Cimento **12**(6), 884–888 (1954). DOI 10.1007/BF02781855. URL http://link.springer.com/10.1007/BF02781855

14. G.P. Parent, P. Roman, On the matrix formulation of the theory of partial polarization in terms of observables, Nuovo Cimento **15**, 370–388 (1960). DOI https://doi.org/10.1007/BF02902573. URL https://link.springer.com/article/10.1007/BF02902573

15. K. Blum, *Density matrix theory and applications*, 3rd edn. No. 64 in Springer series on atomic, optical and plasma physics (Springer, Heidelberg, Germany; New York, NY, USA, 2012). ISBN 978-3-642-20560-6. (Originally published 1981.)

16. J. Von Neumann, R.T. Beyer, N.A. Wheeler, *Mathematical foundations of quantum mechanics*, new edn. (Princeton University Press, Princeton, NJ, USA, 2018). ISBN 978-0-691-17856-1978-0-691-17857-8. (Originally published: Princeton University Press, 1996, 1955, and in German by Springer, Berlin, Germany 1932; OCLC: on1004934776.)

17. Y.S. Kim, E.P. Wigner, Entropy and Lorentz transformations, Physics Letters A **147**(7), 343–347 (1990). DOI 10.1016/0375-9601(90)90550-8. URL http://linkinghub.elsevier.com/retrieve/pii/0375960190905508

18. J. Eisert, M. Cramer, M.B. Plenio, *Colloquium* : Area laws for the entanglement entropy, Reviews of Modern Physics **82**(1), 277–306 (2010). DOI 10.1103/RevModPhys.82.277. URL https://link.aps.org/doi/10.1103/RevModPhys.82.277

19. Y.S. Kim, M.E. Noz, Entropy and Temperature from Entangled Space and Time, Physical Science International Journal **4**(7), 1015–1039 (2014). DOI 10.9734/PSIJ/2014/10336. URL http://www.sciencedomain.org/abstract.php?iid=533&id=33&aid=4834

20. R.P. Feynman, *Statistical Mechanics: a set of lectures*. Advanced book classics (Westview Press, Boulder, CO, USA, 1998). ISBN 978-0-201-36076-9. (Originally published 1972 by Benjamin Cummings, Reading, MA, USA; OCLC: ocm60679997.)

21. D. Han, Y.S. Kim, M.E. Noz, Illustrative example of Feynman's rest of the universe, American Journal of Physics **67**(1), 61–66 (1999). DOI 10.1119/1.19192. URL http://aapt.scitation.org/doi/10.1119/1.19192

22. Y.S. Kim, M.E. Noz, Dirac Matrices and Feynman's Rest of the Universe, Symmetry **4**(4), 626–643 (2012). DOI 10.3390/sym4040626. URL http://www.mdpi.com/2073-8994/4/4/626/

23. Y.S. Kim, M.E. Noz, Symmetries Shared by the Poincaré Group and the Poincaré Sphere, Symmetry **5**(3), 233–252 (2013). DOI 10.3390/sym5030233. URL http://www.mdpi.com/2073-8994/5/3/233/

24. Y.S. Kim, M.E. Noz, *Phase space picture of quantum mechanics: group theoretical approach*. No. 40 in Lecture notes in physics series (World Scientific Publishing Co., Singapore; Hackensack, NJ, USA, 1991). ISBN 978-981-02-0360-3,978-981-02-0361-0. URL https://doi.org/10.1142/1197

25. M. Hübner, Explicit computation of the Bures distance for density matrices, Physics Letters A **163**(4), 239–242 (1992). DOI 10.1016/0375-9601(92)91004-B. URL https://linkinghub.elsevier.com/retrieve/pii/037596019291004B

26. P.M. Alsing, C. Cafaro, D. Felice, O. Luongo, Geometric Aspects of Mixed Quantum States Inside the Bloch Sphere, Quantum Reports **6**(1), 90–109 (2024). DOI 10.3390/quantum6010007. URL https://www.mdpi.com/2624-960X/6/1/7

27. S.M. Carroll, *Spacetime and geometry: an introduction to general relativity* (Addison Wesley, San Francisco, 2004). ISBN 9780805387322. OCLC: ocm53245141

28. A. Zee, *Einstein gravity in a nutshell*. In a nutshell (Princeton University Press, Princeton, NJ, USA, 2013). ISBN 978-0-691-14558-7

29. S. Başkal, Y.S. Kim, Lorentz group in ray and polarization optics, in *Mathematical Optics: Classical, Quantum and Computational Methods*, ed. by V. Lakshminarayanan, M. L. Calvo, T. Alieva (Taylor and Francis, Boca Raton, FL, USA, 2013), 303–349. ISBN 978-1-4398-6961-1. URL https://www.taylorfrancis.com/books/9781439869611

30. S. Başkal, Y. Kim, M. Noz, *Mathematical Devices for Optical Sciences*. (IOP Publishing, Bristol, UK, 2019). ISBN 978-0-7503-1612-5. URL https://dx.doi.org/10.1088/2053-2563/aafe78. (OCLC: 1034620988.)

31. S. Başkal, Y.S. Kim, M.E. Noz, *Physics of the Lorentz Group (Second Edition): Beyond high-energy physics and optics* (IOP Publishing, Bristol, UK, 2021). DOI 10.1088/978-0-7503-3607-9. ISBN 978-0-7503-3607-9. URL https://iopscience.iop.org/book/978-0-7503-3607-9

Chapter 15
Lorentz Group in Classical Optics

Abstract In this chapter, Jones vector formalism is considered for studying the polarization properties of the two-component electric field vector. A general form for the attenuator is introduced, and thereof attenuations, phase-shifts, and the rotation of the polarization axes are reformulated. Their combined effects are investigated and it is shown that they amount to a two-by-two representation of the six parameter $SL(2, c)$ group. In connection with these combined effects, the Wigner rotation is also reviewed. Ray transfer matrices are introduced and their equi-diagonalization process is explained. It is shown how the results of this process facilitate computations when a large number of cycles is required to account for the behaviour of a beam subjected to the effects of a series of optical elements. It is also noted that the Wigner and Bargmann decompositions can be explained in terms of the decomposition properties of the $ABCD$ matrix. This matrix as a general element of the $Sp(2)$ group can be composed by assembling some specific combinations of lens and translation matrices. Concrete physical examples of $ABCD$ matrices governing the propagation of rays in a camera, in a laser cavity, in multi-layers and in an optically active medium are presented. It is observed that they serve as analog computers for the essential features of Wigner's little groups dictating the internal space–time symmetries of relativistic particles.

Throughout this book we have seen that the Poincaré group and its subgroups, namely the Lorentz group and little groups have significant importance in many branches of physics. The coherent and squeezed states of light [1, 2] are the representations of these subgroups in quantum optics [3, 4]. We have considerably investigated several important properties of quantum light in earlier chapters. A crude viewpoint may suggest that classical optics has no or very little bearing with the quantum. On the contrary, we are currently witnessing more and more examples of quantum phenomena in the macroscopic world. One example is the Josephson

S. Başkal et al., *Theory and Applications of the Poincaré Group*, Fundamental Theories of Physics 217, https://doi.org/10.1007/978-3-031-64376-7_15

junction having applications in quantum computing [5]. Linear quantum computing is another case in point, where it uses classical and quantum optical elements. In this realm, qubits can be encoded to the two modes of the photon states, while operations on qubits are implemented using linear classical optical elements, such as beam splitters, phase shifters, mirrors, and wave guides, in circuits for computations [6, 7].

Classical ray optics is a fairly old subject, yet we should not assume that novel physics can progress without observations using optical instruments. Historically, the science of optics was developed without using matrices or Fourier transformations [8]. On the other hand from a mathematical perspective, physics pertinent to modern classical optics can now be represented by two-by-two matrices [9, 10, 11, 4]. We observe that optical activities, as well as optical components like lenses, polarizers, interferometers, lasers, and multi-layers may all be described in terms of Wigner's little group as well as the Lorentz group.

It is already a well established knowledge that optical activities can perform rotations, and when modulated attenuations can perform symmetry operations [12]. Conversely, the propagation of light through optical components, along with those components themselves can serve as analog computers for internal space-time symmetries of elementary particles. Thus the Lorentz group as a fundamental mathematical device in many branches of physics that we have discussed throughout this book, has established itself as a powerful means of describing optical phenomena and constitutes one of the most interesting features of Wigner's frontiers in physics.

In Sect. 15.1, we first examine the Jones vector formalism as the traditional approach for studying the polarization properties of the two-component electric field vector. We install a general attenuator to the Jones-matrix formalism, and thereof we study attenuations, phase-shifts, and the rotation of the polarization axes. We investigate their combined effects and show that altogether they amount to a two-by-two representation of the six parameter $SL(2, c)$ group, which is locally isomorphic to the Lorentz group. In connection with these combined effects, the Wigner rotation is also reviewed.

We conclude this section by introducing new optical filters and examine their possible applications. It is also pointed out that these new filters share the same mathematical construct of the space-time symmetries for massless particles, namely the $E(2)$-like subgroup of the Poincaré group, that we have extensively studied in Chap. 10.

In Sect. 15.2, we are interested in mathematical properties of the $ABCD$ matrices whose components consist of optical elements which govern the propagation of the beam. We give a detailed account about the equi-diagonalization of the $ABCD$ matrices. We see how the results of this process facilitate computations when a large number of cycles is required to account for the behaviour of beam subjected to the effects of a series of optical elements. Compositions of the $ABCD$ matrix are obtained by assembling suitable combinations of lens and translation matrices alone.

In Sect. 15.3, we provide physical examples for $ABCD$ matrices governing the propagation of the rays in a camera, in a laser cavity, in multi-layers, and in an optically active medium. We observe that they serve as analog computers for the essential features of Wigner's little groups dictating the internal space–time symmetries of relativistic particles.

15.1 Polarization Optics

Polarization is roughly defined to be a characteristic of transverse waves, most notably the electromagnetic waves, in which the geometrical orientation of the oscillations is specified. Polarized light can be described with a two component Jones vector and the calculus of two-by-two Jones matrices. In this section we shall lay out the details of this matrix calculus and show how the subgroups of the Poincaré group come into play. Specifically we shall be employing the groups $SL(2, c)$ and $E(2)$ and those that are isomorphic to them.

15.1.1 Jones Vector

Consider a plane wave propagating along the z-direction, and having polarizations along the x- and y-directions. In this case, the electric field vector is written as

$$E_x = a \cos (kz - \omega t + \phi_1), \quad E_y = b \cos (kz - \omega t + \phi_2), \tag{15.1}$$

where a and b are the amplitudes which are real and positive numbers, and ϕ_1 and ϕ_2 are the phases of the x- and y-components, respectively. The benefits in the usage of this form is not limited to classical optics, but is also applicable to coherent and squeezed states of light [3, 13].

In this formalism the two transverse components of the electric field are combined into one column vector with the exponential form by replacing the cosinusoidal function with

$$\begin{pmatrix} E_x \\ E_y \end{pmatrix} = \begin{pmatrix} a \exp \{i(kz - \omega t + \phi_1)\} \\ b \exp \{i(kz - \omega t + \phi_2)\} \end{pmatrix}. \tag{15.2}$$

This column matrix is called the Jones vector [14, 15, 16, 17].

The ratio of the transverse electric field components

$$\frac{E_y}{E_x} = \left(\frac{b}{a}\right) e^{i(\phi_2 - \phi_1)} \tag{15.3}$$

determines the content of polarization. By using

$$r = \frac{b}{a}, \quad \phi = \phi_2 - \phi_1 \tag{15.4}$$

Eq. (15.3) can be expressed as one complex number

$$w = re^{i\phi} . \tag{15.5}$$

The degree of polarization as measured by these two real numbers, are the amplitude ratio and the phase difference, respectively. The transformation of this complex number takes place when the light beam goes through an optical filter. In general its transmission properties are not isotropic.

The Jones-matrix formalism begins with the projection operator [18]:

$$\begin{pmatrix} 1 & 0 \\ 0 & 0 \end{pmatrix} . \tag{15.6}$$

In the standard textbooks [16, 19], this matrix is applied to the Jones vector of Eq. (15.2). Its effect is to completely eliminate the y-component while keeping the x-component of the electric field. Here, in order to avoid this oversimplification of realistic observations, we shall replace this projection operator by an attenuation matrix which will eventually take the form of a two-by-two squeeze matrix.

Another basic element in the conventional formalism gives the phase difference between the x- and y-components with two different values of the index of refraction along the two orthogonal directions. This filter is expressed as

$$\begin{pmatrix} e^{-i\delta_1} & 0 \\ 0 & e^{-i\delta_2} \end{pmatrix} = e^{-i(\delta_1+\delta_2)/2} \begin{pmatrix} e^{-i\delta/2} & 0 \\ 0 & e^{i\delta/2} \end{pmatrix} , \tag{15.7}$$

with $\delta = \delta_1 - \delta_2$. Not being an observable, the overall phase factor $e^{-i(\delta_1+\delta_2)/2}$ can safely be deleted. Then we denote the phase shifter as

$$P(\delta) = \begin{pmatrix} e^{-i\delta/2} & 0 \\ 0 & e^{i\delta/2} \end{pmatrix} . \tag{15.8}$$

Furthermore, the rotation matrix

$$R(\theta) = \begin{pmatrix} \cos(\theta/2) & -\sin(\theta/2) \\ \sin(\theta/2) & \cos(\theta/2) \end{pmatrix} \tag{15.9}$$

is also needed, since the x- and y-axes are not always the polarization axes.

The attenuation coefficient along one transverse direction can be different from that along the other direction. For this purpose, we consider the matrix

$$\begin{pmatrix} e^{-\eta_1} & 0 \\ 0 & e^{-\eta_2} \end{pmatrix} = e^{-(\eta_1+\eta_2)/2} \begin{pmatrix} e^{\eta/2} & 0 \\ 0 & e^{-\eta/2} \end{pmatrix} . \tag{15.10}$$

The exponential factor $e^{-(\eta_1+\eta_2)/2}$ decreases both components at the same rate and has no affect on the state of polarization. Therefore, the polarization effect is determined only by the squeeze matrix

$$B(\eta) = \begin{pmatrix} e^{\eta/2} & 0 \\ 0 & e^{-\eta/2} \end{pmatrix} . \tag{15.11}$$

This way the projection operator of Eq. (15.6) is superseded with a more realistic operator where it will be referred to as an attenuator. This matrix becomes the projection operator of Eq. (15.6) if η_1 is very close to zero and η_2 becomes infinitely large. The conventional projection operator is therefore a special case of the above attenuation matrix.

The systematic combination of the three components given in Eq. (15.8), Eq. (15.9), and Eq. (15.11) constitute the traditional Jones-matrix formalism. From Chap. 3, we have already seen that the type of mathematical operation given in Eq. (15.11) is a Lorentz boost acting on spinors. Thus, the Jones-matrix formalism will be expanded by replacing the projection operator of Eq. (15.6) by the above squeeze operator $B(\eta)$.

15.1.2 The SL(2, c) Content of Polarization

As was noted in Sect. 8.1.1 of Chap. 8, the generators K_3 and J_2 are needed for the squeeze matrix of Eq. (15.11). The repeated application leads to the commutation relation for these two generators:

$$[J_2, K_3] = iK_1. \tag{15.12}$$

Indeed, J_2, K_1, and K_3 form a closed set of commutation relations for the $Sp(2)$ subgroup of $SL(2, c)$ [20]. This three-parameter subgroup has been extensively discussed in connection with squeezed states of light [21, 3, 22, 4].

If we consider only the phase shifters, the mathematics is basically repeated applications of J_1 and J_2, resulting in applications also of J_3, where their explicit two-by-two matrix forms are given in Table 3.1 of Chap. 3. Thus, the phase-shift filters form an $SU(2)$ or $O(3)$-like subgroup of the group $SL(2, c)$.

If we use both the attenuators and phase shifters, the result is the full $SL(2, c)$ group with six parameters. The transformation matrix which we previously saw in Eq. (3.25) of Chap. 3, can be rewritten as

$$G = \begin{pmatrix} \alpha & \beta \\ \gamma & \delta \end{pmatrix}, \tag{15.13}$$

with the condition that the determinant of G be one: $\alpha\delta - \gamma\beta = 1$. The successive use of two such matrices yields

$$\begin{pmatrix} \alpha_2 & \beta_2 \\ \gamma_2 & \delta_2 \end{pmatrix} \begin{pmatrix} \alpha_1 & \beta_1 \\ \gamma_1 & \delta_1 \end{pmatrix} = \begin{pmatrix} \alpha_2\alpha_1 + \beta_2\gamma_1 & \alpha_2\beta_1 + \beta_2\delta_1 \\ \gamma_2\alpha_1 + \delta_2\gamma_1 & \gamma_2\beta_1 + \delta_2\delta_1 \end{pmatrix}. \tag{15.14}$$

The most general form of the polarization transformation is the application of this algebra to the column matrix of Eq. (15.2).

The generators of the rotation or the phase-shifters are Hermitian. Thus, they form a unitary subset of the G matrices of Eq. (15.13). Repeated applications of the phase-shifters represents the three-parameter rotation-like subgroup. The generators J_2, K_1 and K_3 are all imaginary and they generate the real $Sp(2)$ group. Thus, the real subset of the G matrices represents the attenuation filters and their repeated applications. Let us review the algebraic properties of the operators we have in hand. The phase-shifter of Eq. (15.8) can be written as

$$P(\delta) = \exp\left(-i\delta J_1\right),\tag{15.15}$$

with

$$J_1 = \frac{1}{2}\begin{pmatrix} 0 & 1 \\ 1 & 0 \end{pmatrix}.\tag{15.16}$$

The rotation operator of Eq. (15.9) takes the form:

$$R(\theta) = \exp\left(-i\theta J_2\right),\tag{15.17}$$

with

$$J_2 = \frac{1}{2}\begin{pmatrix} 0 & -i \\ i & 0 \end{pmatrix},\tag{15.18}$$

and finally the squeeze operator of Eq. (15.11) can also be written in the exponential form:

$$B(\eta) = \exp\left(-i\eta K_3\right),\tag{15.19}$$

with

$$K_3 = \frac{i}{2}\begin{pmatrix} 1 & 0 \\ 0 & -1 \end{pmatrix}.\tag{15.20}$$

These generators do not form a closed set of commutation relations among themselves, but rather their commutation relations produce three more generators. As discussed in Chap. 3, these six generators form the Lie algebra for the six-parameter Lorentz group. The two-by-two representation given is locally isomorphic to the six-parameter $SL(2, c)$.

15.1.3 The Effects of Squeeze, Phase Shift, and Rotation of Axis

The squeeze matrix of Eq. (15.11) did not receive the due attention in the literature as that of the phase shift matrix $P(\delta)$ of Eq. (15.8), which has a well-known effect on the Jones vector.

The combined effects of these two matrices will be examined in this section. It can be noted that both are diagonal and commute.

We apply the squeeze matrix of Eq. (15.11) to the Jones vector, to obtain

$$\begin{pmatrix} e^{\eta/2} & 0 \\ 0 & e^{-\eta/2} \end{pmatrix}\begin{pmatrix} E_x \\ E_y \end{pmatrix} = \begin{pmatrix} e^{\eta/2}E_x \\ e^{-\eta/2}E_y \end{pmatrix}.\tag{15.21}$$

This squeeze transformation increases one amplitude, while contracting the other so that the product of the amplitudes remains invariant. This squeeze transformation is illustrated in Figure 15.1.

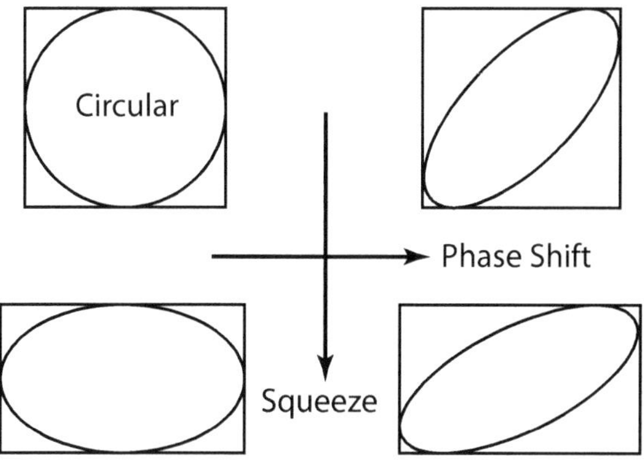

Fig. 15.1: Squeeze and phase shift. The effect of both squeeze and phase shift transformations are elliptic deformations, yet they are performed in a different way [12].

In order to explain phase shifts, we start with electric field of the form

$$\begin{pmatrix} E_x \\ E_y \end{pmatrix} = \begin{pmatrix} \exp(ikz) \\ \exp[i(kz - \pi/2)] \end{pmatrix},$$
(15.22)

whose real part is denoted by $\mathrm{Re}(E_x, E_y)$ as (x, y),

$$\begin{pmatrix} x \\ y \end{pmatrix} = \begin{pmatrix} \cos(kz) \\ \sin(kz) \end{pmatrix},$$
(15.23)

which corresponds to a circular polarization with

$$x^2 + y^2 = 1.$$
(15.24)

When the phase shift matrix is applied, the resulting vector is

$$\begin{pmatrix} x \\ y \end{pmatrix} = \begin{pmatrix} \cos(kz + \delta/2) \\ \sin(kz - \delta/2) \end{pmatrix}.$$
(15.25)

This can also be written as

$$\begin{pmatrix} x \\ y \end{pmatrix} = \begin{pmatrix} \cos(kz - \pi/4 + \alpha) \\ \cos(kz - \pi/4 - \alpha) \end{pmatrix},$$
(15.26)

with

$$\alpha = \frac{\delta}{2} + \frac{\pi}{4} \, . \tag{15.27}$$

Then

$$x + y = 2(\cos\alpha)\cos(kz - \pi/4) \, ,$$

$$x - y = 2(\sin\alpha)\sin(kz - \pi/4) \, , \tag{15.28}$$

and

$$\frac{(x+y)^2}{4(\cos\alpha)^2} + \frac{(x-y)^2}{4(\sin\alpha)^2} = 1 \, . \tag{15.29}$$

This is an elliptic polarization.

In order to illustrate effect of the rotation matrix $R(\theta)$ to the polarized beams, let us start with the circularly polarized wave

$$\begin{pmatrix} E_x \\ E_y \end{pmatrix} = \begin{pmatrix} 1 \\ -i \end{pmatrix} e^{(ikz - i\omega t)} \, , \tag{15.30}$$

and as before, denote the real part of $\mathrm{Re}(E_x, E_y)$ as (x, y), then

$$\begin{pmatrix} x \\ y \end{pmatrix} = \begin{pmatrix} \cos(kz - \omega t) \\ \sin(kz - \omega t) \end{pmatrix} \, . \tag{15.31}$$

This leads to the familiar equation for the circle

$$x^2 + y^2 = 1 \, . \tag{15.32}$$

If the polarization coordinate is the same as the x- and y-coordinate where the electric field components take the form of Eq. (15.2), the attenuator matrix is directly applicable to the column vector of Eq. (15.2). If the polarization coordinate is rotated by an angle $\theta/2$, by the matrix of Eq. (15.9) the phase shifter takes the form

$$P(\theta, \delta) = R(\theta)P(\delta)R(-\theta)$$

$$= \begin{pmatrix} \cos(\delta/2) + i\sin(\delta/2)\cos\theta & i\sin(\delta/2)\sin\theta \\ i\sin(\delta/2)\sin\theta & \cos(\delta/2) - i\sin(\delta/2)\cos\theta \end{pmatrix} \tag{15.33}$$

When the polarization coordinate system is rotated by $45°$ the phase shifter matrix takes the form

$$Q(\delta) = \begin{pmatrix} \cos(\delta/2) & i\sin(\delta/2) \\ i\sin(\delta/2) & \cos(\delta/2) \end{pmatrix} \, . \tag{15.34}$$

This phase shifter is applied to the Jones vector of Eq. (15.31) to yield

$$\begin{pmatrix} [\cos(\delta/2) + \sin(\delta/2)]\cos(kz - \omega t) \\ i[\sin(\delta/2) - \cos(\delta/2)]\sin(kz - \omega t) \end{pmatrix} \tag{15.35}$$

with

$$\cos(\delta/2) = \cos\left(\left[\delta/2 + \pi/4\right] - \pi/4\right) ,$$

$$\sin(\delta/2) = \cos\left(\left[\delta/2 + \pi/4\right] + \pi/4\right) . \tag{15.36}$$

To simplify further, we consider

$$\cos(\delta/2) + \sin(\delta/2) = \sqrt{2}\cos\left(\delta/2 + \pi/4\right) ,$$

$$\cos(\delta/2) - \sin(\delta/2) = \sqrt{2}\sin\left(\delta/2 + \pi/4\right) . \tag{15.37}$$

After the phase shift, the Jones vector becomes

$$\begin{pmatrix} \left[\sqrt{2}\cos\alpha\right]\cos(kz - \omega t) \\ \left[\sqrt{2}\sin\alpha\right]\sin(kz - \omega t) \end{pmatrix} , \tag{15.38}$$

with

$$\alpha = \frac{\delta}{2} + \frac{\pi}{4} . \tag{15.39}$$

Then the x and y components will satisfy the equation

$$\frac{x^2}{(\sqrt{2}\cos\alpha)^2} + \frac{y^2}{(\sqrt{2}\sin\alpha)^2} = 1 . \tag{15.40}$$

This is an elliptic polarization. These steps are illustrated in Figure 15.2.

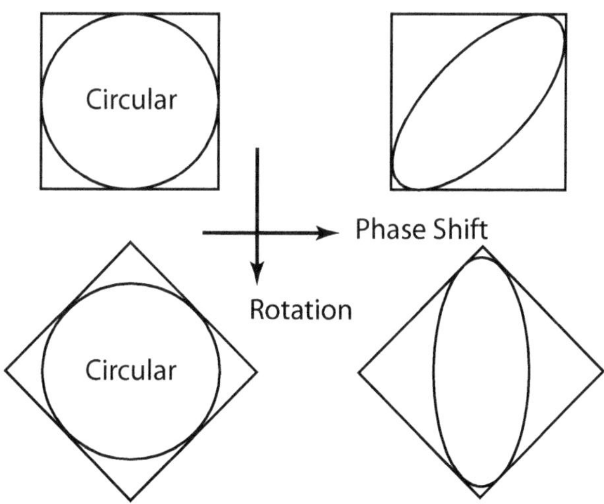

Fig. 15.2: Phase shift and rotation. They are rotated by 45° [12].

The squeeze operation of Eq. (15.11) is relatively simple. It changes the amplitudes, and it commutes with the phase shift matrix. Their combined effect is illustrated in Figure 15.1.

The combined effects of filters is not necessarily restricted to those that are analyzed above. It is possible to consider rotations of the squeeze matrix.

$$B(\theta,\eta) = R(\theta)B(\eta)R(-\theta),\qquad(15.41)$$

which leads to

$$B(\theta,\eta) = \begin{pmatrix} \cosh(\eta/2) + \sinh(\eta/2)\cos\theta & \sinh(\eta/2)\sin\theta \\ \sinh(\eta/2)\sin\theta & \cosh(\eta/2) - \sinh(\eta/2)\cos\theta \end{pmatrix}.\qquad(15.42)$$

If the squeeze angle becomes $45°$, we use the notation $S(\eta)$ for this special case, and write

$$S(\eta) = \begin{pmatrix} \cosh(\eta/2) & \sinh(\eta/2) \\ \sinh(\eta/2) & \cosh(\eta/2) \end{pmatrix}.\qquad(15.43)$$

We have extensively made use of this matrix through out this book in connection with the group $SL(2, c)$. We have seen in Sect. 7.2 of Chap. 7 that this matrix transformation leaves the Hamiltonian of the harmonic oscillator function H_- invariant.

Calculations above may give the impression that, two squeeze transformations carried out in two different directions yield another squeeze transformation. However, the result is not just one other squeeze, but another squeeze matrix followed by a rotation [23, 24, 25], which is expressed as:

$$B(\theta,\lambda)B(0,\eta) = B(\phi,\xi)R(\omega),\qquad(15.44)$$

where $R(\omega)$ is a rotation around the z-axis by ω. The relation between the parameters ϕ,ξ,ω and θ,λ,η are explicitly calculated in Sect. 8.1 of Chap. 8. $R(\omega)$ is referred to as the Wigner rotation in the literature [26, 3, 4]. In the context of Wigner's little groups, this rotation does not change the four-momentum of a massive particle, nevertheless it does affect the spin [27]. Eq. (15.44) constitutes the kinematical origins of the well-known Thomas precession [28]. It is also interesting to note that a close relation between Thomas precession and Berry's phase can be established through this particular property of the Lorentz group [29, 30].

15.1.4 New Filters

We should note at this point that the Poincaré group has another set of three-parameter subgroup. It is like the two-dimensional Euclidean group, $E(2)$. Let us consider the generators J_3, N_1 and N_2, with

$$N_1 = K_1 - J_2,\qquad N_2 = K_2 + J_1,\qquad(15.45)$$

where

$$N_1 = \begin{pmatrix} 0 & i \\ 0 & 0 \end{pmatrix}, \qquad N_2 = \begin{pmatrix} 0 & 1 \\ 0 & 0 \end{pmatrix}. \tag{15.46}$$

These matrices satisfy closed set of commutation relations:

$$[J_3, N_1] = iN_2, \quad [J_3, N_2] = -iN_1, \quad [N_1, N_2] = 0. \tag{15.47}$$

As shown in Sect. 3.3 of Chap. 3 these commutation relations are like those for the two-dimensional Euclidean group consisting of two translations and one rotation around the origin. This group has been studied extensively in connection with the space-time symmetries of massless particles, where J_3 and the two N generators correspond to the helicity and gauge degrees of freedom respectively [31].

The physics of J_3 is well known through the phase shifter given in Eq. (15.8). If the angle δ is $\pi/2$, the phase shifter becomes a quarter-wave shifter, which is expressed as

$$Q = P(0, \pi/2) = \begin{pmatrix} e^{-i\pi/4} & 0 \\ 0 & e^{i\pi/4} \end{pmatrix}. \tag{15.48}$$

Then J_2 and K_1 are the quarter-wave conjugates of J_1 and K_2 respectively:

$$J_2 = QJ_1Q^{-1}, \qquad K_1 = -QK_2Q^{-1}. \tag{15.49}$$

Consequently,

$$N_2 = QN_1Q^{-1}. \tag{15.50}$$

The generators N_i lead to the following transformation matrices:

$$T_1(\gamma) = \exp(-i\gamma N_1) = \begin{pmatrix} 1 & \gamma \\ 0 & 1 \end{pmatrix}, \tag{15.51}$$

$$T_2(\gamma) = \exp(-i\gamma N_2) = \begin{pmatrix} 1 & -i\gamma \\ 0 & 1 \end{pmatrix}. \tag{15.52}$$

It is clear that T_1 is the quarter-wave conjugate of T_2. T_1 transforms the column matrix of Eq. (15.2) as

$$\begin{pmatrix} 1 & \gamma \\ 0 & 1 \end{pmatrix} \begin{pmatrix} E_x \\ E_y \end{pmatrix} = \begin{pmatrix} E_x + \gamma E_y \\ E_y \end{pmatrix}. \tag{15.53}$$

This new filter superposes the y component of the electric field to the x component with an appropriate constant, but it leaves the y component invariant.

Let us examine how this is achieved. The generator N_1 consists of J_2 which generates rotations around the y-axis, and K_1 which generates a squeeze along the 45° axis. Physically, J_2 generates optical activities. Thus, the new filter consists of a suitable combination of these two operations. In both cases, we have to take into account the overall attenuation factor. This can be measured by the attenuation of the y component which is not affected by the symmetry operation of Eq. (15.53).

It is possible to produce optical filters by starting from an optically active material. Then we can introduce an asymmetry in attenuation by either mechanical or electrical means. Another approach would be to pile up alternately the J_2-type and K_1-type layers. In either case, it is interesting to note that the combination of these two effects produces a special effect predicted from the Lorentz group.

The $E(2)$-like symmetry includes transformations generated by J_1. The transformation matrix is

$$P(0, \delta) = \begin{pmatrix} e^{-i\delta/2} & 0 \\ 0 & e^{i\delta/2} \end{pmatrix} . \tag{15.54}$$

This matrix is given in Eq. (15.8) and its physics is well understood. Let us apply this matrix to T_1 from left and from right. Then

$$P(0, \delta)T_1(\gamma) = \begin{pmatrix} e^{-i\delta/2} & e^{-i\delta/2}\gamma \\ 0 & e^{i\delta/2} \end{pmatrix} ,$$

$$T_1(\gamma)P(0, \delta) = \begin{pmatrix} e^{-i\delta/2} & e^{i\delta/2}\gamma \\ 0 & e^{i\delta/2} \end{pmatrix} . \tag{15.55}$$

This leads to

$$P(0, -\delta)T_1(\gamma)P(0, \delta) = T_1(\delta, \gamma) = \begin{pmatrix} 1 & e^{i\delta}\gamma \\ 0 & 1 \end{pmatrix} . \tag{15.56}$$

We can of course obtain the $P(0, -\delta)$ filter by rotating the $P(0, \delta)$ around the z-axis by 90 degrees. It is thus possible to add a phase factor to the γ variable using phase shifters of the type $P(0, \delta)$.

15.1.5 Possible Applications of the New Filter

The three-dimensional rotation group occupies an important place in many different branches of physics. The group $Sp(2)$ also is useful in a number of fields including optics [3, 32, 33]. Thus, traditional attenuation and phase-shift filters may be useful in constructing analog computers performing the symmetry operations of these groups.

The group $E(2)$ is somewhat new in optics [34]. However, as was shown in Chaps. 4 and 10, it deals with translations and rotations on a flat surface. This filter may therefore be useful as a computational device for recording and reading two-dimensional maps. Let us rewrite the T_1 matrix of Eq. (15.51) as

$$\begin{pmatrix} 1 & \beta \\ 0 & 1 \end{pmatrix} . \tag{15.57}$$

This matrix has an interesting algebraic property:

$$\begin{pmatrix} 1 & \beta_1 \\ 0 & 1 \end{pmatrix}\begin{pmatrix} 1 & \beta_2 \\ 0 & 1 \end{pmatrix} = \begin{pmatrix} 1 & \beta_1 + \beta_2 \\ 0 & 1 \end{pmatrix} , \tag{15.58}$$

which can be used for converting multiplication into addition, like the logarithmic function. This is one of the most basic operations in computational machines.

As for a better immediate application, let us consider lens optics. It is a trivial laboratory operation to rotate a given filter around the z-axis by 90 degrees. If we rotate the matrix of Eq. (15.57), the result is the matrix of the form

$$\begin{pmatrix} 1 & 0 \\ \beta & 1 \end{pmatrix}. \tag{15.59}$$

This form together with the original form of Eq. (15.57) serve as lens and translation matrices respectively in para-axial optics. Indeed, a system of polarization filters can serve as an analog computer for a multi-lens system.

Furthermore, the matrix of the form given in Eq. (15.57) represents a *shear* transformation. This is one of the basic deformations in engineering applications.

15.2 Ray Optics and ABCD Matrices

A two-by-two matrix associated with an optical element, the ray transfer matrix or the so-called $ABCD$-matrix, is widely used for describing the element's effect on several types of rays and beams. They prove to be indispensable tools for resonators [17], problems with varying radial indices, and most notably for lasers [35].

15.2.1 ABCD Matrices

The Lorentz group is generated by six matrices. The elements of the $ABCD$ matrices, however, are always real and unimodular (i.e., determinant = 1). As we saw earlier a group with such a restriction can be generated by J_2, K_3, and K_1. These generators form a closed set of commutation relations given by

$$[J_2, K_1] = -iK_3, \quad [J_2, K_3] = iK_1, \quad [K_1, K_3] = iJ_2, \tag{15.60}$$

where

$$J_2 = \frac{i}{2}\begin{pmatrix} 0 & -1 \\ 1 & 0 \end{pmatrix}, \quad K_3 = \frac{i}{2}\begin{pmatrix} 1 & 0 \\ 0 & -1 \end{pmatrix}, \quad K_1 = \frac{i}{2}\begin{pmatrix} 0 & 1 \\ 1 & 0 \end{pmatrix}, \tag{15.61}$$

and form the Lie algebra for the $Sp(2)$ group as discussed in Chap. 8 [36, 37, 12].

This representation of the $Sp(2)$ group consists of a rotation about the origin, a squeeze along the x-direction and another squeeze along axes rotated by $45°$, respectively. Another representation of the $Sp(2)$ group which has two shear transformations and a squeeze transformation [38, 39] can also be obtained from these generators. In what follows, we shall use a rotation $R(\theta)$, a squeeze $B(\eta)$, and a shear matrix $T(\gamma)$ of the form given in Eqs. (15.9), (15.11), and (15.51), respectively [40].

Specific arrangements of optical materials along with those materials themselves determine the elements of the $ABCD$ matrices. Before discussing the practical applications of the $ABCD$ matrices in the following sections, we shall first dwell on its mathematical properties.

15.2.2 Equi-Diagonalization of the ABCD Matrices

Diagonalization is a common procedure in order to put a matrix into its simplest possible form while keeping its characteristic properties unaltered. Since diagonalization may not always be achieved, the next best option is to transform them so that they have equi-diagonal elements. For example, in Eq. (15.51), although the triangular matrix cannot be diagonalized, the two diagonal elements are equal.

In dealing with the $ABCD$ matrices, whose components are determined by the optical instrument in hand, we often encounter matrices that are not equi-diagonal. To make the diagonal elements equal we start from the matrix

$$[ABCD] = \begin{pmatrix} A & B \\ C & D \end{pmatrix}, \tag{15.62}$$

where A and D are not necessarily equal to each other. We have two different ways of transforming $ABCD$ matrices to have equal diagonal elements.

Using the matrix $B(\eta)$ from Eq. (15.11) we apply the transformation

$$B(\eta) \, [ABCD] \, B(\eta) = \begin{pmatrix} e^{\eta/2} & 0 \\ 0 & e^{-\eta/2} \end{pmatrix} \begin{pmatrix} A & B \\ C & D \end{pmatrix} \begin{pmatrix} e^{\eta/2} & 0 \\ 0 & e^{-\eta/2} \end{pmatrix}. \tag{15.63}$$

This yields the equi-diagonal form

$$[ABCD]_S = \begin{pmatrix} \sqrt{AD} & B \\ C & \sqrt{AD} \end{pmatrix} \tag{15.64}$$

with

$$e^{\eta} = \sqrt{D/A}. \tag{15.65}$$

The squeeze matrix is symmetric and Hermitian. The transformation of Eq. (15.63) is a determinant-preserving transformation, but it is not a similarity transformation. This form of equi-diagonalization will prove to be useful in the following Sect. 15.3.1.

If we now consider a rotation of the $ABCD$ matrix

$$R(\theta) \, [ABCD] \, R(-\theta)$$
$$= \begin{pmatrix} \cos(\theta/2) & -\sin(\theta/2) \\ \sin(\theta/2) & \cos(\theta/2) \end{pmatrix} \begin{pmatrix} A & B \\ C & D \end{pmatrix} \begin{pmatrix} \cos(\theta/2) & \sin(\theta/2) \\ -\sin(\theta/2) & \cos(\theta/2) \end{pmatrix}, \tag{15.66}$$

we obtain

$$[ABCD]_R = \begin{pmatrix} A' & B' \\ C' & D' \end{pmatrix}, \tag{15.67}$$

with

$$A' = \frac{1}{2}[A(1 + \cos\theta) + D(1 - \cos\theta) - (C + B)\sin\theta],$$

$$D' = \frac{1}{2}[A(1 - \cos\theta) + D(1 + \cos\theta) + (C + B)\sin\theta]. \tag{15.68}$$

To make the two diagonal elements equal, the rotation angle should satisfy

$$\tan\theta = \frac{A - D}{A + B}. \tag{15.69}$$

The rotation matrix is anti-symmetric and unitary. The transformation of Eq. (15.66), unlike that of Eq. (15.63), is a similarity transformation.

When an $ABCD$ matrix has been equi-diagonalized, it can be put into one of the forms of the two-by-two matrices W_i', given in Table 15.1, by calculating the appropriate relations between the group parameters and the physical parameters of the optical system in hand [40, 41]. The Lorentz boosted version of Wigner matrices are defined by

$$W_i' = B(\eta) \, W_i \, [B(\eta)]^{-1}, \tag{15.70}$$

where W_i is one of the three single-parameter matrices $R(\theta), T(\gamma)$, and $S(\eta)$, whose generic forms are given in Eqs. (15.9), (15.51), and (15.43), respectively.

Table 15.1: The two-by-two representations of the Wigner matrices and their boosted forms. They are the repesentations of little groups for massive, massless and imaginary mass particles.

Particle Mass	Wigner Matrix W_i	Transformed Wigner Matrix W_i'
Massive	$R(\theta) = \begin{pmatrix} \cos(\theta/2) & -\sin(\theta/2) \\ \sin(\theta/2) & \cos(\theta/2) \end{pmatrix}$	$R'(\theta) = \begin{pmatrix} \cos(\theta/2) & -e^{\eta}\sin(\theta/2) \\ e^{-\eta}\sin(\theta/2) & \cos(\theta/2) \end{pmatrix}$
Massless	$T(\gamma) = \begin{pmatrix} 1 & \gamma \\ 0 & 1 \end{pmatrix}$	$T'(\gamma) = \begin{pmatrix} 1 & e^{\eta}\gamma \\ 0 & 1 \end{pmatrix}$
Imaginary Mass	$S(\lambda) = \begin{pmatrix} \cosh(\lambda/2) & \sinh(\lambda/2) \\ \sinh(\lambda/2) & \cosh(\lambda/2) \end{pmatrix}$	$S'(\lambda) = \begin{pmatrix} \cosh(\lambda/2) & e^{\eta}\sinh(\lambda/2) \\ e^{-\eta}\sinh(\lambda/2) & \cosh(\lambda/2) \end{pmatrix}$

It is now possible to enjoy all the mathematical devices developed for the two-by-two representation of Wigner's little groups in earlier Chap. 3.

It should be noted that these three W_i' matrices form different classes with different traces which are smaller than, equal to, and greater than 2, respectively. As is shown in [42], it is possible to make continuations from one to the other through the tangential continuity.

Since the two-by-two $ABCD$ matrices can be written as a similarity transformation of one of the three possible W_i matrices given in Table 15.1, they can further take a convenient form by the following transformation

$$[ABCD]_R$$
$$= R(\sigma)B(\eta) \; W \; B(-\eta)R(-\sigma) = [R(\sigma)B(\eta)] \; W \; [R(\sigma)B(\eta)]^{-1} . \quad (15.71)$$

This will prove to be very useful when repeated applications of an $ABCD$ matrix are required. In case such a necessity arises, we have

$$[ABCD]^N = [R(\sigma)B(\eta)] \; W^N \; [R(\sigma)B(\eta)]^{-1} \qquad (15.72)$$

as we shall see in Sects. 15.3.2 and 15.3.3.

It is possible to combine all three different classes of Eq. (15.70) into one analytic expression using the relation between two momentum preserving transformations

$$B(\eta) \; W_i \; B(\eta)^{-1} = R(\alpha)S(-2\chi)R(\alpha) \qquad (15.73)$$

which was discussed in detail in Sect. 8.1.2 of Chap. 8. The forms of the rotation matrix $R(\alpha)$ and the squeeze matrix $S(\chi)$ are given in Eq. (15.9) and Eq. (15.43), respectively. The matrix multiplication on the right hand side of Eq. (15.73) yields

$$W_B' = \begin{pmatrix} (\cosh \chi) \cos \alpha & -\sinh \chi - (\cosh \chi) \sin \alpha \\ -\sinh \chi + (\cosh \chi) \sin \alpha & (\cosh \chi) \cos \alpha \end{pmatrix} . \qquad (15.74)$$

Here, both α and χ are independent variables. They can be expressed in terms of the η, θ, λ, and γ parameters of the matrices W_i' listed in Table 15.1. These relations are already calculated in Sect. 8.1.2 of Chap. 8. The matrix W_B' proves to be useful when the transition from one form to another within the set of matrices W_i' is required. Furthermore, this particular equivalence between the Bargmann and Wigner forms of the $ABCD$ matrix turns out to be convenient when dealing with cascaded systems having N cycles for the light beam to perform.

15.2.3 Recomposition of the ABCD Matrices

As to the recomposition of the $ABCD$ matrices, the best example for its demonstration can be derived from para-axial lens optics, where we borrow the lens and translation matrices from the forthcoming Sect. 15.3.1:

$$L(f) = \begin{pmatrix} 1 & 0 \\ -1/f & 1 \end{pmatrix} \quad \text{and} \quad T(d) = \begin{pmatrix} 1 & d \\ 0 & 1 \end{pmatrix}. \tag{15.75}$$

Here f is the focal length of the lens and d is the distance between lenses [17]. It is readily seen that both L and T convert multiplications into additions

$$L(f_1 + f_2) = L(f_1)\,L(f_2) \qquad \text{and} \qquad T(d_1 + d_2) = T(d_1)\,T(d_2) \tag{15.76}$$

if computations are restricted to L-type or T-type matrices.

A one-lens system consists of a TLT chain, while a two-lens system has the form $TLTLT$. The chain becomes longer when more lenses are included. However, the final result will be one $ABCD$ matrix with three independent real parameters. As mentioned earlier, this is a representation of the $Sp(2)$ group, whose most general form can be expressed as

$$W''(\phi, \rho, \eta) = R(\phi)B(\eta)R(\rho) \tag{15.77}$$

or more explicitly as

$$W''(\phi, \rho, \eta) = \begin{pmatrix} \cos\phi & -\sin\phi \\ \sin\phi & \cos\phi \end{pmatrix} \begin{pmatrix} e^\eta & 0 \\ 0 & e^{-\eta} \end{pmatrix} \begin{pmatrix} \cos\rho & -\sin\rho \\ \sin\rho & \cos\rho \end{pmatrix}, \tag{15.78}$$

by considering Bargmann decomposition given in Sect. 8.1.2 of Chap. 8. Let us note that $R(\rho)$ can be decomposed as $R(\rho) = R(-\phi)R(\theta)$. Thus in terms of matrices, the last matrix in Eq. (15.78) becomes

$$R(\rho) = \begin{pmatrix} \cos\phi & \sin\phi \\ -\sin\phi & \cos\phi \end{pmatrix} \begin{pmatrix} \cos\theta & -\sin\theta \\ \sin\theta & \cos\theta \end{pmatrix}, \tag{15.79}$$

with $\rho = \theta - \phi$. Finally, we have

$$W''(\phi, \rho, \eta) = R(\phi)B(\eta)R(-\phi)R(\theta) \tag{15.80}$$

and instead of the angle ρ, now θ is an independent parameter.

The matrix W'' can be obtained from the product of two matrices, one symmetric S and the other orthogonal R as:

$$W''(\phi, \theta, \eta) = S(\phi, \eta)R(\theta) \quad \text{with} \quad S(\phi, \eta) = R(\phi)B(\eta)R(-\phi). \tag{15.81}$$

This symmetric matrix S takes the form [43]:

$$S(\phi, \eta) = \begin{pmatrix} \cosh\eta + (\sinh\eta)\cos(2\phi) & (\sinh\eta)\sin(2\phi) \\ (\sinh\eta)\sin(2\phi) & \cosh\eta - (\sinh\eta)\cos(2\phi) \end{pmatrix}. \tag{15.82}$$

Our procedure is to write S and R separately as L and T chains. Let us consider first the rotation matrix that can be obtained through the multiplication of L-type and T-type matrices as [44]:

$$R(\theta) = \begin{pmatrix} 1 & -\tan(\theta/2) \\ 0 & 1 \end{pmatrix} \begin{pmatrix} 1 & 0 \\ \sin\theta & 1 \end{pmatrix} \begin{pmatrix} 1 & -\tan(\theta/2) \\ 0 & 1 \end{pmatrix}. \tag{15.83}$$

This expression is in the form of TLT, but it can also be written in the form of LTL. If we take the transpose and change the sign of θ, R becomes

$$R'(\theta) = \begin{pmatrix} 1 & 0 \\ \tan(\theta/2) & 1 \end{pmatrix} \begin{pmatrix} 1 & -\sin\theta \\ 0 & 1 \end{pmatrix} \begin{pmatrix} 1 & 0 \\ \tan(\theta/2) & 1 \end{pmatrix}. \tag{15.84}$$

Both R and R' have the same matrix form, but are decomposed in different ways.

As for the two parameter symmetric matrix of Eq. (15.82), we start with the form $LTLT$ [38]:

$$S(a, b) = \begin{pmatrix} 1 & 0 \\ b & 1 \end{pmatrix} \begin{pmatrix} 1 & a \\ 0 & 1 \end{pmatrix} \begin{pmatrix} 1 & 0 \\ a & 1 \end{pmatrix} \begin{pmatrix} 1 & b \\ 0 & 1 \end{pmatrix}, \tag{15.85}$$

which can be combined into one symmetric matrix:

$$S = \begin{pmatrix} 1 + a^2 & b(1 + a^2) + a \\ b(1 + a^2) + a & 1 + 2ab + b^2(1 + a^2) \end{pmatrix}. \tag{15.86}$$

By comparing Eq. (15.82) and Eq. (15.86), we can compute the parameters a and b in terms of η and ϕ. The result is

$$a = \pm\sqrt{(\cosh\eta - 1) + (\sinh\eta)\cos(2\phi)},$$

$$b = \frac{(\sinh\eta)\sin(2\phi) \mp \sqrt{(\cosh\eta - 1) + (\sinh\eta)\cos(2\phi)}}{\cosh\eta + (\sinh\eta)\cos(2\phi)}. \tag{15.87}$$

The matrix in Eq. (15.82) can also be written in a $TLTL$ form:

$$S' = \begin{pmatrix} 1 & b' \\ 0 & 1 \end{pmatrix} \begin{pmatrix} 1 & 0 \\ a' & 1 \end{pmatrix} \begin{pmatrix} 1 & a' \\ 0 & 1 \end{pmatrix} \begin{pmatrix} 1 & 0 \\ b' & 1 \end{pmatrix}. \tag{15.88}$$

Then the parameters a' and b' are

$$a' = \pm\sqrt{(\cosh\eta - 1) - (\sinh\eta)\cos(2\phi)},$$

$$b' = \frac{(\sinh\eta)\sin(2\phi) \mp \sqrt{(\cosh\eta - 1) - (\sinh\eta)\cos(2\phi)}}{\cosh\eta - (\sinh\eta)\cos(2\phi)}. \tag{15.89}$$

The difference between the two sets of parameters ab and $a'b'$ is the sign of the parameter η. This sign change means that the squeeze operation is in the direction perpendicular to the original direction. In choosing ab or $a'b'$, we will also have to take care of the sign of the quantity inside the square root to be positive. If $\cos(2\phi)$ is sufficiently small, both sets are acceptable. On the other hand, if the absolute value of $(\sinh\eta)\cos(2\phi)$ is greater than $(\cosh\eta - 1)$, only one of the sets, ab or $a'b'$, is valid.

We can now combine S and R matrices in order to construct the $ABCD$ matrix. In so doing, the number of matrices is reduced by one as:

$$SR = \begin{pmatrix} 1 & 0 \\ b & 1 \end{pmatrix} \begin{pmatrix} 1 & a \\ 0 & 1 \end{pmatrix} \begin{pmatrix} 1 & 0 \\ a & 1 \end{pmatrix} \begin{pmatrix} 1 & b - \tan(\theta/2) \\ 0 & 1 \end{pmatrix}$$
$$\times \begin{pmatrix} 1 & 0 \\ \sin\theta & 1 \end{pmatrix} \begin{pmatrix} 1 & -\tan(\theta/2) \\ 0 & 1 \end{pmatrix} . \tag{15.90}$$

We can also combine making the product $S'R'$, which becomes

$$S'R' = \begin{pmatrix} 1 & b' \\ 0 & 1 \end{pmatrix} \begin{pmatrix} 1 & 0 \\ a' & 1 \end{pmatrix} \begin{pmatrix} 1 & a' \\ 0 & 1 \end{pmatrix} \begin{pmatrix} 1 & 0 \\ b' + \tan(\theta/2) & 1 \end{pmatrix}$$
$$\times \begin{pmatrix} 1 & -\sin\theta \\ 0 & 1 \end{pmatrix} \begin{pmatrix} 1 & 0 \\ \tan(\theta/2) & 1 \end{pmatrix} . \tag{15.91}$$

To reduce SR of Eq. (15.90), two adjoining T matrices were combined into one T matrix. Similarly, two L matrices were combined into one, to simplify $S'R'$ of Eq. (15.91).

In both cases, there are six matrices, consisting of three T and three L matrices. Now, we have come to the conclusion that the minimum number of L and T-type matrices required for the the most general form of the $ABCD$ matrix is six.

In para-axial optics, we often encounter special forms of the $ABCD$ matrix. For instance, the squeeze matrix $S(\phi, \eta)$ in Eq. (15.82) is for pure magnification [45]. A special case of the decomposition given for S and S' in Eq. (15.85) and Eq. (15.88) respectively, is with $\phi = 0$. However, if η is positive, the set $a'b'$ is not acceptable because the quantity in the square root in Eq. (15.89) becomes negative. For the ab set, we have

$$a = \pm \left(e^\eta - 1\right)^{1/2}, \qquad b = \mp e^{-\eta} \left(e^\eta - 1\right)^{1/2} . \tag{15.92}$$

The L and T matrices of Eq. (15.75) are products of Iwasawa decompositions (again see Sect. 8.1.2 of Chap. 8). Since the most general form of the $ABCD$ matrix is obtained by implementing the inverse process of the Iwasawa decomposition, using solely the products of L and T matrices, we have the recomposition of the $ABCD$ matrix.

15.3 Physical Examples Using ABCD Matrices

We shall look through some concrete physical examples to demonstrate the mathematical tools we developed for the $ABCD$ matrices. We shall see that the $ABCD$ matrices serve as analog computers for the essential features of Wigner's little groups dictating the internal space–time symmetries of relativistic particles.

15.3.1 Ray Optics Applied to Cameras

A lens with focal length f and the propagation of the ray by an amount d comprise the basic optical arrangement for a camera. We can write the lens matrix and the translation matrix as given in Eq. (15.75). These matrices are applicable to the two-dimensional space of $(y, m)^T$, where y measures the height of the ray from the propagation axis and m is the slope of the ray.

Suppose the object and the image are d_1 and d_2 distances away from the lens respectively, then the system is represented by an $ABCD$ matrix as

$$[ABCD] = T(d_2)L(f)T(d_1) \tag{15.93}$$

or more explicitly as

$$[ABCD] = \begin{pmatrix} 1 & d_2 \\ 0 & 1 \end{pmatrix} \begin{pmatrix} 1 & 0 \\ -1/f & 1 \end{pmatrix} \begin{pmatrix} 1 & d_1 \\ 0 & 1 \end{pmatrix}. \tag{15.94}$$

The result of the multiplication above gives the camera matrix

$$[ABCD] = \begin{pmatrix} 1 - d_2/f & d_1 + d_2 - d_1 d_2/f \\ -1/f & 1 - d_1/f \end{pmatrix}. \tag{15.95}$$

The focal condition stated as

$$\frac{1}{d_1} + \frac{1}{d_2} = \frac{1}{f} \tag{15.96}$$

follows from the vanishing of the upper right component of the $ABCD$ matrix [39].

Since both d_1 and d_2 must be longer than f for camera optics, it is more convenient to work with the negative of the matrix given in Eq. (15.95) having positive diagonal elements. The variables are redefined

$$x_1 = d_1/f, \qquad x_2 = d_2/f \tag{15.97}$$

so that all components become dimensionless. Then the camera matrix can be written as

$$\begin{pmatrix} x_2 - 1 & (x_1 - 1)(x_2 - 1) - 1 \\ 1 & x_1 - 1 \end{pmatrix}, \tag{15.98}$$

where we have applied the focal condition

$$\frac{1}{x_1} + \frac{1}{x_2} = 1. \tag{15.99}$$

To study the formula of the camera matrix Eq. (15.98) as a representation of the Poincaré group, the matrix must first be brought into an equi-diagonal form. In order to preserve the focal condition of Eq. (15.99), the off-diagonal elements should remain invariant. For this purpose we carry out the Hermitian transformation given in Eq. (15.63), where

$$e^\eta = \sqrt{(1 - x_2)/(1 - x_1)}. \tag{15.100}$$

Then the diagonal elements become

$$\sqrt{(1-x_1)\,(1-x_2)}\,.\tag{15.101}$$

The camera matrix should take the form

$$\begin{pmatrix} \cos(\theta/2) & -e^{\eta}\sin(\theta/2) \\ e^{-\eta}\sin(\theta/2) & \cos(\theta/2) \end{pmatrix}\tag{15.102}$$

when the diagonal elements are smaller than one. Setting

$$e^{-\eta}\sin(\theta/2) = 1\tag{15.103}$$

we obtain

$$\begin{pmatrix} \cos(\theta/2) & -\sin^2(\theta/2) \\ 1 & \cos(\theta/2) \end{pmatrix}\,.\tag{15.104}$$

For the diagonal elements greater than one, the camera matrix should be given by

$$\begin{pmatrix} \cosh(\lambda/2) & e^{\eta}\sinh(\lambda/2) \\ e^{-\eta}\sinh(\lambda/2) & \cosh(\lambda/2) \end{pmatrix}\tag{15.105}$$

with

$$e^{-\eta}\sinh(\lambda/2) = 1\,,\tag{15.106}$$

leading to

$$\begin{pmatrix} \cosh(\lambda/2) & \sinh^2(\lambda/2) \\ 1 & \cosh(\lambda/2) \end{pmatrix}\,.\tag{15.107}$$

The focusing process produces a transition from

$$\begin{pmatrix} (1-\xi^2)/2 & -\xi^2 \\ 1 & (1-\xi^2)/2 \end{pmatrix} \quad \text{to} \quad \begin{pmatrix} (1+\xi^2)/2 & \xi^2 \\ 1 & (1+\xi^2)/2 \end{pmatrix},\tag{15.108}$$

passing through the point $\xi = 0$, where these two matrices are glued together. Here, we used the usual addition formulae for the relations between ξ and λ or θ. When ξ gradually approaches to zero, the matrix gradually becomes an element of an $E(2)$-like group. Therefore it is fair to say that while the camera is in the process of focusing, the equi-diagonalized $ACBD$ matrix is in the process of group contraction [39, 46]. In terms of its upper right component, specifically the component where the focal condition is located, we observe a tangential continuity at $\xi = 0$ depicted in Fig. 15.3. We shall see in more detail, a similar series of continuous changes taking place in the forthcoming Sect. 15.3.4.

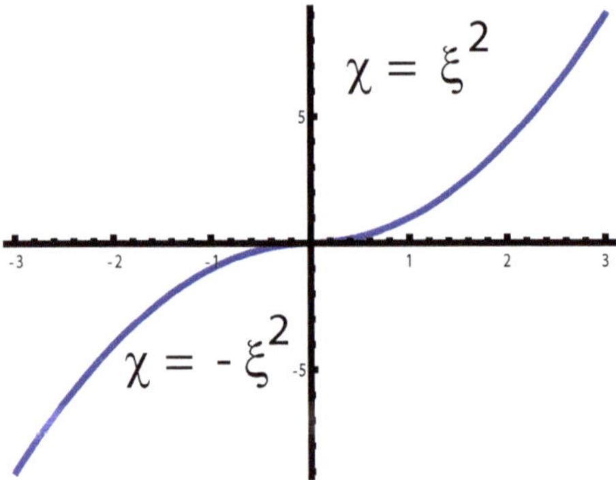

Fig. 15.3: Tangential continuity. At the point $\xi = 0$ the functions $\chi = -\xi^2$ and $\chi = \xi^2$ have a common tangent, however higher derivatives are not continuous [4].

15.3.2 Laser Cavities

Consider a laser cavity consisting of two identical concave mirrors separated by a distant d as shown in Fig. 15.4. The basic matrices are the mirror and translation matrices

$$M(R) = \begin{pmatrix} 1 & 0 \\ -2/R & 1 \end{pmatrix} \quad \text{and} \quad T(d) = \begin{pmatrix} 1 & d \\ 0 & 1 \end{pmatrix}, \tag{15.109}$$

where R and d are the radius of the mirror and the mirror separation, respectively. This form is extensively used in the laser literature [47, 48, 35, 49, 17].

A round trip of one beam can be expressed by the $ABCD$ matrix as

$$[ABCD] = \begin{pmatrix} 1 & 0 \\ -2/R & 1 \end{pmatrix}\begin{pmatrix} 1 & d \\ 0 & 1 \end{pmatrix}\begin{pmatrix} 1 & 0 \\ -2/R & 1 \end{pmatrix}\begin{pmatrix} 1 & d \\ 0 & 1 \end{pmatrix}. \tag{15.110}$$

In a laser cavity, this process is expected to repeat N times, with N taking large values. This issue can be dealt with by putting the sequence of matrices in Eq. (15.110) in an equi-diagonal form. For this purpose we rewrite this matrix as

$$[ABCD] = \begin{pmatrix} 1 & -d/2 \\ 0 & 1 \end{pmatrix}\begin{pmatrix} 1 & d/2 \\ 0 & 1 \end{pmatrix}\begin{pmatrix} 1 & 0 \\ -2/R & 1 \end{pmatrix}\begin{pmatrix} 1 & d/2 \\ 0 & 1 \end{pmatrix}^2$$

$$\times \begin{pmatrix} 1 & 0 \\ -2/R & 1 \end{pmatrix}\begin{pmatrix} 1 & d/2 \\ 0 & 1 \end{pmatrix}\begin{pmatrix} 1 & d/2 \\ 0 & 1 \end{pmatrix}. \tag{15.111}$$

This translates the system by $-d/2$ using a translation matrix given in Eq. (15.109). The $ABCD$ matrix of Eq. (15.111) can then be written as

$$[ABCD] = \begin{pmatrix} 1 & -d/2 \\ 0 & 1 \end{pmatrix}\left[\begin{pmatrix} 1 - d/R & d - d^2/2R \\ -2/R & 1 - d/R \end{pmatrix}\right]^2\begin{pmatrix} 1 & d/2 \\ 0 & 1 \end{pmatrix}. \tag{15.112}$$

fig. (a)

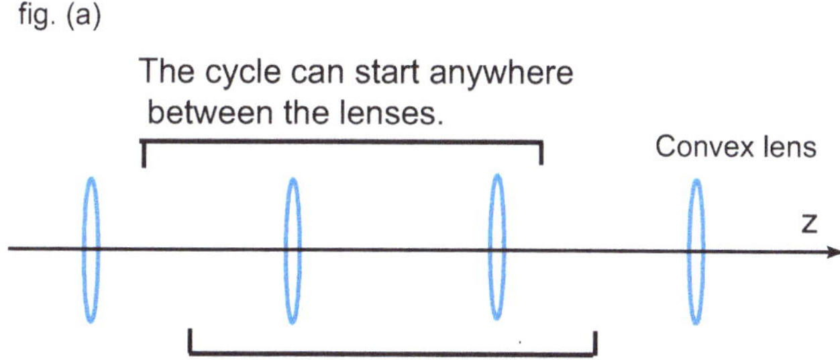

fig. (b)

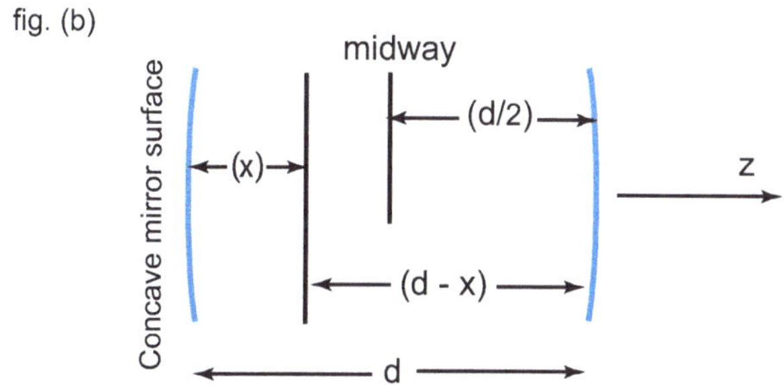

Fig. 15.4: Optical beams in the laser cavity. fig. (a) illustrates that beams going though a series of convex lenses are analogous to the reflected beams in a cavity. fig. (b) describes the reflection of the beams from the surfaces of two concave mirrors, when they originate from an arbitrary distance x [12].

This sequence of matrices contains a similarity transformation, thus it is sufficient to consider only the core matrix which is of the form

$$\begin{pmatrix} 1 - d/R & d - d^2/2R \\ -2/R & 1 - d/R \end{pmatrix}.$$
(15.113)

This can then be written as

$$\begin{pmatrix} \sqrt{d} & 0 \\ 0 & 1/\sqrt{d} \end{pmatrix} \begin{pmatrix} 1 - d/R & 1 - d/2R \\ -2d/R & 1 - d/R \end{pmatrix} \begin{pmatrix} 1/\sqrt{d} & 0 \\ 0 & \sqrt{d} \end{pmatrix}.$$
(15.114)

This part of the $ABCD$ matrix can be decomposed into

$$E\, C^2\, E^{-1},$$
(15.115)

where

$$C = \begin{pmatrix} 1-d/R & 1-d/2R \\ -2d/R & 1-d/R \end{pmatrix} \quad \text{and} \quad E = \begin{pmatrix} 1 & -d/2 \\ 0 & 1 \end{pmatrix} \begin{pmatrix} \sqrt{d} & 0 \\ 0 & 1/\sqrt{d} \end{pmatrix} . \tag{15.116}$$

Since the C matrix contains only dimensionless numbers, we rewrite it as

$$C = \begin{pmatrix} \cos(\gamma/2) & e^{\eta} \sin(\gamma/2) \\ -e^{-\eta} \sin(\gamma/2) & \cos(\gamma/2) \end{pmatrix} , \tag{15.117}$$

with

$$\cos(\gamma/2) = 1 - \frac{d}{R} \quad \text{and} \quad e^{\eta} = \sqrt{\frac{2R-d}{4d}} . \tag{15.118}$$

In this case d and R are both positive, and the restriction on them is that d be smaller than $2R$. This stability condition is frequently mentioned in the literature [48, 49].

The equi-diagonal matrix is now C^2, and it can be expressed as

$$\begin{pmatrix} \cos(\gamma) & e^{\eta} \sin(\gamma) \\ -e^{-\eta} \sin(\gamma) & \cos(\gamma) \end{pmatrix} . \tag{15.119}$$

The equi-diagonal matrix of Eq. (15.119) is now in the familiar form

$$B(\eta) R(2\gamma) B(-\eta) . \tag{15.120}$$

This form is the same as that of the Wigner decomposition for particle physics where a particle is Lorentz boosted to its rest frame, rotated without changing the momentum, and then Lorentz boosted back to the original state. Hence, the $ABCD$ matrix becomes

$$[ABCD] = [E\, B(\eta)]\, R(2\chi)\, [E\, B(\eta)]^{-1} . \tag{15.121}$$

Now, it is possible to express the repeated application of this process as

$$[ABCD]^N = [E\, B(\eta)]\, R(2N\chi)\, [E\, B(\eta)]^{-1} . \tag{15.122}$$

15.3.3 Beam Propagation through Multi-Layers

When an optical beam goes through a cascaded periodic medium with two different refractive indexes, it is partially transmitted and partially reflected at the surfaces of the boundaries. Since it is necessary to maintain the continuity of the Poynting vector we define the electric fields for the optical beams in the first and second media respectively as

$$E_1^{(\pm)} = \frac{1}{\sqrt{n_1}} \exp\left(\pm i k_1 z - \omega t\right),$$

$$E_2^{(\pm)} = \frac{1}{\sqrt{n_2}} \exp\left(\pm i k_2 z - \omega t\right). \tag{15.123}$$

The incoming and reflected rays are represented by the superscript $(+)$ and $(-)$ respectively.

The two-by-two $ABCD$ matrix relates the optical beams of the second medium to those of the first by

$$\begin{pmatrix} E_2^{(+)} \\ E_2^{(-)} \end{pmatrix} = \begin{pmatrix} A & B \\ C & D \end{pmatrix} \begin{pmatrix} E_1^{(+)} \\ E_1^{(-)} \end{pmatrix}.$$

(15.124)

The components of the $ABCD$ matrix are determined by the transmission coefficients and the phase shifts that the beams experience while going through the media [11, 36].

As the beam passes through the first medium to the second, it is possible to use the boundary matrix given by Azzam and Bashara [11] and by [50, 51, 52]. We can write the matrix

$$S(\sigma) = \begin{pmatrix} \cosh(\sigma/2) & \sinh(\sigma/2) \\ \sinh(\sigma/2) & \cosh(\sigma/2) \end{pmatrix},$$

(15.125)

where the σ parameter is given by

$$\cosh\left(\frac{\sigma}{2}\right) = \frac{n_1 + n_2}{2\sqrt{n_1 n_2}}, \qquad \sinh\left(\frac{\sigma}{2}\right) = \frac{n_1 - n_2}{2\sqrt{n_1 n_2}}.$$

(15.126)

Here n_1 and n_2 are the refractive indexes of the first and second medium respectively. From Fig. 15.5 it can be deduced that the boundary matrix for the beam going from the second medium should be $S(-\sigma)$.

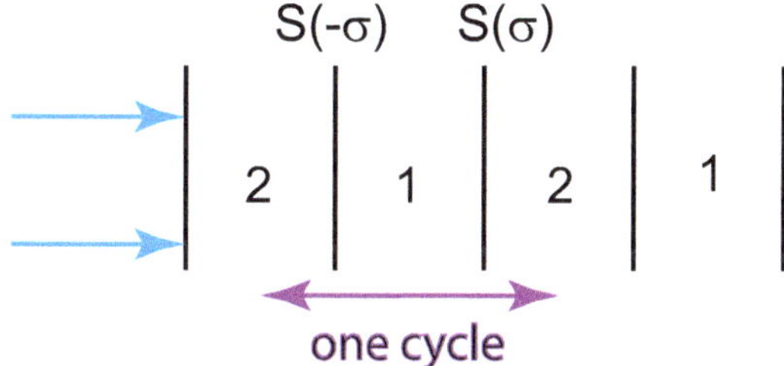

Fig. 15.5: Multi-layers. The resulting ABCD matrix can be made tractable when the beam is prepared to start the cycle from the middle of the second medium [36, 53].

Since the beam experiences phase shifts as it propagates, the phase-shift matrix is to be introduced as

$$P(\delta_1) = \begin{pmatrix} e^{-i\delta_1/2} & 0 \\ 0 & e^{i\delta_1/2} \end{pmatrix}$$

(15.127)

for the beam going through the first medium. There is a similar expression for $P\,(\delta_2)$ for the second medium. The wave number and thickness of the medium determine the phase shift δ.

We choose to start one complete cycle from the midpoint of the second medium. Such a strategy makes the calculations more tractable, when the beam passes through a cascaded system composed of a large number of periodic media. So, we consider

$$M_I = P\,(\delta_2/2)\,S(\sigma)P\,(\delta_1)\,S(-\sigma)P\,(\delta_2/2)\,, \tag{15.128}$$

as is illustrated in Fig. 15.5. If the resulting matrix is composed of real elements, then it may be manageable when dealing with N-cycles. For this purpose we first consider the matrices

$$C = \frac{1}{2}\begin{pmatrix} 1+i & 1+i \\ -(1+i) & 1-i \end{pmatrix} \quad \text{and} \quad C^{-1} = \frac{1}{2}\begin{pmatrix} 1-i & -(1+i) \\ 1-i & 1+i \end{pmatrix}, \tag{15.129}$$

which bring $P(\delta)$ and $S(\sigma)$ to

$$C\,P(\delta)\,C^{-1} = R(\delta) = \begin{pmatrix} \cos(\delta/2) & -\sin(\delta/2) \\ \sin(\delta/2) & \cos(\delta/2) \end{pmatrix} \tag{15.130}$$

and

$$C\,S(\sigma)\,C^{-1} = B(\sigma) = \begin{pmatrix} e^{\sigma/2} & 0 \\ 0 & e^{-\sigma/2} \end{pmatrix} \tag{15.131}$$

respectively, by a similarity transformation.

If we apply this similarity transformation to each of the matrices in Eq. (15.128), then M_I becomes

$$M_R = R\,(\delta_2/2)\,B(\sigma)R\,(\delta_1)\,B(-\sigma)R\,(\delta_2/2)\,, \tag{15.132}$$

where all the matrices are real. We observe that the core of the above equation is familiar and from Table 15.1 it is written as

$$B(\sigma)R\,(\delta_1)\,B(-\sigma) = \begin{pmatrix} \cos(\delta_1/2) & -e^{\sigma}\sin(\delta_1/2) \\ e^{-\sigma}\sin(\delta_1/2) & \cos(\delta_1/2) \end{pmatrix}. \tag{15.133}$$

Because the rotation matrix $R\,(\delta_2/2)$ is at the beginning and at the end of Eq. (15.132), it is not evident that the entire sequence can be expressed as a similarity transformation. This issue can be resolved if we first write Eq. (15.133) as a Bargmann decomposition that we have discussed in detail in Sect. 8.1.2 of Chap. 8:

$$R(\alpha)S(-2\chi)R(\alpha)\,. \tag{15.134}$$

The result of the above multiplication is provided in Eq. (15.74), where the relation between the parameters δ_1, σ and α, χ is found to be

$$\cos(\delta_1/2) = (\cosh\chi)\cos\alpha \quad \text{and} \quad e^{2\sigma} = \frac{(\cosh\chi)\sin\alpha + \sinh\chi}{(\cosh\chi)\sin\alpha - \sinh\chi}\,. \tag{15.135}$$

The entire sequence of matrices in Eq. (15.128) can be written as another Bargmann decomposition

$$M_B = R(\alpha + \delta_2/2)S(-2\chi)R(\alpha + \delta_2/2)\,. \tag{15.136}$$

It is possible now to convert this expression to another Wigner decomposition [53]

$$M_W = B(\eta)R(\theta)B(-\eta)\,, \tag{15.137}$$

where

$$\cos(\theta/2) = (\cosh\chi)\cos(\alpha + \delta_2/2)$$
$$e^{2\eta} = \frac{(\cosh\chi)\sin(\alpha + \delta_2/2) + \sinh\chi}{(\cosh\chi)\sin(\alpha + \delta_2/2) - \sinh\chi}\,. \tag{15.138}$$

With the form of M_W, when the beam propagates through a periodic system of N-cycles, we obtain

$$(M_W)^N = B(\eta)R(N\theta)B(-\eta)\,, \tag{15.139}$$

which has now become tractable, by virtue of the successive implementations of the Wigner and Bargmann decompositions.

15.3.4 Optical Activities

A substance is said to be optically active if it has the ability to rotate the plane of polarization of a beam of light that passes through it. Therefore, optical activities can be described by the real rotation and squeeze matrices given in Eq. (15.9) and Eq. (15.10), respectively. We assume here that the optical ray propagates along the z-direction, and that the polarization rotates on the xy-plane. If the attenuation along the x-direction is different from that along the y-direction, this results in an asymmetric attenuation and the rotation around the z-axis.

Thus as the ray propagates along the z-direction, the polarization goes through the rotation

$$R(\chi z) = \begin{pmatrix} \cos(\chi z) & -\sin(\chi z) \\ \sin(\chi z) & \cos(\chi z) \end{pmatrix}\,. \tag{15.140}$$

Here we have explicitly included the z-direction and for simplicity we use χ for $\theta/2$. This matrix is applicable to the x and y components of the polarization, and the rotation angle increases as z increases.

The optical ray is usually attenuated due to absorption by the medium through which it travels. The attenuation coefficient, however, could be different in each of the transverse directions. Hence if the rate of attenuation along the x-direction is different from that along y-axis, this asymmetric attenuation can be described by squeeze matrix, where we use μ for $\eta/2$

$$\begin{pmatrix} e^{-\mu_1 z} & 0 \\ 0 & e^{-\mu_2 z} \end{pmatrix} = e^{-\lambda z} \begin{pmatrix} e^{\mu z} & 0 \\ 0 & e^{-\mu z} \end{pmatrix}, \tag{15.141}$$

with

$$\lambda = \frac{\mu_2 + \mu_1}{2} \qquad \text{and} \qquad \mu = \frac{\mu_2 - \mu_1}{2}. \tag{15.142}$$

The overall attenuation is given by the exponential factor $e^{-\lambda z}$. The matrix

$$\begin{pmatrix} e^{\mu z} & 0 \\ 0 & e^{-\mu z} \end{pmatrix} \tag{15.143}$$

performs a squeeze transformation where the x component of the polarization is expanded, while the y component is contracted. This is therefore a squeeze along the x-direction.

The squeeze, however, can be in the direction which makes an angle ϕ with the x-axis and not along the x-direction. Let us call this matrix S. The squeeze matrix then becomes

$$S(\phi, \mu z) = \begin{pmatrix} \cos\phi & -\sin\phi \\ \sin\phi & \cos\phi \end{pmatrix} \begin{pmatrix} e^{\mu z} & 0 \\ 0 & e^{-\mu z} \end{pmatrix} \begin{pmatrix} \cos\phi & \sin\phi \\ -\sin\phi & \cos\phi \end{pmatrix}, \tag{15.144}$$

which can be compressed to one matrix

$$\begin{pmatrix} \cosh(\mu z) + \cos(2\phi)\sinh(\mu z) & \sin(2\phi)\sinh(\mu z) \\ \sin(2\phi)\sinh(\mu z) & \cosh(\mu z) - \cos(2\phi)\sinh(\mu z) \end{pmatrix}. \tag{15.145}$$

If $\phi = 45^o$, this matrix becomes

$$S(\pi/4, \mu z) = \begin{pmatrix} \cosh(\mu z) & \sinh(\mu z) \\ \sinh(\mu z) & \cosh(\mu z) \end{pmatrix}. \tag{15.146}$$

In the following discussion, this form of squeeze matrix will be used as $S(\mu z)$ for the above expression, without referring to the angle $\pi/4$. Thus, if the squeeze is made along the x-axis, it is expressed as

$$S(0, \mu z) = R(-\pi/4, \mu z) S(\mu z) R(\pi/4, \mu z). \tag{15.147}$$

When this squeeze is followed by the rotation of Eq. (15.140), the net effect is

$$e^{-\lambda z} \begin{pmatrix} \cos(\chi z) & -\sin(\chi z) \\ \sin(\chi z) & \cos(\chi z) \end{pmatrix} \begin{pmatrix} \cosh(\mu z) & \sinh(\mu z) \\ \sinh(\mu z) & \cosh(\mu z) \end{pmatrix}, \tag{15.148}$$

where z is considered to be on a macroscopic scale, maybe measured in centimeters. However, this is not an accurate description of the optical process.

This happens in a microscopic scale of z/N where N is a very large number. This becomes accumulated into the macroscopic scale of z after the N repetitions. Therefore the transformation matrix takes the same form

$$M(\chi, \mu, z) = \left[e^{-\lambda z/N} S(\mu z/N) R(\chi z/N) \right]^N . \tag{15.149}$$

In the limit of large N, this quantity becomes

$$e^{-\lambda z} \left[\begin{pmatrix} 1 & \mu z/N \\ \mu z/N & 1 \end{pmatrix} \begin{pmatrix} 1 & -\chi z/N \\ \chi z/N & 1 \end{pmatrix} \right]^N . \tag{15.150}$$

Since $\chi z/N$ and $\mu z/N$ are very small, we have

$$M(\chi, \mu, z) = e^{-\lambda z} \left[\begin{pmatrix} 1 & 0 \\ 0 & 1 \end{pmatrix} + \begin{pmatrix} 0 & -(\chi - \mu) \\ (\chi + \mu) & 0 \end{pmatrix} \frac{z}{N} \right]^N . \tag{15.151}$$

For large N, this matrix can be written as

$$M(\chi, \mu, z) = e^{-\lambda z} e^{Uz} , \tag{15.152}$$

with

$$U = \begin{pmatrix} 0 & -(\chi - \mu) \\ (\chi + \mu) & 0 \end{pmatrix} . \tag{15.153}$$

We now are interested in calculating the exponential form e^{Uz} by making a Taylor expansion. In order to proceed we need to compute U^N. This can be tractable if U is expressed as one of the three forms given below.

If $\chi > \mu$ then off-diagonal elements have opposite signs, and we can write U as

$$U = k \begin{pmatrix} 0 & -e^{2\zeta} \\ e^{-2\zeta} & 0 \end{pmatrix} \tag{15.154}$$

with

$$k = \sqrt{\chi^2 - \mu^2} \quad \text{and} \quad e^{2\zeta} = \sqrt{(\chi + \mu)/(\chi - \mu)} \tag{15.155}$$

or

$$\chi = k \cosh(2\zeta) \quad \text{and} \quad \mu = k \sinh(2\zeta) . \tag{15.156}$$

Now

$$U = k \begin{pmatrix} e^{\zeta} & 0 \\ 0 & e^{-\zeta} \end{pmatrix} \begin{pmatrix} 0 & -1 \\ 1 & 0 \end{pmatrix} \begin{pmatrix} e^{-\zeta} & 0 \\ 0 & e^{\zeta} \end{pmatrix} , \tag{15.157}$$

with e^{ζ} given in Eq. (15.155). This is a similarity transformation with respect to a squeeze matrix $B(\zeta)$. The role of this squeeze matrix is quite different from that of Eq. (15.143). It does not depend on z.

We can write U^N as

$$k^N \begin{pmatrix} e^{\zeta} & 0 \\ 0 & e^{-\zeta} \end{pmatrix} \begin{pmatrix} 0 & -1 \\ 1 & 0 \end{pmatrix}^N \begin{pmatrix} e^{-\zeta} & 0 \\ 0 & e^{\zeta} \end{pmatrix} , \tag{15.158}$$

and

$$\exp\left[kz\begin{pmatrix}0 & -1\\ 1 & 0\end{pmatrix}\right] = \begin{pmatrix}\cos(kz) & -\sin(kz)\\ \sin(kz) & \cos(kz)\end{pmatrix}. \tag{15.159}$$

Then the exponential form e^{Uz} becomes

$$\begin{pmatrix}e^{\zeta} & 0\\ 0 & e^{-\zeta}\end{pmatrix}\begin{pmatrix}\cos(kz) & -\sin(kz)\\ \sin(kz) & \cos(kz)\end{pmatrix}\begin{pmatrix}e^{-\zeta} & 0\\ 0 & e^{\zeta}\end{pmatrix} \tag{15.160}$$

and the transformation matrix of Eq. (15.152) takes the form

$$M(\chi,\mu,z) = e^{-\lambda z}\begin{pmatrix}\cos(kz) & -e^{2\zeta}\sin(kz)\\ e^{-2\zeta}\sin(kz) & \cos(kz)\end{pmatrix}, \tag{15.161}$$

with k and e^{ζ} given in Eq. (15.155).

If $\chi < \mu$ then off-diagonal elements have the same signs, and we can write U as

$$U = k\begin{pmatrix}0 & e^{2\zeta}\\ e^{-2\zeta} & 0\end{pmatrix} \tag{15.162}$$

with

$$k = \sqrt{\mu^2 - \chi^2} \qquad \text{and} \qquad e^{2\zeta} = \sqrt{(\chi+\mu)/(\mu-\chi)} \tag{15.163}$$

or

$$\chi = k\sinh(2\zeta) \qquad \text{and} \qquad \mu = k\cosh(2\zeta). \tag{15.164}$$

We can also perform similar calculations as we did above and obtain

$$M(\chi,\mu,z) = e^{-\lambda z}\begin{pmatrix}\cosh(kz) & e^{2\zeta}\sinh(kz)\\ e^{-2\zeta}\sinh(kz) & \cosh(kz)\end{pmatrix}, \tag{15.165}$$

with k and e^{ζ} given in Eq. (15.163).

If $\mu = \chi$ then the upper right element of Eq. (15.151) has to vanish, then U is

$$U = \begin{pmatrix}0 & 0\\ 2\chi & 0\end{pmatrix}, \tag{15.166}$$

with the property $U^2 = 0$. The Taylor expansion truncates and the transformation matrix takes the form

$$M(\chi,z) = \begin{pmatrix}1 & 0\\ 2\chi z & 1\end{pmatrix}. \tag{15.167}$$

As μ becomes larger, it travels from $\mu < \chi$ to $\mu > \chi$ while the U matrix of Eq. (15.154) has to go through the above triangular matrix to take the form in Eq. (15.162). This process involves a non-analytical but a tangential continuity discussed in [46, 54, 42], which can be visualized by the Fig. 15.6, when the sine function in the upper right component of $M(\chi,\mu,z)$ in Eq. (15.161) makes a transition to a hyberbolic function, sinh, in Eq. (15.165). At $kz = 0$, we see that they have a common tangent.

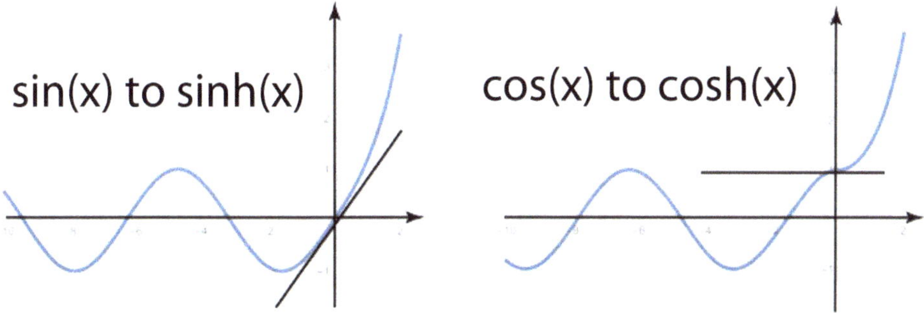

Fig. 15.6: Tangential continuity. The continuation of the trigonometric functions sine and cosine to the hyperbolic functions sinh and cosh when their respective arguments are zero. At that point they have a common tangent, however higher derivatives are not continuous [4].

15.3.5 One Final Remark

We are by now aware of the fact that improvements in the area of optical sciences are responsible for the majority of significant scientific breakthroughs. We don't need to be reminded how the spectrometers developed in the nineteenth century paved the way for quantum mechanics by revealing the distinct spectra of atoms. The optical phase shifting experiments of the Michelson interferometer initiated a line of research that eventually culminated in special relativity. Relativistic quantum mechanics emerged from the unification of these two modern concepts. Light is, in fact, our most significant source of knowledge about our universe from the very small to the very large scale using microscopes and/or telescopes composed of various optical elements that we have discussed in this chapter. It is inspiring to observe how the applications of the Poincaré group are integrated in this formidable area of science and facilitate many of its challenging computations.

References

1. S.M. Barnett, P.M. Radmore, *Methods in theoretical quantum optics*, reprint edn. No. 15 in Oxford series in optical and imaging sciences (Clarendon Press, Oxford, UK, 2005). ISBN 978-0-19-856361-7. (Originially published 1997; OCLC: 316132663.)
2. E. Goldin, *Waves and photons: an introduction to quantum optics*. Wiley series in pure and applied optics (John Wiley and Sons, New York, NY, USA, 1982). ISBN 978-0471085928
3. Y.S. Kim, M.E. Noz, *Phase space picture of quantum mechanics: group theoretical approach*. No. 40 in Lecture notes in physics series (World Scientific Publishing Co., Sin-

gapore; Hackensack, NJ, USA, 1991). ISBN 978-981-02-0360-3,978-981-02-0361-0. URL https://doi.org/10.1142/1197

4. S. Başkal, Y. Kim, M. Noz, *Mathematical Devices for Optical Sciences.* (IOP Publishing, Bristol, UK, 2019). ISBN 978-0-7503-1612-5. URL https://dx.doi.org/10.1088/2053-2563/aafe78. (OCLC: 1034620988.)

5. A.F. Kockum, F. Nori, Quantum Bits with Josephson Junctions, in *Fundamentals and Frontiers of the Josephson Effect*, vol. 286, ed. by F. Tafuri (Springer International Publishing, Cham, 2019), 703–741. DOI 10.1007/978-3-030-20726-7_17. ISBN 9783030207243, 9783030207267. URL http://link.springer.com/10.1007/978-3-030-20726-7_17

6. E. Knill, R. Laflamme, G.J. Milburn, A scheme for efficient quantum computation with linear optics, Nature **409**(6816), 46–52 (2001). DOI 10.1038/35051009. URL https://www.nature.com/articles/35051009

7. P. Kok, W.J. Munro, K. Nemoto, T.C. Ralph, J.P. Dowling, G.J. Milburn, Linear optical quantum computing with photonic qubits, Reviews of Modern Physics **79**(1), 135–174 (2007). DOI 10.1103/RevModPhys.79.135. URL https://link.aps.org/doi/10.1103/RevModPhys.79.135

8. F.A. Jenkins, H.E. White, *Fundamentals of optics* (McGraw-Hill Primis Custom Publishing, New York, NY, USA, 2001). ISBN 978-0-07-256191-3. (OCLC: 871773448.)

9. M. Herzberger, *Modern Geometrical Optics.* Pure and Applied Mathematics: Interscience (Interscience Publishers, New York, NY, USA, 1958). ISBN 978-1114379947. URL https://books.google.com.tr/books?id=ctsNAQAAIAAJ. (Iccn 58009218.)

10. K. Halbach, Matrix Representation of Gaussian Optics, American Journal of Physics **32**(2), 90–108 (1964). DOI 10.1119/1.1970159. URL https://pubs.aip.org/ajp/article/32/2/90/1043235/Matrix-Representation-of-Gaussian-Optics

11. R.M.A.G. Azzam, N.M. Bashara, *Ellipsometry and polarized light*, 4th edn. North-Holland personal library (Elsevier, Amsterdam, NL, 1999). ISBN 978-0-444-87016-2. (Originally published 1977; OCLC: 247501433.)

12. S. Başkal, Y.S. Kim, Lorentz group in ray and polarization optics, in *Mathematical Optics: Classical, Quantum and Computational Methods*, ed. by V. Lakshminarayanan, M. L. Calvo, T. Alieva (Taylor and Francis, Boca Raton, FL, USA, 2013), 303–349. ISBN 978-1-4398-6961-1. URL https://www.taylorfrancis.com/books/9781439869611

13. A.S. Chirkin, D.Y. Parashchuk, A.A. Orlov, Quantum theory of two–mode interactions in optically anisotropic media with cubic nonlinearities: Generation of quadrature– and polarization–squeezed light, Quantum Electronics **23**(10), 870–874 (1993). DOI 10.1070/QE1993v023n10ABEH003182. URL http://stacks.iop.org/1063-7818/23/i=10/a=A05?key=crossref.63264a54ba8e432397266848f6398b50

14. R.C. Jones, A New Calculus for the Treatment of Optical Systems I Description and Discussion of the Calculus, Journal of the Optical Society of America **31**(7), 488–493 (1941). DOI 10.1364/JOSA.31.000488. URL https://www.osapublishing.org/abstract.cfm?URI=josa-31-7-488

15. R.C. Jones, A New Calculus for the Treatment of Optical Systems V A More General Formulation, and Description of Another Calculus, Journal of the Optical Society of America **37**(2), 107–110 (1947). DOI 10.1364/JOSA.37.000107. URL https://www.osapublishing.org/abstract.cfm?URI=josa-37-2-107

16. E. Hecht, *Optics*, 5th edn. (Pearson, London, UK, 2015). ISBN 978-0133977226. (Originally published: Addison-Wesley, Reading, MA, USA, 1974.)

17. B.E.A. Saleh, M.C. Teich, *Fundamentals of photonics*, 2nd edn. Wiley series in pure and applied optics (Wiley Interscience; A John Wiley & Sons, Inc., Publication, New NY, USA, Hoboken, NJ, USA, 2007). ISBN 978-0-471-35832-9. (Originaly published 1991.)

18. F.L. Pedrotti, L.S. Pedrotti, *Introduction to optics*, 3rd edn. (Cambridge Univeristy Press, Cambridge. UK, 2017). ISBN 978-1108428262. (Originally published: 1987 Prentice-Hall, Englewood Cliffs, NJ, USA.)

19. E. Hecht, Note on an Operational Definition of the Stokes Parameters, American Journal of Physics **38**(9), 1156–1158 (1970). DOI 10.1119/1.1976574. URL http://aapt.scitation.org/doi/10.1119/1.1976574

20. M. Kitano, T. Yabuzaki, Observation of Lorentz-group Berry phases in polarization optics, Physics Letters A **142**(6-7), 321–325 (1989). DOI 10.1016/0375-9601(89)90373-3. URL http://linkinghub.elsevier.com/retrieve/pii/0375960189903733

21. H.P. Yuen, Two–photon coherent states of the radiation field, Physical Review A **13**(6), 2226–2243 (1976). DOI 10.1103/PhysRevA.13.2226. URL https://link.aps.org/doi/10.1103/PhysRevA.13.2226

22. B. Yurke, S.L. McCall, J.R. Klauder, SU(2) and SU(1,1) interferometers, Physical Review A **33**(6), 4033–4054 (1986). DOI 10.1103/PhysRevA.33.4033. URL https://link.aps.org/doi/10.1103/PhysRevA.33.4033

23. D. Han, Y.S. Kim, M.E. Noz, Linear canonical transformations of coherent and squeezed states in the Wigner phase space, Physical Review A **37**(3), 807–814 (1988). DOI 10.1103/PhysRevA.37.807. URL https://link.aps.org/doi/10.1103/PhysRevA.37.807

24. D. Han, E.E. Hardekopf, Y.S. Kim, Thomas precession and squeezed states of light, Physical Review A **39**(3), 1269–1276 (1989). DOI 10.1103/PhysRevA.39.1269. URL https://link.aps.org/doi/10.1103/PhysRevA.39.1269

25. S. Başkal, Y.S. Kim, Rotations associated with Lorentz boosts, Journal of Physics A: Mathematical and General **38**(29), 6545–6556 (2005). DOI 10.1088/0305-4470/38/29/009. URL http://stacks.iop.org/0305-4470/38/i=29/a=009?key=crossref.e99c952d1a8a20610d56358b253d04b4

26. D. Han, Y.S. Kim, D. Son, Thomas precession, Wigner rotations, and gauge transformations, Classical and Quantum Gravity **4**(6), 1777–1783 (1987). DOI 10.1088/0264-9381/4/6/029. URL http://stacks.iop.org/0264-9381/4/i=6/a=029?key=crossref.cdf832b6cc75bd1798168dbe7d84aa28

27. S. Başkal, Y.S. Kim, Wigner rotations in laser cavities, Physical Review E **66**(2), 026,604 –6 – 026,604 –6 (2002). DOI 10.1103/PhysRevE.66.026604. URL https://link.aps.org/doi/10.1103/PhysRevE.66.026604

28. J.D. Jackson, *Classical Electrodynamics*, 3rd edn. (John Wiley and Sons, New York, NY, USA, 1999). ISBN 978-0-471-30932-1. (Originally published 1962.)

29. H. Mathur, Thomas precession, spin-orbit interaction, and Berry's phase, Physical Review Letters **67**(24), 3325–3327 (1991). DOI 10.1103/PhysRevLett.67.3325. URL https://link.aps.org/doi/10.1103/PhysRevLett.67.3325

30. Ö.F. Dayi, E. Kilinçarslan, A semiclassical kinetic theory of Dirac particles and Thomas precession, Physics Letters B **749**, 119–124 (2015). DOI 10.1016/j.physletb.2015.07.059. URL https://linkinghub.elsevier.com/retrieve/pii/S0370269315005699

31. D. Han, Y.S. Kim, D. Son, E(2)–like little group for massless particles and neutrino polarization as a consequence of gauge invariance, Physical Review D **26**(12), 3717–3725 (1982). DOI 10.1103/PhysRevD.26.3717. URL https://link.aps.org/doi/10.1103/PhysRevD.26.3717

32. R. Abraham, E. Marsden, J, *Foundations of mechanics*, 2nd edn. (AMS Chelsea Pub./American Mathematical Society, Providence, RI, USA, 2008). ISBN 978-0-8218-4438-0. (Originally published 1978; OCLC: ocn191847156.)

33. V. Guillemin, S. Sternberg, *Symplectic techniques in physics*, reprinted edn. (Cambridge Univ. Press, Cambridge, UK, 2001). ISBN 978-0-521-38990-7. (Originally published 1984; OCLC: 248721606.)

34. Y.S. Kim, L. Yeh, E(2)–symmetric two–mode sheared states, Journal of Mathematical Physics **33**(4), 1237–1246 (1992). DOI 10.1063/1.529701. URL http://aip.scitation.org/doi/10.1063/1.529701

35. A.E. Siegman, *Lasers* (University Science Books, Mill Valley, CA, USA, 1986). ISBN 9780935702118,9780198557135

36. E. Georgieva, Y.S. Kim, Iwasawa effects in multilayer optics, Physical Review E **64**(2), 26,602 –1 – 26,602 – 6 (2001). DOI 10.1103/PhysRevE.64.026602. URL https://link.aps.org/doi/10.1103/PhysRevE.64.026602

37. L.L. Sánchez-Soto, J.J. Monzón, A.G. Barriuso, J.F. Cariñena, The transfer matrix: A geometrical perspective, Physics Reports **513**(4), 191–227 (2012). DOI 10.1016/j.physrep.2011.10.002. URL http://linkinghub.elsevier.com/retrieve/pii/S0370157311002560

38. S. Başkal, Y.S. Kim, Shear representations of beam transfer matrices, Physical Review E **63**(5), 056,606 –1 – 056,606 – 6 (2001). DOI 10.1103/PhysRevE.63.056606. URL https://link.aps.org/doi/10.1103/PhysRevE.63.056606

39. S. Başkal, Y.S. Kim, Lens optics as an optical computer for group contractions, Physical Review E **67**(5), 056,601 –1 – 056,601 – 8 (2003). DOI 10.1103/PhysRevE.67.056601. URL https://link.aps.org/doi/10.1103/PhysRevE.67.056601

40. S. Başkal, Y.S. Kim, ABCD matrices as similarity transformations of Wigner matrices and periodic systems in optics, Journal of the Optical Society of America A-Optics Image Science and Vision **26**(9), 2049–2054 (2009). DOI 10.1364/JOSAA.26.002049. ISBN 1084-7529

41. S. Başkal, Y.S. Kim, One analytic form for four branches of the ABCD matrix, Journal of Modern Optics **57**(14–15), 1251–1259 (2010). DOI 10.1080/09500340903576433. URL http://www.tandfonline.com/doi/abs/10.1080/09500340903576433

42. S. Başkal, Y.S. Kim, M.E. Noz, *Physics of the Lorentz Group (Second Edition): Beyond high-energy physics and optics* (IOP Publishing, Bristol, UK, 2021). DOI 10.1088/978-0-7503-3607-9. ISBN 978-0-7503-3607-9. URL https://iopscience.iop.org/book/978-0-7503-3607-9

43. D. Han, Y.S. Kim, M.E. Noz, Wigner rotations and Iwasawa decompositions in polarization optics, Physical Review E **60**(1), 1036–1041 (1999). DOI 10.1103/PhysRevE.60.1036. URL https://link.aps.org/doi/10.1103/PhysRevE.60.1036

44. A.W. Lohmann, Image rotation, Wigner rotation, and the fractional Fourier transform, Journal of the Optical Society of America A **10**(10), 2181–2186 (1993). DOI 10.1364/JOSAA.10.002181. URL https://www.osapublishing.org/abstract.cfm?URI=josaa-10-10-2181

45. A. Gerrard, J.M. Burch, *Introduction to matrix methods in optics* (Dover, New York, NY USA, 1994). ISBN 978-1-62198-651-5. URL http://app.knovel.com/hotlink/toc/id:kpIMMO0001/introduction-to-matrix. (Originally published 1975 John Wiley and Sons, New York, NY, USA.)

46. S. Başkal, Y.S. Kim, Lens optics and the continuity problems of the ABCD matrix, Journal of Modern Optics **61**(2), 161–166 (2014). DOI 10.1080/09500340.2014.880524. URL http://www.tandfonline.com/doi/abs/10.1080/09500340.2014.880524

47. A. Yariv, *Quantum electronics*, 3rd edn. (Wiley, Hoboken, NJ, USA, 1989). ISBN 978-0-471-60997-1. (Originally published 1975.)

48. H.A. Haus, *Waves and fields in optoelectronics*. Prentice-Hall series in solid state physical electronics (Prentice-Hall, Englewood Cliffs, NJ, USA, 1984). ISBN 978-0-13-946053-1

49. J. Hawkes, I. Latimer, *Lasers: theory and practice*. Prentice-Hall international series in optoelectronics (Prentice Hall, Upper Saddle River, NJ, USA, 1995). ISBN 978-0-13-521493-0

50. J.J. Monzón, L.L. Sánchez-Soto, Fully relativistic–like formulation of multilayer optics, Journal of the Optical Society of America A **16**(8), 2013–2018 (1999). DOI 10.1364/JOSAA.16.002013. URL https://www.osapublishing.org/abstract.cfm?URI=josaa-16-8-2013

51. J.J. Monzón, L.L. Sánchez-Soto, Fresnel formulas as Lorentz transformations, Journal of the Optical Society of America A **17**(8), 1475–1481 (2000). DOI 10.1364/JOSAA.17.001475. URL https://www.osapublishing.org/abstract.cfm?URI=josaa-17-8-1475

52. D. Dragoman, Polarization optics analogy of quantum wavefunctions in graphene, Journal of the Optical Society of America B **27**(7), 1325–1331 (2010). DOI 10.1364/JOSAB.27.001325. URL https://www.osapublishing.org/abstract.cfm?URI=josab-27-7-1325

53. E. Georgieva, Y.S. Kim, Slide-rule-like property of Wigner's little groups and cyclic S matrices for multilayer optics, Physical Review E **68**(2), 026,606 –1 – 026,606 – 12 (2003). DOI 10.1103/PhysRevE.68.026606. URL https://link.aps.org/doi/10.1103/PhysRevE.68.026606

54. S. Başkal, Y.S. Kim, M.E. Noz, Wigner's Space-Time Symmetries Based on the Two-by-Two Matrices of the Damped Harmonic Oscillators and the Poincaré Sphere, Symmetry **6**(3), 473–515 (2014). DOI 10.3390/sym6030473. URL http://www.mdpi.com/2073-8994/6/3/473/

Appendix A
Elements of Group Theory

Abstract Since there are many excellent textbooks on group theory, this Appendix contains only those aspects of group theory that are essential for understanding the theory and application of the Poincaré group. Usually physicists' first exposure to group theory is with the three-dimensional rotation group in quantum mechanics. This group is used throughout this book to illustrate the abstract concepts. Among other examples contained in this Appendix, the two-dimensional Euclidean group or $E(2)$ is used extensively. The $E(2)$ group consists of two translations on the xy-plane and rotations around the z-axis. Since 1960, group theory has become an indispensable tool in theoretical physics in connection with the quark model. However, the trend now is that there is more interest in studying space-time coordinate transformations and in constructing explicit representations of non-compact groups, particularly those which are neither simple nor semisimple. Because it describes the fundamental space-time symmetries in the four-dimensional Minkowski space, the Poincaré group occupies an important place in this trend. We start with definitions of standard terms used in group theory. We then illuminate vector spaces and representations followed by a discussion of the properties of matrices. Finally, we discuss Schur's Lemma.

Since there are already many excellent textbooks and lecture notes on group theory, it is not necessary to give another full-fledged treatment of group theory in this book. We are interested here in only those aspects of group theory which are essential for understanding the theory and applications of the Poincaré group. We shall not present proofs of theorems if they are readily available in standard textbooks [1, 2, 3, 4, 5, 6] and many others. Physicists learn group theory not by proving theorems but by working out examples. The purpose of this Appendix and of the entire book is to present a systematic selection of examples from which the reader can formulate his/her own concepts.

S. Başkal et al., *Theory and Applications of the Poincaré Group*, Fundamental Theories of Physics 217, https://doi.org/10.1007/978-3-031-64376-7

Physicists' first exposure to group theory takes place through the three-dimensional rotation group in quantum mechanics. Since 1960, group theory has become an indispensable tool in theoretical physics in connection with the quark model. During the past fifty or more years, the trend has been toward more abstract group theory, with emphasis on constructing unitary representations of compact Lie groups which are simple or semisimple in connection with constructing multiplet schemes for quarks and other fundamental particles.

However, the present trend, as is manifested in the study of supersymmetry [7] and the Kaluza-Klein theory [8], is that we are becoming more interested in studying space-time coordinate transformations and in constructing explicit representations of non-compact groups, particularly those which are neither simple nor semisimple. Because it describes the fundamental space-time symmetries in the four-dimensional Minkowski space, the Poincaré group occupies an important place in this trend. The purpose of this book is to discuss the representations of the Poincaré group which is non-compact and which is neither simple nor semisimple.

The mathematical theorems connected with non-compact groups are somewhat beyond the grasp of most physics students. Fortunately, however, calculations involved in the study of the Poincaré group are simple enough to carry out with only a basic knowledge of group theory. The examples selected in this book are aimed at building a bridge between the abstract concepts and explicit calculations.

It is assumed that the reader is already familiar with the three-dimensional rotation group. This group is used throughout this book to illustrate the abstract concepts. Among other examples contained in this Appendix, also in Appendix B, and in Chap. 2, the two-dimensional Euclidean group or $E(2)$ is used extensively. As we note in Chap. 2, there are quite a variety of reasons for its extensive usage. In addition to those, Gilmore [5] uses the two-parameter Lie group consisting of multiplications and additions of real numbers, which is very similar to the $E(2)$ group. His book contains a very comprehensive coverage of Lie groups, and is thus one of the most popular books among students. By using the $E(2)$ group, we can establish a bridge between Gilmore's textbook and what we do in this book.

In Sect. A.1, we give definitions of standard terms used in group theory. In Sect. A.2, subgroups are reviewed. In Sect. A.3, vector spaces and representations are examined. In Sect. A.4, equivalence classes, orbits, and little groups are discussed. In Sect. A.5, the properties of matrices are given, and Sect. A.6 contains a discussion of Schur's Lemma. In Sect. A.7, we list further examples in the form of exercises and problems.

A.1 Definition of a Group

A set of elements g forms a group G if there is a multiplication (i.e., a group operation) defined for any two elements a and b of G according to the group axioms:

1. Closure: for any a and b in G, $a \cdot b$ is in G, where $\cdot$ is the group operation generally referred to as group multiplication.

2. Associative law: $(a \cdot b) \cdot c = a \cdot (b \cdot c)$, where c is also in G.
3. Identity: there exists a unique identity element e such that $e \cdot a = a \cdot e = a$ for all a in G.
4. Inverse: for every a in G, there exists an inverse element, denoted a^{-1}, such that $a^{-1} \cdot a = a \cdot a^{-1} = e$.

The third axiom is a consequence of the other three.

The number of elements in the group is called the order of G. The order can be either finite or infinite. It is assumed that the reader is familiar with some examples of groups, such as the group S_3 which consists of permutations of three objects. The group most familiar to physicists is $O(3)$, namely the three-dimensional rotation group without space inversions, sometimes referred to as the *proper rotation group*. The order of S_3 is six, while the order of the three-dimensional rotation group is infinite.

Groups in which the commutative law for group multiplication holds are called Abelian groups. The group consisting of rotations around the origin on the xy-plane or $O(2)$ is Abelian. The group of translations in three-dimensional space is also Abelian. S_3 and $O(3)$ are not Abelian groups.

Two groups are isomorphic if there is a one-to-one mapping F from one group (G) to the other (H) that preserves group multiplication, i.e. $F(g_1 g_2) = F(g_1)F(g_2)$, where $h_1 = F(g_1)$ and $h_2 = F(g_2)$. S_3 is isomorphic to the symmetry group of an equilateral triangle. This group consists of three rotations (by $0°$, $120°$, and $240°$) and three flips, each of which results in interchange of two vertices. The groups are *homomorphic* if the mapping is onto but not one-to-one. The three-dimensional rotation group is homomorphic to the group $SU(2)$ which consists of two-by-two unitary matrices which are unimodular (the determinant is equal to one).

There are many examples of various groups discussed in standard textbooks. One example which will be useful in this book is the *four group*, consisting of four elements, e, a, b, c, where e is the unit element. This group can have two distinct multiplication laws:

$$(A) \qquad a^2 = b, \qquad ab = c = a^3, \qquad a^4 = b^2 = e, \qquad (A.1)$$

with the multiplication table:

e	a	b	c
a	b	c	e
b	c	e	a
c	e	a	b

$$(B) \qquad a^2 = b^2 = c^2 = e, \qquad ab = c, \qquad ac = b, \qquad bc = a, \qquad (A.2)$$

with the table:

e	a	b	c
a	e	c	b
b	c	e	a
c	b	a	e

It is clear that type A is cyclic and is isomorphic to the group consisting of $1, i, -1$, and $-i$. The multiplication law for the four group of type B is identical to that for the Pauli spin matrices up to the phase factor. This group will be useful in understanding space-time reflection properties [9].

   Another example which will play an important role throughout this book is the two-dimensional Euclidean group. This group, consists of rotations and translations in the xy-plane, resulting in the linear transformation:

$$x = x_0 \cos\theta - y_0 \sin\theta + u\,,$$
$$y = x_0 \sin\theta + y_0 \cos\theta + v\,. \tag{A.3}$$

This is the rotation of the coordinate system by angle θ followed by the translation of the origin to (u, v). In this book we are using the coordinate system (t, z, x, y). This coordinate system is used to represent the Lorentz group $O(3, 1)$ in Minkowski space. In matrix notation, the linear transformation of the $E(2)$ group takes the form

$$\begin{pmatrix} 1 \\ 1 \\ x \\ y \end{pmatrix} = \begin{pmatrix} 1 & 0 & 0 & 0 \\ 0 & 1 & 0 & 0 \\ 0 & u & \cos\theta & -\sin\theta \\ 0 & v & \sin\theta & \cos\theta \end{pmatrix} \begin{pmatrix} 1 \\ 1 \\ x_0 \\ y_0 \end{pmatrix}. \tag{A.4}$$

The determinant of the above transformation matrix is 1, and the inverse transformation takes the form

$$\begin{pmatrix} 1 \\ 1 \\ x_0 \\ y_0 \end{pmatrix} = \begin{pmatrix} 1 & 0 & 0 & 0 \\ 0 & 1 & 0 & 0 \\ 0 & -u' & \cos\theta & \sin\theta \\ 0 & -v' & -\sin\theta & \cos\theta \end{pmatrix} \begin{pmatrix} 1 \\ 1 \\ x \\ y \end{pmatrix}, \tag{A.5}$$

where

$$u' = u \cos\theta + v \sin\theta\,,$$
$$v' = -u \sin\theta + v \cos\theta\,.$$

Like matrices representing the three-dimensional rotation group, the matrix in Eq. (A.4) contains three parameters. While similar to the rotation group in some aspects, $E(2)$ also shares many characteristics with the Poincaré group also known as the inhomogeneous Lorentz group. Indeed, this group will play a very important role both in illustrating mathematical theorems and in constructing representations of the Poincaré group.

A.2 Subgroups, Cosets, and Invariant Subgroups

A set of elements H, contained in a group G, forms a subgroup of G, if all elements in H satisfy group axioms. Even permutations in S_3 form a subgroup. Rotations around the z-axis form a subgroup of the rotation group.

If H is a subgroup of G, the set gH with g in G is called the left coset of H. Similarly, the set Hg is called the right coset of H. The number of right cosets and the number of left cosets are either both infinite, or both finite and equal. The number of cosets is called the index of H in G. If the order of G is finite, it is the product of the order and the index of H. In other words, if the order of G is a, and the order and index of H are b and c respectively, then a = bc. Two cosets are either identical or disjoint. The group G therefore is a union of all left or right cosets, or can be expanded in terms of cosets. This leads us to the concept of coset space in which cosets form the entire group. The coset space is usually written as G/H.

S_3 is the union of the two cosets of the even permutation subgroup. One coset is the even permutation subgroup itself, and the other is the set of even permutations followed or preceded by a transposition of two objects resulting in odd permutations. The coset space consists of the set of even permutations and the set of odd permutations. The order of the subgroup in this case is 3, and the index is 2, while the order of S_3 is 6.

Rotations around the z-axis form a subgroup of the three-dimensional rotation group. Cosets in this case are rotations of the z-axis followed or preceded by the rotation around the z-axis. The coset space consists of all possible directions the rotation axis can take or the source of a unit sphere. In this case, the order of G, the order of H, and the index are all infinite.

A right coset of H in general need not be identical to the left coset. There are certain subgroups N in which every right coset is a left coset, satisfying

$$gN = Ng \quad \text{or} \quad gNg^{-1} = N \quad \text{for all} \quad g \quad \text{in} \quad G. \tag{A.6}$$

Subgroups N having this property are called normal or invariant subgroups. The set of even permutations in S_3 is an invariant subgroup. The group of rotations around the z-axis is not an invariant subgroup of the three-dimensional rotation group.

If one multiples two cosets of an invariant subgroup N, the result itself is a coset. Since

$$Nn = nN = N \tag{A.7}$$

for every n in N, we have

$$NN = N. \tag{A.8}$$

For g_1 and g_2 in G,

$$\begin{aligned}
(Ng_1)(Ng_2) &= N(g_1N)g_2 \\
&= N(Ng_1)g_2 = (NN)(g_1g_2) \\
&= N(g_1g_2). \tag{A.9}
\end{aligned}$$

The set of cosets of an invariant subgroup is therefore a group, with N as the identity element. This group is called the quotient group of G with respect to N and is written G/N.

If a group does not contain invariant subgroups, it is called a simple group. $O(3)$ is a simple group. If a group contains invariant subgroups which are not Abelian, it is called a semisimple group. S_3 is semisimple group. $E(2)$ contains an Abelian invariant subgroup, and is therefore neither simple nor semisimple.

Let us look at $E(2)$ closely. This group has two subgroups. One of them is the rotation subgroup consisting of matrices:

$$R(\theta) = \begin{pmatrix} 1 & 0 & 0 & 0 \\ 0 & 1 & 0 & 0 \\ 0 & 0 & \cos\theta & -\sin\theta \\ 0 & 0 & \sin\theta & \cos\theta \end{pmatrix}, \tag{A.10}$$

and the other is the translation subgroup represented by:

$$T(u,v) = \begin{pmatrix} 1 & 0 & 0 & 0 \\ 0 & 1 & 0 & 0 \\ 0 & u & 1 & 0 \\ 0 & v & 0 & 1 \end{pmatrix}. \tag{A.11}$$

If we use the notation $F(u,v,\theta)$ for the matrix of Eq. (A.4), then

$$F(u,v,\theta) = T(u,v)R(\theta). \tag{A.12}$$

Both T and R are closed under their respective group multiplications. They are Abelian subgroups. The matrix in Eq. (A.4) is a product of the two matrices: $T(u,v)R(\theta)$. It is easy to verify that T is an invariant subgroup, while R is not.

As we can see in S_3 and $E(2)$, a group can be generated by two subgroups H and K:

$$G = KH. \tag{A.13}$$

This means that an element of H is to be multiplied from left by an element of K. This is usually written as

$$g = kh = (k,h). \tag{A.14}$$

The multiplication law for g is then

$$\begin{aligned} g_2 g_1 &= (k_2 h_2)(k_1 h_1) \\ &= k_2 (h_2 k_1 h_2^{-1}) h_2 h_1. \end{aligned} \tag{A.15}$$

If $(h_2 k_1 h_2^{-1})$ belongs to K, the above product can be written as

$$g_2 g_1 = (k_2 h_2 k_1 h_2^{-1}, h_2 h_1). \tag{A.16}$$

The multiplication of two groups given in Eq. (A.16) is called the semi-direct product. The subgroup K of G in this case has to be an invariant subgroup in order to satisfy the requirement of Eq. (A.14).

If every element in H commutes with every element in K, then we say that K and H commute with each other. In that case, the product becomes

$$
\begin{aligned}
g_2 g_1 &= (k_2 h_2)(k_1 h_1) \\
&= (k_2 k_1, h_2 h_1).
\end{aligned}
\tag{A.17}
$$

If this multiplication law is satisfied, it is said that G is a direct product of H and K. The addition of orbital and spin angular momenta in quantum mechanics is an example of a direct product of the $O(3)$ and $SU(2)$ groups.

S_3 is a semi-direct product of the even permutation group and the two-element group consisting of the identity and a transposition of the first two elements. $E(2)$ is a semi-direct product of the translation and rotation subgroups. Because the translation subgroup is an invariant subgroup, T and R can still be separated as

$$
F(u_2, v_2, \theta_2) F(u_1, v_1, \theta_1) = T(u', v') R(\theta_2 + \theta_1),
\tag{A.18}
$$

where

$$
\begin{aligned}
u' &= u_2 + u_1 \cos \theta_2 - v_1 \sin \theta_2, \\
v' &= v_2 + u_1 \sin \theta_2 + v_1 \cos \theta_2.
\end{aligned}
\tag{A.19}
$$

Indeed, this property plays an important role in constructing representations of the Poincaré group as seen in this book.

A.3 Equivalence Classes, Orbits, and Little Groups

An element a of G is said to be *conjugate* to the element b if there exists an element g in G such that $b = gag^{-1}$. It is easy to show that conjugacy is an equivalence relation, i.e.,

1. $a \sim a$ (reflexive),
2. $a \sim b$ implies $b \sim a$ (symmetric),
3. $a \sim b$ and $b \sim c$ implies $a \sim c$ (transitive)

For this reason, a and b are said to belong to the same *equivalence class*. Thus the elements of G can be divided into equivalent classes of mutually conjugate elements.

The class containing the identity element consists of just one element, because $geg^{-1} = e$ for all g in G. The even and odd permutations in S_3 form separate classes. All rotations by the same angle around the different axes going through the origin in three-dimensional space belong to the same equivalence class in the rotation group.

The subgroup H of G is said to be conjugate to the subgroup K if there is an element g such that $K = gHg^{-1}$. If H is an invariant subgroup, then it is conjugate to itself. As was noted in Sect. A.2, even permutations form an invariant subgroup in S_3. In the three-dimensional rotation group, the subgroup of rotations around any given axis is conjugate to the $O(2)$-like subgroup consisting of rotations around the z-axis.

Let p be a point of the vector space X. The maximal subgroup $G(p)$ of G which leaves p invariant, i.e., $G^{(p)}p = p$, is called the *little group* of G at p. In the three-dimensional x, y, z space, rotations around the z-axis form the $O(2)$-like little group of $O(3)$ at $(0, 0, 1)$. This little group is conjugate to the little group at $(1, 0, 0)$ which consists of rotations around the x-axis.

The set of points V that can be reached through the application of G on a single point p in X is called the *orbit* of G at p. Two orbits are either identical or disjoint. The orbit of $O(3)$ at $(0, 0, R)$ is the surface of the sphere with radius R centered around the origin. The orbit of $O(2)$ at the point $(x = a, y = 0)$ on the xy-plane is the circumference of a circle with radius a.

The coset space $G/G^{(p)}$ is therefore identical to orbit V. As was noted in Sect. A.2, the coset space $O(3)/O(2)$ is the surface of a unit sphere in three-dimensional space.

Let us see how these concepts are helpful in understanding the $E(2)$ group. The little group of $E(2)$ at $x = y = 0$ is the rotation subgroup represented by Eq. (A.10) satisfying the relation

$$
\begin{pmatrix} 1 \\ 1 \\ 0 \\ 0 \end{pmatrix} = \begin{pmatrix} 1 & 0 & 0 & 0 \\ 0 & 1 & 0 & 0 \\ 0 & 0 & \cos\theta & -\sin\theta \\ 0 & 0 & \sin\theta & \cos\theta \end{pmatrix} \begin{pmatrix} 1 \\ 1 \\ 0 \\ 0 \end{pmatrix}. \tag{A.20}
$$

The orbit of $E(2)$ at the origin is the entire plane, because every point on the plane can be reached through a translation of the origin:

$$
\begin{pmatrix} 1 \\ 1 \\ u \\ v \end{pmatrix} = \begin{pmatrix} 1 & 0 & 0 & 0 \\ 0 & 1 & 0 & 0 \\ 0 & u & 1 & 0 \\ 0 & v & 0 & 1 \end{pmatrix} \begin{pmatrix} 1 \\ 1 \\ 0 \\ 0 \end{pmatrix}. \tag{A.21}
$$

The coset space $E(2)/O(2)$ is therefore the entire plane.

Rotations by the same angle around two different points on the plane belong to the same equivalence class in the $E(2)$ group. For instance, the rotation around the origin is represented by the matrix of Eq. (A.20). The rotation around (u, v) is

$$\begin{pmatrix} 1 & 0 & 0 & 0 \\ 0 & 1 & 0 & 0 \\ 0 & u & 1 & 0 \\ 0 & v & 0 & 1 \end{pmatrix} \begin{pmatrix} 1 & 0 & 0 & 0 \\ 0 & 1 & 0 & 0 \\ 0 & 0 & \cos\theta & -\sin\theta \\ 0 & 0 & \sin\theta & \cos\theta \end{pmatrix} \begin{pmatrix} 1 & 0 & 0 & 0 \\ 0 & 1 & 0 & 0 \\ 0 & -u & 1 & 0 \\ 0 & -v & 0 & 1 \end{pmatrix}$$

$$= \begin{pmatrix} 1 & 0 & 0 & 0 \\ 0 & 1 & 0 & 0 \\ 0 & -u\cos\theta & v\sin\theta + u & \cos\theta & -\sin\theta \\ 0 & -u\sin\theta - v\cos\theta - v & \sin\theta & \cos\theta \end{pmatrix} . \tag{A.22}$$

Clearly, the rotation around the (u, v) is conjugate to the rotation around the origin. In matrix language, the above operation is known as a *similarity* transformation.

Let us consider another example. The group consisting of four-by-four Lorentz transformation matrices is called the Lorentz group. We can now consider a four-vector P and the maximal subgroup which leaves this four-vector invariant. This subgroup is the little group of the Lorentz group at P. If P is the four-momentum of a free massive particle, and if this particle is at rest, the little group is the three-dimensional rotation group. If we perform Lorentz transformations on this, then the four-momentum traces the hyperbolic surface:

$$P_0^2 - |\mathbf{P}|^2 = M^2 , \tag{A.23}$$

where M, P_0, and $\mathbf{P}$ are the mass, energy, and momentum of the particle respectively. This hyperbolic surface is the orbit. The matrices of the little group are Lorentz-boosted rotation matrices. The little groups and orbits of the Lorentz group are studied in more detail in Chaps. 2 and 3.

A.4 Representations and Representation Spaces

A set of linear transformation matrices homomorphic to the group multiplication of G is called a *representation* of the group. The word *homomorphic* is appropriate here because the matrices need not have a one-to-one correspondence with the group elements. In addition, there can be more than one set of matrices forming a representation of the group. For instance, the three-dimensional rotation group can be represented by two-by-two, three-by-three, or n-by-n matrices, where n is an arbitrary integer.

$GL(n, c)$ and $GL(n, r)$ are the groups of non-singular n-by-n complex and real matrices respectively. $SL(n, c)$ is an invariant subgroup of $GL(n, c)$ with determinant one, and is called the *unimodular group*. These matrices perform linear transformations on a linear vector space V containing vectors of the form:

$$X = (x_1, x_2, x_3, ..., x_n) . \tag{A.24}$$

$U(n)$ is the unitary subgroup of $GL(n, c)$ which leaves the norm

$$(|x_1|^2 + |x_2|^2 + \cdots + |x_n|^2)^{1/2} \tag{A.25}$$

invariant. $O(n)$ is a subgroup of $U(n)$ in which all elements are real, and $SO(n)$ is the unimodular subgroup of $O(n)$.

$U(n, m)$ is the *pseudo-unitary group* applicable to the $(n+m)$-dimensional vector space which leaves the quantity

$$[(|x_1|^2 + |x_2|^2 + \cdots + |x_n|^2) - (|y_1|^2 + |y_2|^2 + \cdots + |y_m|^2)] \tag{A.26}$$

invariant. $O(n, m)$ is the real subgroup of $U(n, m)$. The group of four-by-four Lorentz transformation matrices is $O(3, 1)$.

In addition to homogeneous linear transformations, there are inhomogeneous linear transformations. The $E(2)$ transformation defined in Eq. (A.3) is an inhomogeneous linear transformation in that the u and v variables are added to x and y respectively after a rotation. Inhomogeneous linear transformations can also be represented by matrices as can be seen in the the case of $E(2)$. Properties of this kind of matrix transformation are not well known to physicists. While avoiding general theorems on this subject, we study some special cases of the inhomogeneous transformations, including those of the inhomogeneous Lorentz transformations.

We shall use the symbol L to denote a representation of G. L therefore consists of matrices acting on the vector space V. A closed subspace W of V will be said to be *invariant* if the action of L on any element in W for all elements in W transforms into an element belonging to W. If we restrict the operations of L to W, we obtain a new representation L^W whose vector space is W. We call L^W a subrepresentation of L. If a representation does not contain subrepresentations, it is said to be *irreducible*.

Let us consider the $n = 2$ state hydrogen atom consisting of spinless proton and electron. This energy state has four degenerate states with two different values of the angular momentum quantum number ℓ. There is only one state for $\ell = 0$ which is usually called the s state. There are three different states for the p state with $\ell = 1$. Under rotations, the s state remains invariant, and the p state wave functions undergo homogeneous linear transformations. The p state wave functions never mix with the s state. Thus we say that the $n = 2$ hydrogen wave functions can be divided into two invariant subspaces, and the rotation matrices become reduced to the three-by-three matrix for the p state and a trivial one-by-one matrix for the s state.

In addition to vectors, there are tensors. For example, the second rank tensor is formed from the *direct product* of two vector spaces. Let V and V' be two invariant vector spaces with n and m components respectively, and let L and L' be representations of the same group. The direct product of these two spaces results in a set of $x_i y_j$, where $i = 1, 2, ..., n$ and $j = 1, 2, ..., m$. The question of whether these elements form an nm-dimensional vector space to which nm-by-nm matrices are applicable, whether these matrices are homomorphic to the multiplication law of the group G, and whether this matrix is reducible is one of the prime issues in representation theory. We are already familiar with some aspects of this through our experience with the rotation group.

Let us consider two spin-1/2 particles. Because each particle can have two different spin states, the dimension of the resulting space is four. However, this space can be divided into one corresponding to the spin-0 state and three for the spin-1 states. The rotation matrix for each of the spinors is two-by-two. The direct product of the two two-by-two matrices result in a four-by-four matrices which can be reduced to a block diagonal form consisting of one three-by-three matrix and one trivial one-by-one matrix. In quantum mechanics, this procedure is known as the calculation of Clebsch-Gordan coefficients [10, 11, 12]. Calculations of Clebsch-Gordan coefficients occupy a very important part of mathematical physics, and the literature on this subject is indeed extensive. For this reason, we shall not elaborate further on this problem.

If L consists of finite-dimensional matrices acting on finite-dimensional vector spaces, such as those of $GL(n, c)$ and its subgroups, the representation is said to be finite-dimensional. If L is reducible, the matrices can be brought to a block diagonal form. In addition to finite-dimensional matrices, we have to consider the possibility of the size of the representation matrices becoming infinite. In this case, the representation is called infinite-dimensional. The technique of handling infinite-dimensional matrices is far more complicated than that for the finite-dimensional case. We discuss some of the infinite-dimensional matrices in chapters throughout the book without going into full-fledged representation theory.

The ultimate purpose of physics is to calculate numbers which can be measured in laboratories. Therefore we have to convert the concept of abstract group theory into measurable numbers. This is only possible through construction of representations or matrices. The most common practice in constructing such matrices is solving the eigenvalue equations. Solving the time-independent Schrödinger equation is a process of constructing representations. Wave functions correspond to vector spaces. Construction of vector spaces often precedes that of the matrices representing the group. Therefore we often call this *representation space*. The spherical harmonics with a given value of ℓ form a representation space for $(2\ell + 1)$-by-$(2\ell + 1)$ matrices representing the three-dimensional rotation group.

There are many different methods of constructing representations, and many of them are based on the techniques of deriving new representations of the same or a different group starting from known representations. The most common practice is the above-mentioned direct product of two representations. As we demonstrated in the case of $E(2)$, the semidirect product is also used often in physics, especially in the study of the Poincaré group. This has been elaborated on in many chapters in this book.

It is also quite common to use the exponentiation of matrices. Let us consider a set of matrices A. Its exponentiation results in another set of matrices of the form

$$B = e^A = \sum_{n=0}^{\infty} A^n / n! . \tag{A.27}$$

This method is quite convenient when the matrices A take a simpler form than that of B. This point will be studied in greater detail in Appendix B.

The use of complex numbers is often helpful in constructing representations. Unitary or orthogonal matrices are much easier to deal with than other forms of matrices. For this reason, we sometimes study a pseudo-unitary matrix by converting it to a unitary matrix. The best known example is the group of Lorentz transformations. Lorentz transformation matrices constitute a four-by-four pseudo-orthogonal representation of $O(3, 1)$. However, if we change the time variable t to (it), the transformation matrix becomes orthogonal. We discuss this method in Chap. 3.

Another useful method in constructing representations is the method of group contraction. The surface of the earth appears flat most of the time. However, it is a spherical surface. We are therefore led to the idea that the $E(2)$ group which deals with transformations on a flat surface is an approximation of those on a spherical surface. Therefore, $E(2)$ may be a limiting case of the rotation group. This process and its physical relevance is discussed in Chap. 4.

Most of the matrix calculations are straight-forward. However, sometimes, it is easier to do the same calculation with different mathematics. For instance, the matrix algebra of the $E(2)$ group which we carried out using four-by-four matrices throughout this Chapter can be translated into the algebra of the complex variable $z = x + iy$. Another example is the $SL(2, c)$ group. This group consists of two-by-two matrices of the form

$$W = \begin{pmatrix} a & b \\ c & d \end{pmatrix}, \tag{A.28}$$

with the condition: $ad - bc = 1$. Here, a, b, c, and d are complex numbers. The algebra of this matrix can be that of the conformal transformation [13]:

$$w = \frac{az + b}{cz + d}. \tag{A.29}$$

Quite often in physics, we use the differential form for operators. For instance, momentum and angular momentum operators are written in terms of differential operators. Where do these operators stand in group theory? This question will be addressed in Appendix B on Lie groups.

When we construct representations, we are constructing matrices. Then it is important to know to what vector or representation space these matrices are applied. Indeed, the construction of a representation cannot be separated from the construction of its representation space. In dealing with this vector space, it is convenient to use the basis vectors. Every vector in the vector space can be written as a linear combination of the basis vectors. The basis vectors in the three-dimensional Cartesian system are usually discussed in Freshmen physics. The spherical harmonics constitute the basis vectors for a given integer ℓ representation of the rotation group.

There are many other forms for basis vectors. Let us for example consider an arbitrary polynomial in the complex variable z:

$$F_n(z) = a_0 + a_1 z^1 + a_2 z^2 + \cdots + a_n z^n. \tag{A.30}$$

This is a linear combination of z^k, with $k = 0, 1, 2, ..., n$. Thus each z^k can be regarded as a basis vector. Thus a linear transformation on this representation space should result in rearrangement of the coefficients $a_0, a_1, ..., a_n$. This point is discussed in detail in Chap. 3.

In terms of the basis vectors, a similarity transformation is a rearrangement of the basis vectors. In the three-dimensional rotation group, a similarity transformation is a rotation of the coordinate system. In the two-dimensional Euclidean space, a similarity transformation with translation matrix results in a translation of the origin. Again in the three-dimensional rotation group, it is more convenient to use the combinations $(x+iy)$, $(x-iy)$, and z, instead of the x, y, and z coordinate system. This is also a similarity transformation.

A.5 Properties of Matrices

Matrices are in general rectangular, i.e., their number of rows does not have to be the same as that of columns. We shall in fact be using one of them in Appendix B. However, in order to represent a group, every matrix in the set has to have an inverse. For this reason, matrices representing a group have to be square and non-singular. This means that the determinant of the matrix cannot vanish.

It is easy to study a matrix if it is of diagonal form. Therefore it is of prime importance to see whether any given matrix can be diagonalized by a similarity transformation or a conjugate operation. Any matrix M defined over a complex vector space can be brought to diagonalized form by a similarity transformation provided that M commutes with its Hermitian conjugate, i.e.,

$$MM^{\dagger} = M^{\dagger}M . \tag{A.31}$$

Diagonal elements of the diagonalized matrices are the eigenvalues of the matrices.

Those matrices with which physicists are most familiar are unitary and Hermitian matrices. Unitary matrices, U, satisfy the condition:

$$U^{\dagger} = U^{-1}, \qquad \text{or} \qquad U^{\dagger}U = I . \tag{A.32}$$

Hermitian matrices are of the type:

$$H = H^{\dagger} . \tag{A.33}$$

Both Hermitian and unitary matrices can be brought to a diagonal form through a similarity transformation:

$$SHS^{-1} = \text{Diagonal} , \tag{A.34}$$

and

$$SUS^{-1} = \text{Diagonal} . \tag{A.35}$$

S in this case is a unitary matrix. The eigenvalues of unitary matrices have unit modulus. The eigenvalues of Hermitian matrices are real. It is possible to write a unitary matrix as

$$U = \exp(-iH) . \tag{A.36}$$

There are also anti-Hermitian matrices:

$$A = -A^\dagger . \tag{A.37}$$

The exponentiation of the anti-Hermitian matrix is not unitary. However, an anti-Hermitian matrix commutes with its Hermitian conjugate, and therefore can be diagonalized by a similarity transformation.

In dealing with inhomogeneous linear transformations, we often use triangular matrices. For example, if we choose the bases to be $x + iy$ and $x - iy$ on the two-dimensional Euclidean plane as discussed in Sect. A.4, then the rotation matrix becomes diagonal:

$$R(\theta) = \begin{pmatrix} 1 & 0 & 0 & 0 \\ 0 & 1 & 0 & 0 \\ 0 & 0 & e^{-i\theta} & 0 \\ 0 & 0 & 0 & e^{i\theta} \end{pmatrix}, \tag{A.38}$$

and the translation matrix becomes

$$T(u,v) = \begin{pmatrix} 1 & 0 & 0 & 0 \\ 0 & 1 & 0 & 0 \\ 0 & u+iv & 1 & 0 \\ 0 & u-iv & 0 & 1 \end{pmatrix}. \tag{A.39}$$

This translation matrix is a lower triangular matrix in the sense that it has only zero elements above the main diagonal. The semidirect product of the rotation and translation becomes the triangular matrix:

$$T(u,v)R(\theta) = \begin{pmatrix} 1 & 0 & 0 & 0 \\ 0 & 1 & 0 & 0 \\ 0 & u+iv & e^{-i\theta} & 0 \\ 0 & u-iv & 0 & e^{i\theta} \end{pmatrix}. \tag{A.40}$$

An upper triangular matrix has only zero elements below the main diagonal. Throughout this book, by triangular matrix, we shall mean an upper triangular or a lower triangular matrix depending on the context.

Matrix multiplication of two triangular matrices yields a triangular matrix. Thus there can be groups of triangular matrices, provided an inverse exists for each matrix.

Unlike the above-mentioned unitary, Hermitian, and anti-Hermitian matrices, triangular matrices cannot be brought to diagonal form by a similarity transformation. However, every non-zero matrix B defined over a complex vector space is unitarily equivalent (similarity transformation with the transformation matrix being unitary) to an upper triangular matrix if not a diagonal matrix. The diagonal elements of this triangular matrix are the eigenvalues of B.

Among triangular matrices, there are those with zero diagonal elements, such as

$$T_0(3) = \begin{pmatrix} 0 & a & b \\ 0 & 0 & c \\ 0 & 0 & 0 \end{pmatrix}.$$ (A.41)

This matrix has the property that

$$[T_0(3)]^3 = 0.$$ (A.42)

In general, an n-by-n triangular matrix with zero diagonal elements yields

$$[T_0(n)]^n = 0.$$ (A.43)

For this reason, the exponentiation of the T_0 matrix becomes

$$\exp(T_0) = \sum_{k=0}^{n-1} \frac{1}{k!} (T_0)^k.$$ (A.44)

The series truncates!

In physics, the size of the matrix is an important factor in carrying out calculations, and also in illustrating mathematical theorems. As the size of the matrix becomes larger, the problem turns out to be more complicated. For this reason, we seek constantly the smallest matrix with which we can do our calculations. There are two important methods which are most commonly employed. One is to seek whether the matrices can be reduced to block diagonal form. This method is called the construction of irreducible representation which we shall discuss in Sect. A.6. Another method commonly used in physics is the method of covering group. The best known example of this method is to use the two-by-two unitary unimodular matrices or $SU(2)$ for the rotation group. We discuss this procedure in more detail in Chapters throughout the book.

In addition, there are infinite-by-infinite matrices applicable to infinite-dimensional vector spaces. This kind of matrix is quite common in physics. Fourier transformations, which cannot be separated from quantum mechanics, are applications of this infinite-dimensional matrix. The so-called S matrix which relates the initial and final states in scattering processes is also an infinite-dimensional matrix. As we see in other chapters, certain representations of the Lorentz group are infinite-dimensional.

A.6 Schur's Lemma

We discussed in Sect. A.4 the fact that a set of matrices or linear transformations on a vector space V_m, homomorphic to the group multiplication of g, constitutes a representation of the group. The dimension of the vector space is denoted by m. We shall denote this set of matrices or linear transformations as $L(s)$ where the parameter s may have a domain which is finite or infinite, denumerable or continuous. For the permutation group S_3, the domain of s is denumerable and finite, while the domain is continuous and infinite for the three-dimensional rotation group.

If we consider $L'(s)$ to be another set of matrices which also forms a representation of the same group over another vector space V'_n, then the two representations are said to be equivalent if V_m can be mapped onto V'_n by a one-to-one linear transformation. This means that any two vectors related by the transformation $L(s)$, e.g., x and $x' = L(s)x$, map into related vectors y and $y' = L'(s)y$ with the same value of s. Then m must equal n, otherwise there would not be any linear one-to-one mapping from V_m to V'_n. The equivalence mapping is accomplished as usual by a non-singular matrix S:

$$x = Sy, \tag{A.45}$$

and hence

$$L'(s) = S^{-1}L(s)S. \tag{A.46}$$

In Sect. A.4 we noted also that the set of matrices or linear transformation $L(s)$ is called reducible if there exists an invariant subspace W of V, i.e., if every vector of W maps into a vector of W under each of the transformations L. If there exists no invariant subspace of V except V and the null vector, the set $L(s)$ is said to be irreducible. The set is reducible if it can be decomposed into several irreducible systems. The corresponding vector space V is then a direct sum of irreducible invariant linear subspaces.

The construction of nonequivalent irreducible representations is one of the fundamental problems in the representation theory of groups. In addition, a practical method must be devised for decomposing a reducible representation into irreducible representations. Schur's Lemma, which consists of the following two theorems, yields via the second theorem, a practical method for determining if a given representation is irreducible.

The *first theorem* consists of the following. Let $L(s)$ and $L'(s)$ be two irreducible representations of the group G defined on the vector spaces V_m and V'_n respectively. Let S be a (rectangular) matrix of n rows and m columns, mapping V into V', i.e., for every s,

$$SL(s) = L'(s)S. \tag{A.47}$$

Then either $S = 0$ or S is nonsingular. If S is nonsingular, $m = n$ and $L(s)$ and $L'(s)$ are equivalent. That this is true can be seen from the following.

Consider the matrix S of rank r. Then all the transformations Sx where x is a vector in V form a linear subspace y in $V'[y = Sx]$ of dimension r which by our theorem is invariant under $L'(s)$. Since $L'(s)$ is irreducible, either $y = 0$ (then $r = 0$ and $S = 0$) or $y = V'$ (then $r = m$ and $n \geq m$). The vectors x of V for which $0 = Sx$ constitute a linear subspace x_1 of V of dimension $(n - r)$ which is invariant under $L(s)$ by Eq. (A.47). Since $L(s)$ is irreducible, either $x = V$ (then $r = 0$ and $S = 0$) or $x = 0$ (then $r = n$ and $m \geq n$). Thus either $S = 0$ or $r = n = m$, and S is nonsingular. Hence $L(s)$ and $L'(s)$ are equivalent. Since we are free to choose the coordinate systems as we like, we choose that coordinate system in which $L(s) = L'(s)$. We use this form of the equivalence statement in what follows.

The theorem which constitutes the second part of Schur's Lemma is extremely useful, because it provides a practical method for determining if a group representation is irreducible. This *second theorem* states that, if $L(s)$ is a representation of the group G on a complex vector space V, $L(s)$ is irreducible if and only if the only transformation $SV \rightarrow V$ such that

$$L(s)S = SL(s) \tag{A.48}$$

for all x in V are those for which S is a multiple of the identity. This says that $S = \lambda I$, where λ is a complex number and I is the identity matrix on V.

We know from linear algebra that a linear operator, such as a matrix, operating on a finite-dimensional complex vector space always has at least one eigenvalue. Suppose that λ is that eigenvalue of S which satisfies Eq. (A.48). Then there is a subspace C_λ of V consisting of vectors which satisfy

$$S\xi = \lambda\xi , \tag{A.49}$$

for all ξ in C_λ. This means that C_λ is a subspace of V with dimension greater than 0. Also, C_λ is invariant under $L(s)$ because

$$SL(s)\xi = L(s)S\xi = \lambda L(s)\xi . \tag{A.50}$$

If $L(s)$ is irreducible, then there is a parameter s such that $x = [L(s)\xi]$ for x in V. Thus the subspace $C_\lambda = V$.

Conversely, suppose that $L(s)$ is reducible. Then we know that V can be decomposed into a direct sum of invariant subspaces. Call one of these V_1. Any vector x in V can also be written uniquely as a direct sum of its vector components in each invariant subspace. Call x_1 the component in V_1. It is always possible to define a projection operator P on V such that $Px = x_1$ in V_1. Then $PL(s)x = L(s)Px = L(x)x_1$. Since, however, P cannot be a multiple of identity, $L(s)$ must be reducible.

A.7 Exercises and Problems

Exercise 1. Construct the maximal set of commuting operators for S_3.

For the system of three similar objects labeled as 1, 2, 3 respectively, we can perform six permutations. First, there are three permutations of the form

$$(12), \quad (23), \quad (31), \tag{A.51}$$

where each number is replaced by the succeeding number in the bracket, while the first one goes to the last position. In addition, there are two permutations of the form

$$(123), \quad (132), \tag{A.52}$$

The above five permutations together with the identity form the six permutations which can be performed on the three objects.

As Dirac did in his Eq. (13) of Sect. 55 in his book entitled *Principles of Quantum Mechanics* [14], we construct the following operators for the three-body system:

$$X_1 = I,$$
$$X_2 = [(12) + (23) + (31)]/3,$$
$$X_3 = [(123) + (132)]/2, \tag{A.53}$$

where I is the identity operator. X_1 can also be written as

$$X_1 = [I + (12) + (23) + (31) + (123) + (132)]/6. \tag{A.54}$$

The above operators commute with every permutation, and therefore with one another. They form the maximal set of commuting operators, the eigenvalues of these operators specify the representation.

Exercise 2. S_3 deals with three objects. Therefore it must be possible to construct three-by-three matrices performing permutations. Construct the multiplication table.

The identity matrix is

$$I = \begin{pmatrix} 1\ 0\ 0 \\ 0\ 1\ 0 \\ 0\ 0\ 1 \end{pmatrix}. \tag{A.55}$$

The even permutations (123) and (132) are respectively

$$A = \begin{pmatrix} 0\ 0\ 1 \\ 1\ 0\ 0 \\ 0\ 1\ 0 \end{pmatrix}, \qquad B = \begin{pmatrix} 0\ 1\ 0 \\ 0\ 0\ 1 \\ 1\ 0\ 0 \end{pmatrix}. \tag{A.56}$$

The odd permutations (23), (13), and (12) are

$$C = \begin{pmatrix} 1\ 0\ 0 \\ 0\ 0\ 1 \\ 0\ 1\ 0 \end{pmatrix}, \qquad D = \begin{pmatrix} 0\ 0\ 1 \\ 0\ 1\ 0 \\ 1\ 0\ 0 \end{pmatrix}, \qquad E = \begin{pmatrix} 0\ 1\ 0 \\ 1\ 0\ 0 \\ 0\ 0\ 1 \end{pmatrix}. \tag{A.57}$$

Using these matrices, we can construct the multiplication table as given below in Table A.1 . It is clear from this table that the S_3 is closed under group operation, and that I, A and B form a subgroup.

Exercise 3. Using Schur's Lemma check whether the above three-by-three representation of S_3 is irreducible.

We can start with the most general three-by-three matrix:

$$F = \begin{pmatrix} a\ b\ c \\ d\ e\ f \\ g\ h\ k \end{pmatrix}, \tag{A.58}$$

Table A.1: Group Multiplication Table for S_3. If we multiply one element by another, we end up with an element in the group. I, A and B form a subgroup.

	A	B	C	D	E
A	B	I	D	E	C
B	I	A	E	C	D
C	E	D	I	B	A
D	C	E	A	I	B
E	D	C	B	A	I

and impose the condition that

$$AF = FA, \qquad BF = FB, \qquad \ldots, \tag{A.59}$$

for all five nontrivial matrices given in Eqs. (A.56) and (A.57). Then the result is that F is a multiple of I or

$$G = \begin{pmatrix} 1 & 1 & 1 \\ 1 & 1 & 1 \\ 1 & 1 & 1 \end{pmatrix}, \qquad \text{or} \qquad H = \begin{pmatrix} 0 & 1 & 1 \\ 1 & 0 & 1 \\ 1 & 1 & 0 \end{pmatrix}. \tag{A.60}$$

We could have calculated these matrices using the expressions for X_2 and X_3 of Eq. (A.53). The above matrices commute with all six permutations, but are not diagonal. What went wrong? The answer to this question is simple. H in the above expression is a linear combination of I and G. G is a singular matrix. Therefore, the only non-singular matrix which commutes with all the elements is the unit matrix or a multiple thereof. Therefore, according Schur's lemma, the matrices given in Exercise 2 form an irreducible representation. In Chap. 12, we discuss representations for which X_2 and X_3 are not singular.

Exercise 4. We noted in Sect. A.5 that a multiplication of two triangular matrices leads to another triangular matrix. Is it possible to construct a triangular matrix by taking a product of diagonalizable non-triangular matrices?

We shall point out that this is possible by giving a concrete example. Let us consider the following three matrices:

$$A = \begin{pmatrix} \cosh(\eta/2) & -\sinh(\eta/2) \\ -\sinh(\eta/2) & \cosh(\eta/2) \end{pmatrix}, \tag{A.61}$$

$$B = [1/\cosh(\eta)]^{1/2} \begin{pmatrix} \cosh(\eta/2) & -\sinh(\eta/2) \\ \sinh(\eta/2) & \cosh(\eta/2) \end{pmatrix}, \tag{A.62}$$

$$C = \begin{pmatrix} [\cosh(\eta)]^{1/2} & 0 \\ 0 & [\cosh(\eta)]^{1/2} \end{pmatrix}. \tag{A.63}$$

If we take the product of these matrices:

$$D = ABC, \tag{A.64}$$

then the result is

$$D = \begin{pmatrix} 1 & -\tanh(\eta) \\ 0 & 1 \end{pmatrix}. \tag{A.65}$$

In mathematics, this operation is called the Iwasawa decomposition [see [15, 16]]. The physics of the above form is discussed in Chaps. 8 and 15.

Exercise 5. One of the most common representations of the three-dimensional rotation group consists of three-by-three orthogonal matrices applicable to a three-component column vector consisting of three real numbers. Show this three-component vector can be represented by a two-by-two unimodular Hermitian matrix consisting also of three-independent real numbers.

Let us consider the Pauli spin matrices. In addition to their well known Hermiticity and commutativity properties, they satisfy the orthogonality condition:

$$\frac{1}{2} Tr(\sigma_i \sigma_j) = \sigma_{ij}. \tag{A.66}$$

Therefore, the Pauli spin matrices can serve as three basis vectors for the rotation group, and a vector $\mathbf{A}$ with three real components A_1, A_2, and A_3, can be written as

$$[A] = \begin{pmatrix} A_3 & A_1 - iA_2 \\ A_1 + iA_2 & -A_3 \end{pmatrix}, \tag{A.67}$$

with

$$\frac{1}{2} Tr([A]^\dagger [B]) = A_1 B_1 + A_2 B_2 + A_3 B_3. \tag{A.68}$$

The rotation of the above matrix is performed by a unitary matrix $U(\theta, \phi, \alpha)$:

$$[A'] = U[A]U^\dagger, \tag{A.69}$$

where

$$U(\theta, \phi, \alpha) = \begin{pmatrix} e^{-i(\phi+\alpha)/2} \cos\frac{\theta}{2} & e^{-i(\phi-\alpha)/2} \sin\frac{\theta}{2} \\ -e^{-i(\phi-\alpha)/2} \sin\frac{\theta}{2} & e^{i(\phi+\alpha)/2} \cos\frac{\theta}{2} \end{pmatrix}. \tag{A.70}$$

It is quite clear that the trace of Eq. (A.68), which represents the inner product, is invariant under the unitary transformation given in Eq. (A.69).

The matrix form for a vector is quite common in physics. If this procedure is generalized to $SU(3)$, the above procedure is the *eightfold-way* representation of elementary particles [17, 18, 19]. This procedure can also be extended to four-vectors in the Minkowski space, as we discuss in Chap. 3.

Problem 1. Write out explicitly the multiplication table associated with the rotation through angles of $0°$, $120°$ and $240°$ and reflections at each vertex of an equilateral triangle. From the multiplication table, determine the number of subgroups.

Problem 2. Using Schur's lemma, show that the following six matrices constitute an irreducible representation of S_3 [[20]].

$$\begin{pmatrix} 1 & 0 \\ 0 & 1 \end{pmatrix}, \qquad \begin{pmatrix} 1 & 0 \\ 0 & -1 \end{pmatrix},$$
$$\begin{pmatrix} -1/2 & \sqrt{3}/2 \\ \sqrt{3}/2 & 1/2 \end{pmatrix}, \quad \begin{pmatrix} -1/2 & -\sqrt{3}/2 \\ -\sqrt{3}/2 & 1/2 \end{pmatrix},$$
$$\begin{pmatrix} -1/2 & \sqrt{3}/2 \\ -\sqrt{3}/2 & -1/2 \end{pmatrix}, \begin{pmatrix} -1/2 & -\sqrt{3}/2 \\ \sqrt{3}/2 & -1/2 \end{pmatrix}. \tag{A.71}$$

Problem 3. The $SL(2, c)$ group consists of complex two-by-two matrices of the form given in Eq. (A.28). Show how this group corresponds to the conformal transformation of Eq. (A.29). Do they have the same algebraic property? Is the correspondence one-to-one?

Problems 4. Show that $SL(2, r)$ consisting of two-by-two real unimodular matrices is unitarily equivalent to the subgroup of $SL(2, c)$ of the form

$$\begin{pmatrix} a & b \\ b^* & a^* \end{pmatrix}. \tag{A.72}$$

Problem 5. Assume that $C = [c_{ij}]$ is any square matrix over a complex vector space. Show that C can be written as $A + iB$, where A and B are Hermitian.

Problem 6. If we take a direct product of two $SU(2)$ spinors, the product is reducible to the symmetric component with spin-1 and the spin-0 antisymmetric state. Construct rotation matrices for this reducible representation. Show that the rotation matrix for the spin-0 state is 1, and that those for the spin-1 states are equivalent to the three-by-three rotation matrix. See Sect. 3.7 of Chap. 3.

Problem 7. In Eq. (A.30), we noted that an arbitrary polynomial in the complex variable z:

$$F_n(z) = a_0 + a_1 z^1 + a_2 z^2 + \cdots + a_n z^n, \tag{A.73}$$

can be regarded as a vector with the basis vectors z^k. Show that the replacement of z by $(z - z_0)$ constitutes a similarity transformation. Calculate the matrix which performs this similarity transformation when $n = 2$. Explain why this matrix has to be triangular. Calculate the inverse of the transformation matrix. Is the inverse matrix also triangular?

Problem 8. Repeat Problem 7 when the transformation is

$$z' = \alpha z - z_0. \tag{A.74}$$

See [5].

Problem 9. Let us consider a quadratic form x defined for the interval $-1 < x < 1$:

$$G(x) = a_0 + a_1 x^1 + a_2 x^2 . \tag{A.75}$$

This form can also be expressed as a linear expansion in terms of the Legendre polynomials:

$$G(x) = b_0 P_0(x) + b_1 P_1(x) + b_2 P_2(x) . \tag{A.76}$$

Now the coefficients b_0, b_1 and b_2 are linear combinations of a's. Show that this is a similarity transformation and calculate the transformation matrix.

Problem 10. Problem 9 is an example of the procedure generally known as the Gramm-Schmidt orthonormalization [5]. Calculate two lowest non-trivial order Hermite and Laguerre polynomials using the Gramm-Schmidt procedure.

Problem 11. Show that $E(2)$ discussed throughout this Appendix is isomorphic to the group of Galilei transformations in two-dimensional space.

Problem 12. We are familiar with the complex $z = (x+iy)$-plane. For a given complex number z, show that $E(2)$ is isomorphic to the group consisting of multiplication by a factor of unit modulus and addition by a complex number.

Problem 13. Show that the matrix of $E(2)$ given in Eq. (A.4) has the same algebraic property as that of

$$\begin{pmatrix} e^{-i\theta/2} & (u - iv)e^{i\theta/2} \\ 0 & e^{i\theta/2} \end{pmatrix} , \tag{A.77}$$

or

$$\begin{pmatrix} e^{-i\theta/2} & 0 \\ (u + iv)e^{-i\theta/2} & e^{i\theta/2} \end{pmatrix} . \tag{A.78}$$

Problem 14. Let us go back to Eq. (A.27) where two matrices A and B are related by

$$B = e^A . \tag{A.79}$$

Show that the determinant of B is $e^{Tr(A)}$.

Problem 15. Addition of numbers can be converted to multiplication if we use an exponential function. Show that the matrix of the form:

$$\begin{pmatrix} 1 & a \\ 0 & 1 \end{pmatrix} \tag{A.80}$$

also converts addition into a matrix multiplication.

Problem 16. Let A, B be two n-by-n matrices. We call

$$[A, B] = AB - BA \tag{A.81}$$

the commutator of the two matrices. Show that

$$[[A, B], C] + [[B, C], A] + [[C, A], B] = 0 . \tag{A.82}$$

This is often called the Jacobi identity.

References

1. M. Hamermesh, *Group theory and its application to physical problems*. Dover books on physics and chemistry (Dover Publications, New York, NY USA, 1989). ISBN 978-0-486-66181-0. (Originally published 1962, Addison-Wesley, Reading MA, USA.)
2. L.S. Pontríaġin, *Topological groups*, 3rd edn. No. v. 1 in Classics of Soviet mathematics (Gordon and Breach Science Publishers, New York. NY, USA, 1986). ISBN 978-2-88124-133-8. (Second Edition published in 1966; First Edition published by Princeton University Press, 1946.)
3. H. Börner, *Representations of groups: with special consideration for the needs of modern physics*, 2nd edn. (North-Holland Publ. [u.a.], Amsterdam, NL, 1970). ISBN 978-0-7204-2040-1,978-0-444-10002-3
4. W. Miller, *Symmetry groups and their applications*. No. 50 in Pure and applied mathematics; a series of monographs and textbooks (Academic Press, New York, NY, USA, 1972). ISBN 978-0-12-497460-9
5. R. Gilmore, *Lie groups, Lie algebras, and some of their applications* (Dover Publications, Mineola, NY, USA, 2005). ISBN 978-0-486-44529-8. (Originally published: 1974, John Wiley and Sons, New York, NY, USA.)
6. A. Zee, *Group theory in a nutshell for physicists*. In a nutshell (Princeton University Press, Princeton, 2016). ISBN 9780691162690
7. E. Witten, Anti-de Sitter space and holography, Advances in Theoretical and Mathematical Physics **2**(2), 253–291 (1998). DOI 10.4310/ATMP.1998.v2.n2.a2. URL http://www.intlpress.com/site/pub/pages/journals/items/atmp/content/vols/0002/0002/a002/
8. R. Kerner, Generalization of the Kaluza-Klein theory for an arbitrary non-abelian gauge group, Annales de l'institut Henri Poincaré. Section A, Physique Théorique **9**(2), 143–152 (1968). URL http://www.numdam.org/item?id=AIHPA_1968__9_2_143_0
9. E.P. Wigner, Unitary representations of the inhomogeneous Lorentz group including reflections, in *Group Theoretical Concepts and Methods in Elementary Particle Physics*, ed. by F. Gürsey (Routledge, Abingdon, Oxfordshire, UK, 1964). ISBN 9780677101408. (Lectures of the Istanbul Summer School of Theoretical Physics, July 16-August 4, 1962, Istanbul, Turkey; OCLC: 948806935.)
10. J.Y. Shapiro, Projection operator techniques for compact groups, Journal of Mathematical Physics **14**(9), 1262–1270 (1973). DOI 10.1063/1.1666477. URL https://doi.org/10.1063/1.1666477
11. J.M. Casilio, M.E. Noz, Analytic formulation of SU(3) vector coupling coefficients for n particles, Computer Physics Communications **5**(5), 365–378 (1973). DOI https://doi.org/10.1016/0010-4655(73)90063-5. URL https://www.sciencedirect.com/science/article/pii/0010465573900635
12. M.E. Noz, J.Y. Shapiro, SU(3) projection operator as a polynomial in the generators, Nuclear Physics B **51**, 309–316 (1973). DOI 10.1016/0550-3213(73)90518-X. URL https://www.sciencedirect.com/science/article/pii/055032137390518X
13. V. Bargmann, Irreducible Unitary Representations of the Lorentz Group, The Annals of Mathematics **48**(3), 568–640 (1947). DOI 10.2307/1969129. URL http://www.jstor.org/stable/1969129?origin=crossref
14. P.A.M. Dirac, *The principles of quantum mechanics*, 4th edn. No. 27 in International series of monographs on physics (Clarendon Press, Oxford University Press, Oxford, UK, 1967). ISBN 978-0-19-852011-5. (Originally published 1930, 2nd edn., 1935, 3rd edn., 1947, 4th edn., 1958, 4th edn., revised 1967.)
15. K. Iwasawa, On Some Types of Topological Groups, The Annals of Mathematics **50**(3), 507–558 (1949). DOI 10.2307/1969548. URL http://www.jstor.org/stable/1969548?origin=crossref
16. R. Hermann, *Lie groups for physicists*, 2nd edn. The mathematical physics monograph series (Benjamin/Cummings Publ. Comp, Reading, MA, USA, 1978). ISBN 978-0-8053-3951-2978-0-8053-3950-5. (Originally pulished, 1966.)

17. M. Gell-Mann, *The Eightfold Way: A theory of strong interaction symmetry*, vol. TID-12608;CTSL-20 (California Institute of Technology, Synchrotron, Laboratory, Pasadena, CA, 1961). DOI 10.2172/4008239

18. M. Gell-Mann, Symmetries of Baryons and Mesons, Physical Review **125**(3), 1067–1084 (1962). DOI 10.1103/PhysRev.125.1067. URL https://link.aps.org/doi/10.1103/PhysRev.125.1067

19. Y. Ne'eman, Derivation of strong interactions from a gauge invariance, Nuclear Physics **26**(2), 222–229 (1961). DOI 10.1016/0029-5582(61)90134-1. URL https://linkinghub.elsevier.com/retrieve/pii/0029558261901341

20. E.P. Wigner, *Group Theory: And its Application to the Quantum Mechanics of Atomic Spectra* (Academic Press, New York, NY, USA, 1959). ISBN 978-0127505503. (Originally published as: Gruppentheorie und ihre Anwendung auf die Quantenmechanik der Atomspektren, Springer Verlag, Braunscheig, Germany 1931.)

Appendix B
Lie Groups and Lie Algebras

Abstract This Appendix discusses continuous groups whose elements depend on a given number of continuous parameters. A continuous group is a Lie group if it is possible to represent it within the framework of a theory of infinitesimal transformations from which the original group can be reconstructed. Since most groups in modern physics are Lie groups, we discuss a selected set of examples to elucidate the basic concepts of Lie groups. Of most interest is the Poincaré group. The rotation group is a subgroup of this group, and we can expand our knowledge from what we know about it. The Poincaré group also has many properties shared by the $E(2)$ group which we continue to use as the prime example. We explain the basic concepts of Lie groups. The generators and how a Lie group is constructed from them is given. The commutation relations for the generators, namely the Lie algebras, determine the multiplication properties of the group. Group properties, such as subgroups and their invariance in terms of the Lie algebra and how they are translated into those of the group are examined. We list many other theorems of Lie groups which we use without proof. The purpose here is to give concrete illustrations for those mathematical theorems.

This Appendix is a continuation of Appendix A on the basic elements of group theory, and is devoted to continuous groups whose elements depend on a given number of continuous parameters. We are already familiar with some aspects of continuous groups from our experience with the three-dimensional rotation group.

Because the parameters are continuous, we are led to ask whether there exists differential and integral calculus for groups with continuous parameters. The question then is whether it is possible to develop the mathematics of converting a continuous group into a theory of infinitesimal transformations from which the original group can be reconstructed. If this process is possible, the group is called a Lie group.

S. Başkal et al., *Theory and Applications of the Poincaré Group*, Fundamental Theories
of Physics 217, https://doi.org/10.1007/978-3-031-64376-7

Because most of the groups in modern physics are Lie groups, there are already many excellent textbooks on Lie groups for physicists, and it is not necessary to prove all the theorems. As in Appendix A, we shall discuss a selected set of examples in order to elucidate the basic concepts of Lie groups. The Lie group in which we are most interested in this book is the Poincaré group. Fortunately, the rotation group is a subgroup of the Poincaré group, and we can expand our knowledge from what we know about the three-dimensional rotation group. The Poincaré group has many properties which are shared by the two-dimensional Euclidean group. For this reason, we continue to use the $E(2)$ group as the prime example.

In Sect. B.1, we explain the basic concepts of Lie groups using the $E(2)$ group as a specific example. The concept of generators is introduced. In Sect. B.2, we give Lie's theorems which enable us to construct a Lie group by using only its generators. It is pointed out that the commutation relations for the generators, which are called Lie algebras, determine the multiplication properties of the group. Section B.3 states group properties, such as subgroups and their invariance in terms of the Lie algebra. Section B.4 explains how these properties of the Lie algebra are translated into those of the group.

In addition, there are many other theorems which we intend to use without proof. In Sect. B.5, we list some of them. Indeed, the purpose of this book is to give concrete illustrations for those mathematical theorems the average physicist does not wish to prove. Section B.6 contains exercises and problems which will further serve as illustrative examples for Lie groups.

B.1 Basic Concepts of Lie groups

Many of the groups in physics are continuous groups. The rotation and translation groups are continuous groups. The Lorentz and Poincaré groups which we study in this book are also continuous groups. Continuous groups depend on a set of continuous parameters. The number of parameters which dictates the form of the transformation matrix is not in general equal to the dimension of the vector space to which the matrix is applied. For example, the spinor rotation matrices have three parameters while acting on a two dimensional space.

A Lie group is a connected component of a continuous group which can be continuously connected to the identity element. The group $O(3)$ is only a continuous group while the three-dimensional rotation group without space inversions, $SO(3)$, the proper rotation group, is a Lie group. In this book we will refer to the Lie group $SO(3)$ simply as $O(3)$. As stated in Appendix A, this is the group most familiar to us as physicists. The proper rotation group is unimodular, that is its determinant is equal to one.

In studying Lie groups, we are interested not only in linear transformations on coordinates, but also in transformations of functions which depend on the coordinate variables. For example, the rotation of the $\ell = 1$ spherical harmonics $Y_1^m(\theta, \phi)$ can be regarded as a coordinate transformation, but the rotation of Y_ℓ^m with $\ell > 1$ is the same transformation of a function depending on the coordinate variables.

We are interested first in linear transformations on a set of n variables x_0^i, where $i = 1, \ldots, n$, which may be regarded as the coordinates of a point in a certain space. Consider now the set of equations

$$x^i = f^i(x_0^1, \cdots, x_0^n; \alpha^1, \cdots, \alpha^r), \tag{B.1}$$

in which α^ρ appear as a set of r independent parameters. By omitting indices, we shall write this and similar relations in the form

$$x = f(x_0; \alpha). \tag{B.2}$$

Linear transformations are either homogeneous or inhomogeneous. Linear coordinate transformations in $O(3)$ are homogeneous, thus do not contain translations, while those in $E(2)$ are inhomogeneous.

Let us go back to the coordinate transformations of Eq. (B.1). We shall assume that the f^i have all the required derivatives, and that r is the smallest number of parameters needed to specify the transformations completely and uniquely. The set of transformations f^i will form a group if they obey the following two conditions:

(i) The result of performing successively any two transformations of the set is another transformation belonging to this set. Formally if $x = f(x_0; \alpha)$ and $x' = f(x; \beta)$, then there exists a set of parameters

$$\gamma^\rho = \phi^\rho(\alpha; \beta), \tag{B.3}$$

such that

$$x' = f(x; \beta) = f(f(x_0; \alpha); \beta) = f(x_0; \phi(\alpha; \beta)). \tag{B.4}$$

(ii) Corresponding to every transformation, there exists a unique inverse, which also belongs to the set. This means that given Eq. (B.1), there exists a set of parameters α^{-1} such that
$$x_0 = f(x, \alpha^{-1}). \tag{B.5}$$

Transforming x_0 onto x and then inversely back to x_0, we obtain according to (i) a transformation which belongs to the group and is characterized by the set of parameters α_0:
$$x_0 = f(x_0; \alpha_0). \tag{B.6}$$

This is the identity transformation. Since it imposes no restriction on x_0 or on α_0, we shall take
$$\alpha_0^\rho = 0, \qquad \rho = 1, \cdots, r, \tag{B.7}$$

so that
$$f(x; 0) = x, \tag{B.8}$$

and

$$\phi(\alpha\,;\,0) = \alpha\,. \tag{B.9}$$

The coordinate transformations given in Eqs. (B.1) and (B.6) can be written respectively as

$$x = f(x_0\,,\,\alpha)\,, \tag{B.10}$$

and

$$x = f(x\,,\,0)\,. \tag{B.11}$$

Corresponding to these, there are two ways of expressing a transformation such that the new components of x differ infinitesimally from the old ones:

$$x + dx = f(x_0\,;\,\alpha + d\alpha)\,, \tag{B.12}$$

or

$$x + dx = f(x\,,\,\delta\alpha)\,, \tag{B.13}$$

where $d\alpha$ and $\delta\alpha$ are the infinitesimal increments in the parameter space from α and from the origin $\alpha = 0$, respectively. From the above equations,

$$dx = \frac{\partial f(x_0\,,\,\alpha)}{\partial \alpha^\sigma}\,d\alpha^\sigma\,, \tag{B.14}$$

or

$$dx = \left.\frac{\partial f(x_0\,,\,\alpha)}{\partial \alpha^\sigma}\right|_{\alpha=0}\,\delta\alpha^\sigma\,. \tag{B.15}$$

The basic idea of the Lie group is to derive all the group properties from the behavior of the functions near the origin in the parameter space. We are thus led to look at the second equation more closely. Eq. (B.15) can be written as

$$dx^i = u^i_\sigma(x)\,\delta\alpha^\sigma\,, \tag{B.16}$$

where

$$u^i_\sigma(x) = \left.\frac{\partial f^i(x,\alpha)}{\partial \alpha^\sigma}\right|_{\alpha=0}\,. \tag{B.17}$$

Next, let us consider transformations of a function depending on the coordinate variables, i.e., the same function on the same point only in terms of a different coordinate system,

$$F(f(x_0)) = F_0(x_0)\,, \tag{B.18}$$

or

$$F(x) = F_0(x_0) = F_0(f(x\,;\,\alpha^{-1}))\,. \tag{B.19}$$

This is a reflection of the fact that the transformation of the function is achieved through the inverse transformation of its arguments. We are interested in the operator which changes $F_0(x)$ to $F(x)$:

$$F(x) = S_\alpha F_0(x)\,. \tag{B.20}$$

For example, consider a function:

$$F_0(x_0, y_0) = (x_0 + iy_0)^2 \,, \tag{B.21}$$

defined on the two-dimensional $x_0 y_0$-plane, and an inhomogeneous linear transformation

$$x = x_0 + u \,,$$
$$y = y_0 + v \,. \tag{B.22}$$

The original function of Eq. (B.21) can be described in the form

$$F(x, y) = F_0(x_0, y_0) = [(x - u) + i(y - v)]^2 \,. \tag{B.23}$$

Thus

$$F(x, y) - F_0(x, y) = -2(x + iy)(u + iv) + (u + iv)^2 \,. \tag{B.24}$$

$F(x)$ is clearly different from $F_0(x)$, and this difference is the effect of the S_α operation defined in Eq. (B.20). The corresponding coordinate transformation is given in Eq. (B.22).

The transformation of the coordinate point through Eq. (B.10) or Eq. (B.22) is called the *active transformation*. The transformation of the coordinate system through which the function is transformed as is given in Eq. (B.19) or Eq. (B.23) is called the *passive transformation*. From a purely calculation point of view, the active and passive transformations are the inverse of each other [See Exercise 2 in Sect. B.6].

The infinitesimal transformation of the function F_0 corresponding to the coordinate transformation of Eq. (B.12) or Eq. (B.13) is

$$dF_0(x) = F(x) - F_0(x) = F_0(x_0) - F_0(x) = F_0(x - dx) - F_0(x) \,. \tag{B.25}$$

Thus

$$dF_0(x) = \frac{\partial F_0}{\partial x^i} dx^i = -\delta\alpha^\sigma u^i_\sigma(x) \frac{\partial F_0}{\partial x^i} = -i\delta\alpha^\sigma X_\sigma(x) F_0(x) \,, \tag{B.26}$$

where

$$X_\sigma(x) \equiv -iu^i_\sigma(x) \frac{\partial}{\partial x^i} \,. \tag{B.27}$$

The operators $X_\sigma(x)$ are called the generators of the Lie group.

Let us carry out explicit calculations for the $E(2)$ group. Remember that in this book we are using the coordinate system $x^\mu = (t, z, x, y)$ with $\mu = 0, 3, 1, 2$. This coordinate system is used to represent the Lorentz group $O(3, 1)$ in Minkowski space. For the $E(2)$ group there are two coordinate variables, namely

$$x^1 = x \,, \qquad x^2 = y \,, \tag{B.28}$$

and three parameters:

$$\alpha^1 = \theta, \qquad \alpha^2 = u \,, \qquad \alpha^3 = v \,. \tag{B.29}$$

For small values of the parameters,

$$
\begin{pmatrix} 1 \\ 1 \\ x + dx \\ y + dy \end{pmatrix} = \begin{pmatrix} 1 & 0 & 0 & 0 \\ 0 & 1 & 0 & 0 \\ 0 & \delta u & 1 & -\delta\theta \\ 0 & \delta v & \delta\theta & 1 \end{pmatrix} \begin{pmatrix} 1 \\ 1 \\ x \\ y \end{pmatrix},
\tag{B.30}
$$

and Eq. (B.16) in this case can be written as

$$
\begin{aligned}
dx &= \left(\frac{\partial x}{\partial \theta}\right)\Bigg|_{\theta=0} \delta\theta + \left(\frac{\partial x}{\partial u}\right)\Bigg|_{u=0} \delta u + \left(\frac{\partial x}{\partial v}\right)\Bigg|_{v=0} \delta v \\
&= -y\delta\theta + \delta u,
\end{aligned}
\tag{B.31}
$$

and

$$
\begin{aligned}
dy &= \left(\frac{\partial y}{\partial \theta}\right)\Bigg|_{\theta=0} \delta\theta + \left(\frac{\partial y}{\partial u}\right)\Bigg|_{u=0} \delta u + \left(\frac{\partial y}{\partial v}\right)\Bigg|_{v=0} \delta v \\
&= x\delta\theta + \delta v.
\end{aligned}
$$

The generators are

$$
\begin{aligned}
X_\theta(x) &= -i\left[\left(\frac{\partial x}{\partial \theta}\right)\Bigg|_{\theta=0} \frac{\partial}{\partial x} + \left(\frac{\partial y}{\partial \theta}\right)\Bigg|_{\theta=0} \frac{\partial}{\partial y}\right] \\
&= -i\left(x\frac{\partial}{\partial y} - y\frac{\partial}{\partial x}\right), \\
X_u(x) &= -i\left[\left(\frac{\partial x}{\partial u}\right)\Bigg|_{u=0} \frac{\partial}{\partial x} + \left(\frac{\partial y}{\partial u}\right)\Bigg|_{u=0} \frac{\partial}{\partial y}\right] \\
&= -i\left(\frac{\partial}{\partial x}\right),
\end{aligned}
\tag{B.32}
$$

and

$$
\begin{aligned}
X_v(x) &= -i\left[\left(\frac{\partial x}{\partial v}\right)\Bigg|_{v=0} \frac{\partial}{\partial x} + \left(\frac{\partial y}{\partial v}\right)\Bigg|_{v=0} \frac{\partial}{\partial y}\right] \\
&= -i\left(\frac{\partial}{\partial y}\right).
\end{aligned}
$$

The expressions given in Eq. (B.31) and Eq. (B.32) are Eq. (B.16) and Eq. (B.27) applied to the $E(2)$ group respectively. Both forms need the matrix $u^i_\sigma(x)$, which can be written as

$$
[u^i_\sigma(x)] = \begin{pmatrix} 0 & 1 & -y \\ 1 & 0 & x \end{pmatrix}.
\tag{B.33}
$$

Since $E(2)$ is a three-parameter group operating on a two-dimensional geometrical space, the above matrix is two-by-three. This matrix has no inverse. The generators of $E(2)$ given in Eq. (B.32) satisfy the commutation relations:

$$[X_\theta(x), X_u(x)] = iX_v(x), \qquad [X_\theta(x), X_v(x)] = -iX_u(x),$$
$$[X_u(x), X_v(x)] = 0. \tag{B.34}$$

Let us next consider the transformation of functions defined in the parameter space specified by Eq. (B.3). Following the same procedure as that for deriving Eqs. (B.14) and (B.15), we obtain

$$\alpha + d\alpha = \alpha + \left.\frac{\partial(\alpha\,;\beta)}{\partial\beta^\tau}\right|_{\beta=0} \delta\alpha^\tau. \tag{B.35}$$

Thus $d\alpha$ is a linear combination of the $\delta\alpha$:

$$d\alpha^\rho = \Theta^\rho_\tau(\alpha)\delta\alpha^\tau \tag{B.36}$$

where

$$\Theta^\rho_\tau = \left.\frac{\partial\phi^\rho(\alpha;\beta)}{\partial\beta^\tau}\right|_{\beta=0}. \tag{B.37}$$

Following the same logic as that which led to Eq. (B.26) for $dF_0(x)$, we can write for a function $\Phi(\alpha)$:

$$\begin{aligned}
d\Phi(\alpha) &= -i\delta\alpha^\sigma\Theta^\rho_\sigma(\alpha)\frac{\partial\Phi_0}{\partial\alpha^\rho}\\
&= -i\delta\alpha^\sigma X_\sigma(\alpha)\Phi_0(\alpha),
\end{aligned} \tag{B.38}$$

where $\Theta^\rho_\sigma(\alpha)$ is given in Eq. (B.37). Consequently

$$X_\sigma(\alpha) = -i\Theta^\rho_\sigma(\alpha)\frac{\partial}{\partial\alpha^\rho}. \tag{B.39}$$

Unlike the case for $u^i_\sigma(x)$, the $\Theta^\rho_\sigma(\alpha)$ matrix is a square matrix. Since there are the minimum necessary number of α parameters, this matrix is expected to have an inverse:

$$\Psi^\sigma_\tau\Theta^\tau_\rho = \delta^\sigma_\rho, \tag{B.40}$$

such that

$$\delta\alpha^\sigma = \Psi^\sigma_\tau d\alpha^\tau. \tag{B.41}$$

Let us carry out explicit calculations for the $E(2)$ case [See also Exercise 1 in Sect. B.6]. The Θ^ρ_σ and Ψ^σ_τ matrices take the form

$$\Theta^\rho_\sigma(\alpha) = \begin{pmatrix} 1 & 0 & 0 & 0\\ 0 & 1 & 0 & 0\\ 0 & -v & 1 & 0\\ 0 & u & 0 & 1 \end{pmatrix} \tag{B.42}$$

and

$$\Psi_\tau^\sigma(\alpha) = \begin{pmatrix} 1 & 0 & 0 & 0 \\ 0 & 1 & 0 & 0 \\ 0 & v & 1 & 0 \\ 0 & -u & 0 & 1 \end{pmatrix} . \tag{B.43}$$

From these expressions, we can derive

$$X_\theta(\alpha) = -i\left(\frac{\partial}{\partial\theta} + u\frac{\partial}{\partial v} - v\frac{\partial}{\partial u}\right) ,$$

$$X_u(\alpha) = -i\frac{\partial}{\partial u} , \qquad X_v(\alpha) = -i\frac{\partial}{\partial v} . \tag{B.44}$$

These operators satisfy the commutation relations:

$$\begin{aligned}
[X_\theta(\alpha), X_u(\alpha)] &= iX_v(\alpha) , \qquad [X_\theta(\alpha), X_v(\alpha)] = -iX_u(\alpha) , \\
[X_u(\alpha), X_v(\alpha)] &= 0 .
\end{aligned} \tag{B.45}$$

These commutation relations are identical to those given in Eq. (B.34). Is this an accidental coincidence or a manifestation of a more fundamental theorem? We shall study this point in more detail in Sect. B.2.

B.2 Basic Theorems Concerning Lie Groups

Now combining Eq. (B.16) and Eq. (B.41), we arrive at

$$dx^i = u^i_\sigma(x)\Psi_\tau^\sigma(\alpha)d\alpha^\tau , \tag{B.46}$$

or

$$\frac{\partial x^i}{\partial\alpha^\sigma} = u^i_\rho(x)\Psi_\sigma^\rho(\alpha) . \tag{B.47}$$

This is known as *Lie's first theorem.*

The commutation relations of Eq. (B.34) and Eq. (B.45) for the $E(2)$ case strongly suggest some fundamental theorems concerning the generators. In order to see this possibility, we observe first that the necessary and sufficient condition for the differential equation of Eq. (B.47) to have a unique solution with a given initial condition is that all mixed derivatives be equal:

$$\frac{\partial^2 x^i}{\partial\alpha^\sigma\partial\alpha^\lambda} = \frac{\partial^2 x^i}{\partial\alpha^\lambda\partial\alpha^\sigma} . \tag{B.48}$$

This condition is often called the integrability condition.

Let us then apply this condition to Eq. (B.47):

$$\frac{\partial}{\partial\alpha^\sigma}(u^i_\rho(x)\Psi_\lambda^\rho(\alpha)) = \frac{\partial}{\partial\alpha^\lambda}(u^i_\rho(x)\Psi_\sigma^\rho(\alpha)) . \tag{B.49}$$

The calculation of this equation will require the derivatives of the $u(x)$ function with respect to the α parameters:

$$\frac{\partial u_\rho^i(x)}{\partial \alpha^\sigma} = \frac{\partial u_\rho^i(x)}{\partial x^j}\frac{\partial x^j}{\partial \alpha^\sigma} = \Psi_\sigma^\lambda(\alpha)u_\lambda^j(x)\frac{\partial u_\rho^i(x)}{\partial x^j}. \tag{B.50}$$

Then Eq. (B.49) becomes

$$\Psi_\sigma^\tau(\alpha)\Psi_\lambda^\rho(\alpha)\left\{u_\rho^j\frac{\partial u_\tau^i}{\partial x^j} - u_\tau^j\frac{\partial u_\rho^i}{\partial x^j}\right\}$$

$$= \left\{\frac{\partial \Psi_\tau^\rho(\alpha)}{\partial \alpha^\sigma} - \frac{\partial \Psi_\sigma^\rho(\alpha)}{\partial \alpha^\tau}\right\}u_\rho^i(x). \tag{B.51}$$

From Eq. (B.40) the Ψ matrix has an inverse, so we have

$$u_\nu^j(x)\frac{\partial u_\mu^i(x)}{\partial x^j} - u_\mu^j(x)\frac{\partial u_\nu^i(x)}{\partial x^j}$$

$$= \left\{\Theta_\mu^\sigma(\alpha)\Theta_\nu^\lambda(\alpha)\left(\frac{\partial \Psi_\tau^\rho(\alpha)}{\partial \alpha^\sigma} - \frac{\partial \Psi_\sigma^\rho(\alpha)}{\partial \alpha^\tau}\right)\right\}u_\rho^i(x). \tag{B.52}$$

The left-hand side of the above expression depends only on the x variables. The quantity inside the curly bracket on the right-hand side depends only on the α variables. Furthermore, this equation cannot be completely separated because the $u_\rho^i(x)$ matrix is not a square matrix and does not have an inverse.

This, however, does not discourage us from asking whether the quantity inside the square bracket is constant. In order to determine this, let us replace $f^i(x;\alpha)$ by $\phi^\mu(\beta;\alpha)$. Then Eq. (B.52) can be rewritten as

$$\Theta_\nu^\rho(\beta)\frac{\partial \Theta_\mu^\sigma(\beta)}{\partial \beta^\rho} - \Theta_\mu^\rho(\beta)\frac{\partial \Theta_\nu^\sigma(\beta)}{\partial \beta^\rho}$$

$$= \left\{\Theta_\mu^\sigma(\alpha)\Theta_\nu^\lambda(\alpha)\left(\frac{\partial \Psi_\tau^\rho(\alpha)}{\partial \alpha^\sigma} - \frac{\partial \Psi_\sigma^\rho(\alpha)}{\partial \alpha^\tau}\right)\right\}\Theta_\rho^\sigma(\beta). \tag{B.53}$$

Now the $\Theta_\rho^\sigma(\beta)$ on the right-hand side of the above equation can be moved to the left-hand side. Then the separation of the variables becomes complete. The quantity in the curly bracket is indeed a constant. This allows us to write

$$u_\sigma^j(x)\frac{\partial u_\tau^i(x)}{\partial x^j} - u_\tau^j(x)\frac{\partial u_\sigma^i(x)}{\partial x^j} = iC_{\tau\sigma}^\rho u_\rho^i(x), \tag{B.54}$$

and

$$\Theta_\rho^\nu(\alpha)\frac{\partial \Theta_\tau^\mu(\alpha)}{\partial \alpha^\nu} - \Theta_\tau^\nu(\alpha)\frac{\partial \Theta_\rho^\mu(\alpha)}{\partial \alpha^\nu} = iC_{\rho\tau}^\sigma\Theta_\sigma^\mu(\alpha), \tag{B.55}$$

where the coefficients $C_{\sigma\tau}^\mu$ are constants.

The immediate consequence of the above relations is that the generators of the Lie group $X_\sigma(x)$ defined in Eq. (B.27) and in Eq. (B.39) satisfy the commutation relations:

$$[X_\rho(x), X_\sigma(x)] = iC^\tau_{\rho\sigma}X_\tau(x),$$ (B.56)

and

$$[X_\rho(\alpha), X_\sigma(\alpha)] = iC^\tau_{\rho\sigma}X_\tau(\alpha).$$ (B.57)

The constants $C^\tau_{\rho\sigma}$ are commonly called the *structure constants*. This result is known as *Lie's second theorem*.

Evidently, from Eqs. (B.55) and (B.56),

$$C^\tau_{\rho\sigma} = -C^\tau_{\sigma\rho},$$ (B.58)

and the generators satisfy the Jacobi identity:

$$[[X_\rho, X_\sigma], X_\tau] + [[X_\sigma, X_\tau]X_\rho] + [[X_\tau, X_\rho], X_\sigma] = 0,$$ (B.59)

resulting in

$$C^\mu_{\rho\sigma}C^\nu_{\mu\tau} + C^\mu_{\sigma\tau}C^\nu_{\mu\rho} + C^\mu_{\tau\rho}C^\nu_{\mu\sigma} = 0.$$ (B.60)

The properties of the structure constants given in Eqs. (B.58) and (B.60) constitute *Lie's third theorem*.

As stated in Sect. B.1, the fundamental idea of the theory of Lie groups is to consider only that part of the group which lies near the identity. This is equivalent to considering only infinitesimal transformations. In terms of the generators, an infinitesimal transformation of the group takes the form

$$S_{\delta\alpha} = I - iX_\sigma\delta\alpha^\sigma,$$ (B.61)

where I is the identity operator, and $\delta\alpha^\sigma$ is an infinitesimal quantity defined to be of first order. The commutation relations for the generators given in Eqs. (B.56) and (B.57) determine the algebraic properties of the Lie group. These commutation relations are of course determined by the structure constants.

Let us see how these infinitesimal transformations lead to transformations with finite parameters. Consider

$$S_N = S_N[(S_{N-1})^{-1}S_{N-1}][(S_{N-2})^{-1}S_{N-2}]...[(S_1)^{-1}S_1]I$$
$$= [S_N(S_{N-1})^{-1}][S_{N-1}(S_{N-2})^{-1}]...[S_2(S_1)^{-1}]S_1I.$$ (B.62)

The condition given in Eq. (B.62) defines the multiplication law of the group.

Let us consider the parametrization $\alpha^\mu t$, where t is allowed to vary along the real axis while the α^μ are a fixed set of numbers. The transformations with different values of t form a one-parameter Abelian subgroup. We are particularly interested in the interval $0 \le t \le 1$. This interval can be divided into

$$t = k/N, \qquad \text{with} \qquad k = 0, 1, 2, ..., N,$$ (B.63)

where N can be an arbitrarily large integer. On this interval, the transformation S_N satisfies the multiplication law:

$$S_k S_m = S_{k+m} , \tag{B.64}$$

and

$$S_N = [S_1]^N . \tag{B.65}$$

S_N is therefore determined from S_1 which describes the transformation for small values of t:

$$S_\alpha = \exp\left(-i\alpha^\rho X_\rho\right) = \sum_{k=0}^{\infty} \frac{1}{k!}(-i\alpha^\rho X_\rho)^k . \tag{B.66}$$

In Eq. (B.61), an infinitesimal transformation of the group S was written as the sum of the identity plus an infinitesimal transformation, where $\delta\alpha^\sigma$ is an infinitesimal quantity defined to be of first order. If two such transformations are considered, we get

$$\begin{aligned}
S_{\delta\alpha} S_{\delta\beta} &= (I - i\delta\alpha^\rho X_\rho)(I - i\delta\beta^\sigma X_\sigma) \\
&= I - i\delta\alpha^\rho X_\rho - i\delta\beta^\sigma X_\sigma ,
\end{aligned} \tag{B.67}$$

where the first non-vanishing infinitesimal terms have been written. Thus the operation of multiplication in S corresponds to addition in the infinitesimal group of S. If the first-order quantities vanish, then higher-order quantities must be used. But in order for the u^i in Eq. (B.47) to be the vector space of a transformation such as is given in Eq. (B.1), Eq. (B.47) must be completely integrable. The integrability of Eq. (B.47) implies that in Eq. (B.67) we never need go beyond the second order infinitesimals (second partial derivatives must exist and be continuous in order for Eq. (B.47) to be integrable). Thus we need only to consider *commutators* which are expressions of the form $S_\alpha S_\beta S_\alpha^{-1} S_\beta^{-1}$ and to ask that the corresponding infinitesimal operators of second order,

$$\delta\alpha^\rho \delta\beta^\sigma [X_\rho, X_\sigma] \tag{B.68}$$

be continuous in the linear manifold of the infinitesimal operators. Thus the structure constants given in Eqs. (B.56) and (B.57) come in second order because they are given by the commutator alone, and it is the structure constants which determine the group.

B.3 Properties of Lie Algebras

As we have seen in Sects. B.1 and B.2, the properties of Lie groups are determined by the closed algebra of the generators. We call this the Lie algebra. Let us rewrite the commutation relations for the generators:

$$[X_\rho, X_\sigma] = iC_{\rho\sigma}^\tau X_\tau \qquad \text{with} \qquad \rho, \sigma = 1, 2, \ldots , r . \tag{B.69}$$

If all the generators commute and the structure constants vanish, we say that the algebra is Abelian.

If a subset of generators X_i satisfy the closed algebraic relation

$$[X_i, X_j] = iC_{ij}^k X_k \qquad \text{with} \qquad i, j = 1, \dots, s < r \tag{B.70}$$

we say that X_i's form a *subalgebra* generating a subgroup. If the subset of the algebra has the property that the commutator of any member of the subset with any member of the algebra produces a member of the subset:

$$[X_\rho, X_j] = iC_{\rho j}^k X_k, \tag{B.71}$$

then the subalgebra of X_i is *invariant*, and the subgroup generated by this subalgebra is an invariant subgroup discussed in Sect. A.2 of Appendix A. If all the X_i's commute:

$$[X_i, X_j] = 0, \tag{B.72}$$

we say that X_i's form an *Abelian invariant subalgebra* generating an Abelian invariant subgroup.

Algebras with no invariant subalgebra are called *simple algebras*, just like simple groups. Algebras with no Abelian invariant subalgebras are called semisimple Lie algebras. For example, the generators of the rotation group form a simple algebra. The Lie algebra of $E(2)$ given in Eqs. (B.34) and (B.45) form an algebra which is neither simple nor semisimple.

Next, let us discuss the isomorphism between a Lie algebra A and another algebra A'. We say that the two algebras A and A' are isomorphic to each other if their generators satisfy the same set of commutation relations. This means that to each element X of A, there is an element $F(X)$ of A' such that

$$F([X_\rho, X_\sigma]) = [F(X_\rho), F(X_\sigma)]. \tag{B.73}$$

Lie groups having the same Lie algebra are said to be *locally isomorphic*. The best known example of this isomorphism is the relation between the three-dimensional rotation group and the $SU(2)$ group generated by the Pauli spinors.

It is important to note that the generators of $E(2)$ given in Eq. (B.32) are not the only ones satisfying the commutation relations of Eq. (B.34). We can consider the two-by-two matrices:

$$Y_\theta = \frac{1}{2}\begin{pmatrix} 1 & 0 \\ 0 & -1 \end{pmatrix}, \quad Y_u = \begin{pmatrix} 0 & 1 \\ 0 & 0 \end{pmatrix}, \quad Y_v = \begin{pmatrix} 0 & -i \\ 0 & 0 \end{pmatrix}. \tag{B.74}$$

These matrices also satisfy the commutation relations for $E(2)$ given in Eq. (B.34). Since it was Wigner who first recognized this [1], we take the liberty of calling the group generated by the above matrices $W(2)$. Thus, $E(2)$ is locally isomorphic to $W(2)$.

Once the commutation relations for the generators are written, we can construct sets of matrices which satisfy the commutation relations:

$$M(X_\rho)M(X_\sigma) - M(X_\sigma)M(X_\rho) = iC^\tau_{\rho\sigma}M(X_\tau),\tag{B.75}$$

where $M(X_\rho)M(X_\sigma)$ implies matrix multiplication. Each such set is called a *representation* of the Lie algebra. In this case, each generator of the algebra can be represented by a square matrix of rank k, in which each element of every matrix is an explicit number. The rank of matrix k is also the dimension of the representation. If k is a finite integer, we say that the representation is *finite-dimensional*. If k is infinite, the representation is called an *infinite-dimensional representation*.

One representation which is an immediate consequence of the Jacobi identity given in Eqs. (B.59) and (B.60) is the set of r matrices consisting of structure constants, defined as

$$(M_\rho)^\sigma_\tau = C^\sigma_{\rho\tau},\tag{B.76}$$

then

$$(M_\mu)^\sigma_\tau (M_\nu)^\tau_\rho - (M_\nu)^\sigma_\tau (M_\mu)^\tau_\rho = iC^\tau_{\mu\nu}(M_\tau)^\sigma_\rho .\tag{B.77}$$

This representation is called the regular representation of the Lie algebra.

We are quite familiar with finite-dimensional representations of various groups. All representations of the three-dimensional rotation group are finite-dimensional. The three three-by-three matrices which generate the rotation of the three-dimensional coordinate system form the regular representation of the Lie algebra of the rotation group.

The matrices describing the $E(2)$ transformation in Section B.1 constitute a finite-dimensional representation of the $E(2)$ group, generated by

$$X_\theta = \begin{pmatrix} 0 & 0 & 0 & 0 \\ 0 & 0 & 0 & 0 \\ 0 & 0 & 0 & -i \\ 0 & 0 & i & 0 \end{pmatrix}, \quad X_u = \begin{pmatrix} 0 & 0 & 0 & 0 \\ 0 & 0 & 0 & 0 \\ 0 & -i & 0 & 0 \\ 0 & 0 & 0 & 0 \end{pmatrix}, \quad X_v = \begin{pmatrix} 0 & 0 & 0 & 0 \\ 0 & 0 & 0 & 0 \\ 0 & 0 & 0 & 0 \\ 0 & -i & 0 & 0 \end{pmatrix}.\tag{B.78}$$

These matrices satisfy the commutation relations for the $E(2)$ group which is given in Eq. (B.34). Furthermore, it is easy to show that they form the regular representation of the Lie algebra for $E(2)$.

Since X_u and X_v for $E(2)$ commute with each other, they form an Abelian subalgebra. Since the commutation of X_θ with X_u or X_v leads to X_u or X_v, the subalgebra of X_u and X_v is also an invariant subalgebra. The Lie algebra of the $E(2)$ group is neither simple nor semisimple. In addition to finite-dimensional representations, $E(2)$ has infinite-dimensional representations which we study in Chap. 10.

If the generators are Hermitian, the resulting representation of the Lie group is unitary, as is evident from the equation

$$\left(\exp[-i\alpha^\rho X_\rho]\right)^\dagger = \exp\left[i\alpha^\rho (X_\rho)^\dagger\right] .\tag{B.79}$$

The generators of the rotation group are Hermitian and all the representations of the rotation group are unitary representations. On the other hand, in the case of $E(2)$, X_θ is Hermitian while X_u and X_v are not. Therefore, the representation generated by the three matrices of Eq. (B.78) is not a unitary representation. However, this does not rule out the possibility of the differential form of the generators given in Eq. (B.32) being Hermitian when they are applied to a suitable Hilbert space. This possibility is detailed in Chap. 10.

When we construct representations of the group starting from the generators, we use the exponential expansion. We know how to do this in the case of rotation group. We know also that

$$\exp\left\{-i(\alpha^1 X_1 + \alpha^2 X_2)\right\} \tag{B.80}$$

is not necessarily equal to

$$\exp\left(-i\alpha^1 X_1\right)\exp\left(-i\alpha^2 X_2\right), \tag{B.81}$$

unless X_1 and X_2 commute with each other. This is known as the Baker-Campbell-Hausdorff relation [2].

As for the $E(2)$ case, the matrix $R(\theta)$ defined in Eq. (A.10) can be obtained from

$$\exp\left(-i\theta M(X_\theta)\right) = \sum_{k=0}^{\infty} \frac{(-i\theta)^k}{n!} \left(M(X_\theta)\right)^k \tag{B.82}$$

as in the case of the rotation group. However, for X_u and X_v, the series truncates, and

$$\exp\left(-iuM(X_u)\right) = I - iuM(X_u),$$
$$\exp\left(-ivM(X_v)\right) = I - ivM(X_v). \tag{B.83}$$

This is because $[M(X_u)]^2 = [M(X_v)]^2 = 0$. In general, if any power of a non-zero matrix M vanishes:

$$(M)^n = 0, \tag{B.84}$$

for an integer n, then this matrix is said to be *nilpotent*. The triangular matrices with vanishing diagonal elements which we discussed in Appendix A are nilpotent matrices. We discuss further nilpotent matrices in Chap. 10.

In constructing representations, as in many other mathematical procedures in physics, it is convenient to find a set of variables which remain invariant under group transformations. In the language of Lie algebras, we should find first the maximal set of operators which commute with all the generators of a given Lie group. These operators are often called the *Casimir operators*. Then by Schur's lemma, the necessary and sufficient condition for the representation to be irreducible is that the Casimir operators be multiples of unity. The eigenvalues of the Casimir operators therefore specify the representation.

The Casimir operator for the rotation group is the square of the total angular momentum. It is straightforward to show that the Casimir operator for the $E(2)$ group is the sum of the squares of the generators of translations. As is well known to many physicists, the $SU(3)$ group has two Casimir operators. As we see in Chap. 2, the Lorentz and the Poincaré group have two Casimir operators.

In addition to the explicit forms of the matrices, we can construct representations by finding differential operators and the Hilbert space to which the operators are applicable. This procedure applied to $O(3)$ is well known. The Hilbert space to which the rotation operators are applicable is called the spherical harmonics. As for the $E(2)$ group, we have already constructed the differential operators. We discuss in Chap. 10 the Hilbert space to which the $E(2)$ operators are applicable.

After constructing the Casimir operators, it is convenient to construct representations which are also diagonal in a commuting set of generators. In the case of the rotation group, we can construct representations which are simultaneously diagonal in the Casimir operator and one of the generators. These representations are not diagonal in the rest of the generators. In the case of $E(2)$, we can choose the representations which are diagonal in the Casimir operator and the two commuting translation operators or in the Casimir operator and the rotation operators.

Let X_ρ and Y_ρ be two different representations of the same Lie algebra such that

$$[X_\rho, Y_\sigma] = 0 . \tag{B.85}$$

We can then consider another representation Z_ρ such that

$$Z_\rho = X_\rho + Y_\rho . \tag{B.86}$$

The operators Z_ρ satisfy the same Lie algebra. We say in this case that the representation Z is a direct sum of X and Y. The Casimir operators for each of the X and Y representations remain as the Casimir operators for the resulting Z representation. However, the Z representation can have new Casimir operators. The new Casimir operators commute with every Z_ρ, but do not necessarily commute with X_ρ or Y_ρ. For example, in the rotation group, the total angular momentum is the direct sum of the generators of the $O(3)$ and $SU(2)$ groups, which are the orbital and spin angular momenta respectively. The total orbital angular momentum and the total spin remain as good quantum numbers even if the two angular momenta are coupled. However, the third component of the orbital angular momentum does not commute with the square of the total angular momentum.

The procedure for constructing representations of a semi-direct product of two groups is much more delicate than the case for direct products. As was pointed out before, the $E(2)$ group is a semi-direct product of the rotation and translation groups in a two-dimensional plane. As we see in Chap. 2, the Poincaré group is a semi-direct product of the Lorentz and space-time translation groups.

The techniques of constructing representations of semi-direct products have been discussed extensively in the mathematical literature. However, for a physicist, it is much more profitable to work out concrete examples before attempting to prove theorems. The $E(2)$ and Poincaré groups are good examples for both physicists and mathematicians.

B.4 Properties of Lie Groups

It is not difficult to translate the properties of Lie algebras to those of Lie groups. A Lie group generated by a given Lie algebra is Abelian if the Lie algebra is Abelian. A subalgebra of a Lie algebra generates a subgroup. A Lie group is simple or semisimple if its Lie algebra is simple or semisimple.

The basic difference between Lie groups and Lie algebras is that, in the case of Lie groups, we have to deal with finite values of the group parameters. For this purpose, it is convenient to consider a Euclidean space spanned by the group parameters. This space is often called the *manifold*. The manifold for a one-parameter group is a straight line. The manifold for the group of translations on a two-dimensional plane, as discussed in Appendix A, is a two-dimensional plane. Both the $O(3)$ and $E(2)$ groups have three-dimensional manifolds.

The Lie algebra of a given Lie group is a description of the group in the neighborhood of the origin in the manifold which corresponds to the identity element of the Lie group. Every element of the group is specified by its coordinate position in the manifold. Because this element can be obtained through the process of continuously increasing the parameters from the identity element, there is a continuous line between the origin and the coordinate point for the group. Every point on this line corresponds to an element of the Lie group.

If all elements of a given group can be specified by the coordinate points confined within an finite region in the manifold, this group is said to be *compact*. If a group has no finite boundary in the manifold, it is said to be *non-compact*. $O(3)$ is compact while $E(2)$ is non-compact.

A given Lie algebra may generate different Lie groups having different boundaries in the manifold. For example, in the case of rotation group, the manifold for $O(3)$ is a sphere with radius π, while that for $SU(2)$ is a sphere with radius 2π [3].

If any closed loop in the manifold can be continuously deformed to a point, we say that the group is *simply connected*. Among all possible Lie groups sharing the same Lie algebra, there is only one simply connected group. This group is often called the *universal covering group* . The fact that $SU(2)$ is a universal covering group for the three-dimensional rotation group is discussed in many textbooks on group theory.

Let us study the manifold structure and the universal covering group for the two-dimensional Euclidean group. As was noted repeatedly before, this group has the three parameters u, v and θ. We therefore have to consider a three-dimensional Euclidean space spanned by these parameters. The translation parameters u and v can cover the entire two-dimensional uv-plane for both $E(2)$ and $W(2)$. However, the rotation parameter θ can extend from $-\pi$ to π in the case of $E(2)$, while it extends from -2π to 2π in $W(2)$.

All closed curves in $W(2)$ can be continuously deformed to a point, while this is not always possible for $E(2)$. For this reason, $W(2)$ is the universal covering group for the two-dimensional Euclidean group.

There is a simple manner in which all Lie groups with the same Lie algebra may be obtained from the universal covering group. Let S be a simply-connected Lie group and N be one of its discrete invariant subgroups. Then

$$S_x S_n S_x^{-1} = S_{n'} \,, \tag{B.87}$$

for all S_n, $S_{n'}$ an element of N, and S_x an element of S. Then by Eq. (A.9) of Appendix A, we can form the factor or quotient group:

$$G = S/N \,, \tag{B.88}$$

such that it is a Lie group whose Lie algebra A is isomorphic to the Lie algebra of S. G is multiply connected when N contains more than one element.

In $SU(2)$, the two-by-two identity matrix and its negative are elements of the group. These two elements form a discrete invariant subgroup. We can therefore form a quotient group as is described in Eq. (B.88). This quotient group is $O(3)$. The story is the same for $W(2)$ and $E(2)$.

B.5 Further Theorems of Lie Groups

The theory of Lie groups is a fascinating subject. It is still a developing subject. In general, representations of compact groups are well understood. Those for simple and semisimple Lie groups have also been thoroughly studied. However, mathematicians are still working hard on non-compact groups and on groups which are neither simple nor semisimple. It appears that what we need in the literature are examples, rather than different versions of proofs. We therefore list here without proof some theorems of Lie groups for which examples are discussed in different Chapters in this book.

The most crucial point in studying space-time symmetries of elementary particles is that we have to deal with non-compact groups. In addition, the groups are neither simple nor semisimple. The Poincaré group and some of the subgroups contained in it, are such groups. The basic advantage of this group is that the space-time symmetry or common sense allows us to carry out explicit calculations even if we do not understand relevant mathematical theorems. Therefore, the explicit construction of the representations useful in physics will provide the examples for studying mathematical theorems. The study of the Poincaré group will provide examples for the following mathematical theorems.

(a) *Theorems on the dimensionality of representations.* If a group is compact, it is possible to construct finite-dimensional unitary representations, as in the case of the three-dimensional rotation group.

If a group is non-compact and does not contain Abelian invariant subgroups, its finite-dimensional representations are non-unitary, and its unitary representations are infinite-dimensional. The $E(2)$ group discussed throughout Appendices A and B is non-compact, and its three-by-three matrix representation given in Eq. (A.4) of Appendix A is non-unitary. However, this group contains an Abelian invariant subgroup. For this reason, we cannot apply this theorem directly to the case of $E(2)$ [4, 5]. We have to resort to explicit calculations!

As we see in many Chapters of this book, whether a representation is unitary or not, depends largely on whether there exists a square-integrable Hilbert space in which the generators of the group act as Hermitian operators. This investigation will require a non-trivial amount of calculation for each problem.

(b) *Cartan's criterion on semisimple Lie algebras.* The criterion for determining whether a given group contains an Abelian invariant subgroup, called *Cartan's criterion* is discussed in many textbooks [6, 7]. Cartan's theorem can be formulated in terms of a symmetrical tensor of the second rank which can be constructed from the structure constants:

$$g_{\rho\sigma} = C^{\mu}_{\rho\lambda} C^{\lambda}_{\sigma\mu} . \tag{B.89}$$

Then the necessary and sufficient condition for the group to be semisimple is

$$\det\left(g_{\rho\sigma}\right) \neq 0 . \tag{B.90}$$

Indeed, if we compute the above determinant for $O(3)$, it does not vanish. The determinant vanishes for $E(2)$ [Problem 5 in Sect. B.6]. In Chap. 4, we study this point in more detail when we discuss a contraction of $O(3)$ to $E(2)$.

The distinction between groups which have Abelian invariant subgroups and those which do not is important, because Abelian invariant subgroups, though apparently easier to deal with can actually be more troublesome from the point of view of representations. Let us consider again the $E(2)$ group. As we have seen in Sect. A.2 of Appendix A, the translation subgroup of the $E(2)$ group is an Abelian invariant subgroup. The four-by-four matrix $T(u, v)$ of Eq. (A.11) of Appendix A is one of its possible representations. The T matrix has the pleasant Abelian property that

$$T(u_1, v_1)T(u_2, v_2) = T(u_2, v_2)T(u_1, v_1)$$
$$= T(u_1 + u_2, v_1 + v_2) . \tag{B.91}$$

However, this matrix can never be brought to a diagonal form through a similarity transformation, while most of the mathematical theorems known to physicists are based on diagonalizable matrices.

B.6 Exercises and Problems

Exercise 1. Work out explicitly the parameter transformation of Eq. (B.3) for the three-parameter group of $E(2)$.

If we use the notation

$$\beta^1 = \theta', \quad \beta^2 = u', \quad \text{and} \quad \beta^3 = v', \tag{B.92}$$

in addition to α^1, α^2, and α^3 given in Eq. (B.29) then

$$\begin{pmatrix} 1 & 0 & 0 & 0 \\ 0 & 1 & 0 & 0 \\ 0 & \phi^2(\alpha\,;\beta) & \cos\phi^1(\alpha\,;\beta) & -\sin\phi^1(\alpha\,;\beta) \\ 0 & \phi^3(\alpha\,;\beta) & \sin\phi^1(\alpha;\beta) & \cos\phi^1(\alpha\,;\beta) \end{pmatrix},$$

$$= \begin{pmatrix} 1 & 0 & 0 & 0 \\ 0 & 1 & 0 & 0 \\ 0 & u' & \cos\theta' & -\sin\theta' \\ 0 & v' & \sin\theta' & \cos\theta' \end{pmatrix} \begin{pmatrix} 1 & 0 & 0 & 0 \\ 0 & 1 & 0 & 0 \\ 0 & u & \cos\theta & -\sin\theta \\ 0 & v & \sin\theta & \cos\theta \end{pmatrix}. \tag{B.93}$$

Thus

$$\begin{aligned} \phi^1 &= \theta + \theta', \\ \phi^2 &= u\cos\theta' - v\sin\theta' + u', \\ \phi^3 &= u\sin\theta' + v\cos\theta' + v' \end{aligned} \tag{B.94}$$

Exercise 2. Discuss active and passive transformations of the $E(2)$ group.

Let us define the transformation given in Eq. (A.12) of Appendix A to be active. This transformation first rotates the coordinate point (x, y) by angle ϕ around the origin. It then translates the rotated point by u and v along the x- and y-directions respectively.

On the other hand, if we perform the same rotation on the function

$$g(x, y) = (x + iy)^m = r^m e^{im\phi}, \tag{B.95}$$

using X_θ given in Eq. (B.32), where

$$r = (x^2 + y^2)^{1/2}, \qquad \phi = \tan^{-1}(y/x), \tag{B.96}$$

we get

$$\left(e^{i\theta X_\theta}\right) g(x, y) = r^m e^{im(\phi - \theta)}. \tag{B.97}$$

If we apply the translation operators on the above expression,

$$\begin{aligned} \left(e^{-i(uX_u + vX_v)}\right)\left(e^{i\theta X_\theta}\right) g(x, y) &= (x'' + iy'')^m \\ &= g(x'', y''), \end{aligned} \tag{B.98}$$

where

$$\begin{aligned} x'' &= (x - u)\cos\theta + (y - v)\sin\theta, \\ y'' &= -(x - u)\sin\theta + (y - v)\cos\theta. \end{aligned}$$

The above linear transformation can also be written as

$$\begin{pmatrix} 1 \\ 1 \\ x'' \\ y'' \end{pmatrix} = \begin{pmatrix} 1 & 0 & 0 & 0 \\ 0 & 1 & 0 & 0 \\ 0 & -(u\cos\theta + v\sin\theta) & \cos\theta & \sin\theta \\ 0 & (u\sin\theta - v\cos\theta) & -\sin\theta & \cos\theta \end{pmatrix} \begin{pmatrix} 1 \\ 1 \\ x \\ y \end{pmatrix}. \tag{B.99}$$

The matrix in this expression is precisely the inverse of that of the active transformation matrix of Eq. (A.12) of Appendix A.

Problem 1. Referring to Eq. (B.3), show that for α, β, γ,

$$\phi[\phi(\alpha ; \beta); \gamma] = \phi[\alpha ; \phi(\beta ; \gamma)], \tag{B.100}$$

independent of x. This is an associative property.

Problem 2. Use the identity property to show that $\alpha_0 = \phi(\alpha; \alpha^{-1})$.

Problem 3. Using the Taylor expansion in $\delta\alpha$, show that Eq. (B.14) is indeed identical to Eq. (B.15).

Problem 4. It is essential that the dx^i of Eq. (B.16) be linearly independent. Show that a necessary and sufficient condition for this is that the α^p's are linearly independent.

Problem 5. Calculate Cartan's determinant of Eq. (B.90) for $O(3)$ and for $E(2)$. Show that it is non-zero for $O(3)$, while it vanishes for $E(2)$.

Problem 6. Calculate the generators of $O(3)$ and $O(2,1)$. Construct the Casimir operators. What are the subgroups of these groups? Carry out the same procedure for $O(4)$ and $O(3,1)$.

Problem 7. Consider $SL(2,c)$. Because there are four complex elements and one complex (or two real) constraints, this group is generated by six independent generators. Show that the generators can be chosen as

$$J_i = \frac{1}{2}\sigma_i, \qquad K_i = \frac{i}{2}\sigma_i, \qquad i = 1,2,3, \tag{B.101}$$

where σ_i are the usual Pauli spin matrices. The commutation relations among σ_i are well known. Complete the algebraic relations for all six generators of this group. Show that

$$C_1 = \sum_{i=1}^{3} ((J_iJ_i - K_iK_i), \tag{B.102}$$

and

$$C_2 = \sum_{i=1}^{3} J_iK_i, \tag{B.103}$$

are the Casimir operators.

Problem 8. Show that $O(2,1)$ and $SL(2,r)$ are locally isomorphic to each other. $SL(2,r)$ is a subgroup of $SL(2,c)$ consisting of real matrices.

Problem 9. Using Cartan's criterion, show that $SL(2, c)$ does not have Abelian invariant subgroups. However, the subgroup of $SL(2, c)$ consisting of the matrices of the form

$$\begin{pmatrix} \alpha & \beta \\ 0 & 1/\alpha \end{pmatrix}$$ (B.104)

contains an invariant subgroup. The invariant subgroup in this case is generated by

$$N_1 = \begin{pmatrix} 0 & 1 \\ 0 & 0 \end{pmatrix}, \qquad N_2 = \begin{pmatrix} 0 & -i \\ 0 & 0 \end{pmatrix}.$$ (B.105)

The above *set of generators* commute with the generators of the subgroup represented by Eq. (B.102). However, this set does not commute with all the generators of the $SL(2, c)$ group. Show therefore that the subgroup generated by the above N_1 and N_2 operators is a subgroup but is not an invariant subgroup of $SL(2, c)$.

Problem 10. Consider the differential operators:

$$J_3 = \frac{1}{2}\left(x\frac{\partial}{\partial x} + y\frac{\partial}{\partial y}\right),$$

$$J_1 = \frac{1}{2}\left(x\frac{\partial}{\partial y} + y\frac{\partial}{\partial x}\right),$$

$$J_2 = \frac{1}{2i}\left(x\frac{\partial}{\partial y} - y\frac{\partial}{\partial x}\right).$$ (B.106)

Show that these differential operators satisfy the commutation relations for the three-dimensional rotation group. Construct a vector space in which these operators act like the generators of the rotation group.

Problem 11. Consider two matrices A and B of the same size. They do not necessarily commute with each other. Show that

$$e^{-A}Be^{A} = B + [B, A] + \frac{1}{2!}[[B, A], A] + \cdots.$$ (B.107)

Problem 12. Prove then that

$$e^{A}e^{B} = e^{C},$$ (B.108)

where

$$C = A + B + \frac{1}{2}[A, B] + \frac{1}{12}[A, [A, B]] - \frac{1}{12}[B, [B, A]] + \cdots.$$ (B.109)

This formula is commonly known as the Baker-Campbell-Hausdorff or BCH formula [8, 2], and is discussed extensively in the literature [9]. Among numerous applications in physics, the BCH formula is very useful in studying Thomas precession [6, 10, 11].

Problem 13. Let us consider the coupled first-order differential equations which can be written in the form:

$$i\frac{d\mathbf{x}}{dt} = A\mathbf{x}\,, \tag{B.110}$$

where $\mathbf{x}$ is a column vector and A is a square matrix. If A does not depend on t, the solution of the above differential equation can be written as

$$\mathbf{x}(t) = e^{-iAt}\mathbf{x}(0)\,. \tag{B.111}$$

Using this formula, find the solutions of the following differential equations:

$$\frac{d}{dt}x_1 = x_2 + x_3\,, \qquad \frac{d}{dt}x_2 = x_1 + x_3\,, \qquad \frac{d}{dt}x_3 = x_2 + x_1\,. \tag{B.112}$$

Problem 14. Let us go back to Eq. (B.108). If the matrix A depends on time t, then the solution may be written formally as

$$\mathbf{x}(t) = \exp\left(-i\int_{t_0}^{t} A(t')dt'\right)x(t_0)\,. \tag{B.113}$$

Give a physical interpretation of this formula for an electron in a time-dependent magnetic field. Study the problem in detail when the magnetic field is the superposition of a strong constant magnetic field along the z-axis and a weak sinusoidal field along the x-direction. The above expression is one of the fundamental formulae in the interaction representation which eventually leads us to Feynman diagrams [12]. For a discussion of this problem with a strong oscillating magnetic field, see [13].

Problem 15. We are quite familiar with the spherical harmonics. They can be obtained from the Legendre differential equation or from the irreducible tensor products of Cartesian coordinate variables. Let us take spherical harmonics with $\ell = 1$. Then its transformation property is the same as that of the coordinate transformation. We are also familiar with the differential form of the rotation operators applicable to the spherical harmonics. Do these operators generate active or passive transformations?

Problem 16. $SU(4)$ is the group of four-by-four unitary unimodular matrices. Construct the generators of $SU(4)$. How many generators does this group have? Construct these generators from those of two $SU(2)$ groups, each having three generators. Discuss all possible subgroups of $SU(4)$. What is the physical motivation for studying this group? See [14]. See also [15].

References

1. E.P. Wigner, On Unitary Representations of the Inhomogeneous Lorentz Group, The Annals of Mathematics **40**(1), 149–204 (1939). DOI 10.2307/1968551. URL http://www.jstor.org/stable/1968551?origin=crossref
2. S. Başkal, Y. Kim, M. Noz, *Mathematical Devices for Optical Sciences.* (IOP Publishing, Bristol, UK, 2019). ISBN 978-0-7503-1612-5. URL https://dx.doi.org/10.1088/2053-2563/aafe78. (OCLC: 1034620988.)

3. M. Hamermesh, *Group theory and its application to physical problems*. Dover books on physics and chemistry (Dover Publications, New York, NY USA, 1989). ISBN 978-0-486-66181-0. (Originally published 1962, Addison-Wesley, Reading MA, USA.)

4. S. Weinberg, Feynman Rules for Any Spin, Physical Review **133**(5B), B1318–B1332 (1964). DOI 10.1103/PhysRev.133.B1318. URL https://link.aps.org/doi/10.1103/PhysRev.133.B1318

5. S. Weinberg, Feynman Rules for Any Spin. II. Massless Particles, Physical Review **134**(4B), B882–B896 (1964). DOI 10.1103/PhysRev.134.B882. URL https://link.aps.org/doi/10.1103/PhysRev.134.B882

6. R. Gilmore, *Lie groups, Lie algebras, and some of their applications* (Dover Publications, Mineola, NY, USA, 2005). ISBN 978-0-486-44529-8. (Originally published: 1974, John Wiley and Sons, New York, NY, USA.)

7. A. Zee, *Group theory in a nutshell for physicists*. In a nutshell (Princeton University Press, Princeton, 2016). ISBN 9780691162690

8. Y.S. Kim, M.E. Noz, *Phase space picture of quantum mechanics: group theoretical approach*. No. 40 in Lecture notes in physics series (World Scientific Publishing Co., Singapore; Hackensack, NJ, USA, 1991). ISBN 978-981-02-0360-3,978-981-02-0361-0. URL https://doi.org/10.1142/1197

9. W. Miller, *Symmetry groups and their applications*. No. 50 in Pure and applied mathematics; a series of monographs and textbooks (Academic Press, New York, NY, USA, 1972). ISBN 978-0-12-497460-9

10. N. Salingaros, Particle in an external electromagnetic field, Physical Review D **28**(10), 2473–2476 (1983). DOI 10.1103/PhysRevD.28.2473. URL https://link.aps.org/doi/10.1103/PhysRevD.28.2473

11. N. Salingaros, Relativistic motion of a charged particle, the Lorentz group, and the Thomas precession, Journal of Mathematical Physics **25**(3), 706–716 (1984). DOI 10.1063/1.526179. URL http://aip.scitation.org/doi/10.1063/1.526179

12. S.S. Schweber, *An Introduction to Relativistic Quantum Field Theory* (Dover Books on Physics, Dover Publications, Inc, New York, NY, USA, 2005). ISBN 978-0-486-44228-0. (Originally published 1961, Harper & Row, Publishers, New York, NY, USA.)

13. J.H. Shirley, Solution of the Schrödinger Equation with a Hamiltonian Periodic in Time, Physical Review **138**(4B), B979–B987 (1965). DOI 10.1103/PhysRev.138.B979. URL https://link.aps.org/doi/10.1103/PhysRev.138.B979

14. E. Wigner, On the Consequences of the Symmetry of the Nuclear Hamiltonian on the Spectroscopy of Nuclei, Physical Review **51**(2), 106–119 (1937). DOI 10.1103/PhysRev.51.106. URL https://link.aps.org/doi/10.1103/PhysRev.51.106

15. S. Oneda, Review talk on SU(4) in elementary particle physics, in *Proceeding of the International Symposium on Mathematical Physics January 5-8 1976*, ed. by A. Bohm (Mexico City, Mexico, 1976), Volumn 1 in Memorias Del Simposio de Fisica Matematica, Mexico, D.F., Enero 5 a 8 de 1976, 368. URL https://books.google.se/books?id=gL3vAAAAMAAJ

Index

MIX
Papier aus verantwortungsvollen Quellen
Paper from responsible sources
FSC® C105338